Advanced Bioceramics

Advanced Bioceramics: Properties, Processing, and Applications describes development of bioceramics and biocomposites, which are used in various biomedical applications including bone tissue repair, remodelling and regeneration. It covers the fundamental aspects of materials science and bioengineering, clinical performance in a variety of applications, ISO/ASTM specifications, and opportunities and challenges.

- Offers a comprehensive view of properties and processing of bioceramics.
- Highlights applications in dentistry, orthopaedic and maxillofacial implants, and regenerative and tissue engineering.
- Covers ISO/ASTM specifications such as processing, clinical applications, recycling/reuse and disposal standards.
- Explores health, environmental and ethical issues.

With contributions from eminent editors and recognized authors around the world, this book should serve as an important reference for academics, scientists, researchers, students and practitioners in materials science and biomedical engineering. It is to assist in the design of novel, targeted and personalised bioceramic-based solutions to advanced healthcare.

Advanced Bioceramics
Properties, Processing, and Applications

Edited by
Md Enamul Hoque, Kheng Lim Goh,
and Suresh Sagadevan

CRC Press
Taylor & Francis Group
Boca Raton London New York

CRC Press is an imprint of the
Taylor & Francis Group, an **informa** business

First edition published 2024
by CRC Press
2385 Executive Center Drive, Suite 320, Boca Raton FL 33431

and by CRC Press
4 Park Square, Milton Park, Abingdon, Oxon, OX14 4RN

CRC Press is an imprint of Taylor & Francis Group, LLC

ISBN: 978-1-032-19251-2 (hbk)
ISBN: 978-1-032-19252-9 (pbk)
ISBN: 978-1-003-25835-3 (ebk)

DOI: 10.1201/9781003258353

Typeset in Times
by codeMantra

Contents

SECTION B Properties and Processing

SECTION D ISO/ASTM Specifications

SECTION E Challenges, Issues and Sustainability

Preface

Biomaterials play a significant role in advancing research and development across a range of industries. A thorough understanding of the fundamental properties and processing techniques of these materials is essential for enhancing their applications. Bioceramics are ceramic materials that are designed and used for biomedical applications, particularly in the field of regenerative medicine. There are various types of bioceramics, such as calcium phosphate ceramics, bioactive glasses and alumina-based ceramics, each with their own unique properties and applications. Bioceramics, in particular, have a wide range of potential applications in the medical field due to their biocompatibility and unique physical properties. As bioceramics continue to gain popularity and increase their socioeconomic impact on the environment, this book aims to explore various aspects of bioceramics, including interface engineering and manufacturing techniques, with a focus on the incorporation of new polymeric materials.

Meeting the growing demands of advanced technology requires the development of new materials with unique properties. Nanotechnology tools and techniques can provide a way to manipulate materials at the atomic level, which enables the creation of advanced structures. One of the materials that have gained widespread attention in recent years is the polymer nanocomposite. This is a solid polymer material that consists of a polymer matrix and an inorganic dispersive phase of nanofillers, which are typically less than 100 nm in scale and can have 1–3 dimensions. Polymer nanocomposites are attractive for many applications due to their unique properties and ease of synthesis. The addition of nanomaterials to polymers can significantly enhance the fundamental properties of the polymer, including tensile strength, impact strength, scratch resistance, fire resistance, thermal stability, and thermal and electrical conductivities. These advanced polymer composites have a wide range of potential applications in industries such as electronics, automotive, marine, defence, sports goods, biological equipment and renewable power generation facilities. Multifunctional polymer nanocomposites are expected to replace traditional composites in the market, making it important to focus on and explore the novel aspects of polymer nanocomposite materials. This will enable the advancement of technology and the creation of new and innovative products.

The broad range of topics covered in this book makes it a suitable resource for research scholars, teaching instructors and faculties involved in the research and development of composite sandwich panels. Additionally, professionals and engineers seeking to understand the factors governing material properties, responses and failure behaviour to various mechanical loads can use this book as a reference. The chapters in this book have been written by esteemed scientists and clinicians based on their knowledge of cutting-edge research on biomaterials. By reading this book, the readers can enhance their understanding of the design and methodology required to develop biomaterials with specific physical and biological properties that are suitable for distinct clinical applications.

This book is organised into 24 chapters. Chapters 1–5 cover processing of bioceramics. Chapters 7–9 cover the physical and chemical properties of bioceramics. Chapters 10–19 cover clinical and cosmetic applications of bioceramics, examining the applications of bioceramics. Chapters 20–26 cover an explanation on the key standards and a discussion on the standards used for processing, testing/properties and end-of-life considerations. Chapters 27 and 28 discuss the challenges and prospects of bioceramics, from ethical issues to future developments.

The publisher, editorial manager and their colleagues' assistance and advice in this potential publication process are greatly acknowledged and appreciated.

INTRODUCTION

Bioceramics, which are ceramic materials created for biomedical purposes, particularly in regenerative medicine, come in various forms including calcium phosphate ceramics, bioactive glasses and alumina-based ceramics, with individual characteristics and functions. In the recent past, there has been a growing interest in bioceramics as they are biocompatible materials with ceramic or glassy properties that can be effectively employed in tissue repair and reconstruction using stem cell technologies. The key advantage of these materials is their ability to be fabricated into diverse forms such as powder, coating and bulk, without compromising their functional properties. Bioceramics, including glasses, glass-ceramics, alumina, zirconia, phosphates and hydroxyapatite, can form various nanocomposites, which have potential for application in clinical sectors including orthopaedics, otolaryngology, maxillofacial and plastic surgery, oral surgery, periodontology and tumour therapy. By integrating bioceramic composites with tissue engineering tools, living cells can grow structurally in a three-dimensional manner, and the engineered tissues can serve as neotissues to repair damaged body parts. Glass-ceramics are particularly biostable, and the composites formed with these materials exhibit high safety and biocompatibility at all stages of tissue growth and development. A comprehensive understanding of the synthesis, processing, properties and applications of bioceramics is needed to further advance this field. With continued scientific research in the field of bioceramics, their developments can be transferred to other industries, leading to the development of various medical prostheses and improving the quality of life for many people. The field of green composites in general manufactures products from natural sources or makes use of semisynthetic and natural materials at a discretionary percentage. These particular sectors in recent years have attained significant momentum from both academic and industrial sectors because of the need to develop new products with environmentally friendly nature and are sustainable. For a majority of applications in the biomedical sector, green composites and renewable materials are highly demanded because of the non-toxic, biocompatibility and biostability. However, due to some inherent issues of metallic biomaterials (e.g. heavyweight, corrosion, limited resource and cost), their usage has been limited and can be replaced by biocomposites that offer superior properties with necessary functionality.

SCOPE OF THE PROPOSED BOOK

This book presents an overview of the current and historical research on bioceramics, as well as biocomposites, specifically as they relate to bone tissue repair, remodelling and regeneration. The topics presented in the respective chapters of this book are grouped into five areas: processing, properties, applications, ISO/ASTM specifications, and challenges and prospects. The discussion covers both the fundamental aspects of materials science and bioengineering, and their relation to the clinical performance of orthopaedic and dental implants. The content is designed to provide a comprehensive understanding of bioceramics and biocomposites, which are considered to be among the most dynamic and highly demanded materials in stem cell research and tissue engineering. Some chapters in this book have also addressed the use of bionanomaterials with enhanced functionalities in a variety of fields; their interactions with biological molecules or structures are also covered in detail. In addition, this book addresses important topics related to tissue regeneration, including the design of novel biomaterials and polymer matrices. Overall, this book is a valuable resource for researchers and professionals involved in the development of implantable devices and tissue engineering.

Biocomposites offer unique advantages that go beyond the properties of metallic biomaterials. For instance, when used as internal fixators, metallic implants typically necessitate revision surgery after the healing process, a requirement that can be eliminated by utilising degradable biocomposites. Despite the recognition of numerous biomaterials and systems for medical purposes, there is still potential for the enhancement and advancement of environmentally friendly composites. This book was created to provide cutting-edge information on the development and uses of bioceramics and biocomposites, serving as a foundation for course materials for undergraduate and post-graduate students in materials science, engineering and biomedicine.

Editors

Prof Dr M Enamul Hoque [PhD (NUS, Singapore), PGCHE (Nottingham Uni, UK), FHEA (UK); FIMechE (UK), CEng (UK)]

Dr Md Enamul Hoque is a Professor at the Department of Biomedical Engineering in the Military Institute of Science and Technology (MIST), Dhaka, Bangladesh. Before joining MIST, he served in several leading positions in some other global universities such as Head of the Department of Biomedical Engineering at King Faisal University (KFU), Saudi Arabia; Founding Head of Bioengineering Division, University of Nottingham Malaysia Campus (UNMC); and so on. He received his PhD from the National University of Singapore (NUS), Singapore, in 2007 with a globally prestigious scholarship from the Singapore Government. He also obtained his PGCHE (Post-Graduate Certificate in Higher Education) from the University of Nottingham, UK, in 2015. He is a Chartered Engineer (CEng) certified by the Engineering Council, UK; Fellow of the Institute of Mechanical Engineering (FIMechE), UK; Fellow of Higher Education Academy (FHEA), UK; and Member, World Academy of Science, Engineering and Technology. To date, he has published 109 journal papers, 5 special issues in journals, 11 books, 70 book chapters and 100 international conference presentations/proceedings. His publications have attracted about 3,402 citations with 27 h-index and 69 i-10 index (https://scholar.google.com/citations?user=aRvzzrEAAAAJ&hl=en) as per Google Scholar citation report. He received the Outstanding Nano-scientist Award in the International Workshop on Recent Advances in Nanotechnology and Applications (RANA-2018) held on 7–8 September 2018, AMET University, Chennai, India. His major areas of research interests include (but are not limited to) bioceramics, biomaterials, biocomposites, biopolymers, nanomaterials, biomedical implants, rehabilitation engineering, rapid prototyping technology, stem cells and tissue engineering. As per AD Scientific Index (as on 08 March 2023), he is the best scientist (biomedical engineering) in the country (Bangladesh), 290 in Asia and 1,388 in the world; https://www.adscientificindex.com/scientist/md-enamul-hoque/424943. He is also among the top 2% scientists in the world; https://scientificbangladesh.com/list-of-the-worlds-top-2-scientists-in-bangladesh/; https://elsevier.digitalcommonsdata.com/datasets/btchxktzyw/4

Kheng Lim Goh is a Fellow of the Institute of Mechanical Engineers (UK), Chartered Engineer and Chartered Physicist with the Institute of Mechanical Engineers (UK) and Institute of Physics (UK), respectively. His research interest is in the repair of fibre-reinforced composites used in aerospace and automotive engineering. He holds the position of Associate Professor (Reader in Materials Technology) at the Newcastle Research and Innovation Institute (NewRIIS) and Newcastle University in Singapore, and is affiliated to the Faculty of Science, Agriculture & Engineering, Newcastle University (UK). He leads the Advanced Composites Research (ACR) Group at NewRIIS. Professor Goh has authored and co-authored over 100 papers in peer-reviewed journals that cover a wide range of composite materials, together with international collaborators from Argentina, Bangladesh, Canada, India, Malaysia, Singapore, Sri Lanka and the UK. His research theme underlies an understanding of the physical properties of natural and synthetic materials and implications for designing composites for engineering applications and for repairing damaged composites. He is the author of the authoritative book *Discontinuous-Fibre Reinforced Composites: Fundamentals of Stress Transfer and Fracture Mechanics*' published by Springer.

Dr. Suresh Sagadevan is currently working as a Senior Research Fellow in the Nanotechnology & Catalysis Research Centre (NANOCAT), University of Malaya, Malaysia. He has authored 10 international book series and 20 book chapters. He has published more than 280 research papers in national and international journals. He is a member of many professional bodies at the national/international level. He is the editor/editorial board member/reviewer for a number of high-impact factor journals. He also has two filed patents to his credit. His current spans and brilliant discoveries are focused on the areas of nanofabrication, functional materials, crystal growth, graphene, polymeric nanocomposites, glass materials, thin films, switchable device, electron microscopy and spectroscopy, bio-inspired materials, drug delivery, tissue engineering, cell culture and integration, flexible and transparent electrodes, supercapacitor, optoelectronics, green chemistry and biosensor applications.

Contributors

Nikita Agrawal
Department of Dentistry
Chirayu Medical College
Bhopal, India

Mahbub Ahmed
Southern Arkansas University
Magnolia, Arkansas

Moeen Akhtar
Eindhoven University of Technology
Eindhoven, the Netherlands

M. Azam Ali
Centre for Bioengineering & Nanomedicine (Dunedin Hub), Faculty of Dentistry
Division of Health Sciences
University of Otago
Dunedin, New Zealand

Yashdi Saif Autul
Department of Materials Science and Engineering, Rensselaer Polytechnic Institute (RPI)
Troy, New York, USA

David Bahati
Euromed Research Center, Euromed Polytechnic School
Euromed University of Fes (UEMF)
Fes, Morocco

N. Banerjee
Department of Mechanical Engineering
Calcutta Institute of Technology
Howrah, India

A.S. Bhattacharyya
Department of Nano Science and Technology
Central University of Jharkhand
Jharkhand, India
and
Centre of Excellence in Green and Efficient Energy Technology (CoE GEET)
Central University of Jharkhand
Jharkhand, India

A.R. Biswas
Department of Mechanical Engineering
Calcutta Institute of Technology
Howrah, India

Meriame Bricha
Euromed Research Center, Euromed Polytechnic School
Euromed University of Fes (UEMF)
Fes, Morocco

Vatsal Chauhan
Department of Conservative Dentistry and Endodontics
Peoples College of Dental Sciences
Bhopal, India

S. Chitra
Department of Prosthodontics
Saveetha Institute of Medical and Technical Sciences (SIMATS)
Chennai, India

Muhammad Ifaz Shahriar Chowdhury
Department of Mechanical Engineering
Imperial College London
London, United Kingdom

Canan Dagdeviren
Media Lab
Massachusetts Institute of Technology
Cambridge, Massachusetts

S. Das
Department of Mechanical Engineering
Calcutta Institute of Technology
Howrah, India

Ritambhara Dash
Department of Nano Science and Technology
Central University of Jharkhand
Jharkhand, India

Khalil El Mabrouk
Euromed Research Center, Euromed Polytechnic School
Euromed University of Fes (UEMF)
Fes, Morocco

Sara V. Fernandez
Department of Materials Science and Engineering
Massachusetts Institute of Technology
Cambridge, Massachusetts

Brijesh Gangil
Department of Mechanical Engineering
HNB Garhwal University
Uttarakhand, India

T.K. Ghosh
Department of General Science and Humanities
Calcutta Institute of Technology
Howrah, India

M.L. Gould
Centre for Bioengineering & Nanomedicine (Dunedin)
Division of Health Sciences, Department of Dentistry
University of Otago
Dunedin, New Zealand

Md Enamul Hoque
Department of Biomedical Engineering
Military Institute of Science and Technology (MIST)
Dhaka, Bangladesh

Azhar Hussain
University of Engineering and Technology
Taxila, Pakistan

Muhammad Uzair Iqbal
ACME Enterprises
Sialkot, Pakistan

M.P. Jahan
Miami University
Oxford, Ohio

Muhammad Ramzan Abdul Karim
Ghulam Ishaq Khan Institute of Engineering Sciences and Technology
Swabi, Pakistan

Lovely Khandakar
Department of Microbiology
University of Manitoba
Winnipeg, Canada

Parimala Kulkarni
Department of Dentistry
Chirayu Medical College
Bhopal, India

M. Kumar
Department of Mechanical Engineering
Calcutta Institute of Technology
Howrah, India

S.R. Maity
Department of Mechanical Engineering
NIT Silchar
Assam, India

Nibin K. Mathew
National Centre for Nanoscience and Nanotechnology
University of Madras
Chennai, India

David Orisekeh
Department of Mechanical and Manufacturing Engineering
Miami University
Oxford, Ohio

Divya Panday
Department of Conservative Dentistry and Endodontics
Peoples College of Dental Sciences
Bhopal, India

Sazedur Rahman
Department of Mechanical and Production Engineering
Ahsanullah University of Science and Technology (AUST)
Dhaka, Bangladesh

S. Rajeshkumar
Department of Pharmacology
Saveetha Institute of Medical and Technical Sciences (SIMATS)
Chennai, India

Asif Mahmud Rayhan
Department of Biomedical Engineering
Military Institute of Science and Technology
Dhaka, Bangladesh
and
Department of Mechanical Engineering
Military Institute of Science and Technology
Dhaka, Bangladesh

Tahrima Binte Rouf
Stephenson School of Biomedical Engineering
University of Oklahoma
Norman, Oklahoma

Poornimadevi Sakthivel
Bone Substitutes
Madurai, India

A. Samanta
Department of Mechanical Engineering
Calcutta Institute of Technology
Howrah, India

Pugalanthi Pandian Sankaralingam
Bone Substitutes
Madurai, India

A. Sen
Department of Mechanical Engineering
Calcutta Institute of Technology
Howrah, India

D. Sengupta
Department of Mechanical Engineering
Calcutta Institute of Technology
Howrah, India

R. Senthilkumar
Department of Biotechnology, School of Applied Sciences
REVA University
Bangalore, India

Muhammad Shahid
ACME Enterprises
Sialkot, Pakistan

Mohammad Ahnaf Shahriar
Department of Industrial, Manufacturing, and Systems Engineering
University of Texas at Arlington
Arlington, TX

Samira Islam Shaily
Department of Mechanical Engineering
Military Institute of Science and Technology
Dhaka, Bangladesh

Samer Shamshad
Department of Bacteriology
ICAR-National Institute of Veterinary Epidemiology and Disease Informatics
Bangalore, India
and
Department of Biotechnology, School of Applied Sciences
REVA University
Bangalore, India

Ikra Iftekhar Shuvo
Media Lab
Massachusetts Institute of Technology
Cambridge, Massachusetts

Ram K. Singh
Department of Nano Science and Technology
Central University of Jharkhand
Jharkhand, India

Santosh Kumar Singh
Department of Conservative Dentistry and Endodontics
Peoples College of Dental Sciences
Bhopal, India

U. Srivastava
Department of Mechanical Engineering
Calcutta Institute of Technology
Howrah, India

G.V. Swarnalatha
Department of Biochemistry
Rayalaseema University
Andhra Pradesh, India

Vijayakumar Chinnaswamy Thangavel
Department of Polymer Technology
Kamaraj College of Engineering and Technology
Virudhunagar, India

M. Vidya
Department of Chemistry and Biochemistry
M.S. Ramaiah College of Arts, Science and Commerce
Bangalore, India

Poorti Yadav
Department of Nano Science and Technology
Central University of Jharkhand
Jharkhand, India

Section A

Introduction and Processing

1 Overview of Bioceramics

Muhammad Ifaz Shahriar Chowdhury
Imperial College London

Sazedur Rahman
Ahsanullah University of Science and Technology (AUST)

Yashdi Saif Autul
Rensselaer Polytechnic Institute (RPI)

Md Enamul Hoque
Military Institute of Science and Technology (MIST)

1.1 INTRODUCTION

Thousands of years ago, humankind discovered that fire could irrevocably turn clay into ceramic pots, leading to the emergence of an agricultural civilization with vast improvements in the quality and duration of life. The use of ceramics in ways that enhance the quality of life has undergone yet another revolution in the span of the last four decades. The revolutionary use of ceramics that have been built specifically for repairing and reconstructing sick or injured areas of the body is at the heart of this revolution. Ceramics that are used for this objective are referred to as bioceramics.

A significant number of biomaterials may be derived from ceramic materials. A biomaterial is an artificial substance used to influence, through controlling interactions with biological systems, therapeutic or diagnostic procedures in human or veterinary medicine [1]. The lives of millions of patients have been significantly altered by novel biomaterials developed over the course of the previous century. Biomaterials have been a significant contributor to contemporary health care and will contribute significantly to the treatment of a variety of conditions, including fragility fractures, osteoarthritis, and osteoporosis, as the older population continues to grow [2]. On the other hand, due to the fact that clinical implantology is the principal and widespread application of biomaterials, biomedical disciplines are gaining a greater amount of importance in biomaterials research. A number of research domains, namely physiology, anatomy, histology, and cell and molecular biology, are included in this study. The end objective is to build the right biological relationship between the artificial grafts and the living tissues of the host.

Bioceramics is a term that describes ceramics that have been created to make interactions with biological structures and are utilized in biomedical applications such as body implants, repairs, augmentations, medication delivery vehicles, vaccination adjuvants, or diagnostics [3]. Since the year 1000 AD, when plaster of Paris ($CaSO_4$)

DOI: 10.1201/9781003258353-2

was commonly used for setting fractured bones, bioceramics have been known and used. A major shift in the history of mankind was brought about by ceramics some decades ago, with the invention of functional ceramics in high-temperature superconductors, piezoelectrics, magnets, semiconductors, and other areas of application.

Thermal stability, mechanical strength, biocompatibility, and similarity to bone tissues are all aspects that impact the utilization of bioceramics in various applications [4]. High hardness, heat conduction, poor electrical conductivity, and high melting temperatures characterize ceramics. Ceramic materials have a very high compression strength, but a considerably lower tensile strength than other types of materials when it comes to their mechanical characteristics. They are stiff, with a high Young's modulus, and they break without deforming plastically [5].

In terms of how bioceramics respond inside of a live organism, their major reactions have enabled them to be categorized as virtually bioinert, bioactive, and resorbable (see Figure 1.1) [5]. Bioactive ceramics undergo a series of surface reactions that form a mechanically strong connection with live tissue. First-generation bioceramics are almost entirely bioinert, whereas second-generation bioceramics are bioactive and resorbable. Additionally, it is conceivable to identify glasses with the same compositions that behave differently depending on how they are formed: either as bioactive when created through the sol–gel process or as bioinert if they are made by melting.

The number of uses of biomaterials in the treatment of human illnesses is limitless and unmatched by other treatments and therapies. Bone substitutes, ocular tissue engineering (TE) scaffolds, contact lenses, skin substitutes, artificial arteries, and joint and limb replacements are all included in this category. The worldwide market for biomaterials is

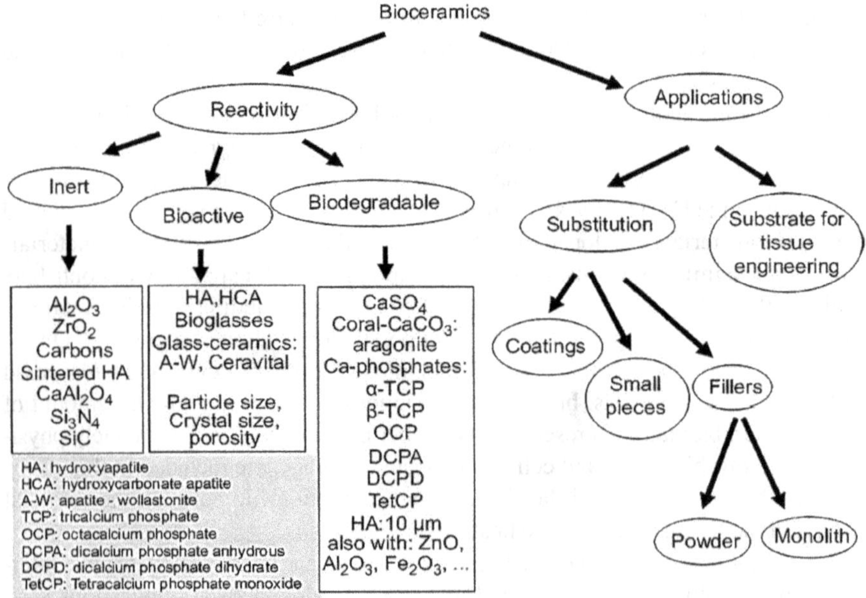

FIGURE 1.1 Bioceramics and their medical utilization. (Reprinted from Ref. [5] with permission from Elsevier.)

expected to reach a value of 149.17 billion U.S. dollars by the end of 2021, up from 70.90 billion U.S. dollars in 2016. This represents a compound annual growth rate of 16% over the period of 2016–2021. The meteoric increase in the number of elderly patients, as well as technical advances in the development and manufacture of implants, the frequency with which hip and knee replacements are performed, and the price of implants will determine the rate of growth. Furthermore, market growth is influenced by the kinds of biomaterials used in therapy, such as polymers, metals, and ceramics, as well as the location [6]. Ceramic biomaterials, such as orthopedic implants, have a high market demand of around 1.5 million each year globally, costing $10 billion [7].

The aim of this chapter is to give an introduction to bioceramics on a more general level. We are going to look at the history of bioceramics, breaking it down into their first, second, and third generations. In addition, the many different kinds of biomaterials and bioceramics, as well as their characteristics, will be spoken about. In conclusion, a variety of applications of bioceramics will be shown. These applications will include TE, orthopedics, dentistry, ophthalmology, cancer treatment, and other fields.

1.2 HISTORICAL BACKGROUND

Bioceramics have been used for the treatment of painful conditions and the restoration of function in calcified tissues (such as teeth and bones) of the human body for millennia in the field of modern medicine. Amadeo Bobbio, for example, uncovered Mayan skulls in 1972 that had missing teeth replaced with nacre replacements, some of which were over 4,000 years old [8].

The first extensively used artificial bioceramics were plaster of Paris (calcium sulfate). German surgeon Themistocles Gluck performed the first ivory knee replacement in 1890. Three weeks later, he replaced a patient's wrist [9]. Dreesman published the first article on the plaster of Paris for bone abnormalities in 1892 [10]. He reported the outcomes of clinical operations on eight patients. Stachow surgically produced flaws in sheep's long bones, which he then repaired with plaster of Paris, according to Peltier [11]. In 1920, U.S. surgeon Fred Houdlette Albee used laboratory-produced calcium phosphate to repair surgically caused abnormalities in rabbit bones [12]. In 1925, Kofmann reported using plaster of Paris to cover an abscessed tibial hollow [11]. Nystrom described four efforts to fix osseous deformities with rivanol, plaster, and oatmeal in 1928 [13]. Petrova prepared plaster of Paris with 10% rivanol and 10% kreolin in the same year and implanted it into both infected and non infected wounds [14].

Edberg reported his 1930 clinical results utilizing Bystrom's plaster of Paris. His initial effort was to fill two huge dentigerous cysts at the tooth's root, but the plaster extruded entirely after 2 weeks. His second use of the filling was for non-malignant calcaneus osteitis. This material was subsequently employed to fill osseous voids created by the tuberculous breakdown [15]. Rock proposed the first use of alumina as a biomaterial in 1933 [16]. Nielson presented the outcomes of 16 diseased flaws and 14 non-infected bone cavities filled with cement in 1944 [17]. In 1952, Hauptli utilized the plaster of Paris on eight patients who had deficiencies as a result of benign tumor excision [14].

Ljubovic and Nikulin published a report on their rabbit-bone-plaster experiment in 1956. They came to the following conclusions: (i) Plaster put in a subperiosteal resection region produces no greater reactivity than would be expected in an untreated fracture; (ii) the new bone that forms in a region where plaster of Paris has been placed is considered to be normal bone; and (iii) as a direct consequence of the plaster, no adverse impacts have been discovered in the tissues that are next to it or in the organs that are different from it [11]. Peltier and his colleagues reported in 1957 that they used tablets of plaster of Paris to cover gaps created in the lateral cortex of distal femurs of 29 mongrel dogs. They found that the tissues that surrounded the implants did not have any adverse reactions to the implants [18]. Peltier described the use of calcium sulfate to treat massive osseous voids in five clinical instances in 1959. At 300°F with dry heat, commercially graded plaster of Paris was sterilized. After being adjusted, tiny tablets of plaster of Paris (commercially graded) were used to fill the spaces [19]. Plaster of Paris naturally has a low tensile strength; therefore, Gourley and Arnold added an epoxy resin to augment the strength of the material. However, they found that despite the increase in strength, the epoxy resin made the plaster physiologically incompatible [20].

The efficiency of calcium hydroxide as a stimulant for the production of ectopic bone was evaluated by the subcutaneous implantation of the compound. Mitchell and Amos (1957) identified a bone that was juvenile and heterotopic [21]. When calcium hydroxide was implanted subcutaneously, the researchers Mitchell and Shankwalker (1958) found that it induced the production of what looked to be juvenile bone. This finding is supported by the studies that they conducted [22].

Smith is recognized for ushering in the "modern" phase of bioceramics, as seen by his successful exploration of Cerosium®, an epoxy resin-coated porous aluminate ceramic. The porosity of that bioceramic was kept at 48% in order to achieve net physical attributes that were extremely similar to those of bone [23]. The first scientific investigation of zirconia's exceptional biological characteristics was published in 1969. The discovery by Hench and colleagues that in rat femurs, a specially composed glass could not be withdrawn forcefully from the transplant region after 6 weeks of an in vivo test appears to be the earliest evidence that implant–bone interface bonding is conceivable [24].

Boutin published a landmark work in 1972 [25], and since then, for femoral balls in hip endoprostheses, alumina has been welcomed as a good bioceramic all over the globe. "Bioceramics" was first used in an abstract in 1971 [26], but it was only in the title of a paper that year that it appeared in print [27]. The years 1972 [28] and 1973 [29] saw the first publications on lunar apatites. Graham and van der Eb's study in 1973 demonstrated the first use of calcium orthophosphates in the consolidation of several genetic elements [30]. With a very straightforward approach, the researchers proved that calcium orthophosphates have the potential to compress DNA and increase the effectiveness of transfection. Furthermore, in the year 1973, the first investigation on the manufacture and implant of porous and resorbable calcium orthophosphate (β-TCP) bioceramics was reported [31]. The present dental use of calcium orthophosphates started in 1975. Specifically, β-TCP was used to treat both surgically produced periodontal abnormalities [32] and pulpless permanent teeth as an adjuvant to apical closure [33]. After that, in 1979, dense HA cylinders were utilized for the purpose of rapidly replacing dental roots in a research study [34]. The first-ever international conference on bioceramics took place on April 26, 1988, in Kyoto, Japan, with participants coming from all over the globe. Figure 1.2 provides a diagrammatic representation of the progression of bioceramics throughout history [35].

The earliest man-made bioceramics that were used to restore teeth and bone were designed with the intention of being bioinert or having low reactivity or contact with living tissues. These bioceramics are often referred to as "first-generation" bioceramics [36]. Alumina (Al_2O_3) and zirconia (ZrO_2) are two more typical bioceramics. They have the characteristics of in vivo biocompatibility, nontoxicity, stability, wear resistance, superior corrosion, and great strength, making them popular biomaterials [5]. Around 1980, the second generation of bioceramics began to bloom. The goal of bioceramic synthesis was to achieve a favorable interaction with the living organism, such as bioactivity and/or resorbability [36]. Crystalline calcium phosphates, bioactive glasses, and glass-ceramics are the most notable examples of bioceramics belonging to the second generation. At the turn of the 21st century, it became abundantly clear that the therapeutic demands placed on biomaterials for implants could not be satisfied alone by bioceramics. More demanding bioceramics were needed. This resulted in the development of third-generation bioceramics, which are utilized as biologically active substances or cell scaffolds (hormones, growth factors, and so on) that may promote the repairment and regeneration of live tissues [37–39]. New forms of sophisticated bioceramics are being investigated, namely mesoporous ordered glasses [40] or star gel [41], mesoporous silica materials [42,43], and organic–inorganic hybrids [44,45]. Figure 1.3 shows various types of bioceramics in different generations.

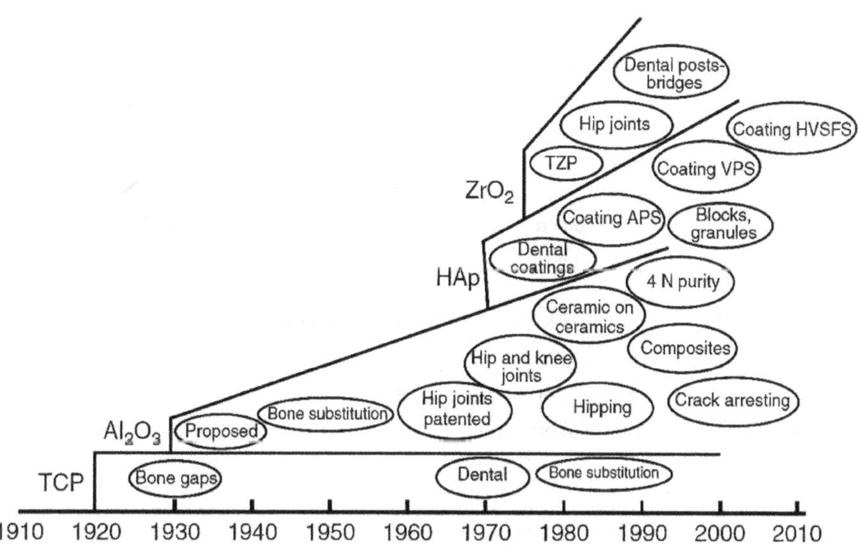

FIGURE 1.2 100 years of clinical application of bioceramics. (Reprinted with permission from Ref. [35], Copyright (2015) John Wiley and Sons.)

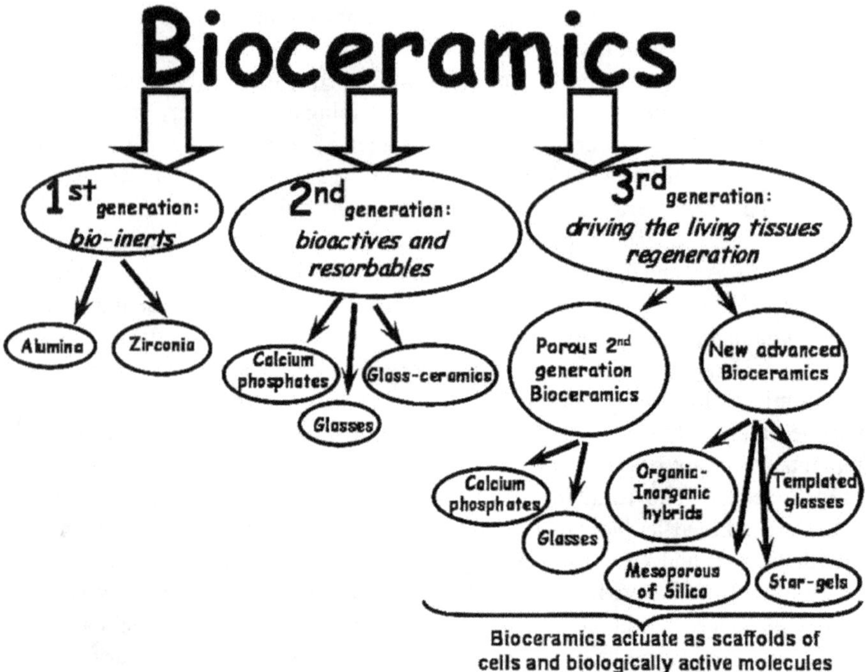

FIGURE 1.3 Evolution of bioceramics. (Reprinted with permission from Ref. [36], Copyright (2007) John Wiley and Sons.)

1.3 BIOMATERIALS

In human or veterinary medicine, a biomaterial is a substance that has been designed to assume a form that may be utilized to regulate the course of any therapeutic or diagnostic operation by controlling interactions with components of biological systems [46]. In 1986, during the European Society for Biomaterials' Consensus Conference, a more exact description of a biomaterial was presented—"a nonviable material used in a medical device, intended to interact with biological systems" [47].

1.4 BIOMATERIALS IN COMPARISON WITH OTHER MATERIALS

Artificial equipment like prosthetic legs and hearing aids is not considered biomaterial since they come into direct touch with the body's outer layer of tissue, the skin [48,49]. Biomaterials differ from other materials in that they can thrive in a biological context without harming themselves or the environment around them. Because of this, there is a clear distinction that needs to be made between biomaterials and biological materials. Biomaterials are defined as materials that are accepted by living tissues and therefore have the potential to be used for tissue replacement, whereas biological materials are defined as materials that are produced by various biological systems but are not accepted by living cells. Biomimetic materials are another potential kind of material that might be used. These materials are not created by living organisms, but they have a constitution, an architecture, and qualities that are analogous to those of biological materials.

1.5 BRIEF CLASSIFICATION OF BIOMATERIALS

It is common practice to categorize materials according to their chemical makeup, and biomaterials are no exception. Materials may be classified as either organic or inorganic. Organic materials are a broad category that includes any chemical compounds that include carbon and that may be found in living organisms. Nucleic acids, proteins, lipids, and carbohydrates are the four groups that make up the organic material classification system. Although each of these substances is composed of atoms of carbon, hydrogen, and oxygen, the relative amounts of these constituents vary from one material category to the next and are determined by the intended use. Inorganic biomaterials do not originate from living organisms, and, with the exception of carbide, cyanide (CN), carbon dioxide (CO_2), carbonate (CO_3), and carbon monoxide (CO), most inorganic biomaterials lack the carbon atom in their chemical structure. The great majority of inorganic biomaterials do not include carbon components in their molecular structure [50]. Organic biomaterials can further be subdivided into lipid-based biomaterials and organic biopolymers. When organic biopolymers are reinforced by a variety of natural fibers (e.g., pineapple leaf fiber (PALF), sisal fiber, etc.) for the purpose of improving mechanical, vibrational, and damping behavior, the resulting materials are referred to as biopolymer composites [51]. On the other hand, inorganic biomaterials can be separated into four distinct categories: biometal, inorganic biopolymers, inorganic biocomposites, and bioceramics. The categorization of biomaterials is shown in Figure 1.4.

1.5.1 Bioceramics

According to Wikipedia, the term ceramic derives from the Greek word κεραμικός (keramikos), which means "of pottery" or "for pottery," and is derived from the word κέραμος (keramos), which means "potter's clay, tile, pottery." The first known reference to the root "ceram-" may be found in the Linear B syllabic writing of the Mycenaean Greek word ke-ra-me-we, which translates to "workers of pottery." A ceramic is a nonmetallic, inorganic substance that is frequently carbide, nitride, or crystalline oxide. Silicon and carbon are common ingredients in ceramics. Ceramics are known for their brittleness, rigidity, and strength in compression, yet they are weak in shear and tension. They are not susceptible to the chemical deterioration that occurs in other materials when exposed to conditions that are either caustic or acidic. Ceramics have a temperature resistance that ranges from 1,000°C to a 1,600°C.

Because of its noncrystalline amorphous nature, glass is often overlooked as a ceramic. Glassmaking, on the other hand, requires multiple phases in the ceramic process, and glass has mechanical characteristics comparable to ceramic materials [52]. As with any other type of biomaterial, bioceramics can be used to replace tissues or joints, as coatings to improve resorbable lattices, or biocompatibility that provides

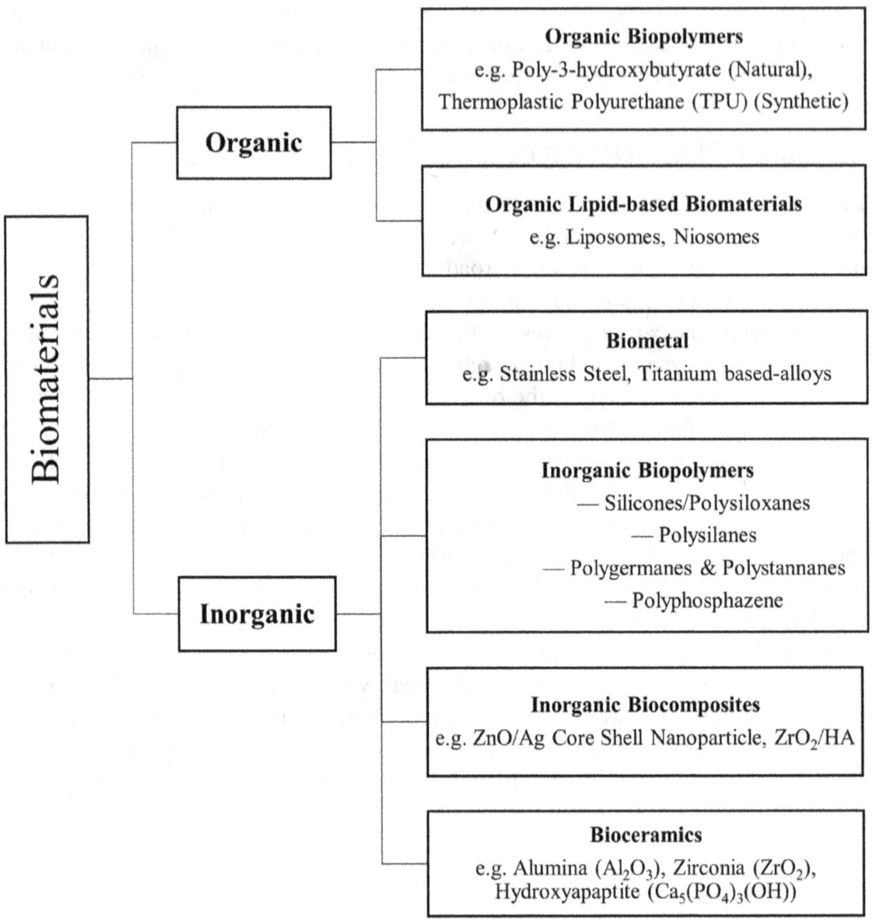

FIGURE 1.4 Classification of biomaterials.

temporary frameworks and structures that are replaced and/or dissolved as the body heals injured tissues [53]. Bioceramics are ceramics that are biocompatible. They are the toughest biomaterials, having a high elastic modulus and toughness [54].

1.6 COMPOSITION OF BIOCERAMICS

In terms of their chemical makeup, a large percentage of inorganic compounds—including metal carbides, metal phosphides, metal nitrides, metal chalcogenides, metal oxides, and metal haloids—are categorized as ceramics. This comprises a wide range of metal salts that include oxygen in their structures (such as silicates, carbonates, acetates, nitrates, phosphates, sulfates, and so on). On the other hand, only a tiny subset of the chemical elements listed in the Periodic Table are employed in the production of bioceramics. To be more specific, bioceramics have the potential to be manufactured using compounds including alumina, zirconia, magnesia, carbon, silica, calcium, and a select few additional chemicals. Therefore, it would seem that all of these compounds, in addition to calcium phosphates and calcium sulfates, as well as glass-ceramics and certain glasses, are true examples of bioceramics. Carbon is classified as a ceramic despite the fact that it is not a compound but rather an element and that the graphite form of carbon is able to conduct electricity. This is because carbon has numerous characteristics that are similar to those of ceramics. At the moment, researchers are looking into new types of sophisticated bioceramics. These bioceramics may consist of certain compositions of organic–inorganic hybrids or organized mesoporous silica materials. Coatings, scaffolds, granules, particles, powders, and/or crystals may be formed from any of these compounds throughout the manufacturing process. In addition, these compounds can be manufactured in bulk in forms that are either porous or dense, depending on the desired characteristics [54].

1.7 PROPERTIES OF BIOCERAMICS

Bioceramics are advantageous in that they may be employed in a vast range of medical applications because they possess a number of characteristics that make them so. The following is an examination of the many characteristics of bioceramics. The typical features of ceramics conformed as dense pieces are shown in Figure 1.5. These qualities are critical in determining the behavior of ceramics when they are used as biomaterials.

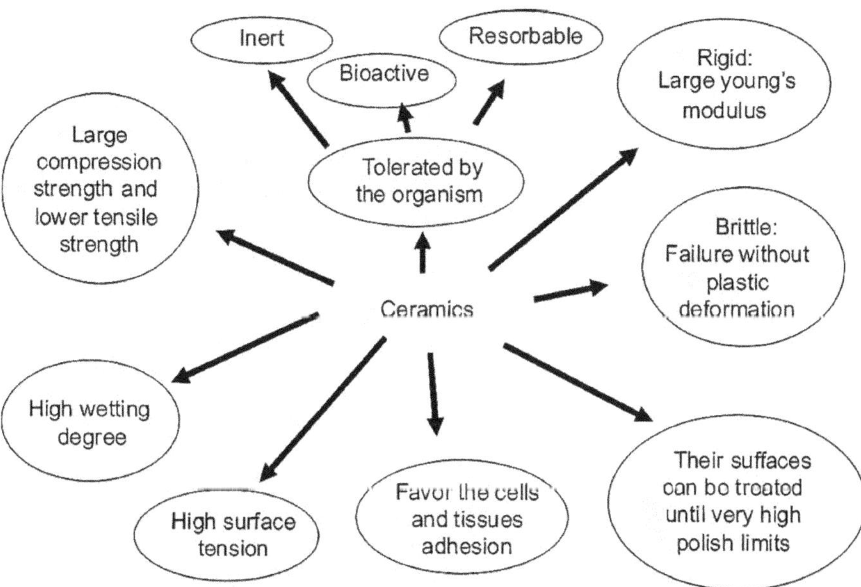

FIGURE 1.5 Ceramic characteristics of implant suitability. (Reprinted from Ref. [5] with permission from Elsevier.)

1.7.1 Porosity

Porosity is a morphological feature that is independent of the material, and it is described as a proportion of open holes in solids. Porous bodies provide a substantially larger surface area, which ensures strong mechanical attachment while also offering locations for chemical interaction between bones and bioceramics on the surface. Porous materials (also known as reticulated materials) have an open-cell aspect that is critical in many applications. Pore connectivity allows for cell movement and dispersion. It is also effective in in vivo blood vessel development that may support bone tissue neo-formation and remodeling [55–57]. As a result, porous bioceramics are readily colonized by cells and bone tissues [58]. Therefore, solid bioceramics with interconnected macroporosity are purposefully introduced [52]. Furthermore, macroporosity may be created artificially by the release of several readily removed substances, which is why the most common method of achieving macroporosity is the use of porogens (pore-creating additives). The porogens are particles, crystals, or threads of volatile or soluble compounds that emit gases at high temperatures. The popular examples include hydrogen peroxide [59], sucrose [60], naphthalene [61], and paraffin [62]. Creating porous 3D bioceramics is analogous to sintering particles, ideally spheres of identical sizes. However, the pores created by this approach are often uneven in size and form and are not completely linked. At a temperature of 250°C and a duration of 24–48 hours, both natural porous materials, namely shells [63] or coral skeletons [63], and artificially manufactured ones [64] can be transformed into porous bioceramics without microstructures being damaged. Furthermore, porous bioceramics may be made by hardening self-setting formulations [59,65]. In addition to that, superporous bioceramics with a porosity of 85% were created as well [66]. Figure 1.6 shows diagrams of several forms of ceramic porosity.

1.7.2 Mechanical Properties of Bioceramics

Modern bioceramics are manufactured with the goal of enhancing the body's innate capacity to heal itself. There must be no lapse in mechanical supporting structure throughout this process, so that mature bone can substitute the contemporary grafts. The strength of dense bioceramics is determined by the grain sizes. Grain sizes that are lower than those of bigger bioceramics result in fewer flaws at the grain borders, making them stronger than the larger ones. As a result, the strength of ceramics is inversely proportional to the square root of grain sizes [68]. Grain sizes, microporosity, and amorphous phase diminish mechanical qualities, whereas tiny grain sizes, low porosity, and high crystallinity increase fracture toughness, tensile and compressive strength, and stiffness. Slow fracture propagation seems to reduce ceramic strength. For mechanical characteristics, composition, porosity, grain boundaries, grain size, and crystallinity seem to be governing factors for bioceramics [69].

By altering the geometry of the pores in porous bioceramics, it is possible to alter the material's strength. It is also important to point out that porous bioceramics have a much lower resistance to fatigue compared to thick bioceramics. It has been found that grain sizes and porosity both have an effect on the fracture route; however, fracture toughness is not affected by the fracture route itself [69].

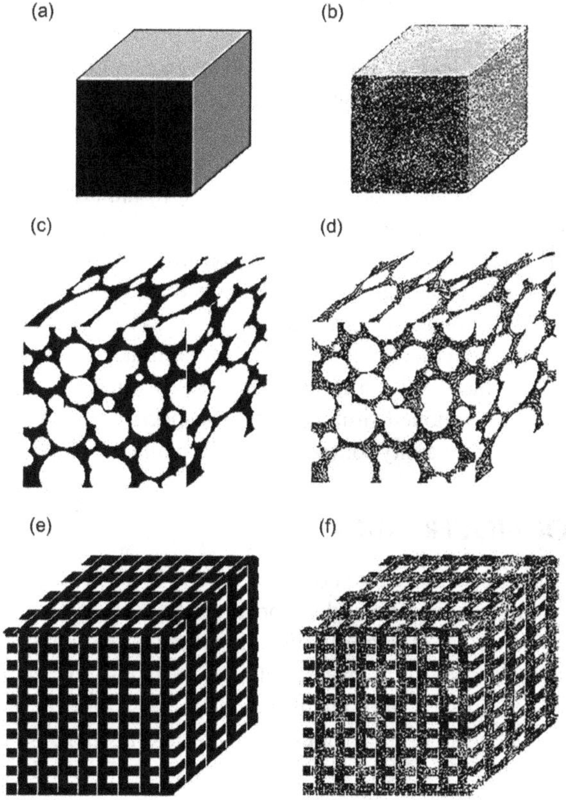

FIGURE 1.6 Variety of porosity in ceramic materials: (a) nonporous, (b) microporous, (c) macroporous (spherical), (d) macroporous (spherical) + microporous, (e) macroporous (3D printing), and (f) macroporous (3D printing) + micropores. (Reprinted from Ref. [67] with permission from Elsevier.)

Bioceramics have limited biological uses due to their high brittleness (which is related to poor break resistance). As a result, methods to increase their dependability are constantly investigated. Specifically, a variety of reinforcements (polymers, metals, and ceramics) have been used to create different hybrid biomaterials and biocomposites [70]. Coating bioceramics with polymers [71] or infiltrating porous structures with polymers is another technique to improve the mechanical characteristics of bioceramics [72].

1.7.3 Possible Transparency

Many ceramic materials have single crystals that are transparent to visible light. Unless colored dopants are added, bioceramics are opaque and white in color due to their polycrystalline structure and random orientation of enormous numbers of tiny crystals. In other circumstances, though, transparency is necessary to deliver some important benefits (for instance, to permit osteogenic differentiation cascade in transmitted light, direct observation of live cells, etc.). This method has been used to generate transparent bioceramics and examine their properties [73]. At a wavelength of 645 nm, they are capable of displaying an optical transmittance of 66% [74]. Some of the preparation techniques comprise spark plasma sintering [75], gel casting coupled with low-temperature sintering [76], and hot isostatic pressing [77]. Temperatures higher than 800°C are required to produce bioceramics that are both fully dense and transparent. Regardless of the process that was used to manufacture the transparent bioceramics, the grain size of the material is always uniform, and there are no pores present in the material. In addition to that, there are also things called transparent bioceramics [78]. It would appear that bioceramics that are transparent to laser light may be effective for minimally invasive surgery. This is because they would allow the laser beam to flow through them and treat the injured tissues that are located beneath. Transparent and translucent bioceramics will be used rarely in medicine, with the exception of the aforementioned situations and prospective eye implants, owing to a lack of porosity and the necessity for transparent implants inside the body.

1.8 TYPES OF BIOCERAMICS

Ceramics are nonmetallic inorganic composites with biological inertness and strong compressive strength, making them ideal for regenerative biomaterials usage and TE scaffolds in hard tissue regeneration. Silicates, graphite, and carbon structures like diamond, selenides, refractory hydrides, sulfides, carbides, and metallic oxides are the ceramics that are utilized in biomedical applications most frequently. Ceramics have a few flaws, such as poor mechanical qualities and brittleness due to their great hardness. Ceramics have low thermal conductivity, poor electrical conductivity, and a high melting point, which are all highly favorable. Because of the rapid progress made in the study of materials, ceramics are now being used in the field of biomedical engineering in the form of future biocomposites and bioceramics [7].

1.8.1 Bioceramics Based on Tissue Interactions

On the basis of how they interact with the tissues, we may divide bioceramics into the following three categories:

1. Resorbable bioceramics or biodegradable or bioabsorbable ceramics (noninert)
2. Bioactive ceramics or surface reactive ceramics (semi-inert)
3. Bioinert ceramics

Figure 1.7 presents several examples of ceramics materials as well as a categorization structure for these materials.

1.8.1.1 Resorbable Bioceramics

Ceramics that are resorbable are made of materials that divide into inorganic chemical species or smaller constituent particles by going through biochemical reactions with the help of a physiological medium. The endogenous tissue in the body will eventually reabsorb these ceramics after they have degraded to their constituent parts.

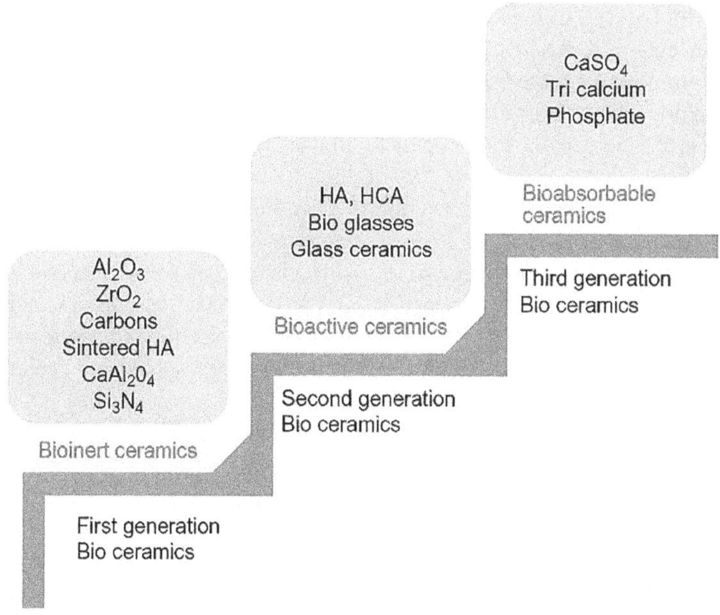

FIGURE 1.7 Classification of bioceramics. (Reprinted from Ref. [7] with permission from Elsevier.)

The pieces of deteriorated ceramics are transported through the metabolic system, where they do not have any negative impact on the body at any point in the process. As an orthopedic implant or TE construct, it deteriorates over time and is eventually substituted by endogenous tissues, allowing for the repair and regeneration of normal, functioning bone in the long term. This occurs whether the construct is an implant or a tissue engineer [7].

The ability of these materials to form osteoid, which refers to their osteoconductive properties or their capability to facilitate the integration of osteoblasts with bone, is an essential characteristic of these materials. Calcium sulfate dihydrate, hydroxyapatite (HA), and calcium phosphates are the most prevalent materials. The rates of material degradation and osteoid development differ depending on the kind of material. Because of the nature of the reabsorption process, the mechanical properties of these materials often undergo severe deterioration. As a consequence, the integrated structure will have a dramatically altered load-bearing capability over the duration of the procedure. TE has made utilization of resorbable scaffolds that have been seeded with cells, which have the ability to act as a synthetic extracellular matrix. Natural ceramics and synthetic resorbable ceramics are two types of ceramics. Calcium phosphate and HA are synthetic bioresorbable ceramics. It crystallizes into salts like HA, the primary constituent of teeth and bone. The exceptional biocompatibility of HA allows it to be linked directly to hard tissues like bone. Coral is a naturally occurring bioresorbable ceramic, and when processed, it may be transformed into HA, a mineral that is biocompatible with physiological conditions, stimulates bone production, and can even be used to mend bone that has been injured. The scaffold made from these materials is utilized to replace bone diseases and deformities as a template [79].

Once combined with the body's tissues, resorbable bioceramics decompose completely and are substituted by the host tissue. In point of fact, the physical collapse of the whole structure is caused by chemical attacks on grain boundaries. They might potentially be reabsorbed by cells like osteoclasts or disintegrate in the fluids that make up the body. When their resorption is taken into consideration, it makes sense to employ them as scaffolds for TE or as systems for the delivery of medications. They are also used in the production of resorbable screws for use in operations to repair the anterior cruciate ligament (ACL). Concerns in this domain include the maintenance of stability prior to the host tissue's capacity for replacement and the matching of desorption rates with repair rates. Calcium phosphates and calcium aluminates are two well-known types of this substance [80].

1.8.1.1.1 Calcium Phosphate Bioceramics

CaPs are chemical compounds that are of special importance for TE applications because of their striking resemblance to the inorganic portion of major normal and pathological calcified tissues of humans [81]. These bioceramics have exceptional biological properties, including biocompatibility, osteoconductivity, and bioresorbability, allowing them to integrate into live tissue using the same mechanisms that are used to remodel bones. CaPs are produced at a low cost, and it is quite simple to ascertain whether or not they are of a quality suitable for medicinal use. Despite this, CaPs are widely used in the biomedical industry as coatings and fillers. This is due to the fact that CaPs have poor mechanical properties, such as resistance to fatigue and low strength [82]. CaPs bioceramics may be found in a variety of forms, including composites with polymers, implant coatings, injectable formulations, dense or porous blocks, and particles [81]. There are other forms that are made specifically for the patient, such as vertebral cage fusion inserts, spine and knee cones, and wedges for tibial opening osteotomy. CaPs are utilized to augment the alveolar ridge, replace teeth, rebuild the maxillofacial area, install orbital implants, increase the size of the hearing ossicles, fuse the spine, and treat bone abnormalities [83].

Based on solubility and acidity, the most well-known CaPs have Ca/P molar ratios ranging from 0.5 to 2. CaP has a higher acidity and greater solubility in water for lower Ca/P molar ratios [84]. CaPs are very slightly soluble in water; however, they are all highly soluble in acids but insoluble in alkaline solutions. CaPs can be arranged in an increasing order of the in-situ degradation rate based on solubility: MCPM > TTCP ≈ α-TCP > DCPD > OCP > β-TCP > HA. Because implants consisting of calcined HA last for years in bone defects after implantation, attention in the biomedical area is typically centered on biphasic CaPs, CDHA, α- and β-TCP, and HA [85]. Below a pH of 4.2, HA exists in a crystalline form and is the CaP with the highest degree of stability and the lowest degree of solubility [82].

1.8.1.2 Bioreactive/Bioactive Ceramics

These are ceramic materials that are capable of establishing a chemical interaction with hard tissue and able to form a connection at the interface. Bioglass and ceravital are ideal examples of these materials, which are employed as composites for hard tissue restoration. Interfacial coatings and bone cement fillers are used in implants to improve biocompatibility and adaptation to physiological conditions. The coating of these compounds on the implant improves osteointegration with the surrounding tissue, reducing stress at the implant tissue contact. The addition of bioactive ceramic coatings to implants increases the surface area available for tissue healing. The creation of a layer atop bioactive ceramics is seen in Figure 1.8, and this layer communicates with cells to initiate the regeneration and repair process [7,86].

Bulk reactive ceramics are considered bioactive bioceramics. They have the ability to chemically attach to tissues. Bioactive ceramics have a wide variety of uses, some of which include the creation of porous scaffolds for bone TE, the coating of metallic substrates, and the biomineralization of bone. Bioactive bioceramics include things like HAs, bioactive glasses, and glass-ceramics.

1.8.1.2.1 Glass-Ceramics and Bioactive Glasses

In the disciplines of dentistry and orthopedics, glass-ceramics and bioactive glasses, in both dense and porous forms, have been developed for use in TE applications [87]. Glass-ceramics are crystalline glasses that have been thermally treated to increase their strength, wear resistance, elastic modulus, and toughness. After implantation, bioactive glasses may eventually change into an amorphous HA or CaP substance, and they may be able to bind to soft and hard connective tissues more quickly than other types of bioceramics. In addition, the ions of Na, P, Ca,

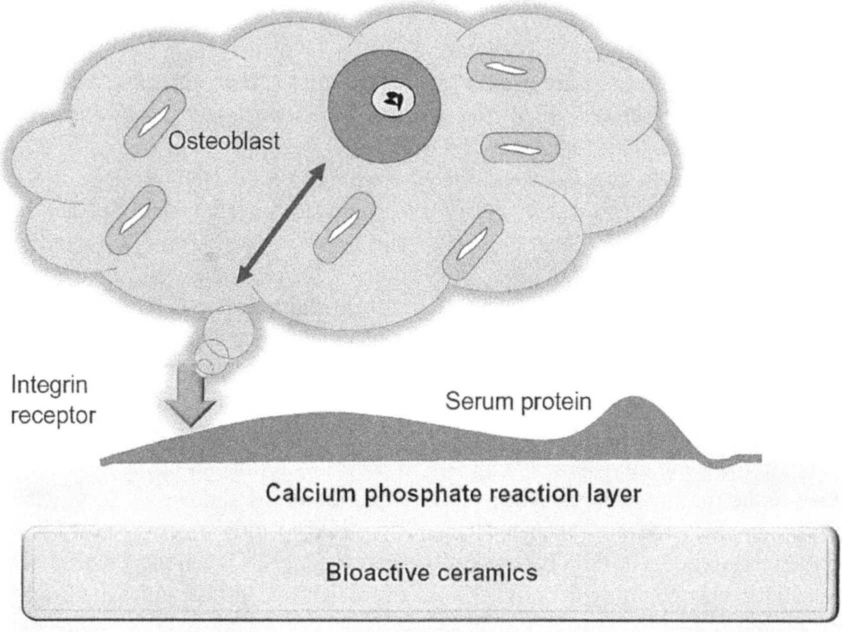

FIGURE 1.8 The cell signaling mechanism when the ceramic came in contact with biological fluids. (Reprinted from Ref. [7] with permission from Elsevier.)

and Si that are generated during the dissolving of particular bioactive glass compounds seem to induce neovascularization and angiogenesis, in addition to the differentiation of mesenchymal stem cells (MSCs) and enzymatic activity [88]. In the early 1970s, Larry Hench made a breakthrough in the field of bioactive glasses for biomedical activities when he invented the 45S5 Bioglass. This development was a watershed point in the history of the field [89]. The composition of bioglass, which is a silica-based bioactive glass, is equivalent to that of a ternary eutectic in the system composed of Na_2O–CaO–SiO_2–P_2O_5. Because there is a coating of hydroxycarbonate apatite on the surface of this kind of glass, it also has the unique quality of promoting bone development at a distance from the point of contact between the bone and the implant [87].

Borate-based and phosphate-based glasses are two further types of bioactive glasses that have been created for use in biomedical applications. Because bioactive phosphate glasses in the Na_2O–CaO–P_2O_5 combination dissolve more quickly in aqueous fluids than silica glasses, they are an excellent choice for use as carriers for complicated chemical compounds and antibacterial ions in chemotherapeutic applications [90]. When submerged in an aqueous phosphate solution, borate-based glasses in the B_2O_3–Na_2O–CaO–P_2O_5 system exhibit quick deterioration rates and may totally convert to apatite, comparable to bioglass, but without the creation of a silica-rich layer [91]. Borate glasses have also been utilized to treat bone infections as medication delivery methods [92]. Microwave irradiation, flame synthesis, sol–gel procedure, and traditional melt quenching are all typical ways to make bioactive glasses [93,94].

1.8.1.3 Bioinert Ceramics

Ceramics that are bioinert are a kind of material that is biocompatible and maintains its mechanical and physical characteristics after implantation. Two noteworthy characteristics of this substance are that it does not cause a biological reaction and that it does not cause cancer in the surrounding tissue. Bioinert ceramics are extensively utilized in orthopedic treatment to provide excellent biostructure support. Because of the better wear properties of these materials, gliding functions are an appropriate use for them. Al_2O_3, ZrO_2, silicon nitrides, and pyrolytic carbon are the most frequent compounds in this group. While maintaining its biocompatibility and high strength, carbon in the form of bioinert ceramics enables fine-tuning of conductivity and lubricity.

Wear resistance and little biological reaction are characteristics of inert bioceramics, which implies they can maintain stiffness, strength, and toughness over time. In addition to being inert, processing, size, and shape are further elements that contribute to the mechanical characteristics of bioceramics. Methods such as statistical distributions and fracture mechanics should be used to forecast crack propagation and the probability of failure behavior at various load levels when inert bioceramics are used for a significantly long time. In the event that bioinert ceramics are used for an extended period of time, biocompatibility would not be an issue. In biomedical applications, inert bioceramics may be utilized as a grouting agent or as a thick structure that attaches to tissue-by-tissue outgrowth that creates a mechanical connection between tissue and bioceramic through tissue ingrowths [7].

1.8.1.3.1 Alumina (Al_2O_3) and Zirconia (ZrO_2) Bioceramics

Alumina (Al_2O_3) and zirconia (ZrO_2) are bioinert ceramics that have been effectively employed in dental repairment and in orthopedics, notably for knee arthroplasty and total hip replacement [95].

The first Al_2O_3-based bioceramics to become available on the commercial market were used for the replacement of the acetabular cup in full hip prostheses and dental implants [96]. Al_2O_3 has a good balance of hardness and abrasion resistance, as well as surface energy and smoothness. This bioceramic has been used as synthetic bone grafts as well as porosity prosthetic devices. This was accomplished by applying a biomimetic coating over alumina in order to create a long-lasting link with the host tissue. Other therapeutic applications for alumina prostheses include segmental bone replacements, corneal replacements, ossicular (middle ear) bone substitutes, maxillofacial reconstruction, alveolar ridge (jaw bone), and bone screws [89].

The addition of zirconia may significantly improve the fracture toughness of alumina ceramics. The resultant composite material has improved toughness and tribological features. This material is also known as alumina-toughened zirconia or zirconia-toughened alumina (ZTA) [97]. The matrix phase of ZTA is alumina, which ranges from 70% to 95%, and the secondary phase is zirconia polycrystals (ZP), which range from 5% to 30%. This composition combines the advantages of monolithic zirconia and alumina. Additionally, ZTA ceramics retain the wear properties of alumina ceramics while also having a low susceptibility to stress-assisted deterioration. This reduces the risk of intrusion and displacement while also increasing the material's stability [97]. Because of their exceptional wear resistance, elastic modulus, high strength, and fracture toughness, zirconia-based bioceramics, specifically TZP, have been widely used in bone TE. This is because zirconia-based bioceramics have a polymorphic crystalline structure that includes cubic, tetragonal, and monoclinic phases [98]. For instance, partly stabilized zirconia materials (including yttria, CaO, and MgO) exhibit fracture toughness of over 8 $MPam^{1/2}$ and flexural strengths of over 1,000 MPa [99]. In addition to its mechanical characteristics, zirconia promotes cellular proliferation in osteogenic routes, as well as osseointegration. Furthermore, zirconia contains radiopaqueness, which assists in the monitoring of radiographs [100]. Zirconia is often used in dentistry because of the fact that it may be colored to correspond with the color of any existing teeth. But the biomedical uses of zirconia require stability of the structure. Various studies have been conducted in this regard. Golieskardi and his colleagues investigated the deterioration, tribological, and microstructural characteristics of a bioceramic that was doped with Al_2O_3 and CeO_2 and stabilized with 3-mol% yttria. The material was intended for use in biomedical applications. Their research centered on the development of stabilized zirconia ceramics that were doped with Al_2O_3 and CeO_2 and were produced using a sintering method that was simple, accessible, and inexpensive [101]. Another study examined zirconia stabilized with 3 mol% yttria (3Y-TZP) and the impact of Al_2O_3 and CeO_2 dopants on tetragonal phase stability and microstructure at 1,250°C–1,550°C. The mechanical characteristics of 3Y-TZP were shown to be temperature and dopant dependent. The optimal sintering temperature for all 3YTZP samples was 1,450°C, achieving 98% of the theoretical density (6.1 g/cm^3) [102]. In one study, MnO_2 and Al_2O_3 were tested to see how well they improved the mechanical characteristics and slowed the breakdown of tetragonal zirconia polycrystalline stabilized with 3-mol% yttria (3Y-TZP) ceramics. Two hours of pressureless sintering at temperatures ranging from 1,250°C to 1,550°C were the experimental conditions. MnO_2 and Al_2O_3 in 3Y-TZP improved mechanical characteristics and ageing resistance. Optimal dopant concentrations were 0.4 wt% MnO_2 and 0.6 wt% Al_2O_3 [103].

The mechanical properties of the materials Al_2O_3 and ZrO_2 are broken down into tabular forms and shown in Tables 1.1 and 1.2, respectively [7].

1.8.2 Bioceramics in Nanotechnology

Bioceramics have the potential to be used either as discrete pieces or as nanobiomaterials in a variety of forms, including bulk (which may include coatings and hollow fiber membranes), fiber nanotubes, and particles. The nanoceramics have unique features—mechanical, chemical, and physical traits of exceptional quality,

TABLE 1.1
Mechanical Characteristics of Al_2O_3 [7]

Properties	Alumina ASTM F603	Alumina ISO 6474
Content (%)	≥99.5%	≥99.5%
Compressive strength (MPa)	4,000	–
Fracture toughness (K1c) MPa	–	5–6
Elastic modulus (GPa)	380	380
Flexural strength (MPa)	45,000	400
Weibull modulus	8	–
Density (g/m³)	>3.94	>3.90
Vickers hardness (GPa)	18	>2,000 (106 psi)
Average grain size (μm)	≤4.5	<7

TABLE 1.2
Mechanical Characteristics of ZrO_2 [7]

Properties	ZrO_2	ZrO_2 (MgO-stabilized)	Yttria-stabilized zirconia ASTM F1873
Content (%)	–	–	≥93.2% + yttria
Compressive strength (MPa)	1,074	–	1,200
Fracture toughness (K1c) MPa	6–15	–	15
Elastic modulus (GPa)	201	200	200
Flexural strength (MPa)	–	–	800
Ultimate strength (MPa)	7500c/420t	634	900
Weibull modulus	–	–	10
Density (g/m³)	6.1	–	>6.0
Vickers hardness (GPa)	12	–	1,200
Average grain size (μm)	–	–	≤0.6

good machinability, and formability—and can easily go through under pressing and sintering processes to take different forms at lower temperatures [7]. As scaffolds in TE, bioceramics are often used as bulk materials for the replacement of grafts. Resorbable bioceramics are often used for these purposes due to the fact that they have a propensity to deteriorate. In research, HA bioceramic scaffolds were chosen for growing adipose-derived stem cells and the influence of various micro- and nano-topographies on osteogenic differentiation, proliferation, and cell adhesion [104]. Calcium silicate scaffolds are also a good substrate for the osteogenic differentiation of MSCs. This is because silicon (Si) ions produced from these structures stimulate angiogenesis and osteogenesis. After strontium is introduced into the structure, the bulk characteristics of the scaffold change. This technique improves the scaffold's characteristics, such as inhibiting bone resorption, making it more suitable for treating bone defects [105]. Calcium phosphate scaffolds manufactured by 3D printing, which are often utilized as synthetic bone transplant alternatives, are another example of widespread bioceramic utilization [106].

Coatings made of bioceramics have been used in a variety of research. Coating medical devices and TE structures may help enhance their biocompatibility, osteoconductivity, and long-term durability. This is the goal of the coating process. As a coating material, bioceramics such as calcium phosphate, hydroxylapatite, and carbon are some examples of materials that find widespread use. Some examples of these structures include prosthetic heart valves [107]. Besides their application as a bulk material, bioceramics are of interest as a biomaterial because of their particle state. Bioceramics, much like particle biopolymers, have the potential to be employed as vehicles for the delivery of drugs. Insoluble and porous glass beads are an excellent transporter for therapeutic chemicals such as chemotherapeutics and radioactive isotopes for the treatment of cancer. Antigens, antibodies, and enzymes can also be transported effectively using glass beads. Graphene oxide nanoparticles are an innovative and exciting kind of bioceramic that has the potential to be employed in applications involving drug delivery [108]. One more illustration of this is a research project that targeted bone abnormalities by attaching the medication risedronate, which is used to treat osteoporosis, to the surface of HA nanoparticles [109].

Hallowed micro- and nanotubes are other sorts of bioceramics. Nanotubes made of carbon were among the first to be discovered when the field of nanotechnology was just getting started. They are hollow cylinders with high aspect ratios and have a diameter in the range of 0.7–2 nm. These hollow cylinders may either have several walls or just one wall. This form of bioceramic is useful for medication and gene delivery as well as tissue regeneration because of its electrical and thermal conductivity and large surface area. They have also been shown to be effective in hyperthermia treatment [110]. Both the size and wall number of graphene oxide nanotubes and carbon nanotubes are distinct from one another. Mediators of cell growth differentiation and biosensors are the most common applications [108]. Halloysite ($Al_2Si_2O_5(OH)_4$-$2H_2O$) is a new nanotube-structured material that occurs naturally. This aluminosilicate has a structure that resembles hollow tubes with two layers and sizes ranging from 15 to 100 nm. This hollow area might be exploited as a nanoreactor, as well as a cell attachment and proliferation substrate for poorly soluble medicines [111]. TiO_2 nanotubes are another kind of new structure that has a variety of applications in biological research. These tubular structures are made by anodizing a Ti substrate and have been used in orthopedic implants, HA modulation, antibacterial substrates, biosensors, and drug delivery [112].

The breakdown of bioresorbable CaPs in the biological system, which occurs via both cell-mediated and physical mechanisms, ultimately results in the production of new bone, as seen in Figure 1.9. The Ca/P ratio determines the solubility of calcium phosphate in physiological media. If there is a drop in the proportion of calcium to phosphorus, then the solubility of CaP will go up. Furthermore, the incorporation of additional metallic ions into HA, such as Sr^{2+} and Mg^{2+}, results in an increase in the compound's solubility. Crystallinity, particle size, and density might all be made smaller in order to increase soluble capacity [113]. To produce fibrous scaffolds that most closely resemble the natural ECM, nanofibers of bioceramics such as CaPs, C, TiO_2, Al_2O_3, and ZrO_2 can be employed. Electrospinning has been shown to be the most flexible method for the production of nanofibrous bioceramics [114].

1.9 APPLICATIONS OF BIOCERAMICS

Not only may bioceramics be used in a variety of medical fields, such as orthopedics, but they can also be used in dentistry. In the field of medicine, applications of

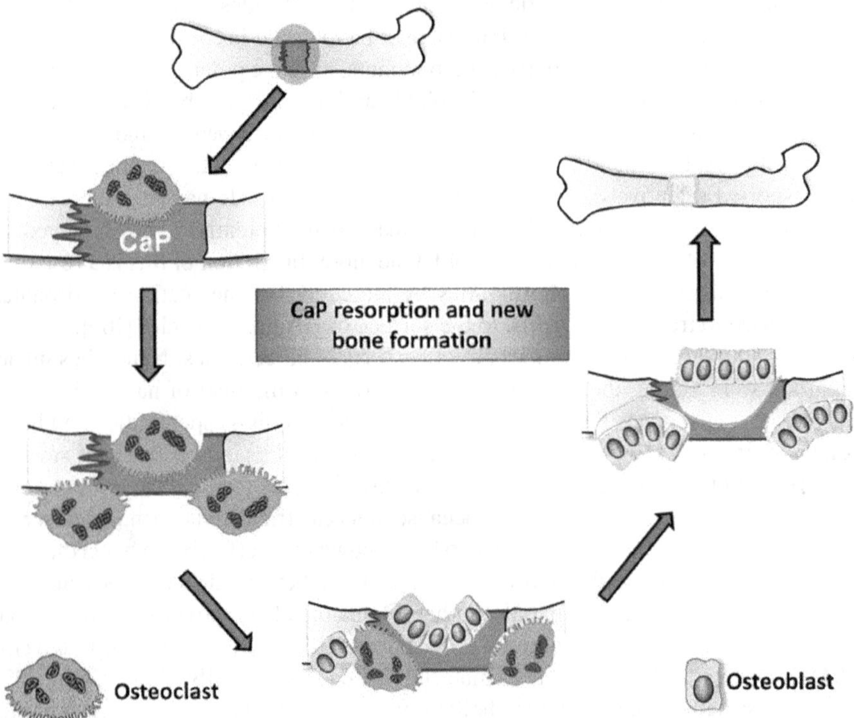

FIGURE 1.9 The osteoclastic resorption of bone using TE techniques. (Reprinted from Ref. [113] with permission from Elsevier.)

bioceramics include prosthetics for the knees, hips, ligaments, and tendons [115], and dental [116], as well as stabilization of bony dental arches [117], implantation [118], maxillofacial reconstruction [119], and alveolar ridge augmentation [120]. Recent advancements in the utilization of ceramics in TE of bones include the introduction of bone space fillers after the removal of tumors and spinal fusion [121]. The many applications of bioceramics found in the human body are shown in Figure 1.10 [122].

1.9.1 Tissue Engineering and Orthopedic Usage

Due to high bone regenerating potentials, bone defects regenerate rapidly with appropriate therapy. A key accomplishment in the field of bone regeneration is the development of scaffolds that can contain MSCs and growth factors. Materials that may be used for bone scaffolding include bioglass, HA, and TCPs either by themselves or in composites. Surgeons are given synthetic bone transplant materials in the form of putty, particle, and porous 3D scaffolds [123]. Dental procedures such as maxillofacial reconstruction, periodontal pocket reduction, and alveolar ridge augmentation can all benefit from the use of these short-term bone gap fillers. Figure 1.11 shows the use of bioceramics in the study and practice of bone TE [124].

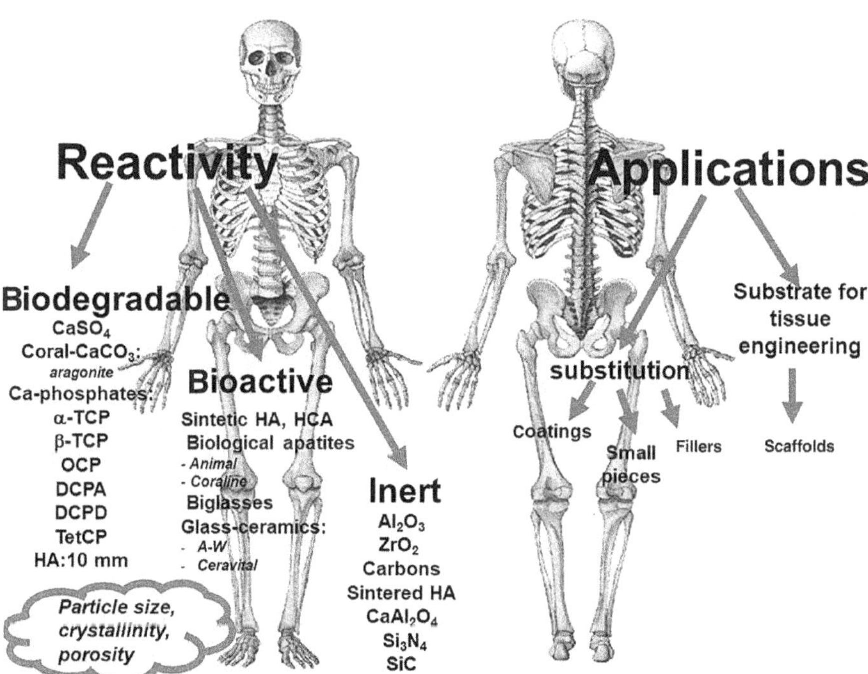

FIGURE 1.10 Clinical usage of bioceramics in the human body (TCP, tricalcium phosphate; OCP, ortho-calcium phosphate; DCPA, dicalcium phosphate anhydrous; DCPD, dicalcium phosphate dihydrate; TetCP, tetra-calcium phosphate; HA, hydroxyapatite; HCA, carbonated hydroxyapatite). (Reprinted with permission from Ref. [122] Copyright 2019, De Gruyter.)

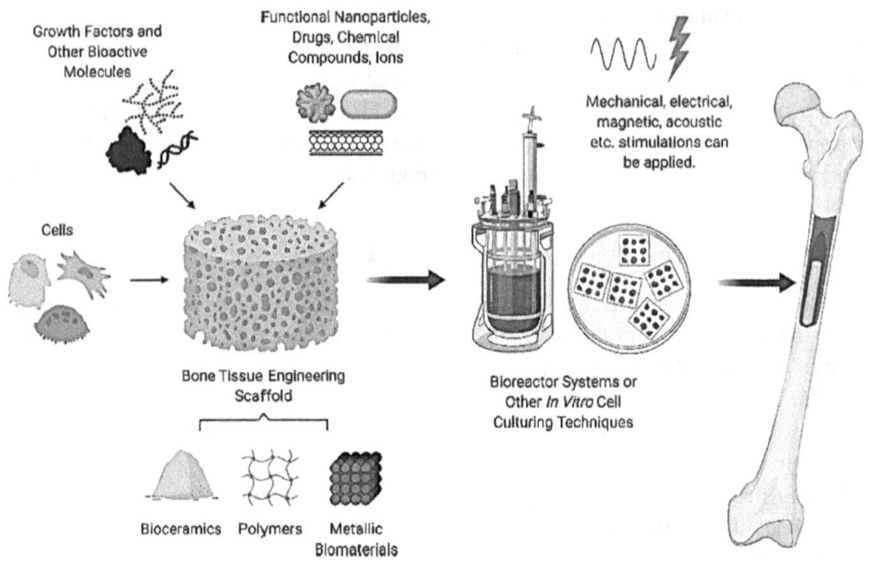

FIGURE 1.11 Incorporation of bioceramics into the process of tissue engineering for bone. (Reprinted from Ref. [124] with permission from Elsevier.)

1.9.1.1 Tissue Engineering and Drug Delivery

In spite of the fact that ceramics are often used in orthopedics and dentistry, the ceramic scaffold has a number of drawbacks, including the fact that it degrades very slowly, has poor tensile strength, and has brittleness. Composite scaffolds that are becoming more popular are ones that replicate the natural structure of bone. The addition of polymer to ceramics may allow for changes in degradability and physical/chemical characteristics. In drug delivery and TE systems, a variety of biodegradable and biocompatible polymers are used [125].

Due to their osteoinductive and vascularization qualities, growth factors such as vascular endothelial growth factor (VEGF), basic fibroblast growth factor (bFGF), transforming growth factor, and bone morphogenic protein-2 (BMP-2) are often used in bone TE scaffolds. Poly(a-hydroxy ester)-based polymers such as poly-caprolactone (PCL), poly-(lactic-co-glycolic) acid (PLGA), polyglycolic acid (PGA), and polylactic acid (PLA) have been the subject of substantial study in the field of drug delivery systems. If polymers are employed to transport pharmaceuticals, the breakdown rate may be customized by modifying the lactic/glycolic ratio (in the case of PLGA), polymerization degree, and molecular weight. A hydrolysis process is used to degrade this ester-backbone polymer. Acid breakdown products of polymers may cause aseptic tissue inflammation, and the hydrophobicity of the products may have a substantial impact on cells that are capable of infiltrating the scaffolds. As a consequence of this, it would be beneficial to integrate the advantages of the two biomaterials in a polymer and bioceramic composite scaffold that is clearly defined. Extensive research is being done on integrating BCP, TCP, and HA as ceramic components in a PCL, PLGA, or PLLA matrix to achieve different degrading and mechanical characteristics.

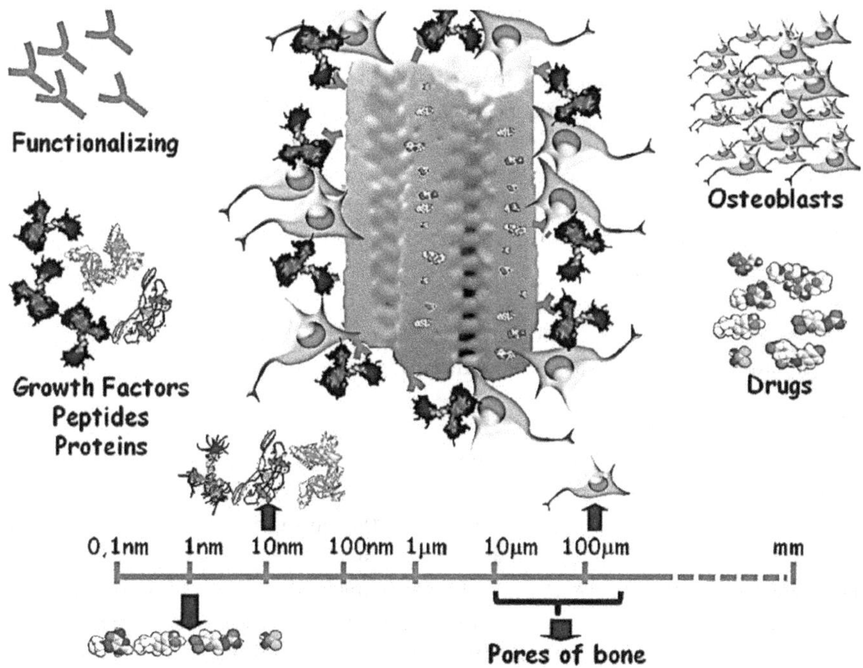

FIGURE 1.12 Mesoporous-based 3D scaffolds for drug delivery. (Reprinted from Ref. [126] with permission from The Royal Society of Chemistry, Copyright 2012.)

The process of bone regeneration, as well as local drug distribution, is shown in Figure 1.12 using mesoporous-based 3D scaffolds [126].

1.9.1.2 Bone-Filling Bioceramics

Bioglass products are surgically implanted into bony defect sites in orthopedics and dentistry, where new bone ingrowth develops after progressive resorption [123]. Bioglass in particle form is preferred by dentists and surgeons over monoliths because it is easier to fill a bone deficiency with it. PerioGlas (NovaBone Products LLC, Alachua, Florida) was the first commercially available particle bioglass. In "guided tissue regeneration," PerioGlas is used in combination with polymeric membranes. Bonalive (Bonalive Biomaterials, Turku, Finland) and Biogran (BIOMET 3i, Palm Beach Gardens, Florida) are two commercial bioglass products that have been utilized as bone grafts in dentistry and orthopedics [127].

1.9.1.3 Bone Constructs Using Tissue Engineering

Because there is a shortage of adequate autograft and allograft materials, tissue-engineered bone constructs are gaining favor as a means to satisfy increased demand due to their capacity to promote bone healing. The scaffolds that are employed in bone regeneration are often constructed of porous materials. These materials provide mechanical strength throughout the process of regenerating and repairing injured or diseased bone, a process that ultimately results in the bone's degeneration. Since the beginning of this decade, researchers have been consistently developing new advancements in bone TE. Their goals have been to create new materials by utilizing new processing methods and then to evaluate those materials for their mechanical support, inhibited angiogenesis and osteogenesis capacities, and their prospective usage [128].

Cell proliferation and adhesion were significantly improved when human mesenchymal stromal cells and osteoblasts were grown on graphene films [129]. With only 0.5 wt% graphenes, a 3D porous graphene/nano-bioglass 58S composite scaffold for bone TE showed compressive strength, improved fracture toughness as well as biocompatibility, and promising bioactivity, indicating its possible usage for TE of bones [130]. In comparison with pure bioglass, the spark plasma sintering process produced bioglass graphene nanoplatelets with greater electrical conductivity, sintering temperature, and viscosity without affecting bioactivity [131].

1.9.1.4 Arthroplasty

In the 1960s, Sir John Charnley did ground-breaking work in the field of full hip arthroplasty, which opened the door for reconstructive techniques to be used in other parts of the musculoskeletal system [132]. In the early 1970s, Pierre Boutin proposed bioceramics in orthopedics to address osteolysis surrounding the implant, which caused inflammation due to polyethylene wear [133]. In an attempt to lower aseptic loosening rates and enhance wear characteristics, ceramic components were originally introduced as a substitute bearing surface in the early 1970s [134]. Polycrystalline glass-ceramic, solid-state sintered ceramic, vitrified ceramic, plasma-sprayed polycrystalline ceramic, and glass are examples of bioceramics utilized for joint replacement or surgical reconstruction. The bioinert characteristic of a bioceramic is seen as an undeniable advantage for use as frictional pairs in artificial joint. Inside the body, bioinertness aids in preserving its properties. The length of time that these bioceramic applications will need to be implanted in the body is estimated by taking into consideration the fact that they are not constrained by factors such as age or prevalent diseases [133].

Pure alumina and pure zirconia were used to create the first generation of bioceramics. Lower purity and large grain sizes (>5 mm) were found in high-temperature sintered pure alumina. Because of poor fixation of the acetabular component, early failure was documented owing to increased fracture rates and unexpected roughening. To reduce grain size, MgO and CaO were added to ZrO_2 and Al_2O_3 in the second generation of bioceramics. The third generation of bioceramics went through hot isostatic pressing and hipped processing as well as pre-sterilization loading, increased purity, and grain size optimization so that it could give appropriate material properties. In comparison with the generations that came before it, the fourth generation of bioceramics, also known as Al_2O_3 matrix composite material, was introduced for medical use in the early 2000s and has shown significant improvements in terms of grain size, grain uniformity, and wear resistance. The toughness of fourth-generation ceramics was improved by including a certain amount of SrO and tetragonal ZrO_2 crystals in the composition. A small quantity of Cr_2O_3 was incorporated to improve hardness, which had been impaired due to the inclusion of zirconium [134].

Total hip replacement alumina bioceramics are mostly used as cups in Al_2O_3-on-Al_2O_3 fusion and as femoral heads articulating against polyethylene in total hip arthroplasty. Because of their fragile nature and expensive cost, utilization of first- and second-generation bioceramics was initially restricted owing to the increased risk of fracture. Fracture risk has decreased significantly in third- and fourth-generation bioceramics due to advancements in manufacturing techniques such as increased purity, density, grain size and dispersion, and enhanced quality control [133].

Since 1984, alumina has been a recommended material for arthroplasty (ISO 6474). Alumina has superior frictional qualities against itself, with a linear wear rate that is 4,000 times lower than metal on polyethylene. Al_2O_3 adhesive wear is reduced by lubrication between the fluid layer and hydrophilic Al_2O_3 surfaces. The grain separation and third-body wear in an Al_2O_3-on-Al_2O_3 combination can be reduced to a minimum by maintaining a clearance of about 50 nm between the elements. The creation of a small quantity of wear debris leads to a considerable biological reaction, according to both in vitro investigations and the examination of recovered implants [133].

Because of its exceptional mechanical qualities, ZrO_2 stabilized by yttrium has been the subject of much research for application in the biomedical sector, despite the fact that it is very difficult to sinter. The use of ZrO_2 for hip head replacement was suggested in 1969, and no detrimental bone or muscle integration or mechanical wear or behavior was detected. Solid ZrO_2 samples were declared noncytotoxic (and nonmutagenic) in 1990. ZrO_2 stabilized by yttrium has been utilized with success in assembling ball heads for replacing the hip completely. A properly constructed Morse taper makes it possible for perfect attachment of the bioceramic ball to the femoral stem. This helps to prevent the head from being subjected to forces that are not desired, and it also helps to prolong the life of the implant [133]. The majority of ceramic femoral heads now have a Morse taper that ranges from 12 to 14 mm, which allows them to be used with the vast majority of metal femoral stems. The bioceramic femoral heads have a fracture rate of 0.03% ZrO_2 and 0.02% for Al_2O_3, which indicates that there is no longer a problem with the fracturing of bioceramic heads. The fact that bioceramic heads are anywhere from six to ten times more costly than metal heads, however, means that cost is still an issue [133,134].

1.9.1.5 Instruments for Orthopedic Fixation

Bioceramic-based bone grafts are a suitable bone graft extender in lumber spine fusion when there is an osteoinductive stimulus present, such as a local bone graft. Because of the material's ease of sterilization, safety profile, inertness, and form flexibility, it is an excellent alternative to standard bone grafting extenders for lumbar spine fusions of the vertebrae. Despite this, its use is restricted since it is easily fractured, it has a low tensile strength, and it cannot be used without an osteoinductive adjunct [135].

Patients displayed a spinal fusion rate of 100% using apatite wollastonite-enriched glass-ceramic, which was the highest rate ever recorded. Other fusion rates are (in ascending order)—synthetic HA (86.7%) (Osteoset; Tennessee, Memphis), calcium sulfate (86.7%), coralline HA (86.9%) (Pro-Osteon 500, Pro-Osteon 200; Biomet, Warsaw, IN), β-TCP (92.5%) (Vitoss; Stryker, Kalamazoo, MI), and dense HA block (96.2%) [135].

Silicon nitride is a commercial ceramic that is robust and strong enough to be used as an implant material. In 1986, clinical research was carried out in Australia with the participation of thirty patients with the purpose of determining its viability as a lumbar spine arthrodesis device [132]. Since 2008, it has been successfully implanted as a fusion cage for thoracolumbar and cervical spine arthrodesis procedures. In the future, silicon nitride might be employed as an articulation member [132].

1.9.2 Cancer Treatment

For cancer treatment, radiation therapy is employed. However, in this way, radiation is needlessly exposed to normal cells, which might cause damage. Chemotherapy and immunotherapy, for example, are not affected by this limitation. The development of a treatment for cancer that eliminates just cancer cells while leaving normal tissue intact and enabling it to heal after treatment is an obvious goal. The vast majority of instances involving radiation therapy result in the destruction of healthy tissue, while the cancerous tumor gets an insufficient dose of radiation. This is especially true for cancers that have spread far into the body. Ceramic microspheres, on the other hand, could be able to help. $17Y_2O_3$–19 Al_2O_3–$64SiO_2$ glass microspheres having a diameter of 20–30 µm have been claimed to be beneficial for in situ irradiation of tumors. Yttrium-89 (^{89}Y), which has a half-life of 64.1 hours, gets radioactivated into producing β-rays when it comes into contact with neutrons. When injected into cancer's target organ, they get stuck within tiny blood capillaries in tumors, blocking the tumor's nutritional supply, and resulting in a significant localized dosage of β-rays. The dosage of β-rays that can be extracted from these glass microspheres, however, is insufficient due to the low yttrium concentration (17 mol%) of the glass. To boost the yttrium concentration in the microspheres, Kawashita and his colleagues employed a thermal plasma flame produced by high-frequency induction. This resulted in 20- to 30-µm-diameter cubic Y_2O_3 microspheres [136]. Concurrently, they used the plasma flame approach to create microspheres made mostly of YPO_4 crystals with some Y_2O_3 crystals.

The process of destroying cancer cells by heating them up with the application of a magnetic field is called hyperthermia. Cancer treated with hyperthermia involves heating tumors to temperatures ranging from 43°C to 47°C. This technique tries to kill cancer cells by subjecting them to higher temperatures, but it also has the potential to cause damage to normal cells if the elevated temperature is not confined to the area of the tumor that is being treated. The most significant drawback of hyperthermia is how difficult it is to both attain and sustain high temperatures at the tumor location. As a consequence of this, implementing this therapy for deep-seated cancers might be challenging from a clinical point of view. This treatment, when combined with other cancer therapies, such as radiation therapy and chemotherapy, will provide excellent results. By combining a bioactive sol–gel glass mixture with a magnetic glass-ceramic, Ruiz- Hernández was able to generate a biphasic material [137]. The magnetic glass-ceramic material in consideration was sufficient, according to the results of initial thermal experiments, to attain hyperthermia temperatures range for the purpose of bone tumor hyperthermia treatment. In order to accomplish this objective, a number of different ferromagnetic glass-ceramic systems have been developed, including magnetite/HA composite, Fe_2–O_3–CaO–ZnO–SiO_2–B_2O_3, SiO_2–Na_2O–CaO–P_2O_5–FeO–Fe_2O_3, FeO–Fe_2O_3–CaO–SiO_2, Li_2–O–MnO_2–CaO–P_2O_5–SiO_2, and ZnO–Fe_2O_3–CaO–SiO_2 [138].

1.9.3 Coatings

In modern medical practice, implants are often fabricated from metals such as titanium alloys, cobalt-chromium alloys, or stainless steel. Unfortunately, the vast majority of these implants are devoid of any bioactivity; as a result, a variety of bioactive implant surface coatings are used in order to accomplish osseointegration [139]. These coating materials should possess osteoinductivity, osteoconductivity, and biocompatibility. These materials should be mechanically stable and should not separate from the surface of the implants under physiologically stressed conditions. Implant surface coating materials should be biocompatible, osteoconductive, and osteoinductive, with appropriate mechanical stability under physiological stress without separating from the implant surface. They should also be antibacterial to limit the possibility of prosthetic infection [140]. Titanium niobium nitride, titanium nitride, zirconium nitride, diamond-like carbon, HA, and calcium phosphates are all examples of bioceramic surface coatings [132,141].

Bioceramics have tough metallic substrates that help them to avoid any fracture caused by brittleness. As bioceramics are hard and resistive to wear, they minimize fragmental wear production. By serving as a buffer between the metallic implant and human bone, this material helps to lower the risk of allergic responses caused by soluble metal ions. These ions include Ni^{+2}, Cr^{+3}, and Co^{+2}. All of these properties of bioceramics coating combinedly help to articulate prostheses efficiently [139]. Bioceramic coating materials of thinner dimensions have been shown to pit, scrape, or even delaminate in vivo, resulting in prosthesis failure [132]. There is only 2% of hip and knee joint prosthesis cases where the joint is loosened without any attack of diseases for a timeline of 10 years [139].

1.9.3.1 Coatings of Calcium Phosphates

Within the calcium phosphate family, only HA and tricalcium phosphate have been subjected to extensive testing as orthopedic implants. To avoid difficulties associated with poly(methyl methacrylate) (PMMA) use, coating of HA on sockets and the femoral prosthesis is a common fixing method. After an average of 8.1 years of follow-up, just 0.3% of the 324 implants in the multicenter study that was conducted in the United States required femoral revision. However, it is yet unknown whether or not HA enhances prosthetic fixation in comparison with bone cement [133].

In order to achieve the needed fixation, it seems that the type of HA present in the metal substrate, chemical composition, surface roughness, and coating thickness are all crucial aspects. Because of its brittle character and low tensile strength, the practical applicability of HA as a bone-graft replacement has been severely restricted. On the therapeutic use of ceramic bone-graft replacements, there is only a limited amount of information accessible. The use of tricalcium phosphate in spinal fusion led to outcomes that were comparable to those obtained with autogenous bone [133].

1.9.3.2 Surface Coatings of HA

HA has shown promise as a candidate for use as a coating in bone implantation, both in medical trials and in in vitro models. It is a bioactive ceramic material having osteoconductivity and has been demonstrated to enhance osteoblastic differentiation and adhesion. HA causes spontaneous deposition of apatite on the surface of the implant owing to an ion exchange mechanism that takes place between the HA coating and the physiological fluids that surround it [142]. When HA coatings, which are doped with fluoride, are used on the surface of the metallic implants, they promote cell differentiation and proliferation. But it takes a longer time to form a new coating of bone. Moreover, the force of adhesion is significantly decreased between the coatings and the substrate [143]. When HA coating fails to attach to a coated implant, resorption and delamination in the implant are observed [144]. Ion-beam-aided deposition, electrophoretic deposition, sputter coating, dip coating, pulsed laser deposition, plasma spraying, thermal spraying, sol–gel, and hot isostatic pressing are some of the new implant coating methods that use HA [144].

1.9.3.3 Nitride Coatings

TiN coatings were launched in the late 1980s for the purpose of full joint arthroplasty. These implants were subjected to clinical investigation after a period of 10 years. The thickness of Ni coatings (nanocrystalline) that were used ranged from 1 to 15 μm. At the present time, these coatings are referred to as femoral knee components or hip resurfacing implants, which are used for total hip arthroplasty [132,145].

1.9.4 Dentistry

Bioceramics, such as $Li_2O_5Si_2$-reinforced ceramics (Mg, Fe)Al_2O_4 (spinel), Al_2O_3, and ZrO_2, can be employed for the purpose of replacing metals in permanent prostheses of teeth [146]. Study showed that ZrO_2 reinforced ceramics had the greatest average load-bearing capabilities, followed by Al_2O_3, while $Li_2O_5Si_2$ had the lowest. Reinforced ceramics are only useful for building a three-unit bridge or an anterior crown repair that is only resistant to chewing loads on the front teeth [147].

Alumina cores are a popular choice for anterior porcelain veneer crown cores [148]. The Al_2O_3-reinforced core porcelain systems Procera AllCeram core, In-Ceram Zirconia core, In-Ceram Alumina core, and In-Ceram Spinell core are a few examples of commercially available porcelain system that can be used in a thickness of 0.5 mm [149]. Zirconia restorations are recommended for fixed dental prostheses that support teeth or implants [150–152].

Zirconia with alumina oxide outperformed all other ceramic-fixed dental prostheses in terms of initial and long-term strength [153]. Zirconia is more appealing than other materials due to its opaque nature and its ability to mimic the color of natural teeth. However, if translucency is essential, a surface coating of Al_2O_3 or $Li_2O_5Si_2$ may be used instead [154]. Software such as CAM or CAD is used in the manufacturing process of the fitting framework of ZrO_2 prosthesis [155].

Bioceramics are insoluble in tissue fluids and are able to be completely bonded chemically to teeth surfaces. They are 100% osteoconductive, radiopaque, and biocompatible. They are easy to handle and able to seal hermetically. Because of these physical features, they have found widespread usage in endodontics [156].

Calcium phosphate is employed for an apical barrier, hard tissue development apexification, and pulp capping. Sealer made of calcium phosphate has a lower risk as compared to AH26 sealers that are based on epoxy and ZnO-eugenol sealers. Clinical experiments and animal testing show HA's effectiveness in endodontic operations such as pulp capping, apical barrier formation, regenerative endodontics, periapical defects, and furcation perforation repair [156]. Torabinejad and White developed the MTA filling material with the intention of using it as a retrograde filler for the repair of dental pulp and tissues [157]. In 1995, MTA was patented and approved for use in endodontics [158]. MTA has gained favor in surgical and nonsurgical endodontic applications because of its bioactivity, biocompatibility, and effective endodontic sealability [159,160]. The commercial calcium silicate-based root-end filling cements that are most commonly found include EndoSequence Root Repair Material (Brasseler, USA), BioAggregate (Innovative BioCeramix Inc., Canada), Biodentine (Septodont, USA), calcium-enriched mixture (CEM) cement, White ProRoot MTA (Dentsply Sirona, USA), and Gray ProRoot MTA (Dentsply Sirona, USA) [161]. CEM cement, a novel endodontic filling material, was released in 2006 [156]. This cement's first setting time flow and film thickness are all good [162]. Its sealing performance is equivalent to that of MTA, despite the fact that it has a quicker aqueous setting time than MTA. CEM cement is preferable to MTA cement because it has an antibacterial impact that is comparable to that of calcium hydroxide [163]. The therapeutic applications of CEM cement are similar to those of MTA cement. CEM has a performance that is comparable to that of MTA when it is put as a pulp capping agent or used for repairing furcation perforation [162,164].

The field of dental implantology has seen the introduction of ZrO_2 implants as an alternative to titanium ones. They are biocompatible and have less ion leaching than metallic implants [165]. Zirconia is an appealing candidate for use as an implant material due to its minimal affinity for plaque, biocompatibility, high fracture toughness, high strength, and tooth-like color it has [166]. Furthermore, zirconium oxide causes less bone resorption and inflammation than titanium [165,166]. Moreover, it is difficult to perform modifications on the surface of ZrO_2 implants. As a consequence, osseointegration does not occur efficiently because the surfaces of implants are flat. For these reasons, the clinical use of ZrO_2 as dental implants is restricted. Sandblasting titanium implants surface with spherical ZrO_2 particles may help increase their osseointegration [165].

Enamel on teeth is the hardest tissue in the human body because it is highly mineralized and contains around 97% HA. The protection and upkeep of tooth enamel are the primary emphases of modern dental care. Enamel is a tissue that does not regenerate once it has emerged from the tooth; hence this tissue is considered to be permanent. In the oral cavity, enamel may be exposed to acids in one of two ways: first, it can be eroded by acidic meals or beverages, which is referred to as acid erosion; second, the crystallites of enamel can be partially demineralized during the caries process in dental plaque. Because of their closeness in composition to natural enamel, calcium phosphates have been singled out as potentially useful biomimetic solutions for the remineralization of tooth enamel [167]. In the realm of oral care, calcium phosphates such as HA, amorphous CaP, β-TCP, and are some of the more common types of calcium phosphates that are used [167,168].

Patients who use kinds of toothpaste containing HA had improved periodontal health, as well as enamel remineralization and higher acidic resistance. This is because the HA helps form a protective layer on the surface of the enamel, which helps prevent tooth decay and gum disease [169–171]. However, mouthwash containing HA was shown to decrease gingival index and plaque [172]. In situ investigations have shown that HA particles inhibit earlier biofilm formation on the surface of enamel and induce initial carious lesions to remineralize [173,174]. The usage of HA-enriched mouthwash has been shown to lower plaque and gingival index [172]. Fluoride-based dentifrices have been shown to be more effective in reversing the effects of white spot lesions and enamel deterioration than amorphous—calcium phosphate complexes (CPP-ACP)—supplemented toothpaste, according to the findings of a number of different in situ investigations [175,176]. Hegde et al. discovered that chewing gums containing amorphous calcium phosphate enhanced saliva buffering capacity [177]. Because of its remineralization impact, calcium silicates have the potential to be utilized to treat enamel white spot lesions and dentin hypersensitivity [178]. According to in vitro experiments, applying calcium silicate cement on the surface of dentin instantly lowers the permeability of dentin by penetrating into tubules of dentin and producing a thick, dense layer of apatite on the surface of the dentin [179,180]. It was observed that the remineralization of demineralized enamel created by 1,000 ppm fluoride treatment was equivalent to that generated by tricalcium silicate therapy. The use of these two together, on the other hand, leads to a substantially larger reduction in enamel demineralization than other therapies [181].

Bioglass obstructs dentinal tubules by being deposited on dentin surfaces and causing carbonated HA to develop. Only a few studies on bioglass's ability to remineralize enamel have been published. Burwell et al. found that a 10-day enamel treatment with a bioglass/5,000 ppm fluoride combination produced much better remineralization than 5,000 ppm fluoride alone [178]. Porcelain veneers have been a remedy for unsightly anterior teeth since the early 1980s. Clinical investigations have shown that porcelain veneers have a satisfactory clinical performance due to the fact that they do not have any adverse effects on gingival health, that they are very satisfying to patients, and that they have wonderful aesthetics [182]. Dental ceramics are constantly improving, giving dentists more options for creating highly esthetic and effective ceramic veneers. Sintered feldspathic and pressable ceramic porcelain are examples of ceramic materials that have adequate translucency. These ceramics have the potential to be used in the production of dental veneers in the form of thin layers. Ceramics with a glassier or noncrystalline microstructure have a greater degree of transparency than ceramics with a crystalline microstructure [183].

Traditional dental veneers are constructed of feldspar porcelain, which is a kind of porcelain that is burned at temperatures higher than 870°C and is composed of kaolin, quartz, and feldspar. As a result of the brittle nature of feldspar-based ceramics, more crystalline ceramics with better mechanical qualities have been invented. Examples of such ceramics are Al_2O_3 and ZrO_2. Ceramics with a high crystalline structure are often utilized as core materials in prosthetic dentistry. On the other hand, ceramics based on feldspar are typically used for veneers [182].

After the glass ionomer cement (GIC) has been set, the incorporation of apatite crystals (in nanometer size) into the fluoroaluminosilicate glass powder not only improves the mechanical characteristics of the GICs but also results in an increase in bioactivity as well as fluoride elution. The incorporation of HA into GICs was shown to increase their chemical stability as well as their insolubility and binding strength with teeth [184,185]. In apical surgery, bioceramics demonstrated much greater clinical success rates than other retrograde filling materials, ranging from 86.4% to 95.6% [161].

1.9.5 Ocular Implant

1.9.5.1 Al_2O_3-Based Prosthesis

There is a lot of potential in using Al_2O_3, which is highly pure, instead of synthetic HA or coralline [186]. Al_2O_3 has been used in the field of orthopedics for many decades now due to its non-reactive nature, biocompatibility, and excellent mechanical qualities [187]. In the year 2000, the United States Food and Drug Administration gave their approval to the porous form of Al_2O_3, and it is now being sold under the trade name "Bioceramic Implant" [188].

1.9.5.2 Porous Hydroxyapatite-Based Prosthesis

Perry unveiled the porous coralline HA sphere in the middle of the 1980s, and it was not until the early 1990s that it was used in a clinical setting. Following primary enucleation, it rapidly became the most preferred kind of ocular implant, and the brand name "Bio-eye sphere" was used to sell the product to consumers [189]. However, coralline HA was pricey; hence, synthetic HA became popular as orbital implants as a result of the fact that they not only were affordable but also facilitated easy pegging-drilling in comparison with Bio-eye. As a direct consequence of this, there was a reallocation of market share between the two distinct forms of HA [186].

The connected porosity design of a HA ocular implant, in contrast to that of a silicon ocular implant, makes it possible for fibrovascular ingrowth of host tissues. This might possibly reduce the risk of infection, migration, and extrusion [190]. Ocular implants made of HA are first coated in a protective substance and then inserted into the orbit [191,192]. Without completely removing the implant, exposed HA ocular implants may be efficiently treated using patched grafts of oral mucosa, dermis, and sclera. This eliminates the need for extensive surgery [193].

Guthoff et al. developed a composited eye implant known as "quasi-integrated implants." It was made up of two parts: the front half was made of synthetic porous HA, while the back section was composed of silicon rubber [194]. It allowed for transverse suturing of the ocular muscles anterior to the implant, which has the potential to increase the implant's stability and mobility [195].

1.9.5.3 Silicon-Based Prosthesis

Silicon is a physiologically and chemically inert substance that is flexible, simple to manipulate, and inexpensive. Porous or solid silicon episcleral implants are the only devices that are medically authorized for use in scleral buckling during retinal detachment surgery and are available for commercial purchase [196]. Nonporous silicon orbital implants were marketed around the end of the 1980s. These were attached to four rectus muscles after being wrapped around the central portion of the muscular cone. In the absence of pegging, they are still an excellent option. Fewer prosthetic movements are possible with these devices than with pegged or mounded ones, and the orbital implant has a tendency to migrate over the course of its lifetime [197].

Orbital implants made of silicon spheres with non-porosity are indicated for use in newborns and children between the ages of preschool and kindergarten, as well as in situations of serious gunshot wounds to the orbit. Silicon orbital implants, which are commercially available in the USA under the brand names "Flexiglass eye" and "Flexiglass system," had a much lower incidence of pre- and post-pegging complications such as pyogenic granuloma and hypo-ophthalmos than other types of implants [198].

1.9.5.4 Glass-Based Prosthesis

Glass is a noncrystalline ceramic material that is composed of oxides and has a significant history of application in the area of medicine [186]. In 1985, Mules successfully implanted the earliest non-integrated orbital glass device after being eviscerated [199]. The Mules implant was a fragile glass spherical with a hollow interior that, upon impact, had a high chance of shattering and, when subjected to temperature fluctuations, posed the danger of imploding. Glass orbital implants are no longer used since they have been replaced by newer, superior materials for orbital implants [188].

1.9.6 Otolaryngologic Applications

Bioceramics such as glass, Ceravital, Bioverit, Macor, Al_2O_3, bioglass, and HA are used in the manufacturing process of passive middle ear implants [200]. In studies that were carried out in the year 2000, HA was shown to be the bioceramic that was favored the most in the USA for the replacement of ossicle chains [201]. Titanium, on the other hand, was far more popular in Europe than ceramics; in fact, titanium was utilized in 68% of instances, while HA was only used in 24% of cases [202]. Plastipore is a lightweight titanium implant that is commercially accessible. Many otologists find that these implants are much easier to manipulate than the more unwieldy and weighty ceramics for prostheses of the middle ear. Nevertheless, results from medical trials have indicated that the acoustic advantage of titanium is not very substantial. This is because various other variables influence postoperative hearing [203]. The choice of implant material is significantly influenced not only by the preferences of individual patients but also by the collective expertise of surgeons, which is why hybrid prostheses like Flex H/A and HAPEX are still utilized in a justifiable manner [200].

1.10 CONCLUSIONS AND PROSPECTS

The development of biomaterials and implants has been significantly aided by bioceramics. Within the human body, bioceramics have been shown to be extraordinarily effective at healing and growing bone more quickly than would be possible by any other method. This has led to the technology's widespread adoption. In spite of the fact that these biomaterials are most often used in dental and orthopedic surgery, there is a chance that they may be suitable for a broad range of important utilization pertaining to TE. Implants and other types of medical devices are increasingly being made out of ceramics, both bioinert and bioactive. In recent years, one of the most widespread advances that have taken place is the use of bio-absorbable ceramics, as well as their application in the field of TE for the purpose of tissue calcification and the efficient regeneration of such tissue. The majority of these projects are still in the research phase and provide optimism for the innovation of novel technologies in the future. The difficult job in TE is to generate bioceramics, which would have the possibility to duplicate the intricate structure of natural calcified tissue. If this objective can be completed, then it may be said that the endeavor was successful. In the field of hard tissue repair, one of the most challenging jobs is the creation of scaffolds for bone TE that also include stem cells. This might be considered one of the toughest undertakings. 3D printing and CAD principles should be used in the field of medicine regeneration for the purpose of designing scaffolds and implants in order to simulate the anatomical organization of the tissue. Nanoceramics have a greater capacity for interaction with cells and may stimulate the process of cell division in a manner that is more favorable to the regeneration of calcified tissue. The activation of osteoblasts and osteoclasts in vivo by nanoceramics leads to the formation of an active form of calcified tissue and the preservation of this form, both of which contribute to an improvement in the performance of orthopedic and dental implants. Investigating the pattern that emerges between the size of the bioceramics and the biological reaction is an important task for the study of the future. It is important to have a thorough understanding of the interactions between tissue and bioceramics, as well as the hierarchical system required for decades of service and the accompanying mechanical strength, in particular, the fatigue limit when subjected to periodic external stress.

ACKNOWLEDGMENTS

The authors would like to thank the Military Institute of Science and Technology (MIST), Dhaka, for overall supporting this work.

REFERENCES

1. Williams DF. On the mechanisms of biocompatibility. *Biomaterials.* 2008;29(20): 2941–53.
2. Naik KS. Chapter 25: Advanced bioceramics. In: Meena SN, Naik MM, editors. *Advances in Biological Science Research* [Internet]. Academic Press; 2019 [cited 2022 May 12]. p. 411–7. Available from: https://www.sciencedirect.com/science/article/pii/B9780128174975000252
3. Ana ID, Satria GAP, Dewi AH, Ardhani R. Bioceramics for clinical application in regenerative dentistry. *Adv Exp Med Biol.* 2018;1077:309–16.
4. Baino F, Vitale-Brovarone C. Bioceramics in ophthalmology. *Acta Biomater.* 2014;10(8):3372–97.
5. Vallet-RegíM, Salinas AJ. 6: Ceramics as bone repair materials. In: Pawelec KM, Planell JA, editors. *Bone Repair Biomaterials* (Second Edition) [Internet]. Woodhead Publishing; 2019 [cited 2022 May 10]. p. 141–78. (Woodhead Publishing Series in Biomaterials). Available from: https://www.sciencedirect.com/science/article/pii/B9780081024515000068
6. *Biomaterials Market Worth $47.5 Billion by 2025* [Internet]. [cited 2022 May 12]. Available from: https://www.marketsandmarkets.com/PressReleases/global-biomaterials.asp

7. Shanmugam K, Sahadevan R. 1: Bioceramics – An introductory overview. In: Thomas S, Balakrishnan P, Sreekala MS, editors. *Fundamental Biomaterials: Ceramics* [Internet]. Woodhead Publishing; 2018 [cited 2022 Apr 21]. p. 1–46. (Woodhead Publishing Series in Biomaterials). Available from: https://www.sciencedirect.com/science/article/pii/B9780081022030000019
8. Bobbio A. The first endosseous alloplastic implant in the history of man. *Bull Hist Dent.* 1972;20(1):1–6.
9. Gluck T. Presentation on modern surgery, In German. *Langenbecks Arch Klin Chir.* 1891;41:187–239.
10. Dreesmann H. Ueber Knochenplombirungl. *DMW-Dtsch Med Wochenschr.* 1893;19 (19):445–6.
11. Peltier LF. The use of plaster of Paris to fill defects in bone. *Clin Orthop.* 1961;21:1–31.
12. Albee FH. Studies in bone growth: Triple calcium phosphate as a stimulus to osteogenesis. *Ann Surg.* 1920;71(1):32–9.
13. Nystrom G. Plugging bone cavities with Kivanol-Plaster porridge. *Acts Chir Scandinav.* 1928;63:296.
14. Bahn SL. Plaster: A bone substitute. *Oral Surg Oral Med Oral Pathol.* 1966;21(5):672–81.
15. Edberg E. Some experiences of filling osseous cavities with plaster. *Acta Chir Scandinav.* 1930;67:313.
16. Rock M. Artificial spare parts for the interior and exterior of the human and animal body, In German. *Dtsch Reichspatent DRP.* 1933;583–9.
17. Nielson A. The filling of infected and sterile bone cavities by means of plaster of Paris. *Acta Chir Scandinav.* 1944;91:17–27.
18. Peltier LF, Bickel EY, Lillo R, Thein MS. The use of plaster of Paris to fill defects in bone. *Ann Surg.* 1957;146(1):61–9.
19. Peltier LF. The use of plaster of Paris to fill large defects in bone. *Am J Surg.* 1959;97:311–5.
20. Gourley IMG, Arnold JP. The experimental replacement of segmental defects in bone with a plaster of Paris-epoxy resin mixture. *Am J Vet Res.* 1960;21:1119.
21. Mitchell DF, Amos ER. Reaction of connective tissue of rats to implanted dental materials. *Int Dent Res.* 1957;35:59–60.
22. Mitchell DF, Shankwalker GB. Osteogenic potential of calcium hydroxide and other materials in soft tissue and bone wounds. *J Dent Res.* 1958;37(6):1157–63.
23. Smith L. Ceramic-plastic material as a bone substitute. *Arch Surg Chic Ill 1960.* 1963;87:653–61.
24. Hench LL, Splinter RJ, Allen WC, Greenlee TK. Bonding mechanisms at the interface of ceramic prosthetic materials. *J Biomed Mater Res.* 1971;5(6):117–41.
25. Boutin P. Total hip arthroplasty with sintered alumina prosthesis, in French. *Rev Chir Orthop.* 1972;58:229–46.
26. Blakeslee KC, Condrate Sr. RA. Vibrational spectra of hydrothermally prepared hydroxyapatites. *J Am Ceram Soc.* 1971;54(11):559–63.
27. Garrington GE, Lightbody PM. Bioceramics and dentistry. *J Biomed Mater Res.* 1972;6(1):333–43.
28. Griffin WL, Åmli R, Heier KS. Whitlockite and apatite from lunar rock 14310 and from Ödegården, Norway. *Earth Planet Sci Lett.* 1972;15(1):53–8.
29. Reed GW, Jovanovic S. Fluorine in lunar samples: Implications concerning lunar fluorapatite. *Geochim Cosmochim Acta.* 1973;37(6):1457–62.
30. Graham FL, van der Eb AJ. A new technique for the assay of infectivity of human adenovirus 5 DNA. *Virology.* 1973;52(2):456–67.
31. Driskell TD, Hassler CR, Tennery VJ, McCoy IR, Clarke WJ. Calcium phosphate resorbable ceramic: A potential alternative for bone grafting. *J Dent Res.* 1973;52:123–31.

32. Nery EB, Lynch KL, Hirthe WM, Mueller KH. Bioceramic implants in surgically produced infrabony defects. *J Periodontol.* 1975;46(6):328–47.
33. Roberts SC, Brilliant JD. Tricalcium phosphate as an adjunct to apical closure in pulpless permanent teeth. *J Endod.* 1975;1(8):263–9.
34. Denissen HW, Groot K de. Immediate dental root implants from synthetic dense calcium hydroxylapatite. *J Prosthet Dent.* 1979;42(5):551–6.
35. Heinmann RB, Lehmann HD. *Bioceramic Coatings for Medical Implants: Trends and Techniques.* Weinheim, Germany: Wiley-VCH Verlag GmbH; 2015. 496 p.
36. Salinas AJ, Vallet-Regí M. Evolution of ceramics with medical applications. *Z Für Anorg Allg Chem.* 2007;633(11–12):1762–73.
37. Hench LL, Polak JM. Third-generation biomedical materials. *Science.* 2002;295(5557):1014–7.
38. Rabkin E, Schoen FJ. Cardiovascular tissue engineering. *Cardiovasc Pathol Off J Soc Cardiovasc Pathol.* 2002;11(6):305–17.
39. Ratner BD, Bryant SJ. Biomaterials: Where we have been and where we are going. *Annu Rev Biomed Eng.* 2004;6:41–75.
40. Shi QH, Wang JF, Zhang JP, Fan J, Stucky GD. Rapid-Setting, Mesoporous, bioactive glass cements that induce accelerated in vitro apatite formation. *Adv Mater.* 2006;18(8):1038–42.
41. Manzano M, Arcos D, Rodríguez Delgado M, Ruiz E, Gil FJ, Vallet-Regí M. Bioactive star gels. *Chem Mater.* 2006;18(24):5696–703.
42. Vallet-Regí M. Ordered mesoporous materials in the context of drug delivery systems and bone tissue engineering. *Chem – Eur J.* 2006;12(23):5934–43.
43. Vallet-Regí M, Ruiz-González L, Izquierdo-Barba I, González-Calbet JM. Revisiting silica based ordered mesoporous materials: Medical applications. *J Mater Chem.* 2006 13;16(1):26–31.
44. Vallet-Regí M, Salinas AJ, Arcos D. From the bioactive glasses to the star gels. *J Mater Sci Mater Med.* 2006;17(11):1011–7.
45. Vallet-Regi M, Arcos D. Nanostructured hybrid materials for bone tissue regeneration. *Curr Nanosci.* 2006;2(3):179–89.
46. Williams DF. On the nature of biomaterials. *Biomaterials.* 2009;30(30):5897–909.
47. El-Meliegy E, van Noort R. History, market and classification of bioceramics. In: El-Meliegy E, van Noort R, editors. *Glasses and Glass Ceramics for Medical Applications* [Internet]. New York, NY: Springer; 2012 [cited 2022 May 10]. p. 3–17. Available from: https://doi.org/10.1007/978-1-4614-1228-1_1
48. Ducheyne P, Healy K, Hutmacher DE, Grainger DW, Kirkpatrick CJ, Ducheyne P, Healy K, Hutmacher DE, Grainger DW, Kirkpatrick CJ, editors. *Comprehensive Biomaterials II* (Second Edition, Vol. 7). Amsterdam, Netherlands: Elsevier; 2017. p. 4858.
49. Ratner BD, Hoffman AS, Schoen FJ, Lemons JE. *Biomaterials Science: An Introduction to Materials in Medicine* (Third Edition). Oxford, UK: Academic Press; 2013. p. 1573.
50. Kiaie N, Aavani F, Razavi M. 2: Particles/fibers/bulk. In: Razavi M, Thakor A, editors. *Nanobiomaterials Science, Development and Evaluation* [Internet]. Woodhead Publishing; 2017 [cited 2022 Feb 23]. p. 7–25. Available from: https://www.sciencedirect.com/science/article/pii/B9780081009635000021
51. Chowdhury MIS, Autul YS, Hoque ME. Free vibration and damping properties of the pineapple leaf fiber- and sisal fiber-based polymer composites. In: *Vibration and Damping Behavior of Biocomposites.* CRC Press; 2022.
52. Dorozhkin SV. Current state of bioceramics. *J Ceram Sci Technol.* 2018;9:353–70.
53. Hench LL, Day DE, Höland W, Rheinberger VM. Glass and medicine. *Int J Appl Glass Sci.* 2010;1(1):104–17.
54. Vallet-Regí M. Evolution of bioceramics within the field of biomaterials. *Comptes Rendus Chim.* 2010;13(1):174–85.

55. Funayama T, Noguchi H, Kumagai H, Sato K, Yoshioka T, Yamazaki M. Unidirectional porous beta-tricalcium phosphate and hydroxyapatite artificial bone: A review of experimental evaluations and clinical applications. *J Artif Organs*. 2021;24(2):103–10.
56. Jin L, Feng ZQ, Wang T, Ren Z, Ma S, Wu J, et al. A novel fluffy hydroxylapatite fiber scaffold with deep interconnected pores designed for three-dimensional cell culture. *J Mater Chem B*. 2013;2(1):129–36.
57. Panzavolta S, Torricelli P, Amadori S, Parrilli A, Rubini K, della Bella E, et al. 3D interconnected porous biomimetic scaffolds: In vitro cell response. *J Biomed Mater Res A*. 2013;101(12):3560–70.
58. De Godoy RF, Hutchens S, Campion C, Blunn G. Silicate-substituted calcium phosphate with enhanced strut porosity stimulates osteogenic differentiation of human mesenchymal stem cells. *J Mater Sci Mater Med*. 2015;26(1):5387.
59. Cheng Z, Zhao K, Wu ZP. Structure control of hydroxyapatite ceramics via an electric field assisted freeze casting method. *Ceram Int*. 2015;41(7):8599–604.
60. Le Ray AM, Gautier H, Bouler JM, Weiss P, Merle C. A new technological procedure using sucrose as porogen compound to manufacture porous biphasic calcium phosphate ceramics of appropriate micro- and macrostructure. *Ceram Int*. 2010;36(1):93–101.
61. Swain SK, Bhattacharyya S. Preparation of high strength macroporous hydroxyapatite scaffold. *Mater Sci Eng C Mater Biol Appl*. 2013;33(1):67–71.
62. Ribeiro GBM, Trommer RM, dos Santos LA, Bergmann CP. Novel method to produce β-TCP scaffolds. *Mater Lett*. 2011;65(2):275–7.
63. Zhang X, Vecchio KS. Conversion of natural marine skeletons as scaffolds for bone tissue engineering. *Front Mater Sci*. 2013;7(2):103–17.
64. Nguyen Xuan Thanh T, Maruta M, Tsuru K, Matsuya S, Ishikawa K. Three-dimensional porous carbonate apatite with sufficient mechanical strength as a bone substitute material. *Adv Mater Res*. 2014;891–892:1559–64.
65. Kunio I, Tsuru K, Pham TK, Maruta M, Matsuya S. Fully-interconnected pore forming calcium phosphate cement. *Key Eng Mater*. 2012;493–494:832–5.
66. Sakamoto M. Development and evaluation of superporous hydroxyapatite ceramics with triple pore structure as bone tissue scaffold. *J Ceram Soc Jpn*. 2010;118(1380):753–7.
67. Habraken WJEM, Wolke JGC, Jansen JA. Ceramic composites as matrices and scaffolds for drug delivery in tissue engineering. *Adv Drug Deliv Rev*. 2007;59(4–5):234–48.
68. Carter CB, Norton MG. *Ceramic Materials: Science and Engineering* (Second Edition). New York, USA: Springer; 2013. 766 p.
69. Pecqueux F, Tancret F, Payraudeau N, Bouler JM. Influence of microporosity and macroporosity on the mechanical properties of biphasic calcium phosphate bioceramics: Modelling and experiment. *J Eur Ceram Soc*. 2010;30(4):819–29.
70. Dorozhkin SV. Calcium orthophosphate-containing biocomposites and hybrid biomaterials for biomedical applications. *J Funct Biomater*. 2015;6(3):708–832.
71. Martínez-Vázquez FJ, Pajares A, Guiberteau F, Miranda P. Effect of polymer infiltration on the flexural behavior of β-tricalcium phosphate robocast scaffolds. *Mater Basel Switz*. 2014;7(5):4001–18.
72. Dressler M, Dombrowski F, Simon U, Börnstein J, Hodoroaba VD, Feigl M, et al. Influence of gelatin coatings on compressive strength of porous hydroxyapatite ceramics. *J Eur Ceram Soc*. 2011;31(4):523–9.
73. Han YH, Kim BN, Yoshida H, Yun J, Son HW, Lee J, et al. Spark plasma sintered superplastic deformed transparent ultrafine hydroxyapatite nanoceramics. *Adv Appl Ceram*. 2016;115(3):174–84.
74. Wang J, Shaw LL. Transparent nanocrystalline hydroxyapatite by pressure-assisted sintering. *Scr Mater*. 2010;63(6):593–6.

75. Yun J, Son H, Prajatelistia E, Han YH, Kim S, Kim BN. Characterisation of transparent hydroxyapatite nanoceramics prepared by spark plasma sintering. *Adv Appl Ceram.* 2014;113(2):67–72.
76. John A, Varma HK, Vijayan S, Bernhardt A, Lode A, Vogel A, et al. In vitro investigations of bone remodeling on a transparent hydroxyapatite ceramic. *Biomed Mater Bristol Engl.* 2009;4(1):015007.
77. Boilet L, Descamps M, Rguiti E, Tricoteaux A, Lu J, Petit F, et al. Processing and properties of transparent hydroxyapatite and β tricalcium phosphate obtained by HIP process. *Ceram Int.* 2013;39(1):283–8.
78. Chaudhry AA, Yan H, Gong K, Inam F, Viola G, Reece MJ, et al. High-strength nanograined and translucent hydroxyapatite monoliths via continuous hydrothermal synthesis and optimized spark plasma sintering. *Acta Biomater.* 2011;7(2):791–9.
79. Paramsothy M, Ramakrishna S. Biodegradable materials for clinical applications: A review. *Rev Adv Sci Eng.* 2015;4(3):221–38.
80. Baino F, Novajra G, Vitale-Brovarone C. Bioceramics and scaffolds: A winning combination for tissue engineering. *Front Bioeng Biotechnol* [Internet]. 2015 [cited 2022 Feb 23];3. Available from: https://www.frontiersin.org/article/10.3389/fbioe.2015.00202
81. Dorozhkin SV. Calcium orthophosphates in nature, biology and medicine. *Materials.* 2009;2(2):399–498.
82. Bohner M. Calcium orthophosphates in medicine: From ceramics to calcium phosphate cements. *Injury.* 2000;31 Suppl 4:37–47.
83. Eliaz N, Metoki N. Calcium phosphate bioceramics: A review of their history, structure, properties, coating technologies and biomedical applications. *Mater Basel Switz.* 2017;10(4):E334.
84. LeGeros RZ, LeGeros JP. Calcium phosphate bioceramics: Past, present and future. *Key Eng Mater.* 2003;240–242:3–10.
85. Daculsi G, Laboux O, Malard O, Weiss P. Current state of the art of biphasic calcium phosphate bioceramics. *J Mater Sci Mater Med.* 2003;14(3):195–200.
86. Ducheyne P, Mauck RL, Smith DH. Biomaterials in the repair of sports injuries. *Nat Mater.* 2012;11(8):652–4.
87. Jones JR. Review of bioactive glass: From Hench to hybrids. *Acta Biomater.* 2013;9(1):4457–86.
88. Gorustovich AA, Roether JA, Boccaccini AR. Effect of bioactive glasses on angiogenesis: A review of in vitro and in vivo evidences. *Tissue Eng Part B Rev.* 2010;16(2):199–207.
89. Greenspan DC. Glass and medicine: The Larry Hench story. *Int J Appl Glass Sci.* 2016;7(2):134–8.
90. Pickup DM, Newport RJ, Knowles JC. Sol-gel phosphate-based glass for drug delivery applications. *J Biomater Appl.* 2012;26(5):613–22.
91. Fu Q, Rahaman MN, Fu H, Liu X. Silicate, borosilicate, and borate bioactive glass scaffolds with controllable degradation rate for bone tissue engineering applications. I. Preparation and in vitro degradation. *J Biomed Mater Res A.* 2010;95A(1):164–71.
92. Xie Z, Cui X, Zhao C, Huang W, Wang J, Zhang C. Gentamicin-loaded borate bioactive glass eradicates osteomyelitis due to Escherichia coli in a rabbit model. *Antimicrob Agents Chemother.* 2013;57(7):3293–8.
93. Balamurugan A, Rebelo A, Kannan S, Ferreira JMF, Michel J, Balossier G, et al. Characterization and in vivo evaluation of sol-gel derived hydroxyapatite coatings on Ti6Al4V substrates. *J Biomed Mater Res B Appl Biomater.* 2007;81B(2):441–7.
94. Kashif I, Soliman AA, Sakr EM, Ratep A. Effect of different conventional melt quenching technique on purity of lithium niobate (LiNbO3) nano crystal phase formed in lithium borate glass. *Results Phys.* 2012;2:207–11.

95. Ghaemi MH, Reichert S, Krupa A, Sawczak M, Zykova A, Lobach K, et al. Zirconia ceramics with additions of Alumina for advanced tribological and biomedical applications. *Ceram Int.* 2017;43(13):9746–52.
96. Kolos E, Ruys AJ. Biomimetic coating on porous alumina for tissue engineering: Characterisation by cell culture and confocal microscopy. *Materials.* 2015;8(6):3584–606.
97. Kurtz SM, Kocagöz S, Arnholt C, Huet R, Ueno M, Walter WL. Advances in zirconia toughened alumina biomaterials for total joint replacement. *J Mech Behav Biomed Mater.* 2014;31:107–16.
98. Pieralli S, Kohal RJ, Jung RE, Vach K, Spies B. Clinical outcomes of zirconia dental implants: A systematic review. *J Dent Res.* 2017;96(1):38–46.
99. Nakamura K, Adolfsson E, Milleding P, Kanno T, Örtengren U. Influence of grain size and veneer firing process on the flexural strength of zirconia ceramics. *Eur J Oral Sci.* 2012;120(3):249–54.
100. Afzal A. Implantable zirconia bioceramics for bone repair and replacement: A chronological review. *Mater Express.* 2014;4(1):1–12.
101. Golieskardi M, Satgunam M, Ragurajan D, Hoque ME, Ng AMH. Microstructural, tribological, and degradation properties of Al2O3- and CeO2-doped 3 mol.% yttria-stabilized zirconia bioceramic for biomedical applications. *J Mater Eng Perform.* 2020;29(5):2890–7.
102. Golieskardi M, Satgunam M, Ragurajan D, Hoque ME, Ng AMH, Shanmuganantha L. Advanced 3Y-TZP bioceramic doped with Al2O3 and CeO2 potentially for biomedical implant applications. *Mater Technol.* 2019;34(8):480–9.
103. Ragurajan D, Golieskardi M, Satgunam M, Hoque ME, Ng AMH, Ghazali MJ, et al. Advanced 3Y-TZP bioceramic doped with Al2O3 and MnO2 particles potentially for biomedical applications: Study on mechanical and degradation properties. *J Mater Res Technol.* 2018;7(4):432–42.
104. Xia L, Lin K, Jiang X, Fang B, Xu Y, Liu J, et al. Effect of nano-structured bioceramic surface on osteogenic differentiation of adipose derived stem cells. *Biomaterials.* 2014;35(30):8514–27.
105. Lin K, Xia L, Li H, Jiang X, Pan H, Xu Y, et al. Enhanced osteoporotic bone regeneration by strontium-substituted calcium silicate bioactive ceramics. *Biomaterials.* 2013;34(38):10028–42.
106. Fahmy MD, Jazayeri HE, Razavi M, Masri R, Tayebi L. Three-dimensional bioprinting materials with potential application in preprosthetic surgery. *J Prosthodont.* 2016;25(4):310–8.
107. Surmenev RA, Surmeneva MA, Ivanova AA. Significance of calcium phosphate coatings for the enhancement of new bone osteogenesis – A review. *Acta Biomater.* 2014;10(2):557–79.
108. Liu J, Cui L, Losic D. Graphene and graphene oxide as new nanocarriers for drug delivery applications. *Acta Biomater.* 2013;9(12):9243–57.
109. Rawat P, Manglani K, Gupta S, kalam A, Vohora D, Ahmad FJ, et al. Design and development of bioceramic based functionalized PLGA nanoparticles of risedronate for bone targeting: In-vitro characterization and pharmacodynamic evaluation. *Pharm Res.* 2015;32(10):3149–58.
110. He H, Pham-Huy LA, Dramou P, Xiao D, Zuo P, Pham-Huy C. Carbon nanotubes: Applications in pharmacy and medicine. *BioMed Res Int.* 2013;2013:e578290.
111. Yuan P, Tan D, Annabi-Bergaya F. Properties and applications of halloysite nanotubes: Recent research advances and future prospects. *Appl Clay Sci.* 2015;112–113:75–93.
112. Cipriano AF, Miller C, Liu H. Anodic growth and biomedical applications of TiO2 nanotubes. *J Biomed Nanotechnol.* 2014;10(10):2977–3003.

113. Roy M, Bandyopadhyay A, Bose S. Chapter 6: Ceramics in bone grafts and coated implants. In: Bose S, Bandyopadhyay A, editors. *Materials for Bone Disorders* [Internet]. Academic Press; 2017 [cited 2022 May 15]. p. 265–314. Available from: https://www.sciencedirect.com/science/article/pii/B9780128027929000069
114. Balu R, Singaravelu S, Nagiah N. Bioceramic nanofibres by electrospinning. *Fibers*. 2014;2(3):221–39.
115. Bunpetch V, Zhang X, Li T, Lin J, Maswikiti EP, Wu Y, et al. Silicate-based bioceramic scaffolds for dual-lineage regeneration of osteochondral defect. *Biomaterials*. 2019;192:323–33.
116. Khalid H, Suhaib F, Zahid S, Ahmed S, Jamal A, Kaleem M, et al. Microwave-assisted synthesis and in vitro osteogenic analysis of novel bioactive glass fibers for biomedical and dental applications. *Biomed Mater Bristol Engl*. 2018;14(1):015005.
117. Zheng Y, Yang Y, Deng Y. Dual therapeutic cobalt-incorporated bioceramics accelerate bone tissue regeneration. *Mater Sci Eng C Mater Biol Appl*. 2019;99:770–82.
118. Furko M, Havasi V, Kónya Z, Grünewald A, Detsch R, Boccaccini AR, et al. Development and characterization of multi-element doped hydroxyapatite bioceramic coatings on metallic implants for orthopedic applications. *Bol Soc Esp Cerámica Vidr*. 2018;57(2):55–65.
119. Pina S, Rebelo R, Correlo VM, Oliveira JM, Reis RL. Bioceramics for osteochondral tissue engineering and regeneration. *Adv Exp Med Biol*. 2018;1058:53–75.
120. Rider P, Kačarević Ž P, Alkildani S, Retnasingh S, Schnettler R, Barbeck M. Additive manufacturing for guided bone regeneration: A perspective for alveolar ridge augmentation. *Int J Mol Sci*. 2018;19(11):E3308.
121. Ma H, Feng C, Chang J, Wu C. 3D-printed bioceramic scaffolds: From bone tissue engineering to tumor therapy. *Acta Biomater*. 2018;79:37–59.
122. Vallet-Regí M. Bioceramics: From bone substitutes to nanoparticles for drug delivery. *Pure Appl Chem*. 2019;91(4):687–706.
123. Gul H, Zahid S, Kaleem M. Bioglass, a new trend towards clinical bone tissue engineering. *Pak Oral Dent J*. 2015;35(4).
124. Jodati H, Yılmaz B, Evis Z. A review of bioceramic porous scaffolds for hard tissue applications: Effects of structural features. *Ceram Int*. 2020;46(10, Part B):15725–39.
125. Yang Y, Kang Y, Sen M, Park S. Bioceramics in tissue engineering. In: Burdick JA, Mauck RL, editors. Biomaterials for Tissue Engineering Applications: A Review of the Past and Future Trends [Internet]. Vienna: Springer; 2011 [cited 2022 Apr 30]. p. 179–207. Available from: https://doi.org/10.1007/978-3-7091-0385-2_7
126. Salinas AJ, Esbrit P, Vallet-Regí M. A tissue engineering approach based on the use of bioceramics for bone repair. *Biomater Sci*. 2012;1(1):40–51.
127. Abbasi Z, Bahrololoom ME, Shariat MH, Bagheri R. Bioactive glasses in dentistry: A review. *J Dent Biomater*. 2015;2(1):1–9.
128. Shadjou N, Hasanzadeh M. Graphene and its nanostructure derivatives for use in bone tissue engineering: Recent advances. *J Biomed Mater Res A*. 2016;104(5):1250–75.
129. Fan H, Wang L, Zhao K, Li N, Shi Z, Ge Z, et al. Fabrication, mechanical properties, and biocompatibility of graphene-reinforced chitosan composites. *Biomacromolecules*. 2010;11(9):2345–51.
130. Gao C, Liu T, Shuai C, Peng S. Enhancement mechanisms of graphene in nano-58S bioactive glass scaffold: Mechanical and biological performance. *Sci Rep*. 2014;4(1):4712.
131. Porwal H, Grasso S, Cordero-Arias L, Li C, Boccaccini AR, Reece MJ. Processing and bioactivity of 45S5 Bioglass((r))-graphene nanoplatelets composites. *J Mater Sci Mater Med*. 2014;25(6):1403–13.
132. McEntire BJ, Bal BS, Rahaman MN, Chevalier J, Pezzotti G. Ceramics and ceramic coatings in orthopaedics. *J Eur Ceram Soc*. 2015;35(16):4327–69.

133. Prakash L. Ceramics in arthroplasty, arthritis and orthopaedics. *Res Arthritis Bone Study*. 2018;1:4.
134. Jaiswal P. The use of ceramics in total hip arthroplasty. *Orthop Rheumatol Open Access J* [Internet]. 2017 [cited 2022 May 11];4(3). Available from: https://juniperpublishers.com/oroaj/OROAJ.MS.ID.555636.php
135. Nickoli MS, Hsu WK. Ceramic-based bone grafts as a bone grafts extender for lumbar spine arthrodesis: A systematic review. *Glob Spine J*. 2014;4(3):211–6.
136. Erbe EM, Day DE. Chemical durability of Y2O3-Al2O3-SiO2 glasses for the in vivo delivery of beta radiation. *J Biomed Mater Res*. 1993;27(10):1301–8.
137. Ruiz-Hernández E, Serrano MC, Arcos D, Vallet-Regí M. Glass-glass ceramic thermoseeds for hyperthermic treatment of bone tumors. *J Biomed Mater Res A*. 2006;79(3):533–43.
138. Shah SA, Hashmi MU, Alam S, Shamim A. Magnetic and bioactivity evaluation of ferrimagnetic ZnFe2O4 containing glass ceramics for the hyperthermia treatment of cancer. *J Magn Mater*. 2010;322(3):375–81.
139. Zhang BGX, Myers DE, Wallace GG, Brandt M, Choong PFM. Bioactive coatings for orthopaedic implants-recent trends in development of implant coatings. *Int J Mol Sci*. 2014;15(7):11878–921.
140. Tobin EJ. Recent coating developments for combination devices in orthopedic and dental applications: A literature review. *Adv Drug Deliv Rev*. 2017;112:88–100.
141. Brunello G, Brun P, Gardin C, Ferroni L, Bressan E, Meneghello R, et al. Biocompatibility and antibacterial properties of zirconium nitride coating on titanium abutments: An in vitro study. *PLoS One*. 2018;13(6):e0199591.
142. Lazarinis S, Kärrholm J, Hailer NP. Effects of hydroxyapatite coating of cups used in hip revision arthroplasty. *Acta Orthop*. 2012;83(5):427–35.
143. Wang J, de Groot K, van Blitterswijk C, de Boer J. Electrolytic deposition of lithium into calcium phosphate coatings. *Dent Mater Off Publ Acad Dent Mater*. 2009;25(3):353–9.
144. Mohseni E, Zalnezhad E, Bushroa AR. Comparative investigation on the adhesion of hydroxyapatite coating on Ti-6Al-4V implant: A review paper. *Int J Adhes Adhes*. 2014;C(48):238–57.
145. Subramanian B, Muraleedharan CV, Ananthakumar R, Jayachandran M. A comparative study of titanium nitride (TiN), titanium oxy nitride (TiON) and titanium aluminum nitride (TiAlN), as surface coatings for bio implants. *Surf Coat Technol*. 2011;21–22(205):5014–20.
146. Raigrodski AJ. Contemporary materials and technologies for all-ceramic fixed partial dentures: A review of the literature. *J Prosthet Dent*. 2004;92(6):557–62.
147. Lüthy H, Filser F, Loeffel O, Schumacher M, Gauckler LJ, Hammerle CHF. Strength and reliability of four-unit all-ceramic posterior bridges. Dent Mater Off Publ Acad Dent Mater. 2005;21(10):930–7.
148. Ozcan M, Vallittu PK. Effect of surface conditioning methods on the bond strength of luting cement to ceramics. *Dent Mater Off Publ Acad Dent Mater*. 2003;19(8):725–31.
149. Heffernan MJ, Aquilino SA, Diaz-Arnold AM, Haselton DR, Stanford CM, Vargas MA. Relative translucency of six all-ceramic systems. Part I: Core materials. *J Prosthet Dent*. 2002;88(1):4–9.
150. Larsson C, Vult von Steyern P, Sunzel B, Nilner K. All-ceramic two- to five-unit implant-supported reconstructions. A randomized, prospective clinical trial. *Swed Dent J*. 2006;30(2):45–53.
151. Piwowarczyk A, Ottl P, Lauer HC, Kuretzky T. A clinical report and overview of scientific studies and clinical procedures conducted on the 3M ESPE Lava All-Ceramic System. *J Prosthodont Off J Am Coll Prosthodont*. 2005;14(1):39–45.
152. Potiket N, Chiche G, Finger IM. In vitro fracture strength of teeth restored with different all-ceramic crown systems. *J Prosthet Dent*. 2004;92(5):491–5.

153. Tinschert J, Natt G, Mohrbotter N, Spiekermann H, Schulze KA. Lifetime of alumina- and zirconia ceramics used for crown and bridge restorations. *J Biomed Mater Res B Appl Biomater.* 2007;80(2):317–21.
154. McLaren EA, Giordano RA. Zirconia-based ceramics: Material properties, esthetics and layering techniques of a new veneering porcelain, VM9. *Quintessence Dent Technol.* 2005;28:99–111.
155. Manicone PF, Rossi Iommetti P, Raffaelli L. An overview of zirconia ceramics: Basic properties and clinical applications. *J Dent.* 2007;35(11):819–26.
156. Utneja S, Nawal RR, Talwar S, Verma M. Current perspectives of bio-ceramic technology in endodontics: Calcium enriched mixture cement – Review of its composition, properties and applications. *Restor Dent Endod.* 2015;40(1):1–13.
157. Lee SJ, Monsef M, Torabinejad M. Sealing ability of a mineral trioxide aggregate for repair of lateral root perforations. *J Endod.* 1993;19(11):541–4.
158. Parirokh M, Torabinejad M. Mineral trioxide aggregate: A comprehensive literature review – Part I: Chemical, physical, and antibacterial properties. *J Endod.* 2010;36(1):16–27.
159. Dammaschke T, Gerth HUV, Züchner H, Schäfer E. Chemical and physical surface and bulk material characterization of white ProRoot MTA and two Portland cements. *Dent Mater Off Publ Acad Dent Mater.* 2005;21(8):731–8.
160. Kim RJY, Shin JH. Cytotoxicity of a novel mineral trioxide aggregate-based root canal sealer [corrected]. *Dent Mater J.* 2014;33(3):313–8.
161. Abusrewil SM, McLean W, Scott JA. The use of bioceramics as root-end filling materials in periradicular surgery: A literature review. *Saudi Dent J.* 2018;30(4):273–82.
162. Asgary S, Eghbal MJ, Parirokh M, Ghanavati F, Rahimi H. A comparative study of histologic response to different pulp capping materials and a novel endodontic cement. *Oral Surg Oral Med Oral Pathol Oral Radiol Endod.* 2008;106(4):609–14.
163. Asgary S, Kamrani FA. Antibacterial effects of five different root canal sealing materials. *J Oral Sci.* 2008;50(4):469–74.
164. Samiee M, Eghbal MJ, Parirokh M, Abbas FM, Asgary S. Repair of furcal perforation using a new endodontic cement. *Clin Oral Investig.* 2010;14(6):653–8.
165. Özkurt Z, Kazazoğlu E. Zirconia dental implants: A literature review. *J Oral Implantol.* 2011;37(3):367–76.
166. Depprich R, Zipprich H, Ommerborn M, Naujoks C, Wiesmann HP, Kiattavorncharoen S, et al. Osseointegration of zirconia implants compared with titanium: An in vivo study. *Head Face Med.* 2008;4:30.
167. Meyer F, Amaechi BT, Fabritius HO, Enax J. Overview of calcium phosphates used in biomimetic oral care. *Open Dent J.* 2018;12:406–23.
168. Khan AS, Syed MR. A review of bioceramics-based dental restorative materials. *Dent Mater J.* 2019;38(2):163–76.
169. Harks I, Jockel-Schneider Y, Schlagenhauf U, May TW, Gravemeier M, Prior K, et al. Impact of the daily use of a microcrystal hydroxyapatite dentifrice on de novo plaque formation and clinical/microbiological parameters of periodontal health. A randomized trial. *PLoS One.* 2016;11(7):e0160142.
170. Lelli M, Putignano A, Marchetti M, Foltran I, Mangani F, Procaccini M, et al. Remineralization and repair of enamel surface by biomimetic Zn-carbonate hydroxyapatite containing toothpaste: A comparative in vivo study. *Front Physiol.* 2014;5:333.
171. Makeeva IM, Polyakova MA, Avdeenko OE, Paramonov YO, Kondrat'ev SA, Pilyagina AA. Effect of long term application of toothpaste Apadent Total Care Medical nanohydroxyapatite. *Stomatologiia (Sofiia).* 2016;95(4):34–6.
172. Hegazy SA, Salama RI. Antiplaque and remineralizing effects of Biorepair mouthwash: A comparative clinical trial. *Pediatr Dent J.* 2016;3(26):89–94.

173. Hannig C, Basche S, Burghardt T, Al-Ahmad A, Hannig M. Influence of a mouthwash containing hydroxyapatite microclusters on bacterial adherence in situ. *Clin Oral Investig.* 2013;17(3):805–14.
174. Kensche A, Holder C, Basche S, Tahan N, Hannig C, Hannig M. Efficacy of a mouthrinse based on hydroxyapatite to reduce initial bacterial colonisation in situ. *Arch Oral Biol.* 2017;80:18–26.
175. Meyer-Lueckel H, Wierichs RJ, Schellwien T, Paris S. Remineralizing efficacy of a CPP-ACP cream on enamel caries lesions in situ. *Caries Res.* 2015;49(1):56–62.
176. Wiegand A, Attin T. Randomised in situ trial on the effect of milk and CPP-ACP on dental erosion. *J Dent.* 2014;42(9):1210–5.
177. Hegde RJ, Thakkar JB. Comparative evaluation of the effects of casein phosphopeptide-amorphous calcium phosphate (CPP-ACP) and xylitol-containing chewing gum on salivary flow rate, pH and buffering capacity in children: An in vivo study. *J Indian Soc Pedod Prev Dent.* 2017;35(4):332–7.
178. Li X, Wang J, Joiner A, Chang J. The remineralisation of enamel: A review of the literature. *J Dent.* 2014;42 Suppl 1:S12–20.
179. Dong Z, Chang J, Deng Y, Joiner A. Tricalcium silicate induced mineralization for occlusion of dentinal tubules. *Aust Dent J.* 2011;56(2):175–80.
180. Mg G, F S, Pd H, G G, P C. Calcium silicate coating derived from Portland cement as treatment for hypersensitive dentine. *J Dent* [Internet]. 2008 Aug [cited 2022 May 12];36(8). Available from: https://pubmed.ncbi.nlm.nih.gov/18538913/
181. Wang Y, Li X, Chang J, Wu C, Deng Y. Effect of tricalcium silicate (Ca(3)SiO(5)) bioactive material on reducing enamel demineralization: An in vitro pH-cycling study. *J Dent.* 2012;40(12):1119–26.
182. Ho GW, Matinlinna JP. Insights on ceramics as dental materials. Part I: Ceramic material types in dentistry. *Silicon.* 2011;3(3):109–15.
183. Pini NP, Aguiar FHB, Lima DANL, Lovadino JR, Terada RSS, Pascotto RC. Advances in dental veneers: Materials, applications, and techniques. *Clin Cosmet Investig Dent.* 2012;4:9–16.
184. Moshaverinia A, Ansari S, Moshaverinia M, Roohpour N, Darr JA, Rehman I. Effects of incorporation of hydroxyapatite and fluoroapatite nanobioceramics into conventional glass ionomer cements (GIC). *Acta Biomater* [Internet]. 2008 Mar [cited 2022 May 11];4(2). Available from: https://pubmed.ncbi.nlm.nih.gov/17921077/
185. Najeeb S, Khurshid Z, Zafar MS, Khan AS, Zohaib S, Martí JMN, et al. Modifications in glass ionomer cements: Nano-sized fillers and bioactive nanoceramics. *Int J Mol Sci.* 2016;17(7):E1134.
186. Baino F, Potestio I. Orbital implants: State-of-the-art review with emphasis on biomaterials and recent advances. *Mater Sci Eng C Mater Biol Appl.* 2016;69:1410–28.
187. Chee E, Kim YD, Woo KI, Lee JH, Kim JH, Suh YL. Inflammatory mass formation secondary to hydroxyapatite orbital implant leakage. *Ophthal Plast Reconstr Surg.* 2013;29(2):e40–42.
188. Catalu CT, Istrate SL, Voinea LM, Mitulescu C, Popescu V, Radu C. Ocular implants-methods of ocular reconstruction following radical surgical interventions. *Romanian J Ophthalmol.* 2018;62(1):15–23.
189. Hornblass A, Biesman BS, Eviatar JA. Current techniques of enucleation: A survey of 5,439 intraorbital implants and a review of the literature. *Ophthal Plast Reconstr Surg* [Internet]. 1995 [cited 2022 May 12];11(2). Available from: https://pubmed.ncbi.nlm.nih.gov/7654621/
190. Gradinaru S, Popescu V, Leasu C, Pricopie S, Yasin S, Ciuluvica R, Ungureanu E. Hydroxyapatite ocular implant and non-integrated implants in eviscerated patients. *J Med Life* [Internet]. 2015 [cited 2022 May 12];8(1). Available from: https://pubmed.ncbi.nlm.nih.gov/25914747/

191. Babar TF, Hussain M, Zaman M. Clinico-pathologic study of 70 enucleations. *Jpma J Pak Med Assoc.* 2009;59(9):612–4.
192. Custer PL. Enucleation: Past, present, and future. *Ophthal Plast Reconstr Surg.* 2000;16(5):316–21.
193. Owji N, Mosallaei M, Taylor J. The use of Mersilene mesh for wrapping of hydroxyapatite orbital implants: Mid-term result. *Orbit Amst Neth.* 2012;31(3):155–8.
194. Jordan DR, Bawazeer A. Experience with 120 synthetic hydroxyapatite implants (FCI3). *Ophthal Plast Reconstr Surg.* 2001;17(3):184–90.
195. Choi HY, Lee JE, Park HJ, Oum BS. Effect of synthetic bone glass particulate on the fibrovascularization of porous polyethylene orbital implants. *Ophthal Plast Reconstr Surg.* 2006;22(2):121–5.
196. Baino F. Scleral buckling biomaterials and implants for retinal detachment surgery. *Med Eng Phys.* 2010;32(9):945–56.
197. Jordan DR, Klapper SR. Controversies in enucleation technique and implant selection: Whether to wrap, attach muscles, and peg? In: Guthoff RF, Katowitz JA, editors. *Oculoplastics and Orbit: Aesthetic and Functional Oculofacial Plastic Problem-Solving in the 21st Century* [Internet]. Berlin, Heidelberg: Springer; 2010 [cited 2022 May 12]. p. 195–209. (Essentials in Ophthalmology). Available from: https://doi.org/10.1007/978-3-540-85542-2_14
198. Baino F, Perero S, Ferraris S, Miola M, Balagna C, Verné E, et al. Biomaterials for orbital implants and ocular prostheses: Overview and future prospects. *Acta Biomater.* 2014;10(3):1064–87.
199. Anderson RL, Thiese SM, Nerad JA, Jordan DR, Tse D, Allen L. The universal orbital implant: Indications and methods. *Adv Ophthalmic Plast Reconstr Surg.* 1990;8:88–99.
200. Beutner D, Hüttenbrink KB. Passive and active middle ear implants. *GMS Curr Top Otorhinolaryngol Head Neck Surg.* 2009;8:Doc09.
201. Goldenberg RA, Emmet JR. Current use of implants in middle ear surgery. *Otol Neurotol Off Publ Am Otol Soc Am Neurotol Soc Eur Acad Otol Neurotol.* 2001;22(2):145–52.
202. Preuss SF, Luers JC, Beutner D, Klussmann JP, Huttenbrink KB. Results of a European survey on current controversies in otology. *Otol Neurotol Off Publ Am Otol Soc Am Neurotol Soc Eur Acad Otol Neurotol.* 2007;28(6):774–7.
203. Truy E, Naiman AN, Pavillon C, Abedipour D, Lina-Granade G, Rabilloud M. Hydroxyapatite versus titanium ossiculoplasty. *Otol Neurotol Off Publ Am Otol Soc Am Neurotol Soc Eur Acad Otol Neurotol.* 2007;28(4):492–8.

2 Processing of Bioceramics by Extrusion and Slip Casting

Sazedur Rahman
Ahsanullah University of Science and Technology (AUST)

Muhammad Ifaz Shahriar Chowdhury
Imperial College London

Mohammad Ahnaf Shahriar
The University of Texas at Arlington

Md Enamul Hoque
Military Institute of Science and Technology

2.1 INTRODUCTION

Ceramics are inorganic crystalline solids with relatively high melting points; therefore, their production and uses necessitate a high temperature. Metals and nonmetals, such as borides, carbides, silicides, and oxides, are combined by ionic or covalent bonding to form ceramics [1,2]. Advanced ceramics are distinguished from traditional ceramics by their superior mechanical and physical properties, as well as the fact that they are made from high-purity synthetic powders with precisely controlled microstructure and properties [3]. In addition, these advanced ceramics include bioceramics as a subcategory. Bioceramics are ceramics that are used to heal and restore sick or injured musculoskeletal components [4–6]. Bioceramics are used to repair and enhance both hard and soft tissues in the skeletal system, which includes bones, joints, and teeth [7–10]. The major purpose of bioceramic material selection, particularly for implant coatings and surface modifications, is to ensure that the metallic implant and surrounding connective tissue are stable [11]. As a result, Fe, Cu, Ce, Yb, Al_2O_3, MnO_2, etc. are sometimes added to bioceramics to make them suitable for a wide range of biological applications [12–15]. The crystalline and amorphous forms of bioceramics are both possible [16]. Alumina, hydroxyapatite (HA), zirconia, glass ceramics, etc. are examples of crystalline forms, while bioglass is an example of an amorphous material. Because of its high biocompatibility, HA is the most significant main bioceramic that may be widely employed in bone tissue engineering and bone defect repair [17–19]. Furthermore, to improve the bonding properties of

 DOI: 10.1201/9781003258353-3

implant materials, bioceramics can be utilized as a coating material for metallic implants, and these coatings are divided into three categories: bioactive, bioinert, and bioresorbable bioceramic coatings [20]. Bioceramics can be used in a number of biomedical applications as rigid or permeable materials with a particular shape, such as implants, prostheses, or prosthetic devices, as well as in powder form to fill the defective areas in the body [21]. Moreover, they come in a variety of phases, including single crystals, polycrystalline, glass, glass ceramics, and composites, and the phase or phases employed are determined by the characteristics and functions required [22]. However, based on bioceramic and tissue attachment processes, bioceramic composites may be classified into four categories, namely, bioinert, bioactive, biodegradable, and porous [16]. Table 2.1 illustrates the mechanical properties of various advanced bioceramics.

Bioceramics are mostly employed in orthopedics, with antimicrobial bioceramics being used to avoid early failure of orthopedic implants [27]. Nonetheless, in addition to bioceramics, biocomposites made up of biodegradable polymeric matrix and bioceramics have shown usefulness in clinics, notably in orthopedics and dentistry [28].

2.2 FABRICATION OF BIOCERAMICS

The manufacturing of bioceramics is divided into many fundamental processes, which are outlined below:

1. Powder processing
2. Shaping or forming of green body
3. Drying and removal of the binder
4. Sintering

This chapter will go through the process of shaping or forming the green body. However, before going on to the forming process, all steps will be briefly discussed.

2.2.1 Powder Processing

Fine, pure, and homogeneous microstructures are required for tough and robust ceramics. As a starting source, pure powders with a wide surface area and a small average size must be employed such that the required microstructure can be achieved. Nevertheless, ceramic powders frequently lack the properties required for the manufacture of sophisticated ceramic components such as bioceramics [29]. As a result, powder treatment or powder processing is necessary before forming and compaction in order to get appropriate properties. The majority of nanocrystalline powders on the market today aren't made up of single-crystal nanometer-sized particles. Crystallites are such particles; however, in most powders, the crystallites are linked together to create bigger units known as "agglomerates" or "aggregates." The distinction between aggregates and agglomerates is not well defined, although in general, aggregate crystallites are more densely and securely linked together, with less intercrystallite porosity, than agglomeration crystallites [30]. Figure 2.1 demonstrates the variation between agglomerated and aggregated powders. Following the powder

TABLE 2.1
Mechanical Properties of some Bioceramics

Bioceramics	Weibull Modulus (m)	Poisson's Ratio	Elastic Modulus (GPa)	Bending Strength (MPa)	Fracture Toughness (MPa $m^{1/2}$)	Hardness (GPa)	Ref.
Cortical bone	–	–	7–30	50–150	2–12	–	[23]
Bioglass 45S5	–	–	35	42	–	458 (HV)	[23]
Alumina bioceramic ($Al_2O_3 > 99.8\%$)	–	–	400	595	5–6	2,300 (HV)	[23]
Sintered hydroxyapatite	–	–	80–110	115–120	1.0	600 (HV)	[23]
Human tooth (enamel)	–	–	9–90	8–35	0.52–1.3	3.2–4.4	[24]
Human tooth (dentin)	–	–	11–20	31–104	2.8–3.1	0.25–0.8	[24]
Pressable ceramic (IPS Empress I)	9.0	–	65 (1.5)	106 (17)	1.2 (0.14)	6.5 (0.4)	[25]
Pressable ceramic (IPS Empress II)	8.0	–	90 (3.7)	303 (49)	3.0 (0.65)	5.5 (0.2)	[25]
Feldspathic porcelain (Vita Mark II)	–	–	69.7	–	1.19 (0.05)	6.3 (0.3)	[26]
Glass-infiltrated alumina ceramic (Vita In-Ceram Alumina)	9.5	0.25	265 (10)	440 (50)	3.6 (0.26)	11 (1.1)	[25]
Full sintered zirconia ceramic (Vita Zirkon)	7.5	0.33	220 (7.5)	840 (140)	7.4 (0.62)	12 (0.2)	[25]
Glass-infiltrated zirconia ceramic (Vita In-Ceram Zirconia)	10.5	0.26	240 (9)	476 (50)	4.9 (0.36)	11 (0.9)	[25]
Glass-ceramic (MGC – fine)	–	–	70.5	–	1.04 (0.04)	4.15 (0.07)	[26]

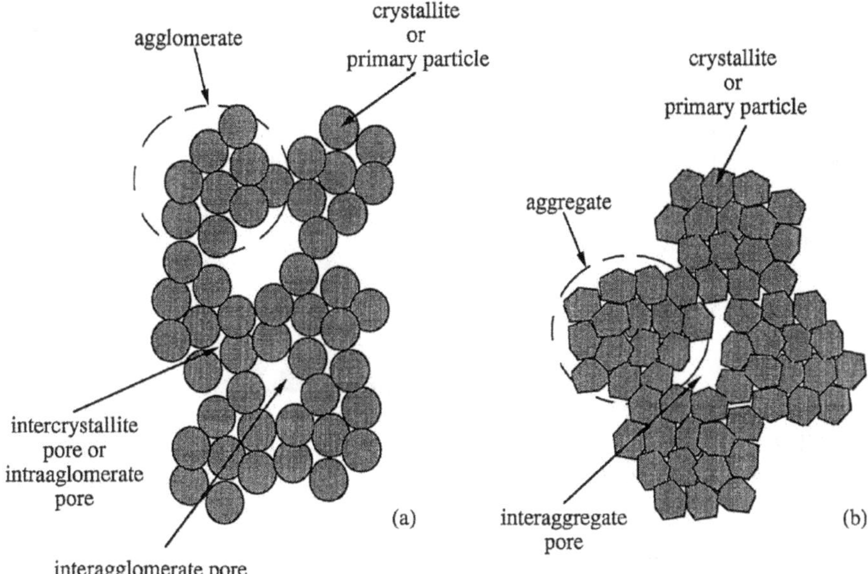

FIGURE 2.1 (a) Agglomerated and (b) aggregated powder categories are depicted in this diagram. (Reprinted from Ref. [30] by permission of the Taylor & Francis Ltd, http://www.tandfonline.com.)

treatment, the next stage in the processing of bioceramics is the shaping or molding of the green body.

Furthermore, the spray-drying technology, which involves atomizing fluids containing precipitated particles at a regulated rate through a nozzle into a hot chamber with a container to seize dry powders, is well suited to the mass manufacture of fine powders with good granular characteristics for molding on a large scale at a reasonable cost. The mechanism has been demonstrated in Figure 2.2.

The wet chemical precipitation process, whose stages are shown in Figure 2.3, can also be used to manufacture HA powders [32].

2.2.2 Shaping or Forming of Green Body

The green body formation process occurs following the powder treatment for the fabrication of bioceramics [33–36]. The powders have been suitably proportioned and pre-consolidated and are ready to be formed into the desired forms in this step [37]. Table 2.2 lists the four basic techniques that are commonly employed to create the green body [29]. Isostatic pressing or uniaxial pressing methods can be used for dry shaping. On the other hand, during the wet shaping process to form the green body, any of the three methods—slip casting, tape casting, or direct casting—can be adopted.

In addition, the plastic shaping process includes methods of extrusion and injection molding. Green body fabrication may also be done in a variety of ways using

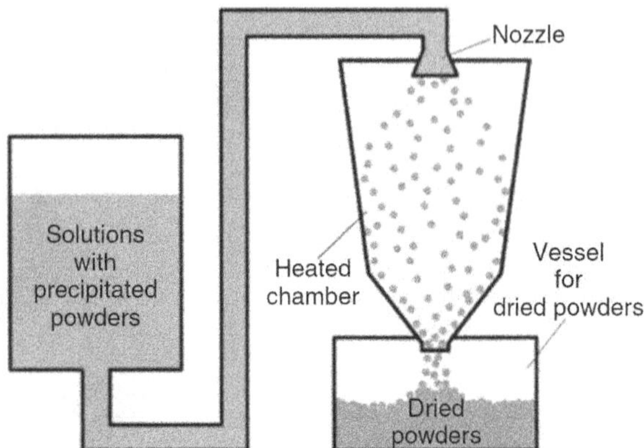

FIGURE 2.2 The spray-drying system is shown schematically. (Reprinted from Ref. [31] with permission from Elsevier.)

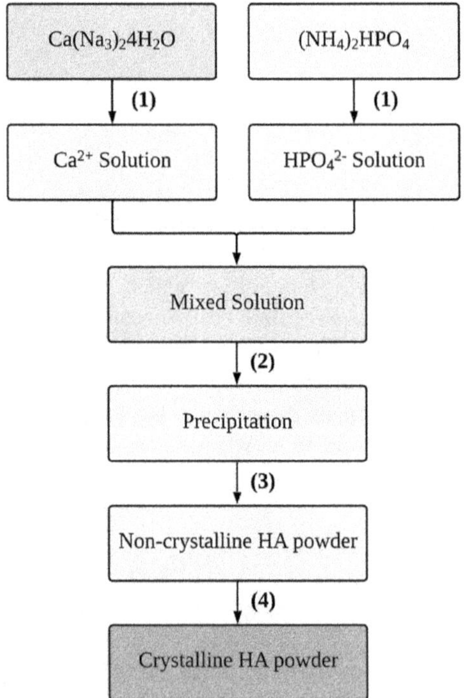

FIGURE 2.3 Hydroxyapatite (HA) powder preparation.

TABLE 2.2
Different Forming/Shaping Methods for the Processing of Bioceramics

Methods	
Isostatic pressing	Dry shaping
Uniaxial pressing	
Slip casting	Wet shaping
Direct casting	
Tape casting	
Injection molding	Plastic shaping
Extrusion	
3D printing	Solid free-form fabrication
Selective laser sintering	
Direct ink jet printing	
Stereolithography	
Robocasting	
Fused deposition	

the solid free-form manufacturing technique. Methodologies of extrusion and slip casting will be discussed in this chapter in order to explore the process of shaping or producing the green body.

2.2.3 Drying and Removal of the Binder

All additives used to form and consolidate the green bodies, such as plasticizers, binders, surfactants, solvents, and dispersants, must be removed from powder compacts before going on to sintering, which is the final stage in bioceramic production. Drying porous bodies and binder removal are two measures that may be used to complete this stage. The stages of porous body drying can be divided into three categories [38], as mentioned below, and the steps are illustrated in Figure 2.4.

1. Initial condition
2. Constant rate period
3. Falling rate period

Furthermore, the binder must be excluded from the formed green body once the porous body has dried. In certain circumstances, removing the binder is considered the most critical step [39]. Moreover, binder removal from thick-walled components is a time-consuming process that might take weeks [40]. Figure 2.5 exhibits the typical debinding of a green part. Binder removal at elevated temperatures is the basis of thermal debinding, and the methods of debinding can be classified into the following groups [39]:

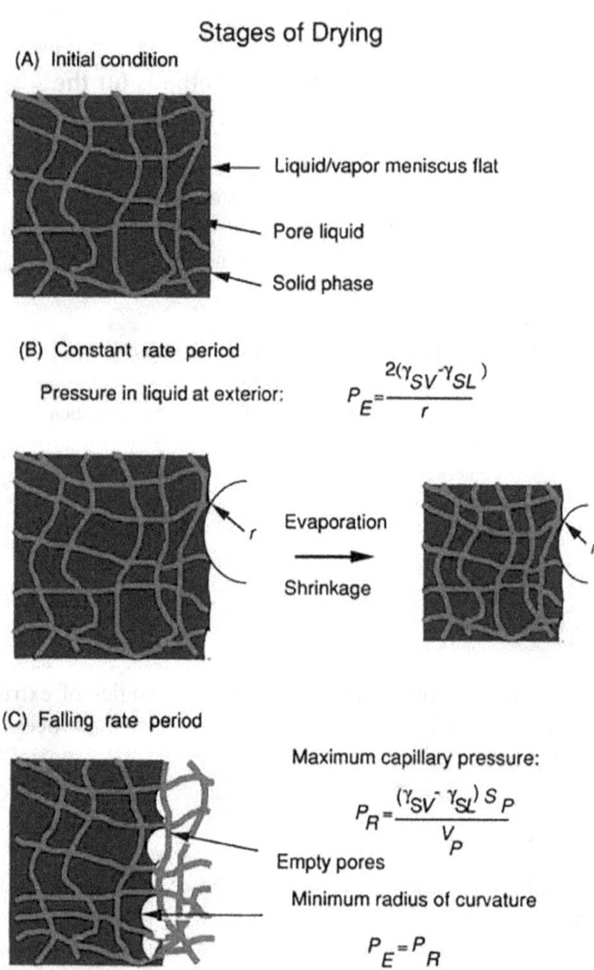

FIGURE 2.4 Depiction of drying process: the solid phase is represented by the black network, while the liquid filling the pores is represented by the shaded region. (Reprinted with permission from Ref. [38]. Copyright (2005) John Wiley and Sons.)

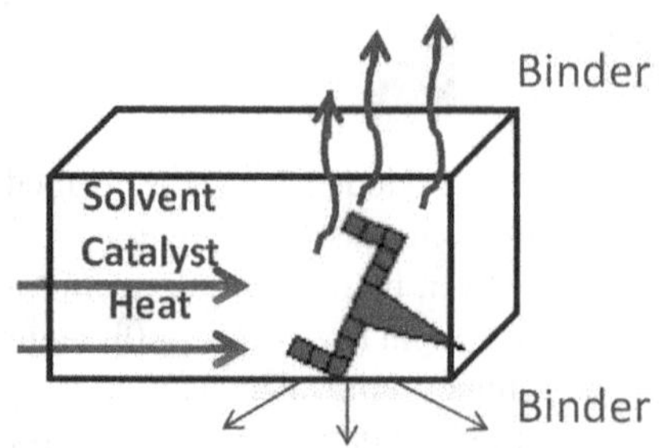

FIGURE 2.5 Debinding of green part. (Reprinted with permission from Ref. [41], An Open Access article distributed under the terms of the Creative Commons Attribution License.)

1. Thermal debinding:
   - debinding with permeation
   - debinding with diffusion
   - wicking
2. Solvent debinding:
   - solvent debinding by immersion
   - supercritical debinding
   - solvent vapor debinding

2.2.4 Sintering

Sintering, which is used to fire ceramic pots and fabricate complicated, high-performance structures like medical implants, serves to fuse particles together into strong, functional shapes. Sintering, on the other hand, is irreversible because the particles give up surface energy associated with tiny particles in order to form bonds between them. Furthermore, before sintering, the particles flow freely. However, after sintering, the particles are linked together to form a solid body [42]. In other words, it is a thermal treatment procedure that binds particles of a pre-shaped green body together to develop a mostly solid structure [43]. Pressure-free and pressure-assisted sintering are two options for sintering, depending on how much external pressure is used during the process [16].

The following techniques are used in the pressure-free sintering process:

- Microwave sintering
- Laser sintering
- Induction heating
- Plasma heating

On the contrary, the following are included in the pressure-assisted sintering process:

- Hot pressing
- Sol-gel
- Spark plasma sintering
- Hot isostatic pressing

The green body undergoes three phases of sintering throughout this process [29]. Figure 2.6 demonstrates all three steps of sintering.

2.3 Green Body Formation by Extrusion

The ceramic green body is formed by pressing the ceramic paste through a nozzle in extrusion [44]. To make the final green body, a straight extruded body with a regulated cross section is trimmed to the necessary length, as demonstrated in Figure 2.7.

Some extruded body constituents are shown in Table 2.3 [44].

In fabricating bioactive composites (PEEK/HA) for bone tissues, HA powder was extruded with a binder, plasticizer, and solvent, followed by 3D printing, and then compression molding was used to permeate polyetheretherketone (PEEK) melt

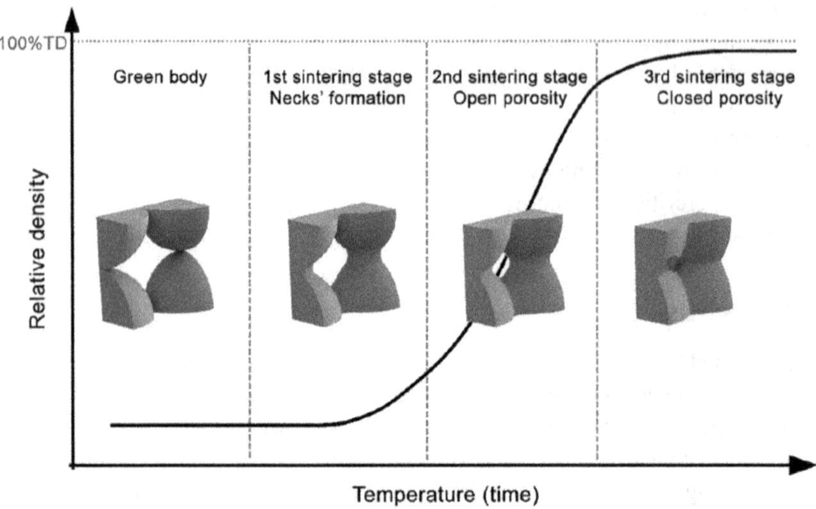

FIGURE 2.6 Three phases of sintering of bioceramic materials. (Reprinted from Ref. [29] with permission from Elsevier.)

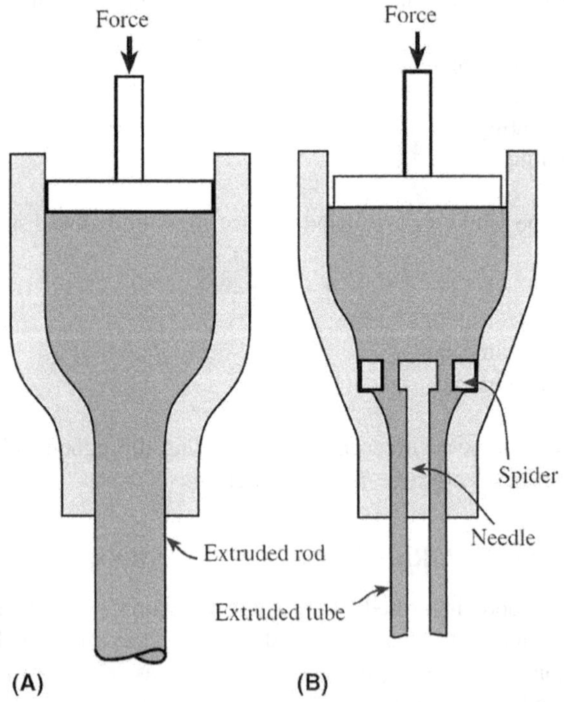

FIGURE 2.7 Schematic diagram of extrusion process for (a) a rod and (b) a tube. (Reprinted by permission from Springer Nature Customer Service Centre GmbH: Springer [44], Copyright 2016.)

TABLE 2.3
Extruded Body Composition Examples (Composition in vol percent)

High Alumina		Refractory Alumina		Electrical Porcelain	
$MgCl_2$	<1	$AlCl_3$ (pH > 8.5)	<1	$CaCl_2$	<1
Methylcellulose	2	Hydroxyethyl cellulose	6	Kaolin	16
Alumina (less than 20 μm)	46	Alumina (less than 20 μm)	50	Quartz (less than 44 μm)	16
Ball clay	4	Water	44	Water	36
Water	48			Feldspar (less than 44 μm)	16
				Ball clay	16

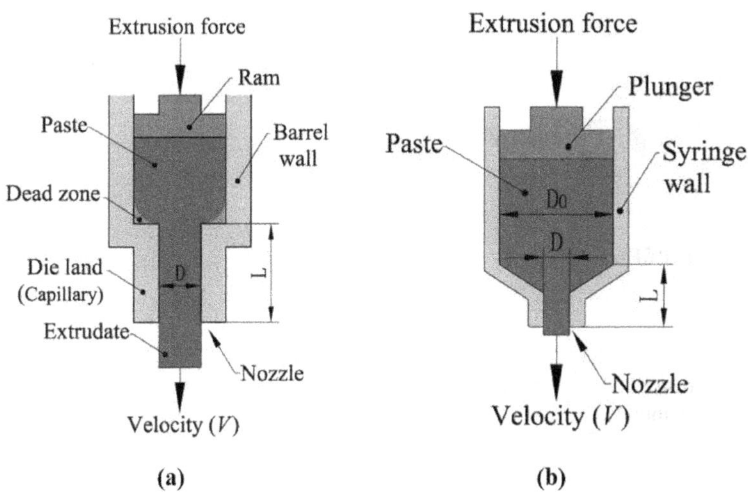

FIGURE 2.8 Cross section of (a) a paste extrusion and (b) a syringe plunger extrusion. (Reprinted from Ref. [45] with permission from Elsevier.)

into the HA scaffolds [45–48]. As indicated in Figure 2.8, the extrusion process can be carried out using either paste extrusion or syringe plunger extrusion [45]. Nonetheless, Figure 2.9 exhibits the steps involved in the manufacture of bioactive PEEK/HA composite. The syringe plunger extrusion model has been presented in Figure 2.10a for better illustration [47]. PEEK is biocompatible with cells and exhibits mechanical qualities similar to the cortical bone, such as elasticity and strength [49,50]. In addition, extrusion heads based on syringes or filaments can be used to make PEEK-based bioceramics.

Calcium phosphate bioceramics are ideal synthetic bone grafting biomaterials because of their exceptional biocompatibility, flexibility in structure design, degradability, plentiful supply, and composition modification [51]. Therefore, the extrusion–microdrilling approach may be utilized to construct 3D interconnected phosphate-based bioceramic scaffolds with channel-like macropores for bone regeneration [52]. In Figure 2.11, the production of green bodies and the construction of calcium phosphate-based bioceramic scaffolds are described in detail.

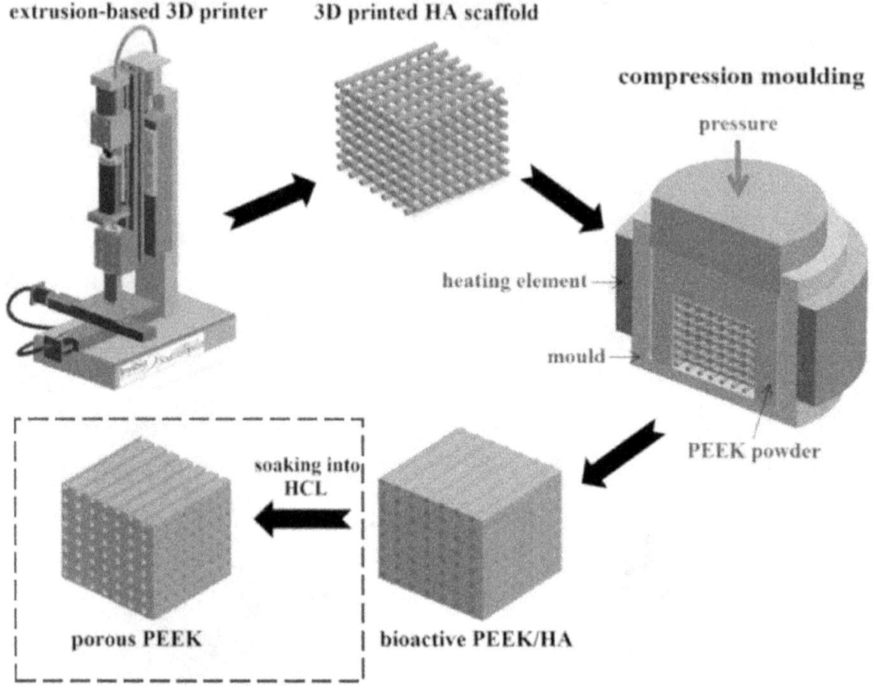

FIGURE 2.9 Procedure for making a bioactive composite (PEEK/HA). (Reprinted with permission from Ref. [48], An Open Access article distributed under the terms of the Creative Commons Attribution License.)

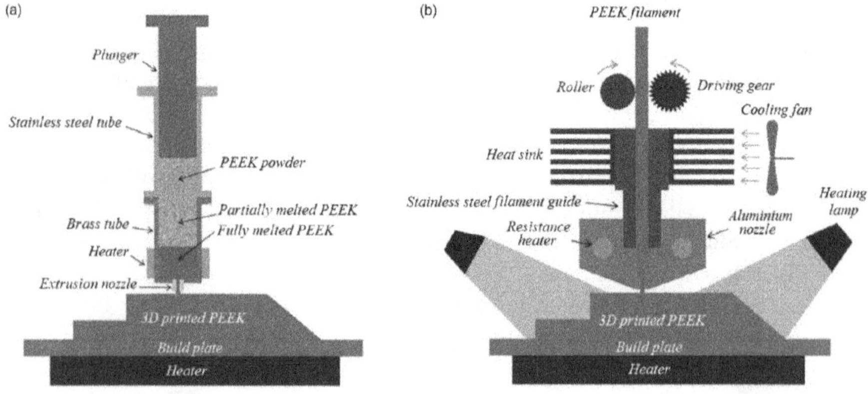

FIGURE 2.10 (a) Syringe-based extrusion head model; (b) filament-based extrusion head model. (Reprinted from Ref. [47] by permission of the Taylor & Francis Ltd, http://www.tandfonline.com.)

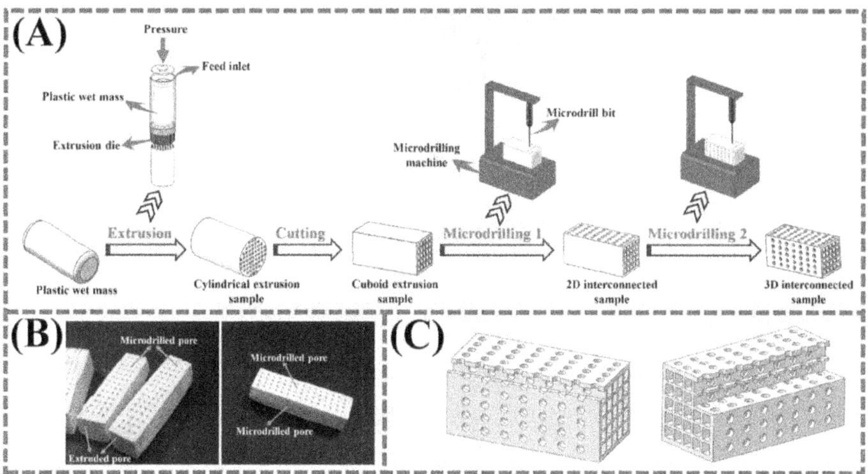

FIGURE 2.11 (a) Extrusion–microdrilling technique for creating 3D-linked green entities is depicted schematically. (b) Digitally photographed samples and (c) diagram of extruded and microdrilled pores. (Reprinted (adapted) with permission from Ref. [52]. Copyright 2020 American Chemical Society.)

Nonetheless, due to inadequacies in porosity, interconnectivity, mechanical characteristics, and biodegradability, clinical usage of Ca–P-based bioceramics is limited [53–56]. As a result, Ca–Si-based bioceramics have gained popularity recently due to their high biocompatibility, osteogenic potential, and degradability [57]. Figure 2.12 demonstrates the fabrication of bioceramics using calcium silicate, the most commonly used Ca–Si-based bioceramic, and strontium, one of the essential components in human bones.

Moreover, because of its simplicity and flexibility in producing complicated geometries, the direct ink writing (DIW) method is an outstanding example of an extrusion process that has received interest in academia and industry [58]. To synthesize bioceramic scaffolds from hardystonite ($Ca_2ZnSi_2O_7$), for instance, DIW with a preceramic polymer and fillers can be employed [59].

Traditional bioceramic/polymer composite manufacturing procedures embed the bioceramic in the polymer matrix, making it challenging to achieve a homogeneous mix [60,61]. As a result, melt extrusion, a type of extrusion approach, has been employed for bone regeneration applications by mixing fillers and polymers [62–66]. Hot melt extrusion is a solvent- and water-free technology that requires fewer processing stages and eliminates the need for time-consuming drying methods since the molten polymer may function as a thermal binder [67]. Figure 2.13 shows a schematic illustration of melt extrusion, and these types of biomaterials have better mechanical strength and can be employed in human regenerative medicine.

Furthermore, bioactive materials such as porous hydroxyapatite (HAp) and tricalcium phosphate (TCP) enhance osteointegration by having the right biological characteristics [68–70]. The term “osteointegration” refers to the direct functional and structural link that exists between organized live bone and the exterior of a

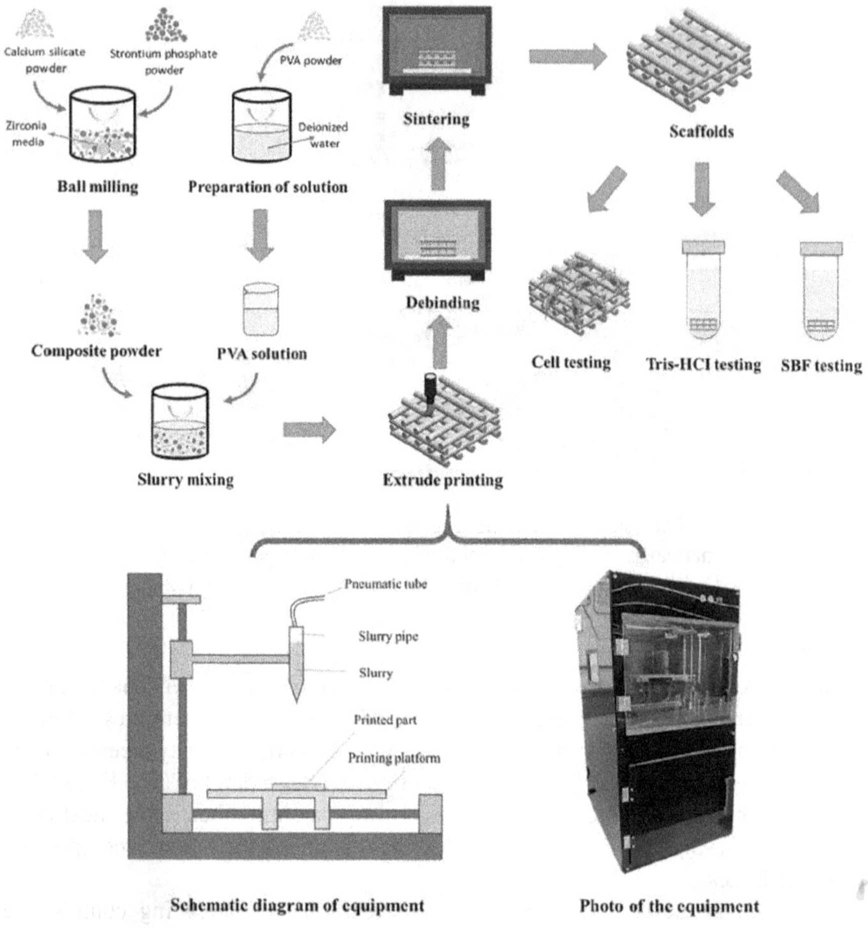

FIGURE 2.12 Ca–Si-based bioceramic fabrication and testing procedures. (Reprinted from Ref. [57] with permission from Elsevier.)

load-bearing implant [71]. Thus, manufacturing polycaprolactone (PCL)/β-TCP bioceramic scaffolds utilizing PCL as polymeric powder and bioceramic powder β-TCP may be done via screw extrusion printing, which is a unique extrusion technology (Figure 2.14) [72].

2.4 GREEN BODY FORMATION BY SLIP CASTING

Slip casting has a lengthy history as a shaping/forming method for both clay-based and nonclay-based ceramics [73]. Wet processing techniques such as slip casting are typically used to provide particles with the necessary flexibility to organize in a suitable position and develop a dense green body with uniform microstructure because dry pressing cannot create a completely compact and homogeneous structure from nanopowders [74,75]. Nonetheless, in order to fabricate high-density green

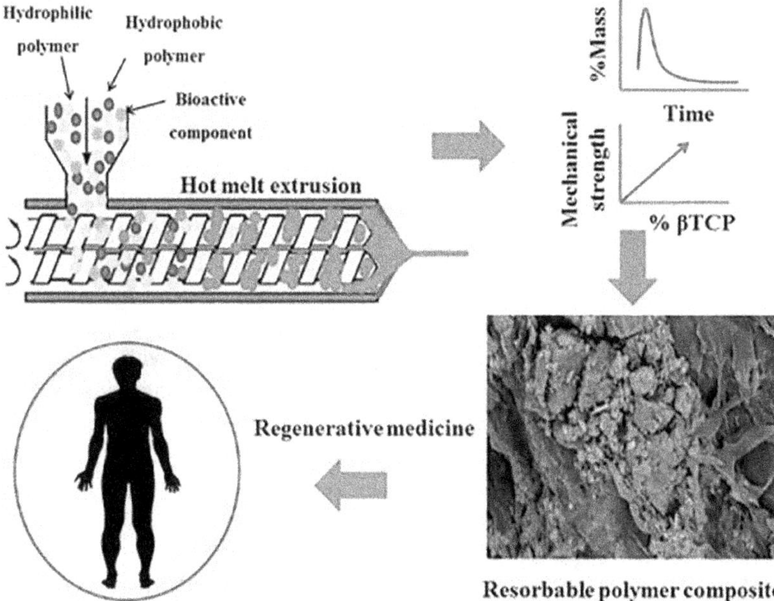

FIGURE 2.13 Hot melt extrusion. (Reprinted from Ref. [67] by permission of the Taylor & Francis Ltd, http://www.tandfonline.com.)

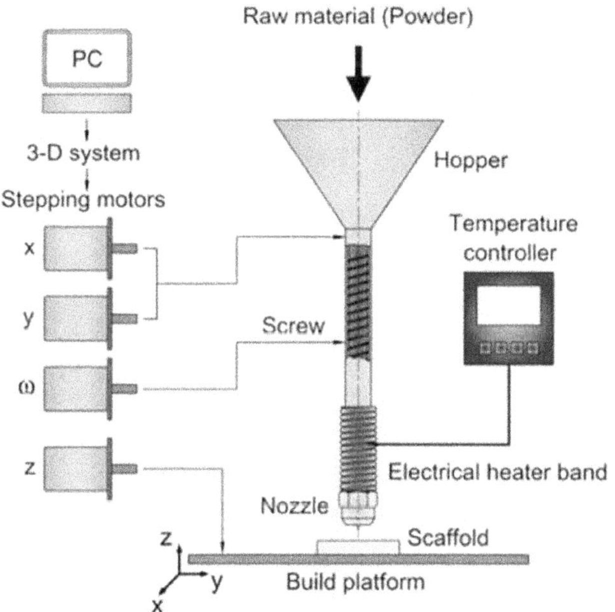

FIGURE 2.14 Schematic illustration of 3D mini-screw extrusion printing. (Reprinted with permission from Ref. [72]. Copyright (2015) John Wiley and Sons.)

bodies by slip casting, controlling the viscosity of the slurry until a solid load with adequate distribution can be obtained is a significant problem [76]. Slip refers to a slurry (a mixture of clay particles suspended in a liquid) of wet powders. Direct pouring into plaster molds can shape a slip with or without organic binders [31]. In this method, the slip is poured into a mold, which is typically made up of plaster of Paris ($2CaSO_4H_2O$) and was created by casting around a model of the required shape that had been suitably enlarged to compensate for the cast ceramic shrinkage during drying and sintering [43]. Slip casting allows for the creation of a wide range of forms and sizes, as well as massive bodies. In addition, slip is dried in molds because plaster molds absorb the water from the slip. During the slip-casting process, additives such as methylcellulose and sodium alginate are utilized, with their weightage ranging from 0.5% to 3.0%. Moreover, water is employed as a solvent, with a concentration of 30% to 60% [31]. Figure 2.15 illustrates a step-by-step procedure of slip casting.

In Figure 2.16, another schematic depiction of the slip-casting process has been presented. Slip casting entails a series of steps. Water travels through the porous plaster by capillary action, leaving a layer of solid on the mold's wall. The excess slip is drained away, and the mold and cast are allowed to dry once a suitable thickness has been cast [43].

Slip casting has become one of the essential forming technologies for the large-scale production of monolithic and composite ceramic components with simple or complicated geometries [77,78]. Now, several bioceramics involving slip casting will be briefly discussed.

Bioceramics made up of bioactive glasses or $Ca_3(PO_4)_2$ are biocompatible and capable of stimulating bone formation and regeneration. Slip casting can be utilized for making the green body of these sorts of bioceramics [79]. Figure 2.17 illustrates stages associated with the production of bioceramic scaffolds built from β-TCP and bioglass, which are bioceramic powders. The slip-casting process can be seen in Step 2 of Figure 2.17. Moreover, the HAp bioceramic was made utilizing the slip-casting technique by mixing a polymeric powder with a slip containing fine HAp particles

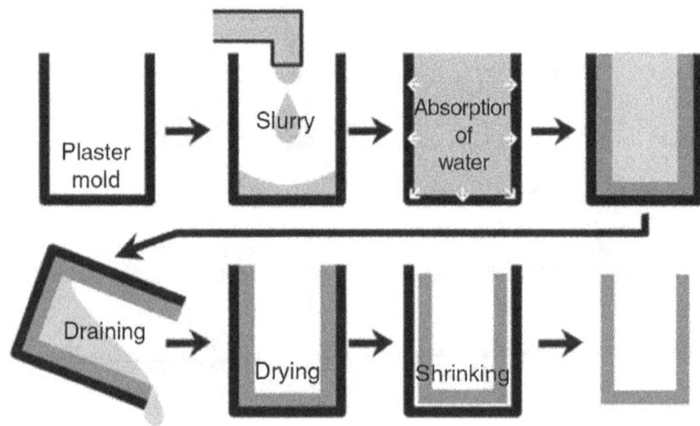

FIGURE 2.15 Step-by-step procedure of slip casting. (Reprinted from Ref. [31] with permission from Elsevier.)

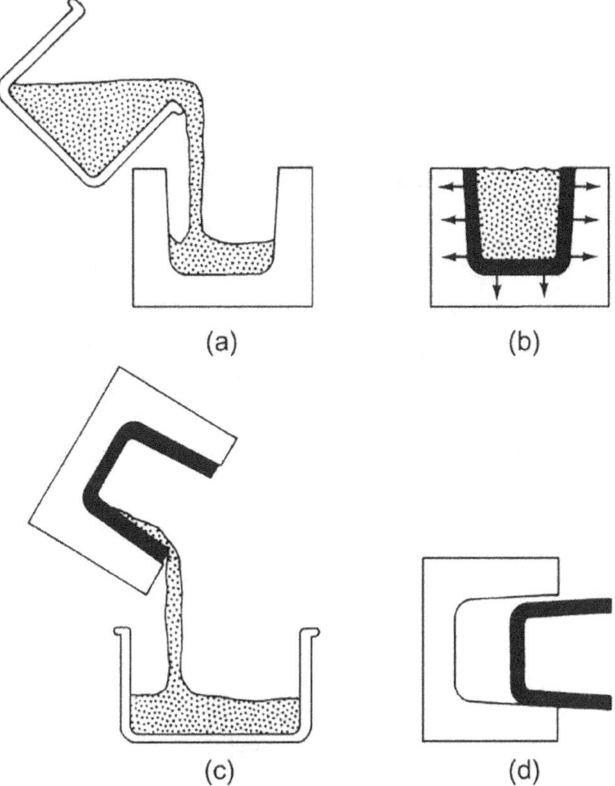

FIGURE 2.16 Slip-casting procedure: (a) pouring the slip into the mold, (b) collecting the liquid to the mold and creating a compact along the walls of the mold, (c) draining the excess slip, and (d) removal of the casting when it is partially dried. (Reprinted from Ref. [29] with permission from Elsevier.)

[80]. In addition, HA powders were prepared for this bioceramic using the following chemical reaction:

$$10Ca(OH)_2 + 6H_3PO_4 \rightarrow Ca_{10}(PO_4)_6(OH)_2 + 18H_2O$$

As the chemical compositions of HA ceramic are quite similar to those of genuine bone's mineral phase, it's an excellent option for bioceramic processing [81]. A flow diagram of the slip-casting process is illustrated schematically to fabricate this well-known hydroxyapatite (HAp or HA) bioceramic in Figure 2.18. Furthermore, Figure 2.19a shows a scanning electron micrograpy (SEM) of a green body that has been processed using the slip-casting method. The photos in Figure 2.19b–d show SEM images of the green body after sintering under various circumstances. Sintering was briefly described in Section 2.2.4.

Due to its near compositional resemblance and great biocompatibility with natural bone, for many years, HA and similar calcium phosphate minerals have been widely

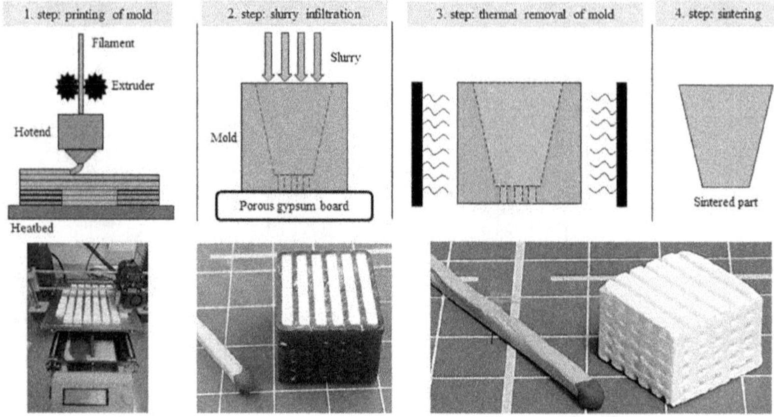

FIGURE 2.17 Bioceramic scaffold fabrication. (Reprinted from Ref. [79] with permission from Elsevier.)

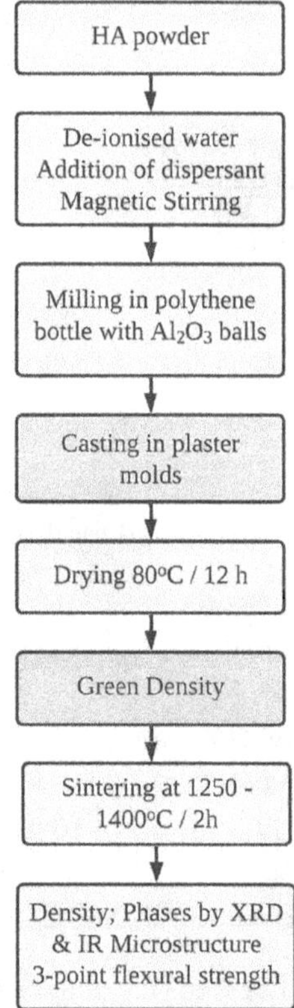

FIGURE 2.18 A schematic flow chart depicts the steps involved in making slip cast HAp bodies. (Reprinted with permission from Ref. [82]. Copyright (2004) John Wiley and Sons.)

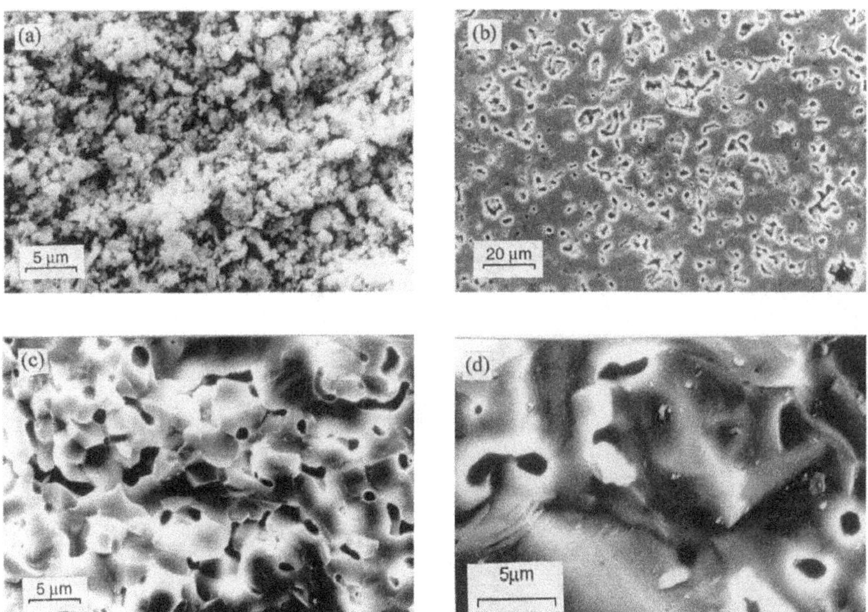

FIGURE 2.19 SEM pictures of (a) HAp slip cast green body fractured surface, (b) sintered HAp (1,400°C): polished and etched surface, (c) sintered HAp (1,350°C): fractured surface, and (d) sintered HAp (1,400°C): fractured surface. (Reprinted with permission from Ref. [82]. Copyright (2004) John Wiley and Sons.)

used as implant materials [83–86]. A combination of CaH_5O_6P and $CaCO_3$ may be used to make fine HA powders, which can then be used to make HA bioceramic using the slip-casting process [87]. Nevertheless, by combining polymethyl methacrylate with HA powder, HAp ceramics with controllable pore properties may be manufactured using the slip-casting process [88]. Figure 2.20 shows images of these bioceramics taken using an SEM.

However, HAp bioceramic scaffolds may be made using either a single or double slip-casting approach [89]. The single slip-casting procedure is used in all of the bioceramics described above. The preparation of bioceramic from hydroxyapatite ($Ca_{10}(PO_4)_6(OH)_2$) powder using the double slip-casting process includes multiple stages, as shown in Figure 2.21.

As illustrated in Figure 2.22, SEM analysis indicates that HAp ceramics are formed of large spherical holes with tiny rectangular pores with connections. However, the pore size of single slip-casting specimens is bigger than that of double slip-casting specimens. In addition, single slip casting produced a specimen with more pores than the double slip-casting approach. The enormous spherical holes have large enough linkages for cells to flow through and thrive, while the microscopic pores are too small for nutrients to flow and stimulate cell proliferation [90,91]. HAp bioceramics created by double slip casting may provide a more acceptable bone-filling material

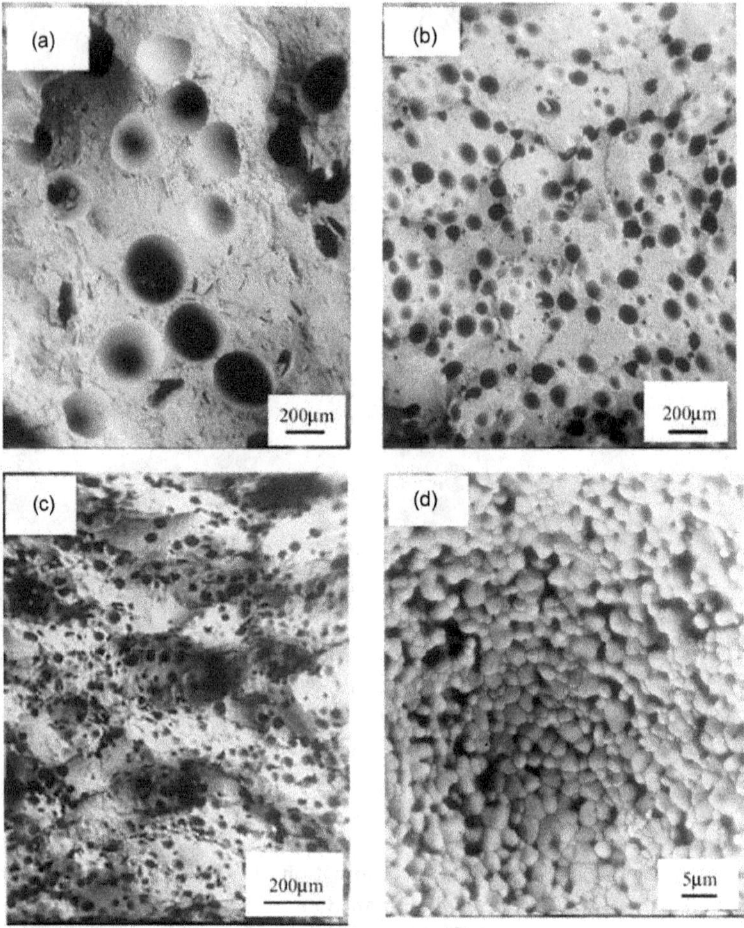

FIGURE 2.20 SEM images of slip cast porous HAp ceramics: porous HAp ceramics with size (a) between 260 and 350 μm, (b) between 120 and 150 μm, (c) between 50 and 75 μm, and (d) porous hydroxyapatite bioceramics with a micropore size between 50 and 75 μm. (Reprinted by permission from Springer Nature Customer Service Centre GmbH: Springer [88], Copyright 1969.)

for biomedical applications because the spherical pore sizes of HAp ceramics manufactured by double slip casting are large enough [89].

After single slip casting, the bioceramic's relative density was low, but after double casting, it was astronomically high. The strength rose dramatically with density owing to the expansion of the total solid area, as predicted, and this tendency is explained by the fact that density increases under load. Figure 2.23 illustrates the link between single and double slip-casting methods in terms of bending or flexural stress.

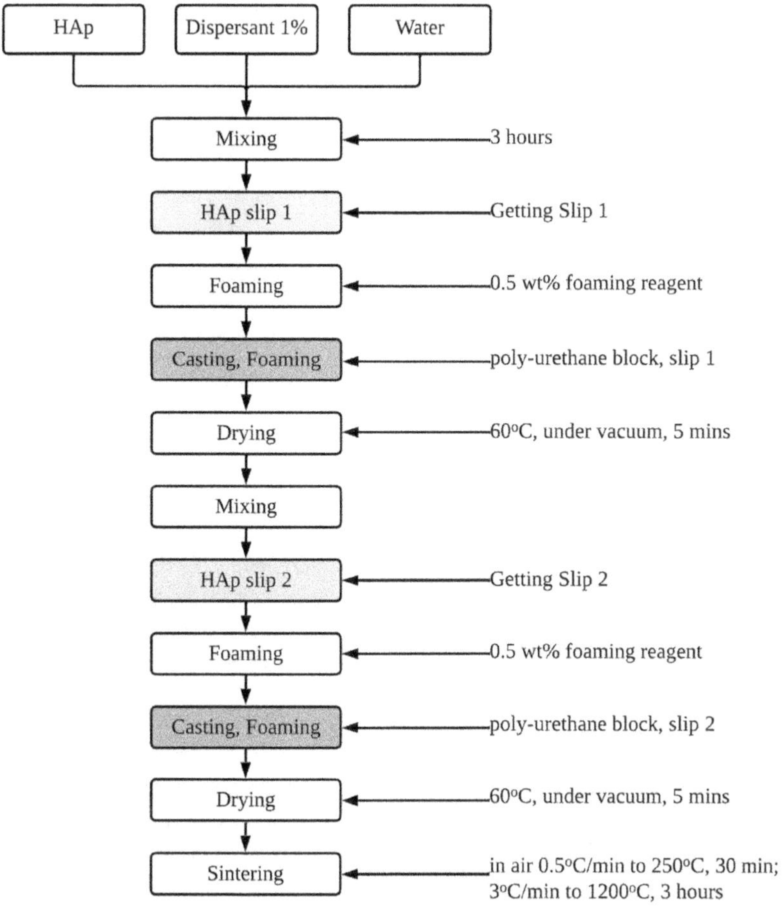

HAp | Dispersant 1% | Water

↓

Mixing ← 3 hours

↓

HAp slip 1 ← Getting Slip 1

↓

Foaming ← 0.5 wt% foaming reagent

↓

Casting, Foaming ← poly-urethane block, slip 1

↓

Drying ← 60°C, under vacuum, 5 mins

↓

Mixing

↓

HAp slip 2 ← Getting Slip 2

↓

Foaming ← 0.5 wt% foaming reagent

↓

Casting, Foaming ← poly-urethane block, slip 2

↓

Drying ← 60°C, under vacuum, 5 mins

↓

Sintering ← in air 0.5°C/min to 250°C, 30 min; 3°C/min to 1200°C, 3 hours

FIGURE 2.21 Bimodal porous HAp ceramics: fabrication process. (Reprinted with permission from [89]. Copyright (2011) John Wiley and Sons.)

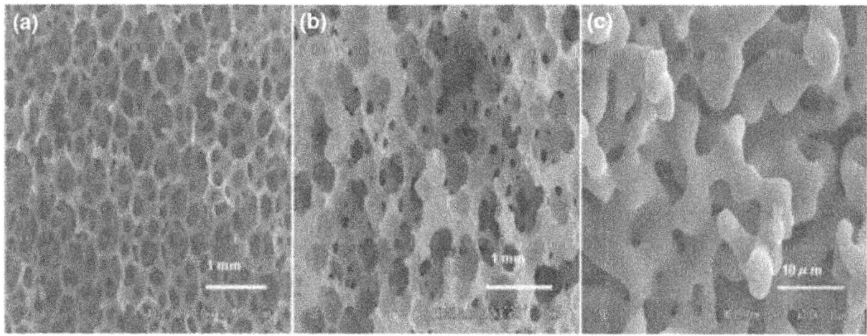

FIGURE 2.22 SEM images of (a) single-casting macropores, (b) double-casting macropores, and (c) microporous HAp scaffolds fabricated by heating foam. (Reprinted with permission from [89]. Copyright (2011) John Wiley and Sons.)

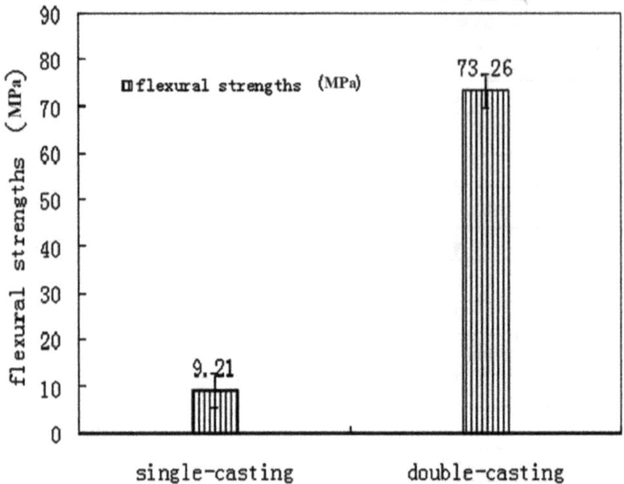

FIGURE 2.23 Relationship between single and double cast HAp porous scaffolds in terms of flexural strength. (Reprinted with permission from Ref. [89]. Copyright (2011) John Wiley and Sons.)

2.5 CONCLUSIONS

Bioceramics are appealing as biological implants because they adhere to bone well and generate little foreign body responses. These advanced ceramics are typically manufactured in three steps—powder shaping, partial drying, and high-temperature firing—to generate a thick material [92,93]. The powder shaping approach for bioceramics, also known as green body shaping, was addressed in this chapter. Moreover, extrusion and slip casting were specifically mentioned among the many processes for producing green bodies, and several examples of bioceramic fabrication involving extrusion and slip casting were provided. When it comes to bioceramics made by extrusion, it's worth noting that the methodology may be used to make not just Ca–P-based bioceramics but also other crystalline ceramics like Ca–Si-based bioceramics. However, HA bioceramic, a Ca-P-based ceramic, is an excellent choice since it is used in a variety of sectors, including medication delivery, bone, dentistry, wound tissue engineering, and orthopedics [94–99]. It's also utilized in nonmedical applications like gas sensors and also to remove nitrate from water. Different processing methods may be used to fabricate bioceramic scaffolds for bone tissue engineering using cockle shell calcium carbonate bioceramic powder also [100]. Furthermore, this chapter contains the single and double slip-casting processes being used to fabricate these HAp bioceramics. Nevertheless, when SEM pictures and the bending stress of bioceramics were compared, the double slip-casting process was found to be more favorable than single casting. Other ways of processing bioceramics will be described in the book's subsequent chapters. Overall, the definition, mechanical properties, and fabrication process of bioceramics were discussed in this chapter, with an emphasis on the production of green bodies via extrusion and slip casting.

ACKNOWLEDGMENTS

The authors are grateful to the Ahsanullah University of Science and Technology, Dhaka, Bangladesh, and the Military Institute of Science and Technology, Dhaka, Bangladesh, for the overall support of this work. This study didn't receive any government or private funds or grants.

REFERENCES

1. Rakshit, R., and A.K. Das. 2019. A review on cutting of industrial ceramic materials. *Precision Engineering* 59 (September): 90–109. https://doi.org/10.1016/j.precisioneng.2019.05.009.
2. Otitoju, A.T., P.U. Okoye, G. Chen, Y. Li, M.O. Okoye, and S. Li. 2020. Advanced ceramic components: Materials, fabrication, and applications. *Journal of Industrial and Engineering Chemistry* 85 (May): 34–65. https://doi.org/10.1016/j.jiec.2020.02.002.
3. Patil, K.C. 1993. Advanced ceramics: Combustion synthesis and properties. *Bulletin of Materials Science* 16 (6): 533–41. https://doi.org/10.1007/BF02757654.
4. Hench, L.L. 1991. Bioceramics: From concept to clinic. *Journal of the American Ceramic Society* 74 (7): 1487–1510. https://doi.org/10.1111/j.1151-2916.1991.tb07132.x.
5. Thamaraiselvi, T., and S. Rajeswari. 2003. Biological evaluation of bioceramic materials – A review trends. *Trends in Biomaterials and Artificial Organs* 18 (November).
6. Doremus, R.H. 1992. Bioceramics. *Journal of Materials Science* 27 (2): 285–97. https://doi.org/10.1007/BF00543915.
7. Raucci, M.G., D. Giugliano, and L. Ambrosio. 2014. Fundamental properties of bioceramics and biocomposites. In *Handbook of Bioceramics and Biocomposites*, edited by Iulian Vasile Antoniac, 1–19. Cham: Springer International Publishing. https://doi.org/10.1007/978-3-319-09230-0_3-1.
8. Best, S.M., A.E. Porter, E.S. Thian, and J. Huang. 2008. Bioceramics: Past, present and for the future. *Journal of the European Ceramic Society, Developments in Ceramic Science and Engineering*: The last 50 years. A meeting in celebration of Professor Sir Richard Brook's 70th Birthday 28 (7): 1319–27. https://doi.org/10.1016/j.jeurceramsoc.2007.12.001.
9. Jodati, H., B. Yılmaz, and Z. Evis. 2020. A review of bioceramic porous scaffolds for hard tissue applications: Effects of structural features. *Ceramics International* 46 (10, Part B): 15725–39. https://doi.org/10.1016/j.ceramint.2020.03.192.
10. Golieskardi, M., M. Satgunam, D. Ragurajan, M.E. Hoque, and A.M.H. Ng. 2020. Microstructural, tribological, and degradation properties of Al_2O_3- and CeO_2-doped 3 mol.% yttria-stabilized zirconia bioceramic for biomedical applications. *Journal of Materials Engineering and Performance* 29, no. 5: 2890–97. https://doi.org/10.1007/s11665-020-04829-3.
11. Depboylu, F.N., P. Korkusuz, E. Yasa, and F. Korkusuz. 2022. Smart bioceramics for orthopedic applications. In *Innovative Bioceramics in Translational Medicine II: Surgical Applications*, edited by Andy H. Choi and Besim Ben-Nissan, 157–86. Springer Series in Biomaterials Science and Engineering. Singapore: Springer. https://doi.org/10.1007/978-981-16-7439-6_8.
12. Ercan, I., O. Kaygili, T. Kayed, N. Bulut, H. Tombuloğlu, T. İnce, F. Al Ahmari, et al. 2022. Structural, spectroscopic, dielectric, and magnetic properties of Fe/Cu Co-doped hydroxyapatites prepared by a wet-chemical method. *Physica B: Condensed Matter* 625: 413486. https://doi.org/10.1016/j.physb.2021.413486.
13. Acar, S., O. Kaygili, T. Ates, S.V. Dorozhkin, N. Bulut, B. Ates, S. Koytepe, F. Ercan, H. Kebiroglu, and A.H. Hssain. 2022. Experimental characterization and theoretical investigation of Ce/Yb Co-doped hydroxyapatites. *Materials Chemistry and Physics* 276: 125444. https://doi.org/10.1016/j.matchemphys.2021.125444.

14. Ragurajan, D., M. Golieskardi, M. Satgunam, M.E. Hoque, A.M.H. Ng, M.J. Ghazali, and A.K. Ariffin. 2018. Advanced 3Y-TZP bioceramic doped with Al_2O_3 and MnO_2 particles potentially for biomedical applications: Study on mechanical and degradation properties. *Journal of Materials Research and Technology* 7, no. 4: 432–42. https://doi.org/10.1016/j.jmrt.2017.05.015.
15. Golieskardi, M., M. Satgunam, D. Ragurajan, M.E. Hoque, A.M.H. Ng, and L. Shanmuganantha. 2019. Advanced 3Y-TZP bioceramic doped with Al_2O_3 and CeO_2 potentially for biomedical implant applications. *Materials Technology* 34, no. 8: 480–89. https://doi.org/10.1080/10667857.2019.1578912.
16. Akin, I., and G. Goller. 2016. Processing technologies for bioceramic based composites. In *Handbook of Bioceramics and Biocomposites*, edited by Iulian Vasile Antoniac, 639–66. Cham: Springer International Publishing. https://doi.org/10.1007/978-3-319-12460-5_14.
17. Feng, C., K. Zhang, R. He, G. Ding, M. Xia, X. Jin, and C. Xie. 2020. Additive manufacturing of hydroxyapatite bioceramic scaffolds: Dispersion, digital light processing, sintering, mechanical properties, and biocompatibility. *Journal of Advanced Ceramics* 9 (3): 360–73. https://doi.org/10.1007/s40145-020-0375-8.
18. Witek, L., Y. Shi, and J. Smay. 2017. Controlling calcium and phosphate ion release of 3D printed bioactive ceramic scaffolds: An in vitro study. *Journal of Advanced Ceramics* 6 (June): 157–64. https://doi.org/10.1007/s40145-017-0228-2.
19. Shen, T., W. Yang, X. Shen, W. Chen, B. Tao, X. Yang, J. Yuan, P. Liu, and K. Cai. 2018. Polydopamine-assisted hydroxyapatite and lactoferrin multilayer on titanium for regulating bone balance and enhancing antibacterial property. *ACS Biomaterials Science & Engineering* 4 (9): 3211–23. https://doi.org/10.1021/acsbiomaterials.8b00791.
20. Vladescu, A., M.A. Surmeneva, C.M. Cotrut, R.A. Surmenev, and I.V. Antoniac. 2016. Bioceramic coatings for metallic implants. In *Handbook of Bioceramics and Biocomposites*, edited by Iulian Vasile Antoniac, 703–33. Cham: Springer International Publishing. https://doi.org/10.1007/978-3-319-12460-5_31.
21. Hayakawa, S., K. Tsuru, and A. Osaka. 2008. 3: The microstructure of bioceramics and its analysis. In *Bioceramics and Their Clinical Applications*, edited by Tadashi Kokubo, 53–77. Woodhead Publishing Series in Biomaterials. Woodhead Publishing. https://doi.org/10.1533/9781845694227.1.53.
22. Hench, L.L., and J. Wilson. 2013. Introduction. In *An Introduction to Bioceramics*, edited by Larry L. Hench, 2nd ed., 1–26. Imperial College Press. https://doi.org/10.1142/9781908977168_0001.
23. Hench, L.L. 1998. Bioceramics. *Journal of the American Ceramic Society* 81 (7): 1705–28. https://doi.org/10.1111/j.1151-2916.1998.tb02540.x.
24. Marshall, G.W., M. Balooch, R.R. Gallagher, S.A. Gansky, and S.J. Marshall. 2001. Mechanical properties of the dentinoenamel junction: AFM studies of nanohardness, elastic modulus, and fracture. *Journal of Biomedical Materials Research* 54 (1): 87–95. https://doi.org/10.1002/1097-4636(200101)54:1<87::aid-jbm10>3.0.co;2–z.
25. Guazzato, M., M. Albakry, S.P. Ringer, and M.V. Swain. 2004. Strength, fracture toughness and microstructure of a selection of all-ceramic materials. Part II. Zirconia-based dental ceramics. *Dental Materials: Official Publication of the Academy of Dental Materials* 20 (5): 449–56. https://doi.org/10.1016/j.dental.2003.05.002.
26. Yin, L., X. Song, Y. Song, T. Huang, and J. Li. 2006. An overview of in vitro abrasive finishing & CAD/CAM of bioceramics in restorative dentistry. *International Journal of Machine Tools & Manufacture* 46 (July): 1013–26. https://doi.org/10.1016/j.ijmachtools.2005.07.045.

27. Riccio, P., M. Zare, D. Gomes, D. Green, and A. Stamboulis. Antimicrobial bioceramics for biomedical applications. In *Innovative Bioceramics in Translational Medicine I: Fundamental Research*, edited by Andy H. Choi and Besim Ben-Nissan, 159–93. Springer Series in Biomaterials Science and Engineering. Singapore: Springer, 2022. https://doi.org/10.1007/978-981-16-7435-8_5.
28. Pina, S., I.K. Kwon, R.L. Reis, and J.M. Oliveira. Biocomposites and bioceramics in tissue engineering: Beyond the next decade. In *Innovative Bioceramics in Translational Medicine I: Fundamental Research*, edited by Andy H. Choi and Besim Ben-Nissan, 319–50. Springer Series in Biomaterials Science and Engineering. Singapore: Springer, 2022. https://doi.org/10.1007/978-981-16-7435-8_11.
29. Trunec, M., and K. Maca. 2014. Chapter 7: Advanced ceramic processes. In *Advanced Ceramics for Dentistry*, edited by James Zhijian Shen and Tomaž Kosmač, 123–50. Oxford: Butterworth-Heinemann. https://doi.org/10.1016/B978-0-12-394619-5.00007-9.
30. Mayo, M. J. 1996. Processing of nanocrystalline ceramics from ultrafine particles. *International Materials Reviews* 41 (3): 85–115. https://doi.org/10.1179/imr.1996.41.3.85.
31. Tanaka, Y, and K Yamashita. 2008. 2: Fabrication processes for bioceramics. In *Bioceramics and Their Clinical Applications*, edited by Tadashi Kokubo, 28–52. Woodhead Publishing Series in Biomaterials. Woodhead Publishing. https://doi.org/10.1533/9781845694227.1.28.
32. Yong, L., M.E. Hoque, and N. Sakinah. Synthesis and characterization of hydroxyapatite bioceramic. *International Journal of Web Engineering and Technology* 3 (1 January 2014): 458.
33. Salomoni, A., A. Tucci, L. Esposito, and I. Stamenkovic. 1994. Forming and sintering of multiphase bioceramics. *Journal of Materials Science: Materials in Medicine* 5 (9): 651–53. https://doi.org/10.1007/BF00120349.
34. Darus, F., R.M. Isa, N. Mamat, and M. Jaafar. 2018. Techniques for fabrication and construction of three-dimensional bioceramic scaffolds: Effect on pores size, porosity and compressive strength. *Ceramics International* 44 (15): 18400–407. https://doi.org/10.1016/j.ceramint.2018.07.056.
35. Bayazit, V., M. Bayazit, and E. Bayazit. 2010. Evaluation of bioceramic materials in biology and medicine. *Digest Journal of Nanomaterials and Biostructures* 7 (April): 211–22.
36. Wu, C., and J. Chang. 2006. A novel akermanite bioceramic: Preparation and characteristics. *Journal of Biomaterials Applications* 21 (2): 119–29. https://doi.org/10.1177/0885328206057953.
37. Richerson, D.W., and W.E. Lee. 2018. Shape-forming processes. In *Modern Ceramic Engineering*, 4th ed. CRC Press.
38. Scherer, G.W. 1990. Theory of drying. *Journal of the American Ceramic Society* 73 (1): 3–14. https://doi.org/10.1111/j.1151-2916.1990.tb05082.x.
39. Trunec, M., J. Cihlar, and K. Nemecek. 1997. Removal of thermoplastic binders from ceramic green bodies. *Removal of Thermoplastic Binders from Ceramic Green Bodies* 41 (2): 67–80.
40. Binner, J.J.G.P., ed. 1990. *Advanced Ceramic Processing and Technology*, vol. 1. Park Ridge: Noyes Publications.
41. Gonzalez-Gutierrez, J., G. Stringari, B. Zupančič, G. Kubyshkina, B. von Bernstorff, and I. Emri. 2012. Time-dependent properties of multimodal polyoxymethylene based binder for powder injection molding*. *Journal of Solid Mechanics and Materials Engineering* 6 (January): 419–30. https://doi.org/10.1299/jmmp.6.419.
42. German, R.M. 2010. 1: Thermodynamics of sintering. In *Sintering of Advanced Materials*, edited by Zhigang Zak Fang, 3–32. Woodhead Publishing Series in Metals and Surface Engineering. Woodhead Publishing. https://doi.org/10.1533/9781845699949.1.3.

43. Rahaman, M.N. 2013. *Sintering of Ceramics*. Boca Raton: CRC Press. https://doi.org/10.1201/b15869.
44. Carter, C.B., and M.G. Norton, eds. 2007. Shaping and forming. In *Ceramic Materials: Science and Engineering*, 412–26. New York, NY: Springer. https://doi.org/10.1007/978-0-387-46271-4_23.
45. Zhong, G., M. Vaezi, P. Liu, L. Pan, and S. Yang. 2017. Characterization approach on the extrusion process of bioceramics for the 3D printing of bone tissue engineering scaffolds. *Ceramics International* 43 (16): 13860–68. https://doi.org/10.1016/j.ceramint.2017.07.109.
46. Vaezi, M., and S. Yang. 2015. A novel bioactive PEEK/HA composite with controlled 3D interconnected HA network. *International Journal of Bioprinting* 1 (1). https://doi.org/10.18063/IJB.2015.01.004.
47. Vaezi, M., and S. Yang. 2015. Extrusion-based additive manufacturing of PEEK for biomedical applications. *Virtual and Physical Prototyping* 10 (3): 123–35. https://doi.org/10.1080/17452759.2015.1097053.
48. Vaezi, M., C. Black, D.M.R. Gibbs, R.O.C. Oreffo, M. Brady, M. Moshrefi-Torbati, and S. Yang. 2016. Characterization of new PEEK/HA composites with 3D HA network fabricated by extrusion freeforming. *Molecules (Basel, Switzerland)* 21 (6): E687. https://doi.org/10.3390/molecules21060687.
49. Edwards, S.L., and J.A. Werkmeister. 2012. Mechanical evaluation and cell response of woven polyetheretherketone scaffolds. *Journal of Biomedical Materials Research. Part A* 100 (12): 3326–31. https://doi.org/10.1002/jbm.a.34286.
50. Kurtz, S.M., and J.N. Devine. 2007. PEEK biomaterials in trauma, orthopedic, and spinal implants. *Biomaterials* 28 (32): 4845–69. https://doi.org/10.1016/j.biomaterials.2007.07.013.
51. Bouler, J.M., P. Pilet, O. Gauthier, and E. Verron. 2017. Biphasic calcium phosphate ceramics for bone reconstruction: A review of biological response. *Acta Biomaterialia* 53 (April): 1–12. https://doi.org/10.1016/j.actbio.2017.01.076.
52. He, F., T. Lu, X. Fang, S. Feng, S. Feng, Y. Tian, Y. Li, F. Zuo, X. Deng, and J. Ye. 2020. Novel extrusion-microdrilling approach to fabricate calcium phosphate-based bioceramic scaffolds enabling fast bone regeneration. *ACS Applied Materials & Interfaces* 12 (29): 32340–51. https://doi.org/10.1021/acsami.0c07304.
53. Chen, Q.Z., I.D. Thompson, and A.R. Boccaccini. 2006. 45S5 bioglass(r)-derived glass-ceramic scaffolds for bone tissue engineering. *Biomaterials* 27 (11): 2414–25. https://doi.org/10.1016/j.biomaterials.2005.11.025.
54. Liu, Z., H. Liang, T. Shi, D. Xie, R. Chen, X. Han, L. Shen, C. Wang, and Z. Tian. 2019. Additive manufacturing of hydroxyapatite bone scaffolds via digital light processing and in vitro compatibility. *Ceramics International* 45 (8): 11079–86. https://doi.org/10.1016/j.ceramint.2019.02.195.
55. Kondo, N., A. Ogose, K. Tokunaga, T. Ito, K. Arai, N. Kudo, H. Inoue, H. Irie, and N. Endo. 2005. Bone formation and resorption of highly purified β-tricalcium phosphate in the rat femoral condyle. *Biomaterials* 26 (28): 5600–5608. https://doi.org/10.1016/j.biomaterials.2005.02.026.
56. Yang, Y., C. He, E. Dianyu, W. Yang, F. Qi, D. Xie, L. Shen, S. Peng, and C. Shuai. 2020. Mg bone implant: Features, developments and perspectives. *Materials & Design* 185 (January): 108259. https://doi.org/10.1016/j.matdes.2019.108259.
57. Zou, A., H. Liang, C. Jiao, M. Ge, X. Yi, Y. Yang, J. Sun, C. Wang, L. Shen, and Y. Li. 2021. Fabrication and properties of CaSiO3/ Sr3(PO4)2 composite scaffold based on extrusion deposition. *Ceramics International* 47 (4): 4783–92. https://doi.org/10.1016/j.ceramint.2020.10.048.
58. Belgin Paul, D.L., A.S. Praveen, and U. Golcha. 2022. Extrusion based 3D-printing of bioceramic structures – A review. In *Composite Materials for Extreme Loading*, edited by Shankar Krishnapillai, Velmurugan R., and Sung Kyu Ha, 221–31. Lecture Notes in Mechanical Engineering. Singapore: Springer. https://doi.org/10.1007/978-981-16-4138-1_17.

59. Zocca, A., G. Franchin, H. Elsayed, E. Gioffredi, E. Bernardo, and P. Colombo. 2016. Direct ink writing of a preceramic polymer and fillers to produce hardystonite (Ca2ZnSi2O7) bioceramic scaffolds. *Journal of the American Ceramic Society* 99 (6): 1960–67. https://doi.org/10.1111/jace.14213.
60. Zhou, H., J.G. Lawrence, and S.B. Bhaduri. 2012. Fabrication aspects of PLA-CaP/PLGA-CaP composites for orthopedic applications: A review. *Acta Biomaterialia* 8 (6): 1999–2016. https://doi.org/10.1016/j.actbio.2012.01.031.
61. Murugan, R., and S. Ramakrishna. 2005. Development of nanocomposites for bone grafting. *Composites Science and Technology* 65 (15–16): 2385–2406. https://doi.org/10.1016/j.compscitech.2005.07.022.
62. Bagheri-Kazemabad, S., D. Fox, Y. Chen, L.M. Geever, A. Khavandi, R. Bagheri, C.L. Higginbotham, H. Zhang, and B. Chen. 2012. Morphology, rheology and mechanical properties of polypropylene/ethylene-octene copolymer/clay nanocomposites: Effects of the compatibilizer. *Composites Science and Technology* 72 (14): 1697–1704. https://doi.org/10.1016/j.compscitech.2012.06.007.
63. Barbieri, D., J.D. de Bruijn, X. Luo, S. Farè, D.W. Grijpma, and H. Yuan. 2013. Controlling dynamic mechanical properties and degradation of composites for bone regeneration by means of filler content. *Journal of the Mechanical Behavior of Biomedical Materials* 20 (April): 162–72. https://doi.org/10.1016/j.jmbbm.2013.01.012.
64. Shikinami, Y., and M. Okuno. 1999. Bioresorbable devices made of forged composites of hydroxyapatite (HA) particles and poly-L-lactide (PLLA): Part I. Basic characteristics. *Biomaterials* 20 (9): 859–77. https://doi.org/10.1016/S0142-9612(98)00241-5.
65. Delabarde, C., C.J.G. Plummer, P-E. Bourban, and J-A.E. Månson. 2011. Accelerated ageing and degradation in poly-l-lactide/hydroxyapatite nanocomposites. *Polymer Degradation and Stability* 96 (4): 595–607. https://doi.org/10.1016/j.polymdegradstab.2010.12.018.
66. Chen, L.J., and M. Wang. 2002. Production and evaluation of biodegradable composites based on PHB-PHV copolymer. *Biomaterials* 23 (13): 2631–39. https://doi.org/10.1016/S0142-9612(01)00394-5.
67. Kenny, E.K., N.M. Gately, J.A. Killion, D.M. Devine, C.L. Higginbotham, and L.M. Geever. 2016. Melt extruded bioresorbable polymer composites for potential regenerative medicine applications. *Polymer-Plastics Technology and Engineering* 55 (4): 432–46. https://doi.org/10.1080/03602559.2015.1098684.
68. Yeo, M.G., W-K. Jung, and G.H. Kim. 2012. Fabrication, characterisation and biological activity of phlorotannin-conjugated PCL/β-TCP composite scaffolds for bone tissue regeneration. *Journal of Materials Chemistry* 22 (8): 3568–77. https://doi.org/10.1039/C2JM14725D.
69. Yeo, M.G., H. Lee, and G.H. Kim. 2011. Three-dimensional hierarchical composite scaffolds consisting of polycaprolactone, β-tricalcium phosphate, and collagen nanofibers: Fabrication, physical properties, and in vitro cell activity for bone tissue regeneration. *Biomacromolecules* 12 (2): 502–10. https://doi.org/10.1021/bm1013052.
70. Shor, L., E.D. Yildirim, S. Güçeri, and W. Sun. 2010. Precision extruding deposition for freeform fabrication of PCL and PCL-HA tissue scaffolds. In *Printed Biomaterials: Novel Processing and Modeling Techniques for Medicine and Surgery*, edited by Roger Narayan, Thomas Boland, and Yuan-Shin Lee, 91–110. Biological and Medical Physics, Biomedical Engineering. New York, NY: Springer. https://doi.org/10.1007/978-1-4419-1395-1_6.
71. LeGeros, R.Z., and R.G. Craig. 1993. Strategies to affect bone remodeling: Osteointegration. *Journal of Bone and Mineral Research: The Official Journal of the American Society for Bone and Mineral Research* 8 Suppl 2 (December): S583–596. https://doi.org/10.1002/jbmr.5650081328.

72. Dávila, J.L., M.S. Freitas, P.I. Neto, Z.C. Silveira, J.V.L. Silva, and M.A. d'Ávila. 2016. Fabrication of PCL/β-TCP scaffolds by 3D mini-screw extrusion printing. *Journal of Applied Polymer Science* 133 (15). https://doi.org/10.1002/app.43031.
73. Rempes, P.E., B.C. Weber, and M.A. Schwartz. 1958. Slip Casting Of Metals, Ceramics, And Cermets. *American Ceramic Society Bulletin (US.)* 37 (July). https://www.osti.gov/biblio/4334061-slip-casting-metals-ceramics-cermets.
74. Mouzon, J., E. Glowacki, and M. Odén. 2008. Comparison between slip-casting and uniaxial pressing for the fabrication of translucent yttria ceramics. *Journal of Materials Science*. https://dx.doi.org/10.1007/s10853-007-2261-y.
75. Letue, L., J. Petit, M.H. Ritti, S. Lalanne, and S. Landais. 2017. Sintering and annealing effects on undoped yttria transparent ceramics. *Materials Chemistry and Physics* 194 (March): 302–7. https://doi.org/10.1016/j.matchemphys.2017.03.046.
76. Chin, C.H., A. Muchtar, C.H. Azhari, M. Razali, and M. Aboras. 2015. Optimization of PH and dispersant amount of Y-TZP suspension for colloidal stability. *Ceramics International* 41 (8): 9939–46. https://doi.org/10.1016/j.ceramint.2015.04.073.
77. Ramachandra Rao, R., H.N. Roopa, and T.S. Kannan. 1999. Dispersion, slip casting and reaction nitridation of silicon-silicon carbide mixtures. *Journal of the European Ceramic Society* 19 (12): 2145–53. https://doi.org/10.1016/S0955-2219(99)00019-9.
78. Herrmann, R. 1989. Slip casting in practice. In *Ceramics Monographs-Handbook of Ceramics, Verlag Schimdt GmbH Freiburg i. Brg, Alemanha*
79. Esslinger, S., and R. Gadow. 2020. Additive manufacturing of bioceramic scaffolds by combination of FDM and slip casting. *Journal of the European Ceramic Society*, 16th European Inter-Regional Conference on Ceramics (CIEC16) 40 (11): 3707–13. https://doi.org/10.1016/j.jeurceramsoc.2019.10.029.
80. Liu, D-M. 1998. Preparation and characterisation of porous hydroxyapatite bioceramic via a slip-casting route. *Ceramics International* 24 (6): 441–46. https://doi.org/10.1016/S0272-8842(97)00033-3.
81. Irfan, M., M. Irfan, S.U. Zaman, M.K.U. Zaman, and N. Muhammad. 2020. Overview of hydroxyapatite; composition, structure, synthesis methods and its biomedical uses 6 (June): 84–99.
82. Rao, R.R., and T.S. Kannan. 2001. Dispersion and slip casting of hydroxyapatite. *Journal of the American Ceramic Society* 84 (8): 1710–16. https://doi.org/10.1111/j.1151-2916.2001.tb00903.x.
83. Daculsi, G. 1998. Biphasic calcium phosphate concept applied to artificial bone, implant coating and injectable bone substitute. *Biomaterials* 19 (16): 1473–78. https://doi.org/10.1016/s0142-9612(98)00061-1.
84. Cao, W., and L.L. Hench. 1996. Bioactive materials. *Ceramics International* 22 (6): 493–507. https://doi.org/10.1016/0272-8842(95)00126-3.
85. Groot, K.de. 1980. Bioceramics consisting of calcium phosphate salts. *Biomaterials* 1 (1): 47–50. https://doi.org/10.1016/0142-9612(80)90059-9.
86. Yamada, S., D. Heymann, J.M. Bouler, and G. Daculsi. 1997. Osteoclastic resorption of calcium phosphate ceramics with different hydroxyapatite/beta-tricalcium phosphate ratios. *Biomaterials* 18 (15): 1037–41. https://doi.org/10.1016/s0142-9612(97)00036-7.
87. Zhang, Y., Y. Yokogawa, X. Feng, Y. Tao, and Y. Li. 2010. Preparation and properties of bimodal porous apatite ceramics through slip casting using different hydroxyapatite powders. *Ceramics International – CERAM INT* 36 (January): 107–13. https://doi.org/10.1016/j.ceramint.2009.07.008.
88. Yao, X., S. Tan, and D. Jiang. 2005. Fabrication of hydroxyapatite ceramics with controlled pore characteristics by slip casting. *Journal of Materials Science. Materials in Medicine* 16 (2): 161–65. https://doi.org/10.1007/s10856-005-5901-2.

89. Zhang, Y., D. Kong, Y. Yokogawa, X. Feng, Y. Tao, and T. Qiu. 2012. Fabrication of porous hydroxyapatite ceramic scaffolds with high flexural strength through the double slip-casting method using fine powders. *Journal of the American Ceramic Society* 95 (1): 147–52. https://doi.org/10.1111/j.1551-2916.2011.04859.x.
90. Sepulveda, P., F. Ortega, M. Innocentini, and V. Pandolfelli. 2000. Properties of highly porous hydroxyapatite obtained by the gelcasting of foams. *Journal of the American Ceramic Society* 83 (December): 3021–24. https://doi.org/10.1111/j.1151-2916.2000.tb01677.x.
91. Chu, T.M., J.W. Halloran, S.J. Hollister, and S.E. Feinberg. 2001. Hydroxyapatite Implants with Designed Internal Architecture. *Journal of Materials Science. Materials in Medicine* 12 (6): 471–78. https://doi.org/10.1023/a:1011203226053.
92. Ducheyne, P., and G.W. Hastings, eds. 1984. *Metal and Ceramic Biomaterials*, vol. 2. Boca Raton, FL: CRC Press.
93. Robert, H.D. 1984. Manufacturing processes of ceramics. In *Metal and Ceramic Biomaterials*. CRC Press.
94. Descamps, M., J.C. Hornez, and A. Leriche. 2009. Manufacture of hydroxyapatite beads for medical applications. *Journal of the European Ceramic Society* 29 (3): 369–75. https://doi.org/10.1016/j.jeurceramsoc.2008.06.008.
95. Balamurugan, A., A.H.S. Rebelo, A.F. Lemos, J.H.G. Rocha, J.M.G. Ventura, and J.M.F. Ferreira. 2008. Suitability evaluation of sol-gel derived Si-substituted hydroxyapatite for dental and maxillofacial applications through in vitro osteoblasts response. *Dental Materials* 24 (10): 1374–80. https://doi.org/10.1016/j.dental.2008.02.017.
96. Sudheesh Kumar, P.T., S. Srinivasan, V-K. Lakshmanan, H. Tamura, S.V. Nair, and R. Jayakumar. 2011. β-chitin hydrogel/nano hydroxyapatite composite scaffolds for tissue engineering applications. *Carbohydrate Polymers* 85 (3): 584–91. https://doi.org/10.1016/j.carbpol.2011.03.018.
97. Mene, R.U., M.P. Mahabole, and R.S. Khairnar. 2011. Surface modified hydroxyapatite thick films for CO_2 gas sensing application: Effect of swift heavy ion irradiation. *Radiation Physics and Chemistry* 80 (6): 682–87. https://doi.org/10.1016/j.radphyschem.2011.02.002.
98. Islam, M., P.C. Mishra, and R. Patel. 2010. Physicochemical characterization of hydroxyapatite and its application towards removal of nitrate from water. *Journal of Environmental Management* 91 (9): 1883–91. https://doi.org/10.1016/j.jenvman.2010.04.013.
99. Rajesh, R., A. Hariharasubramanian, and Y.D. Ravichandran. 2012. Chicken bone as a bioresource for the bioceramic (hydroxyapatite). *Phosphorus, Sulfur, and Silicon and the Related Elements* 187 (8): 914–25. https://doi.org/10.1080/10426507.2011.650806.
100. Hoque, M.E., M. Shehryar, and K.M. Islam. Processing and characterization of cockle shell calcium carbonate (CaCO3) bioceramic for potential application in bone tissue engineering. *Journal of Materials Science and Engineering* 2 (1 January 2013). https://doi.org/10.4172/2169-0022.1000132.

3 Processing of Bioceramics by Pressing and Tape Casting

Muhammad Uzair Iqbal
ACME Enterprises

Azhar Hussain
University of Engineering and Technology

Muhammad Shahid
ACME Enterprises

Muhammad Ramzan Abdul Karim
Ghulam Ishaq Khan Institute of Engineering Sciences and Technology

Moeen Akhtar
Eindhoven University of Technology

3.1 BACKGROUND

Bioceramics are a promising class of ceramics offering a combination of various properties such as mechanical, chemical, thermal, biocompatibility, biodegradability, and bioresorbability [1–5]. They can be used for multiple biomedical applications such as bio-implants, tissue replacements, spinal fusion, bone filling, dental implants, maxillofacial reconstruction, space fillings, and scaffolds [6–10] in the form of nanoparticles, coatings, and cement [11–13]. Many ceramic-based materials [14,15], including alumina, zirconia, zirconia-toughened alumina, silica, and calcium phosphates and their polymorphs, tricalcium phosphate (TCP) and tetracalcium phosphate, have been extensively studied for biomedical applications because of structural similarity and biocompatibility to natural bone. Crystallinity, porosity, composition, etc. are the characteristic features that govern the property of bioceramics [16]. These characteristics can be altered by suitable modification in processing techniques and functionalization with other materials to acquire desired properties for intended applications. Various processing and shaping techniques, i.e., pressing, sol-gel, slip casting, tape casting, gel casting, stereolithography, selective laser

 DOI: 10.1201/9781003258353-4

sintering, bioprinting, and additive manufacturing, are available for fabricating bioceramic products [17–23].

This chapter focuses primarily on the processing of bioceramics by pressing and tape casting. The purpose of the processing of bioceramics is to create a material's form that performs a specified function, such as tissue replacement or bonding, space filling. Bioceramic processing involves the following basic steps:

- Powder treatment
- Shaping of green body
- Drying and binder removal
- Sintering

To improve the particle packing density, the bioceramic powder must be milled and grounded using ball milling or agitation milling, as the packing density of sintered bodies is critical for compaction and densification [24]. Additives such as binders, flocculants, and plasticizers used during consolidation should be removed prior to sintering to produce a porous structure. The powder thus obtained is followed by sintering to get a densified bioceramic product.

3.2 POWDER TREATMENT

In any material synthesis process, metallic, ceramic, or composite, powder is an essential ingredient for batch formulation. Additives such as binders, solvents, and plasticizers play a supportive role in shaping and controlling the bulk density of the green product. After de-binding and consolidation, the powder is the chief component that defines the properties of the final product produced. One must not forget to characterize the powder for its particle size and particle size distribution, and to optimize the packing density prior to material processing.

Many authors and researchers have already discussed the characterization of powder for its size and distribution [24–26]. Powders are characterized by their particle size and size distribution to achieve maximum packing density. Various techniques can do this characterization. Some of the major techniques are sieve/mesh analysis, scanning electron microscopy, static light scattering, X-ray diffraction, dynamic light scattering, etc. Details of these techniques are not in the scope of this chapter but can be taken from References [27–29].

The classical representation of good packing using mixtures of powder particles of different diameters is different [30]. The ideal system is presumed to consist of a framework of large-sized spheres with interstices filled by slightly smaller ones that are of such a size that they touch each of the surrounding large-sized balls. Such a system would ideally have a very small void space, but it would be impossible to attain in practice because small-sized grains are rigidly contained in the space between the large ones and are not free to move. Mixing spheres or particles of different diameters in various proportions to improve the packing and reduce the void space is known as grading and has an important application in the manufacturing of most ceramic bodies. The use of graded materials is desirable for several reasons:

- If only coarse materials are employed, it would require a very large proportion of bond to fill the voids and cement the grains together, and the resultant mass could not be as strong as a better-graded material.
- Porosity and permeability are reduced in a properly graded material, which often has a marked effect on its resistance to corrosion, abrasion, etc.
- The appearance and other qualities of the material may often be improved by removing particles of certain sizes from the material.
- The use of carefully graded material greatly lessens the possibility of spalling, splintering, and cracking when a material or article is exposed to sudden changes in temperature.
- Well-graded raw material has a considerably smaller shrinkage on firing than a body containing appreciable void space.

The void space must be kept to a minimum to compensate for the issues mentioned earlier. This minimum void space can be achieved using a powder system of three different sizes as reported by Westman and Hugill [31]. For ideal packing of the mixture, the ratio of powder particles that results in maximum packing is coarse:medium:fine 70:20:10 by volume, where the diameter ratios of the spheres are 50.5:8:1. Although the individual particles in ceramic raw materials are usually not spherical but angular or subangular, the general principles of grading apply, provided the grains are fairly symmetrical and not elongated in one particular direction. Thus, most ground or crushed materials may be separated into size fractions, which, by suitable blending, will give a close-packed system.

3.3 PRESSING

Pressing is a shape-forming technique used mostly in the ceramic industry and is categorized into two main classes based on the application of pressure: uniaxial pressing/die pressing and isostatic pressing. Each process is further classified into cold and hot pressing based on the working temperature. A detailed overview of each process is given in this section.

3.3.1 Uniaxial Pressing or Die Pressing

For most bioceramic bodies, uniaxial pressing is the preferred method of processing. In this process, punches or pistons impart pressure along one axial direction to a stiff die to compact the powder. The pressure varies between 10 and 12 kpsi. This method is particularly suited to large production because it is easily mechanized. Due to the friction between the die and the ceramic particles, high-body pressing often results in an uneven distribution of ceramic particles [32]. Therefore, low-height or tabular bodies and symmetrical shapes are best suited for uniaxial pressing. A hydraulic press is mainly used for pressing the powder having a load capacity of up to 750 tons. Uniaxial pressing can be cold and hot pressing based on the temperature. Figure 3.1 shows a schematic example of segmented punch uniaxial pressing.

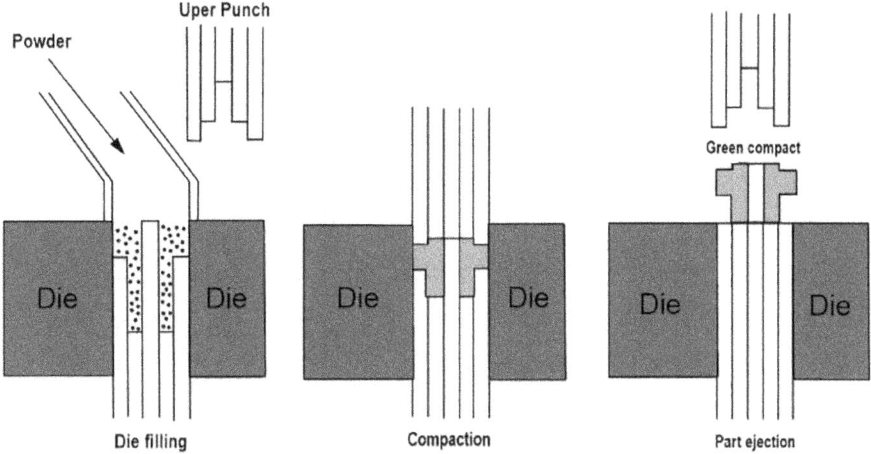

FIGURE 3.1 Schematic example of segmented punch uniaxial pressing.

The uniaxial pressing process completes in three steps:

- **Die filling:**
  A predetermined amount of ceramic powder is fed into the die cavity.
- **Compaction:**
  At this stage, an upper punch presses the powder at a pressure approximately 10–12 kpsi.
- **Part ejection/removal:**
  After compaction, the green body/compact is ejected from the die cavity.

Using this technique, many porous bioceramics, as well as nonporous/bulk bioceramics such as calcium phosphate, calcium sulfate, TCP, alumina, or zirconia-based products, bioactive glasses, hydroxyapatite (HAp), and glass-toughened HAp, can be produced into desired shapes: rods, bars, or other types of products [33–38].

3.3.1.1 Cold Pressing

In this process, a powder mixture is fed in between the closed die, where it is compacted using upper and lower punches with significant pressure. The produced component is removed from the die and quickly heated to remove any remaining lubrication. The component is then sintered in a controlled atmosphere before cooling [39]. The schematic illustration of cold pressing is given in Figure 3.2.

To obtain porous bioceramic structures, bioceramic powders are mixed with sacrificial fillers such as synthetic polymers (polyethylene particles) or natural polymers (rice husk, corn, and potato starches). The mixture is cold pressed to get the desired shape followed by sintering at 1,050°C which results in a porous structure on the firing of fillers at 450°C. Alumina porous scaffolds and bioactive glass-based scaffolds can be produced using this technique [10,36].

Sabree et al. developed the yttria-stabilized zirconia (YSZ)/HAp nanocomposite using cold pressing followed by sintering for orthopedic and dental applications [40].

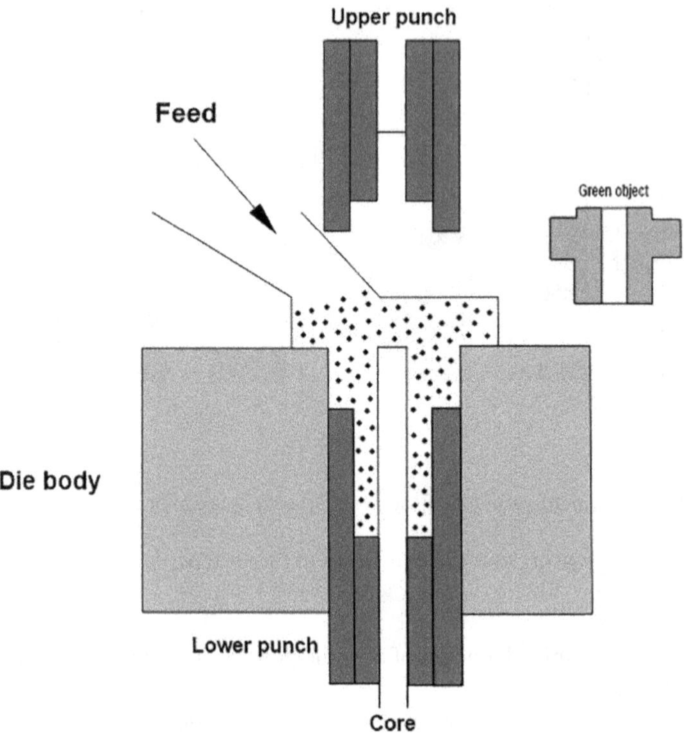

FIGURE 3.2 Schematic illustration of cold pressing.

Zampiva et al. fabricated the magnesium silicon oxide (forsterite) via cold pressing followed by low-temperature sintering presenting high-density and enhanced mechanical properties maintaining a low crystallite size (30 nm) [41].

3.3.1.2 Hot Pressing

Hot pressing can only be used to create simple, solid shapes, such as cubes, squares, and rectangles. In this method, the powder is compacted between two punches in a die under uniaxial pressure at a higher temperature, i.e., sintering temperature. The sintering temperature is lowered because applying pressure avoids the decomposition of ceramic products at high temperatures. Graphite is the ideal die-making material because of its high strength and excellent thermal conductivity, high-temperature creep resistance, minimal thermal expansion, and low cost. Due to graphite's high vapor pressure, an inert environment should be utilized above 2,200°C. However, sometimes, a controlled environment must be maintained at high temperatures.

The most common heat transfer methods for hot pressing are conduction, convection, and radiation. Conventional inductive heating uses a high-frequency electromagnetic field to generate heat within the die, which is then transferred to the product, while the method of "indirect resistance heating" involves electrical energy, which first warms the heating source and then heats a mold. This new approach takes longer to heat up and transfer heat from the die to the chamber and has no bearing

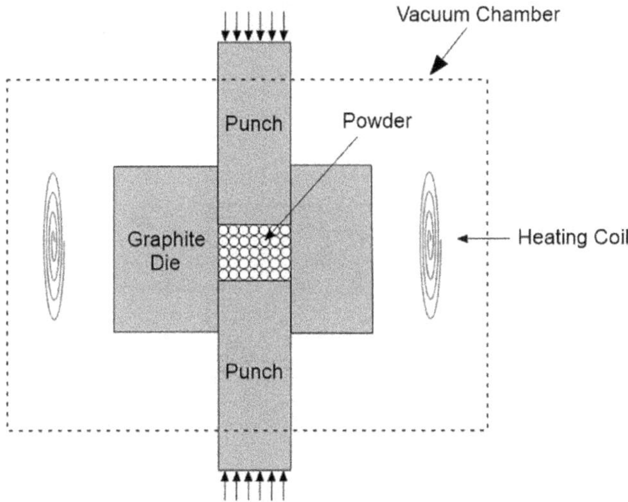

FIGURE 3.3 Schematic illustration of hot pressing.

on thermal conductivity compared to the inductive heating system [42]. A schematic illustration of hot pressing is given in Figure 3.3.

Bioceramics with precise microstructures and high densities are now being produced using hot pressing as part of their manufacturing process as it allows for better part compaction and higher part strength than cold pressing.

Sung et al. successfully developed the HAp/YSZ nanocomposite using hot pressing in a vacuum atmosphere (1,100°C for 1 hour) with improved mechanical properties, making a potential material for load-bearing orthopedic applications [43]. Many nanocomposites such as Al_2O_3–ZrO_2, Si_3N_4–ZrO_2/CeO_2, Al_2O_3–ZrO_2/CeO_2, Ce-TZP/Al_2O_3, etc. were developed with enhanced mechanical properties, opting the hot pressing technique that is mainly employed in joint prostheses [44–47].

3.3.2 Isostatic Pressing

This method applies pressure to a liquid or gaseous medium that surrounds the compacted object in several directions during the powder compaction process. It can be carried out at ambient or elevated temperatures depending on the desired properties. This process is more advantageous to die pressing because of its better uniformity and shape flexibility.

3.3.2.1 Cold Isostatic Pressing (CIP)

This process is carried out at room temperature using a pressure medium like water or oil and is commonly used to prepare porous materials and/or intricate shapes [48]. When the friction between powders/mold walls is sufficient, a green body with homogeneous density is produced. This results in a higher packing homogeneity than in uniaxial/die pressing. A high-pressure pump drives the fluid medium into a high-pressure-resistant steel container. Powders in the elastic mold are subjected to high

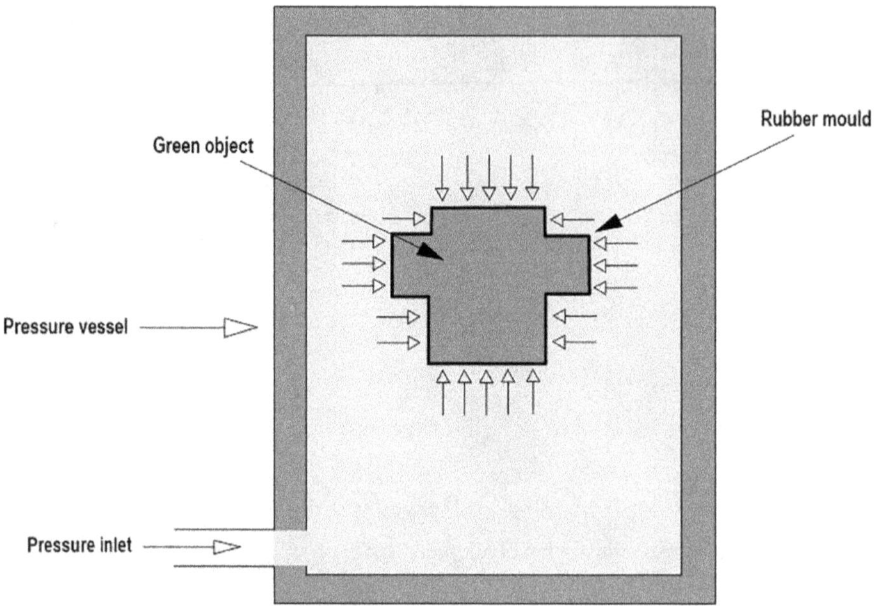

FIGURE 3.4 Cold isostatic pressing schematic.

pressure in the range of 200–300 MPa. A schematic representation of CIP is given in Figure 3.4.

In this process, molds can be made from both natural and synthetic rubbers. On the other hand, rubber is gradually being replaced by plastics because of distortion and wrinkling induced by mineral oil contact. The plasticizer composition and amount can alter the softness and hardness of thermal plastic soft resin, an important moldable material [49]. The following specifications apply to isostatic pressing molds:

- When loading powders, the original geometry must be preserved, with excellent strength and flexibility.
- The powders must not undergo any kind of reaction, physical or chemical.
- Abrasion resistance and ease of machinability are essential.

CIP is commonly used for consolidations of advanced ceramics such as alumina (Al_2O_3), zirconia (ZrO_2), calcium phosphate (CaP), magnesium phosphate (MP), amorphous calcium polyphosphates, zirconium-toughened alumina (ZTA), etc. for various medical and dental applications [50–52].

Rokas et al. successfully developed ZTA/carbon nanotube (CNT) composites opting for CIP followed by sintering relatively higher density with improved mechanical properties suitable for load-bearing orthopedic applications [53]. Thang et al. studied the effect of carbonate content on the mechanical properties and thermal stability of carbonated apatite (CA). Results showed that dense CA with a high carbonate content results in high density and enhanced fracture toughness suitable for artificial bone substitutes [54].

3.3.2.2 Hot Isostatic Pressing (HIP)

Isostatic pressing at a higher temperature using a gas (nitrogen or argon) as a pressure medium is known as hot isostatic pressing (HIP). This process can be carried out without the use of a mold. The part is first compacted using the CIP process and then sintered to close the interconnecting porosity. After that, the sintered (but still porous) component is isostatically pressed at a high temperature without any cans (mold). The working pressure used in this process typically ranges from 15 to 44 kpsi (0.1–0.3 GPa). This process combines pressing and sintering to cause powder particle consolidation and the healing of voids and pores. Heating causes the sample to shrink due to increasing gas (Ar or He) pressure, which results in an isostatic effect. As a result of the uniform pressure distribution during sintering, this approach is particularly useful for processing components with complicated structures. This technique is mainly used for the processing of nonporous bioceramic products. However, HIP equipment is costly.

Bernal et al. developed the Al_2O_3 ceramic with higher density and enhanced mechanical properties (hardness of 19 GPa and toughness of 5.2 MPa$\sqrt{m}$) via HIP [55]. Bal et al. fabricated Si_3N_4-based ceramics via sintering and HIP for hip joint prostheses. Results depict that Si_3N_4-based ceramics have superior mechanical properties as compared to other orthopedic materials such as Al_2O_3, ZTA, and Y-TZP (yttria-stabilized tetragonal zirconia polycrystal) [56]. Multiwalled carbon nanotube–reinforced ZTA composites have been developed using this technique [57].

Typically, a sacrificial polymer binder is mixed or coated with ceramic powder particles, and laser is directed onto the powder to melt and fuse the ceramic particles. The fused ceramic followed by post-processing (warm isostatic pressing and sintering) results in dense zirconia scaffolds [58].

3.4 TAPE CASTING

Tape casting is another material processing technique originally employed in traditional ceramics but has reached a high level of sophistication in bioceramics. Doctor blading is an effective tape casting procedure, where a ceramic powder slurry comprising an organic solvent and other polymer additives is constantly cast onto a nonsticky moving carrier surface such as Teflon. The slurry is spread to the desired thickness, the solvent is evaporated, and the tape is rolled onto a take-up reel for further processing. This method of tape casting is named doctor blade tape casting [59]. The whole tape casting process flow is schematically illustrated in Figure 3.5.

A detailed description of the doctor blade tape casting process, as well as further processing of tapes beyond tape formation, is given below in this section. Moreover, the application of this technique for bioceramic processing is also discussed.

3.4.1 Doctor Blade Tape Casting Process

Doctor blade tape casting is a modern technique that originates from Park patents [60]. In this process, a slip is made to move onto a substrate carrier surface or a polymer carrier. The flexibility and ease of the process urge the ceramic engineers' interest in the manufacturing of thin ceramic sheets as well as functionally graded

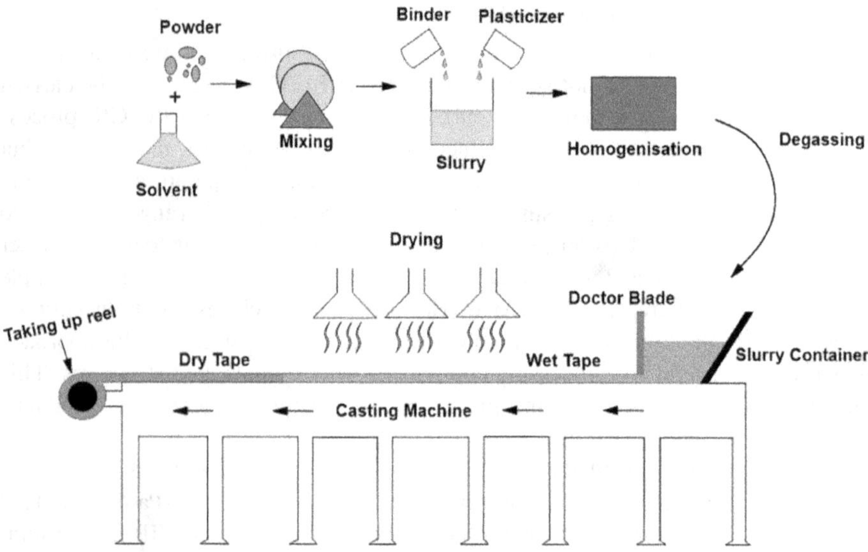

FIGURE 3.5 Process flow diagram of whole tape casting.

and porous structures. However, this process is not as simple as it seems, but certain things such as casting machines, design of doctor blades, and formulation make the process complex.

In this process, the slurry/slip is normally poured into the slurry container just before the doctor blade. The cast carrier is set in motion, and the gap between the doctor blade and the carrier defines the wet thickness of the ceramic sheet to be cast. Other factors such as container depth, carrier speed, slip viscosity, and blade shape may also be considered. The wet sheet is passed through a drying zone, where the solvent is evaporated and left behind a dry sheet that can be rolled for additional processing [61]. A detailed discussion of the process from material selection to the end product is covered in the following section.

3.4.1.1 Material Selection

As tape casting is a fluid-forming process, it requires a powder that behaves as a fluid for its mechanical metering into thin sheets. To get a flow, the powder is suspended into a liquid termed solvent, base, or vehicle. The main purpose of liquid is to homogenize the ingredients thoroughly to create a uniform mixture. This solvent is so frequently used that the process can broadly be referred to as solvent-based casting.

The choice of a **solvent** depends on other additives such as binders, plasticizers, and deflocculants. When you design the composition of ceramic slip, the powder is chosen first and then other additives such as a solvent, binder, plasticizer, etc. The solvent must be chosen carefully, and it should be noted that the chosen solvent could dissolve all other ingredients or additives. To achieve such a degree of solubility, the common practice is to use more than one solvent as a fluid vehicle. Binary solvents are common in tape casting as they have multiple advantages; the most important are increased capacity of dissolution, rheological control cast, and controlled drying speed.

TABLE 3.1
Commonly Used Solvent System for Tape Casting Slip Formation

Xylenes/95% ethanol	MEK/95% ethanol
Xylenes/anhydrous ethanol	Toluene/95% ethanol
Methyl ethyl ketone (MEK)	TCE/95% ethanol
MEK/toluene	TCE/MEK/ethanol
1,1,1 Trichloroethane (TCE)	MEK/95% ethanol/cyclohexanone
MEK/anhydrous ethanol/cyclohexanone	Toluene/ethanol/cyclohexanone
And many more…	

Some of the most common solvent systems that have been effectively used over the years are given in Table 3.1 [62].

Over the past few years, many researchers have taken the idea of **azeotropes or azeotropic mixture** [62]. The azeotropic mixture is a combination of many liquids that acts as a single liquid having a constant boiling point and maintaining the stoichiometric while drying. Solvents such as ethanol/MEK (>46 wt% ethanol) and ethanol/toluene (68 wt% ethanol) are azeotropes at essentially a single point. Criteria for choosing the final solvent depend mainly on cost and time to market. Ethanol and isopropanol are preferred over anhydrous ethanol due to their cost and hydrophobicity. However, the relative concentration of solvent affects slip formability but has no such drastic effect on characteristics and properties.

When deciding on a solvent, another important consideration is the level of regulation around its use. Solvents including 1,1,1 TCE, xylenes, benzene, toluene, and MEK have been under much attention over the years. Certain states or countries have outlawed the use of certain solvents, while others have controlled them so tightly that their usage is impractical. The carcinogenicity of these solvents, as well as their other long-term consequences on the human body, is currently in discussion. Although this book will not get into the nitty-grittiest of the discussion over who's right and who's wrong, the amount of debate and the ever-changing rules imposed on the use and disposal of these solvents and their vapors can have a considerable impact on the expense of utilizing them. Every person who is exposed to industrial chemicals should be aware of the necessary handling practices, storage regulations, and safety protocols for these compounds.

A **homogenizer** is a substance that helps ensure that a system is consistent throughout. The concept of homogenizers in solvent-based tape casting has been around for a while now. It is an organic solvent that helps make a system uniform and acts as a skin retarder. Cyclohexanone is the most commonly used homogenizer.

Surfactants are the additives that mainly alter the particle surface to attain desired properties such as specific chemistry, low/high surface charge, and low/high surface energy. These are the keys for dispersion, density, stability, porosity, green strength, and other measurable properties of the ceramic shape formed. Surfactants are discussed in almost every book on ceramic processing, but here, we may define them as concentrations for the active agents; by this definition, ingredients involved in slip preparation act as a surfactant in one way or the other way except for plasticizers.

Dispersants/deflocculants help in keeping the particles of the slip system apart. Their main function is to hold the primary particles for binders to coat them individually and to disperse the primary particles and keep them in a uniform suspension. There are two ways that the dispersing additive can perform this action: steric hindrance and ionic repellency. Using the dispersant, the vehicle can create a layer of separation by separating and suspending the primary particles. Because of the fluid interparticulate layer's ability to increase particle mobility, a well-dispersed suspension usually has a lower viscosity, which can reduce the amount of solvent resulting, lowering the cost. This solvent reduction helps dry the slip faster with less shrinkage and complete burning of additives, so the final product doesn't get contaminated upon firing during sintering. Deflocculation of these slips is usually excellent. Both aqueous and nonaqueous systems commonly use organic chain molecules to achieve particle–particle separation [63]. The most commonly used dispersants/deflocculants are given in Table 3.2 [62].

Binders are used only in the tape casting process to keep the ceramic body flexible even after drying. The binders should be long-chain polymers that form a film on drying from the solvent system. The conditions for removal before sintering complicate the choosing of binders even more. Because most organics burn with oxygen and heat, removing the binder is simple. Almost all oxide ceramics can be sintered in an oxidizing environment; however, a neutral or decreasing environment is required when working with reactive metal, carbide, or nitride matrix powders. This limits the choice of binders to an organic system that evaporates when heated. Polyvinyl acetate, polyvinyl butyral (PVB), polyvinyl alcohol, polyvinyl chloride, polystyrene, and copolymers are commonly used as oxidizing binders while acrylics are used in a non-oxidizing environment [63].

Aside from the limitations discussed earlier, the binder's cost and its solubility in a cheap, nonflammable, and volatile solvent are even more crucial factors that must be taken into consideration. Some other factors, such as casting surface, final shape thickness, environmental safety, and health issues, must be considered while choosing a binder/solvent system. Toluene or xylene can be utilized in the case of thicker tapes (>0.254 mm); however, for thin tapes, acetone is the preferred solvent, as it is highly volatile. The usage of chlorinated hydrocarbons like trichloroethylene in most tape formulations has been reduced to a minimum because of health issues. Alcohols and

TABLE 3.2
Commonly Used Dispersants/Deflocculants

Linseed oil	Safflower oil
Lanolin fatty acids	Oleic acid
Menhaden fish oil	Synthetic waxy esters
Dibutyl amine	Substituted imidazolines
Polyisobutylene	Citric acid
Linoleic acid	Salts of polyacrylic acids
Salts of methacrylic acids	Stearic acid
Corn oil	Glycerol trioleate

TABLE 3.3
Commonly Used Binders for Tape Casting

Polyvinyl alcohol	Polyvinyl chloride
Vinyl chloride-acetate	Ethylcellulose
Methylcellulose	Polyvinyl butyral
Polyacrylates esters	Polymethyl methacrylate
Polystyrene	Polypropylene carbonate
Polyethylene	Atactic poly(propylene)/poly(butene)
Poly isobutylene	Polypropylene carbonate emulsion
Polytetrafluoroethylene	Hydroxypropyl methylcellulose

TABLE 3.4
Most Commonly Used Plasticizers for Tape Casting

Dimethyl phthalate	Ethyltoluene sulfonamides
(poly)Propylene glycol	Butyl stearate
n-Butyl phthalate	Butyl benzyl phthalate
Triethylene glycol	Propylene carbonate
Mixed esters phthalate	Glycerol
(poly)Ethylene glycol	Polyalkylene glycol
Dipropyl glycol di benzoate glycol	Methyl abietate

their mixtures with toluene, methyl ethyl ketone, and xylene are currently popular safe solvents. Binders in common practice are enlisted in Table 3.3 [62].

The term "**plasticizer**" in tape casting refers to any additive that enhances the green tape's elasticity or plasticity. This is achieved by modifying the polymer binder's glass transition temperature. The plasticizer in green tape allows it to bend or "give" without tearing. If no plasticizer is employed, most polymeric binders generate a stiff, rigid, and brittle sheet. For aqueous and nonaqueous systems, the most prevalent plasticizers are tabulated in Table 3.4 [62].

3.4.1.2 Material Processing

Materials with which you begin limit any process. The tape's quality limits the ability to make a device/product out of it. The basic ingredients and how they have been processed into a slip limit the tape's quality. Easier it is to produce high-quality, high-yield pieces when your slip and tape are superior. Creating a flawless slip isn't the simplest task; for this, pre-batch material processing is mandatory. Pre-batching powder preparation includes cleaning the surface, controlling the surface water or hydration, treating the surface chemically to enhance coupling to the dispersant, protecting the surface chemically from hydration, and adding appropriate dopants to promote sintering. For example, to achieve a uniform coating on the surface of the aluminum oxide particles, soluble salts such as magnesium nitrate are used.

3.4.1.2.1 Slip Preparation

The first stage in making a tape casting slip is dispersion milling of powder with a dispersant/deflocculant in the chosen solvent. The batch size determines the ball mill size. After the selection of material and mill size, the next stage is the loading of dispersant, solvent, wetting agent, and powder into the mill. The standard approach is to dissolve a predefined amount of dispersant in one solvent by stirring or milling. The solvents are then added to the ball mill.

Solid loading in the mill is defined as follows:

$$\%\text{Solid} = \text{Powder weight} / (\text{Powder weight} + \text{solvent weight})$$

Percentage of solid loading in the mill ranges from 20% to 90%, but for excellent dispersion, it is 70%–80%. After solid loading, the mill is sealed and rotated for 4–48 hours. The speed of rotation of the mill must be a fraction (35%–115%) of its critical speed which can be formulated by using the given equation [62].

$$\text{Nc} = \frac{76.6}{\sqrt{D}} \tag{3.1}$$

Here, D is the inner diameter of the mill in feet and Nc is the critical speed in rpm.

The dispersion milling process serves multiple purposes. Of primary importance are the decomposition of aggregates, the reduction of bigger particles to more manageable sizes, and the dispersion of dispersant/deflocculant evenly on each surface of the newly formed particles. This method is used to achieve an easily dispersed, low-viscous liquid slip in the solvent.

Particles should be suspended in a solvent before the plasticizers and binders are added to the mill. First comes the plasticizer and then comes the binder because the binder dissolves more quickly in the plasticizer than in the solvent(s). In some inorganic powder systems, binders like PVB act as secondary dispersants and compete for the powder surfaces. As a result, the milling process separates the dispersion and mixing phases. The slurry gets viscous and the powdered binder dissolves into the slurry during mixing. Most manufacturers opt to dissolve powdered binders before milling, rather than relying on the mill. The binder/plasticizer is typically milled for 24 hours or more into the slurry [63].

The next stage in the preparation of the slip is ousting the entrapped air/gas bubbles from the slurry. Usually, this is done by slow rolling, usually at less than 10 rpm for a minimum of 24 hours or longer. The most popular method involves swirling and vacuuming in a closed container like a desiccator. The vacuum is normally 635–700 mmHg as a high vacuum tends to extract a lot of solvents with bubbles. After this, the characteristics of slip to be monitored are viscosity, specific gravity, and particle size.

Once de-airing is completed, the slip is ready to be tape cast. For the production of defect-free thin sheets, the slurry is filtered using a series of nylon screens. The screen's mesh size is relative to the slip viscosity and particle size of the powder.

3.4.1.2.2 Tape Formation

The filtered slurry is poured into the slurry container just before the doctor blade in the tape casting machine. It has two operational modes: stationary blade–moving carrier and moving blade–stationary carrier. For continuous and mass production, a moving carrier is preferred, where the slip is poured into the doctor blade reservoir and the carrier is set to move. Preset speeds for batch or continuous casting are common. Speed of casting is affected by the drying chamber length on the machine, the tape thickness being cast, and the solvent system employed. Continuous casting speeds typically vary from 15 to 50 cm/min, whereas batch casting speeds often hover around the 20 cm/min level. The doctor blade controls the slurry thickness applied on the substrate surface by adjusting the height of the blade from the carrier surface. Since slip viscosity and casting speed are all interrelated, as is reservoir depth, it's difficult to precisely manage layer thickness.

For evaporation and removal of solvents during drying, the airflow mechanism in the machine is mandatory. Using hot air and heated platens allows for quick drying even when diffusion-controlled drying becomes the method of choice. At this point, diffusion takes over as the liquid slurry begins to harden. In most tape casting processes, laminar airflow is typically advised as the best practice.

In the next step, the dried tape needs to be removed and handled. For batch operations, the cast tape is transported to the drying section. To remove the tape from the carrier, simply use a metal blade and cut it to the desired length. In a continuous operation, the tape is cast and cured while the carrier is moved through and out of the device. The most common way to build a spool is to peel the tape and roll it.

3.4.2 Further Tape Processing

Several processes can be used for the tape cast product to be processed into a functional shape or part. This includes lamination, calendering, blanking, debinding, and sintering of ceramic bodies [63–65]. The whole process flow is given in Figure 3.6.

The green tapes are transferred to the next fabrication steps called calendering and lamination. Calendering is a process in which a flat sheet of material is subjected to high pressure and passed between two or more rollers. This results in flexible, smooth, and flat tapes that could be punched into desired shapes. Calendering of green tapes is done to achieve high green strength, smooth surface finish, and uniform thickness.

Lamination is the process of making a product with numerous layers. Multilayered products are produced by stacking multiple layers of green tapes that result in a single solid as sintered. The lamination process is as follows:

- Using alignment pins, stack the green tapes in the proper order.
- Silicone-coated Mylar must be applied to the substrate to avoid sticking.
- To achieve equilibrium, the stack must be preheated slowly at a low pressure.
- Apply and hold the full laminating pressure (0.2–20 kpsi) for a period ranging from a few seconds to several minutes.

Lamination can also be done by gluing several tapes and applying low pressure for proper adhesion of tapes. Functionally graded components can be designed using this technique.

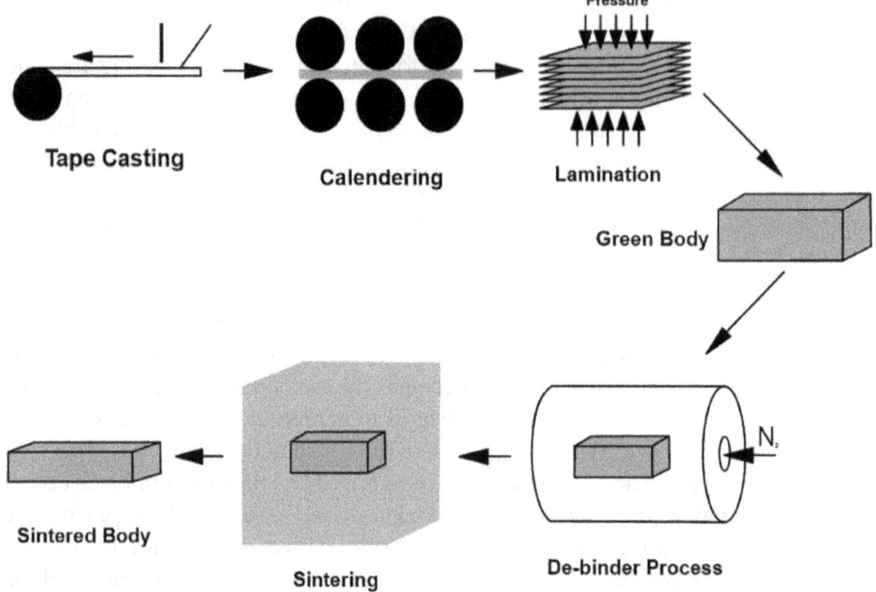

FIGURE 3.6 Flow diagram for further processing of tapes beyond tape casting.

After calendering and lamination, the tape cast materials are processed by punching their intended shape into an area that's been predetermined. Punch-and-die sets are commonly used in hydraulic or mechanical presses in most manufacturing sectors. Punching can be used to complete two distinct and independent tasks, i.e., blanking and hole or via generation.

Blanking is a punching procedure that shapes the intended part's exterior. Depending on the volume of items to be fabricated, usually blanking tools are made of hardened steel or carbide materials. This process can also include large interior holes or other shapes.

The next stage in part manufacturing is to create a hole pattern in a tape cast material. These holes should be used to connect layers in a stacked ceramic package or two surfaces of a substrate. The holes are commonly round but can be of any shape. In addition to the blanking process, numbers/nomenclature of holes can be punched sequentially at the same station or at a different station.

After blanking and hole punching, the green part is subjected to the de-binding process where the cracking of polymer is done at a slow heating rate in an atmosphere-controlled tube furnace from 750°C to 800°C. This typically involves a temperature hold of between 250°C and 650°C and binder removal times of at least 18 hours. After the binder removal, the part is cofired at a temperature range of 1,600°C–1,800°C depending on ceramic formalization. Usually, sintering is done in a controlled atmosphere to avoid oxidation of parts where necessary. After sintering, the part is characterized and finished before its use.

3.4.3 Tape Casting Applications

A variety of bioceramic materials, i.e., highly porous bioceramics, functionally graded structures, transparent bioceramics, porous implant materials, multilayered porous structures, and dense bioceramics, can be produced by tape casting technology. Using this technique, functionally graded materials based on TCP, HAp, and biphasic calcium phosphate [66] having a gradient in composition or structure depending on applications can also be tailored. In moderate stress-bearing applications, extremely porous gradient multilayer components may be used for bone tissue replacement and bone tissue engineering. It is possible to create bone graft materials with variant porosity levels based on the bone cross section. Transparent bioceramics of HAp and YSZ have been fabricated using this method [67–72]. Hofer et al. studied the effect of second-phase particles on the fracture toughness of tape-casted textured alumina. The results showed that the fracture toughness of monolithic textured alumina significantly decreases with the second-phase addition of zirconia [73].

3.5 CONCLUSION AND FUTURE DIRECTIONS

This chapter summarizes various shaping techniques like pressing and tape casting to form a green compact from the bioceramic powder. Shaping is a crucial stage prior to sintering. The non-homogeneities and defects in the green body can't be easily eliminated during sintering. To overcome such issues, powder processing prior to shaping is critical. The shape, size of the object, and manufacturing cost influence the choice of shaping technique. Isostatic pressing is particularly useful for processing intricate, near-net shape and densified bioproducts, while tape casting is beneficial for the fabrication of multilayered and functionally graded bioproducts. A variety of shapes—dense, porous, functionally graded, multilayered porous, and hierarchical scaffolds—can be designed for various implants, orthopedic, and tissue engineering applications. Furthermore, the primary directions for future advancements are:

- To produce high-quality finished goods, bioceramic materials' dimensional accuracy, surface quality, and mechanical qualities should be improved.
- Processing techniques and design approaches should be refined and innovated to obtain pleasing ceramic restorations with minor surface and subsurface damage.
- Tapes can be coined to near-net shape made manually and using a CNC mechanism replicating laminated object manufacturing.

REFERENCES

1. Jayalekshmi AC, Sharma CP. Bioceramics-design, synthesis and biological applications. In *Frontiers in Biomaterials The Design, Synthetic Strategies and Biocompatibility of Polymer Scaffolds for Biomedical Application*, vol. 1, p. 101, 2014.
2. Ebrahimi M, Botelho M, Lu W, Monmaturapoj N. Synthesis and characterization of biomimetic bioceramic nanoparticles with optimized physicochemical properties for bone tissue engineering. *J Biomed Mater Res A*. 2019;107(8):1654–1666.

3. Wei S, Ma JX, Xu L, Gu XS, Ma XL. Biodegradable materials for bone defect repair. *Milit Med Res.* 2020;7(54):1–25.
4. Huang J, Li X, Guo ZX. Biomechanical and biochemical compatibility in innovative biomaterials. *Biocompat Perform Med Dev.* 2020:23–46.
5. Islam MM, Shahruzzaman M, Biswas S, Sakib MN. Rashid TU. Chitosan based bioactive materials in tissue engineering applications – A review. *Bioactive Mater.* 2020;5:164–183.
6. Wang W, Yeung KWK. Bone grafts and biomaterials. Substitutes for bone defect repair: A review. *Bioactive Mater.* 2017;2:224–247.
7. Jitaru S, Hodisan I, Timis L, Lucian A, Bud M. The use of bioceramics in endodontics-literature review. *Clujul Med.* 2016;89:470–473.
8. Sheikh Z, Najeeb S, Khurshid Z, Verma V, Rashid H, Glogauer M. Biodegradable materials for bone repair and tissue engineering applications. *Materials.* 2015;8(9):5744–5794.
9. Denry I, Kelly JR. State of the art of zirconia for dental applications. *Dent. Mater.* 2008;24(3):299–307.
10. Prakasam M, Silvain J-F, Largeteau A. Innovative high-pressure fabrication processes for porous biomaterials – A review. *Bioengineering.* 2021;8:170.
11. Catledge SA, Fries MD, Vohra YK, Lacefield WR, Lemons JE, Woodard S, Venugopalan R. Nanostructured ceramics for biomedical implants. *J Nanosci Nanotechnol.* 2002;2(3–4):293–312.
12. Zhang GB, Myers ED, Wallace GG, Brandt M, Choong FP. Bioactive coatings for orthopaedic implants-recent trends in development of implant coatings. *Int J Mol Sci.* 2014;15(7).
13. Victor SP, Selvam S, Sharma CP. Recent advances in biomaterials science and engineering research in India: A minireview. *ACS Biomater Sci Eng.* 2019;5(1):3–18.
14. Nie L, Chen D, Fu J, Yang S, Hou R, Suo J. Macroporous biphasic calcium phosphate scaffolds reinforced by poly-L-lactic acid/hydroxyapatite nanocomposite coatings for bone regeneration. *Biochem Eng J.* 2015;98:29–37.
15. Shekhawat D, Singh A, Banerjee MK, Singh T, Patnaik A. Bioceramic composites for orthopaedic applications: A comprehensive review of mechanical, biological, and microstructural properties. *Ceram Int.* 2021;47(3):3013–3030.
16. Kuffner BHB, Facci AD, Sachs D, Silva G. Study of the microstructure and mechanical properties of beta tricalcium phosphate-based composites with alumina addition produced by powder metallurgy. *REM.* 2017;70(4):459–464.
17. Poomathi N, Singh S, Prakash C, Subramanian A, Sahay R, Cinappan A, Ramakrishna S. 3D printing in tissue engineering: A state of the art review of technologies and biomaterials. *Rapid Prototyp J.* 2020;26(7):1313–1334.
18. Kiran K, Ramakrishna S. *An Introduction to Biomaterials Science and Engineering.* World Scientific, 2021.
19. Aguilar CR, Pinto UO, Reyes EAA, Juarez RL. Alfonso 16. I. Characterization of β-tricalcium phosphate powders synthetized by sol-gel and mechanosynthesis. *Boletin De La Sociedad Espanola De Ceramica Y Vidrio.* 2018;2018:213–220.
20. Alizadeh-Osgouei M, Li Y, Wen C. A comprehensive review of biodegradable synthetic polymer-ceramic composites and their manufacture for biomedical applications. *Bioactive Mater.* 2019;4:22–36, ISSN 2452-199X.
21. Lin T, Wang X, Jin L, Li W, Zhang Y, Wang A, Peng J, Shao H. Manufacturing of porous magnesium scaffolds for bone tissue engineering by 3D gel-printing. *Mater Design.* 2021;209:109948, ISSN 0264-1275.
22. Peng C, et al. Preparation and properties of SiC ceramics via SLS/CIP process[J]. *J Mater Eng.* 2019;47(3):87–93.
23. Melchels FPW, Feijen J, Grijpma DW. A review on stereolithography and its applications in biomedical engineering. *Biomaterials.* 2010;31(24):6121–6130.

24. Allen. *Particle Size Measurements*, 3rd ed. Chapman and Hall, London, 1981.
25. Malghan SG. Characterization of ceramic powders. In *Engineered Materials Handbook, Vol. 4, Ceramics and Glasses.* edited by SJ Schneider. ASM International, Newbury, OH, pp. 63–74, 1991.
26. Katz RN. Characterization of ceramic powders. In *Treatise on Materials Science and Technology, Vol. 9, Ceramic Fabrication Processes.* edited by EEY Wang. Academic Press, New York, pp. 35–49, 1976.
27. Jillavenkatesa A, et al. *Particle Size Characterization.* National Institute of Standards and Technology, Gaithersburg, Md., 2001.
28. Abbireddy COR, Clayton CRI. A review of modern particle sizing methods. *Proc Inst Civil Eng – Geotech Eng.* 2009;162(4):193–201, doi: 10.1680/geng.2009.162.4.193.
29. Methods of particle size determination – A review https://www.innopharmatechnology.com/products/eyecon2tm/methods-of-particle-size-determination.
30. Dorozhkin SV. Calcium orthophosphate bioceramics. *Ceram Int.* 2015;41(10, Part B):13913–13966, ISSN 0272-8842.
31. Westman AER, Hugill HR. *Amer Ceram Soc.* 1930;13:767.
32. Shen JZ, Kosmač T. *Advanced Ceramics for Dentistry.* Elsevier/Bh, Amsterdam, 2014.
33. Cui W, Cao L, Zhang X. Bioceramics: Materials, properties, and applications. *Ceram Sci Eng.* 2022:65–110.
34. Kuffner BHB, Facci AD, Sachs D, Silva G. Study of the microstructure and mechanical properties of beta tricalcium phosphate-based composites with alumina addition produced by powder metallurgy. *REM.* 2017;70(4):459–464.
35. Jodati H, Yilmaz B, Evis Z. Calcium zirconium silicate (baghdadite) ceramic as a biomaterial. *Ceram Int.* 2020;46(14):21902–21909
36. Wang M, et al. Manufacture of biomaterials. *Encycl Biomed Eng.* 2019:116–134, doi: 10.1016/b978-0-12-801238-3.11027-x.
37. Baino F, et al. Processing methods for making porous bioactive glass-based scaffolds-a state-of-the-art review. *Int J Appl Ceram Technol.* 2019;16(5):1762–1796, doi: 10.1111/ijac.13195.
38. Malik MSRA, et al. Synthesis and mechanical characterization of alumina-based composite material for armor application. *J Ceram Process Res.* 2021;22(2):149–157.
39. Cold pressing and sintering of powders. www.open.edu/openlearn/science-maths-technology/engineering technology/manupedia/cold-pressing-and-sintering-powders
40. Sabree I, Saleh O. Characterization of zirconia-hydroxyapatite nanocomposites for orthopedic characterization of zirconia-hydroxyapatite nanocomposites for orthopedic and dental applications. *Int J Eng Technol.* 2019;8(1.5):554–560.
41. Zampiva RYS. *Ceram Int.* 2017, doi: 10.1016/j.ceramint.2017.08.201.
42. Moustafa S, et al. Hot forging and hot pressing of AlSi powder compared to conventional powder metallurgy route. *Mater Sci Appl.* 2011;02(08):1127–1133, doi: 10.4236/msa.2011.28152.
43. Sung YM, Shin YK, Ryu JJ. Preparation of hydroxyapatite/zirconia bioceramic nanocomposites for orthopaedic and dental prosthesis applications. *Nanotechnology.* 2007;18(6): 1–6.
44. Sigulinski F, Bo˘skovi´c S. Phase composition and fracture toughness of Si_3N_4-ZrO_2 with CeO2 additions. *Ceram Int.* 1999;25:41–47.
45. Verma V, Kumar BVM. Processing of alumina-based composites via conventional sintering and their characterization. *Mater Manuf Process.* 2017;32:21–26.
46. Tanaka, et al. Ce-TZP/Al_2O_3 nanocomposite as a bearing material in total joint replacement. *J Biomed Mater Res.* 2002;63(3):262–270.
47. Rittidech A, Tunkasiri T. Preparation and characterization of Al_2O_3-25 mol% ZrO2 composites. *Ceram Int.* 2012;38S:S125–S129.

48. Liu PS, Chen GF. Making porous metals. *Porous Mater.* 2014:21–112, doi: 10.1016/b978-0-12-407788-1.00002-2. Accessed 4 Nov. 2020.
49. Methods of shape forming ceramic powders [SubsTech]. (2012, May 31). Subtech. https://www.substech.com/dokuwiki/doku.php?id=methods_of_shape_forming_ceramic_powders.
50. Bavya Devi K, Lalzawmliana V, Saidivya M, Kumar V, Roy M, Kumar Nandi S. Magnesium phosphate bioceramics for bone tissue engineering. *Chem Rec.* 2022. https://doi.org/10.1002/tcr.202200136
51. Comeau P, Filiaggi M. A two-stage cold isostatic pressing and gelling approach for fabricating a therapeutically loaded amorphous calcium polyphosphate local delivery system. *J Biomater Appl.* 2017;32(1):126–136, doi: 10.1177/0885328217708639.
52. Amat NF, Muchtar A, Amril MS, Ghazali MJ, Yahaya N. Preparation of presintered zirconia blocks for dental restorations through colloidal dispersion and cold isostatic pressing. *Ceram Int.* 2018;44:6409–6416.
53. Reyes-Rojas A, Dominguez-Rios C, Garcia-Reyes A, Aguilar-Elguezabal A, Bocanegra-Bernal MH. Sintering of carbon nanotube-reinforced zirconia toughened alumina composites prepared by uniaxial pressing and cold isostatic pressing. *Mater Res Exp.* 2018;5:10.
54. Thang LH, Bang LT, Long BD, et al. Effect of carbonate contents on the thermal stability and mechanical properties of carbonated apatite artificial bone substitute. *J Mater Eng Perform.* 2022.
55. Bocanegra-bernal MH, Domínguez-rios C, Garcia-reyes A, Aguilarelguezabal A. Fracture toughness of an a-Al_2O_3 ceramic for joint prostheses under sinter and sinter-HIP conditions. *Int J Refract Metals Hard Mater.* 2009;27(4):722–728.
56. Bal BS, Khandkar A, Lakshminarayanan R, Clarke I, Hoffman AA, Rahaman MN. Testing of silicon nitride ceramic bearings for total hip arthroplasty. *J Biomed Mater Res B Appl Biomater.* 2008;87B(2):447–454.
57. Echeberria J, Ollo J, Bocanegra-bernal MH, Garcia-reyes A, Domínguezrios C. Sinter and hot isostatic pressing (HIP) of multi-wall carbon nanotubes (MWCNTs) reinforced ZTA nanocomposite: Microstructure and fracture toughness. *Int J Refract Metals Hard Mater.* 2010;28(3):399–406.
58. Kumaresan S, Vaiyapuri S, Kang J, Dubey N, Manivasagam G, Yun K, Park S. Perspective chapter: Additive manufactured zirconia-based bio-ceramics for biomedical applications. In *Advanced Additive Manufacturing*. IntechOpen, 2022.
59. Mason TO. Advanced ceramics. *Encycl Britan.* 2016, https://www.britannica.com/technology/advanced-ceramics.
60. Park Jr. J.L. Manufacture of ceramics. US patent 2, 966, 719. 1961.
61. Mistler RE. Tape casting: The basic process for meeting the needs of the electronics industry. *Am Ceram Soc Bull.* 1990;69(6):1022–1026.
62. Mistler, Richard E, Twiname ER. *Tape Casting: Theory and Practice*. Wiley, Hoboken, NJ, Chichester, 2012.
63. Terpstra RA, Pex PPAC, De Vries AH. *Ceramic Processing*. Chapman & Hall, 1995.
64. Eric R. Twiname, tape casting and lamination. In *Encyclopedia of Materials: Technical Ceramics and Glasses*. edited by M Pomeroy. Elsevier, pp. 189–194, 2021.
65. Wang W, et al. Fabrication and properties of tape-casting transparent Ho:Y3Al5O12 ceramic. *Chinese Opt Lett.* 2015;13(5):051404-51407, doi: 10.3788/col201513.051404.
66. Sánchez-Salcedo S, Werner J, Vallet-Regí M. Hierarchical pore structure of calcium phosphate scaffolds by a combination of gel-casting and multiple tape-casting methods. *Acta Biomaterialia.* 2008;4(4):913–922, ISSN 1742-7061.
67. Prakasam M, et al. Fabrication, properties and applications of dense hydroxyapatite: A review. *J Funct Biomater.* 2015;6(4):1099–1140, doi: 10.3390/jfb6041099.

68. Pompe W, et al. Functionally graded materials for biomedical applications. *Mater Sci Eng: A*. 2003;362(1–2):40–60, doi: 10.1016/s0921-5093(03)00580-x.
69. Corbin SF, Zhao-Jie X, Henein H, Apte PS. Functionally graded metal/ceramic composites by tape casting, lamination and infiltration. *Mater Sci Eng A*. 1999;262:192–203.
70. Muthutantri A, Huang J, Edirisinghe M. Novel preparation of graded porous structures for medical engineering. *J Royal Soc Interface*. 2008;5:1459–1467.
71. Sánchez-Salcedo S, et al. Hierarchical pore structure of calcium phosphate scaffolds by a combination of gel-casting and multiple tape-casting methods. *Acta Biomaterialia*. 2008;4(4):913–922, doi: 10.1016/j.actbio.2008.02.005.
72. Yigiterhan O, Tas AC. Manufacture of thin plaques of calcium hydroxyapatite bioceramics by die-pressing and tape- casting," In *IV. Ceramics Congress, Proceedings Book*. Eskisehir, Turkey, pp. 689–695, 1998.
73. Hofer AK, Kraleva I, Prötsch T, Vratanar A, Wratschko M, Bermejo R. Effect of second phase addition of zirconia on the mechanical response of textured alumina ceramics. *J Eur Ceram Soc*. 2022, https://doi.org/10.1016/j.jeurceramsoc.2022.08.058.

4 Processing of Bioceramics by Additive Manufacturing

David Orisekeh and M.P. Jahan
Miami University

NOMENCLATURE

3DP Three-dimensional printing, 3D printing
AM Additive manufacturing
BJ Binder jetting
DIW Direct ink writing
DLP Digital light processing
FDM Fused deposition modeling
HA Hydroxyapatite
LOM Laminated object manufacturing
MJ Material jetting
SLA Stereolithography
SLM Selective laser melting
SLS Selective laser sintering

4.1 INTRODUCTION

Composites are man-made engineering materials consisting of two or more individual constituents that can synergize together having properties better than each individual component. Bioinert and bioactive ceramics (bioceramics) are formed to produce application-specific biocomposites having better bioactivity and mechanical properties. Ceramics are inorganic materials, usually with covalent and ionic bonding, and have a high melting temperature and low electrical and thermal conductivity and are typically hard and brittle in nature.

Over the years, the study of fabrication and processing of bioceramics and their applications has gained interest both from researchers and industries due to the excellent mechanical and biological properties of bioceramics and its wide range of applications. Bioceramics especially find their application as scaffolds for bone tissue engineering and bone tumor therapy. Bioceramics are classes of ceramic materials used for skeletal and dental implants and therapy as well.

Back in the years, much research has been done on bioceramics with traditional methods of preparation such as the hand layoff method, which has many disadvantages.

 DOI: 10.1201/9781003258353-5

The AM method has become a popular method of processing bioceramic materials mostly because of its many advantages over the traditional method.

Casting is one of the earliest manufacturing processes used to create or form parts made up of various types of materials. It could involve liquefaction of solid materials and pouring them into a designed mold of desired shape and allowing them to solidify and harden. It also could involve mixing of different materials at room temperature, pouring them into a mold, and taking the shape of the mold after solidification. Casting is one of the earliest processes used for manufacturing bioceramics, although casting has many limitations and disadvantages. Precise details of shapes and intricate geometry are difficult to attain; measurements and material quality and quantity can be hardly achieved where mass production is required. Bioceramic scaffolds were prepared traditionally by gas foaming, freeze drying, fiber bonding, and particulate leaching. However, this method cannot guarantee accurate dimensions and shapes. In addition, the geometry, pore shape, porosity, and material layer connectivity and cohesion are difficult to control in this process. These limitations have led to the 3D-printing method (Ma et al., 2018) to process bioceramics.

Some of the advantages of 3D printing (3DP) over traditional manufacturing processes of fabricating bioceramic parts are that 3DP is an easier and cheaper process when compared with respect to complexity it can achieve, produces a better layer-by-layer interface and generates better surface finishing, requires minimal human control once the printing has started, reduces wastage of printing materials, and generates a high quality of finished printed products with every smaller detail and intricate feature. Recent progresses in 3DP technology have seen improved printing speed, which resolved the slow printing speed issue with previous 3D printers. The capability of printing multiple similar parts in a single printing bed (based on the print bed dimensions) and improved printing speed have enabled 3D printers to be able to mass produce bioceramic parts.

Although the traditional method of preparing bioceramic scaffolds has some advantages, the disadvantages led to the idea of the 3DP technique. Some of the disadvantages include higher porosity in the scaffolds and difficulty in controlling the internal structure of the scaffolds such as the pore size, pore shape, and interconnectivity of the particles and layers. It is also difficult to form some complex shapes and geometry of the scaffolds using traditional manufacturing processes. AM, since the 1980s, has gained interest from researchers and engineers, and its application is widely used in medical engineering as it could be used to print bio-scaffolds for bone tissue and dental engineering and tumor therapy. AM helps in the easy control and fabrication of objects with a high degree of accuracy. AM begins with designing the desired object in the computer using any computer-aided designing (CAD) tools, such as SolidWorks, Autodesk fusion360, Inventor, etc., making a 3D model and preparing the 3D model for 3DP and post-process (if needed) of the printed part. Getting a 3D printer is not enough to fabricate a complex part with high quality, as some instrumental pieces come between these two, which is known as the slicer software. The software prepares the 3D model in a language or form the 3D printer can understand and execute by generating a G-code. The G-code is a sequential line of

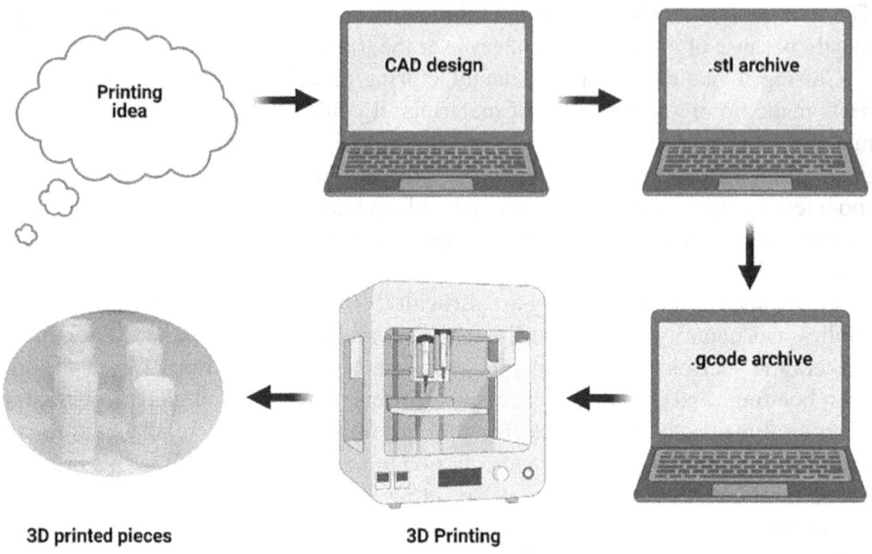

FIGURE 4.1 Basic process of 3D printing (Cano-Vicent et al., 2021).

codes, instructing the printer how to execute it one after the other, i.e., layer-by-layer until the print is completed. Some examples of slicer software include *Slic3r*, *Cura*, *OctoPrint*, *3DPrinterOS*, *IceSL*, etc. Figure 4.1 presents a basic step-by-step method for fabricating a part using AM (Cano-Vicent et al., 2021).

3D printers come in different sizes and purposes; some major types of 3D-printer models are 3D Systems–USA, Admatec–Netherlands, Arcam–Sweden, Aurora–China, BCN3D–Spain, Cytosurge–Switzerland, Materialise NV–Belgium, Zortrax–Poland, Beam–France, Bigrep–Germany, Blueprinter–Denmark, Carima–Korea, CRP Group–Italy, ComeTrue 3D–Taiwan, Nano Dimensions–Israel, and Wombot–Australia (https://io3dprint.com/3d-printer-manufacturers/). There are several 3DP technologies in the market for manufacturing products using a range of materials from polymer to metal to ceramics using different fabricating strategies. Some of the common 3DP technologies are binder jetting (BJ), direct ink writing (DIW), digital light processing (DLP), direct energy deposition (DED), fused deposition modeling (FDM) also known as fused filament fabrication (FFF), laminated object manufacturing (LOM), material jetting/wax casting (MJ), stereolithography (SLA), selective laser melting (SLM), and selective laser sintering (SLS).

This chapter focuses on providing an overview of materials and 3DP processes used for AM of bioceramics. A major focus and application of 3DP of bioceramics is scaffold fabrication for biomedical implants and bone replacement. Therefore, discussions have been added on the related topics using research articles focused on 3DP of bioceramics for scaffold fabrication. The current research trend and future research directions in the areas of 3DP of bioceramics have also been added in this chapter.

4.2 MATERIALS FOR 3D PRINTING OF BIOCERAMICS

4.2.1 Chemical Composition of the Bone

One of the major applications of bioceramics is as scaffolds to repair or replace damaged bone tissues. It is expected that bioceramics will interact with bone tissues and allow the growth of new tissues surrounding the bones. Therefore, most bioceramics have very similar compositions to bones. Bones are known to possess the ability to self-heal, but when the fracture size is large, the natural healing process will not be achieved; thus, the idea and application of bone repair and regeneration set in. Bones are made up of cells in an orderly arranged and large network of proteins and other materials. These materials include majorly of hydroxyapatite (HA) ($Ca_{10}(PO_4)_6(OH)_2$), sodium, carbonate, citrate, magnesium, water, and other minerals (Ma et al., 2018). The organic phase of the bone includes collagen, noncollagenous proteins, and primary bone cells such as osteoblasts, osteocytes, and osteoclasts whereas the inorganic phase of the bone includes HA, water, carbonate, citrate, sodium, magnesium, chlorine, and other elements in a small amount (Ma et al., 2018).

Bone macrostructure can be classified into external cortical and internal cancellous bone. The cortical bone is highly dense with a compressive strength of about 130–190 MPa. It is the load-carrying macrostructure of the bone (Ma et al., 2018). Bone defects include malformation, fracture from injuries, and tumor resections. Treatment includes surgery, bone grafting, inducing osteoinduction using growth factors, implementation of autologous bone for implants, and the use of biomaterials (Jazayeri et al., 2018). Figure 4.2 shows the structural and chemical composition of human bones (Gao et al., 2017).

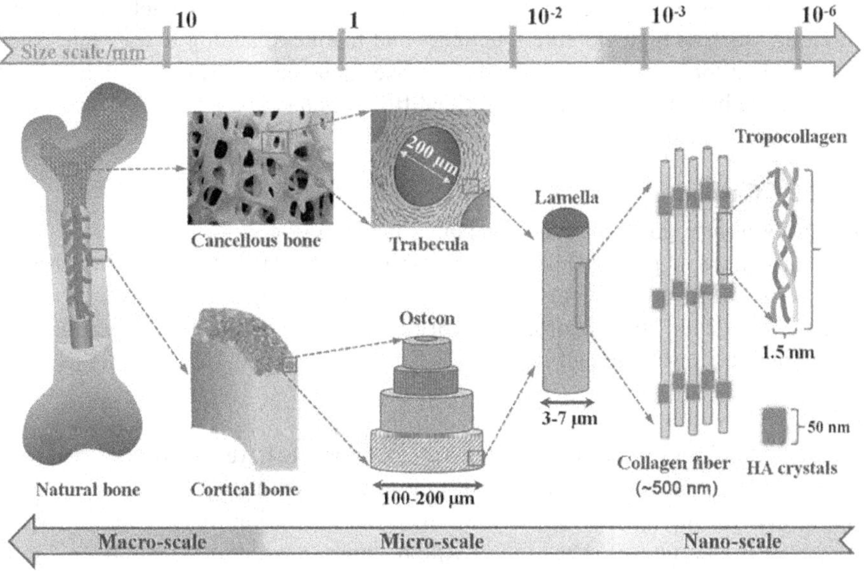

FIGURE 4.2 The structural and chemical composition of bones (Gao et al., 2017).

In 3D-printing of bioceramic scaffolds, the scaffolds' structure, such as dimensions and shapes, geometry, pore shape, porosity, and material layer connectivity and cohesion can be largely controlled to meet desired results and specifications. The scaffold fabrication is faster with very minimal defects and errors (Ma et al., 2018). The materials for scaffolds that can be used for any bone tissue regeneration and bone tumor therapy must first be nontoxic to the human body, they must be able to synergize with the skeletal structure, and they should have good compressive strength and low degradable rate among others. In this review, we shall be looking at some of the most commonly used materials for bone engineering. Materials used for bioceramics can be classified into three categories which are bioinert ceramics, bioactive ceramics, and bioresorbable ceramics.

4.2.2 Bioinert Ceramics

The human immune system naturally attacks foreign substances in the body. Bioinert simply means materials that do not cause any response from the human body. Bioinert ceramics do not interact with the human body, and they have high mechanical strength and chemical stability in vivo. Some examples of bioinert ceramics are zirconia, alumina, titanium nitride, zinc oxide, zinc nitrate, magnesia, titania, yttria, ceria, nickel oxide, iron oxide, barium titanate, europium hydroxide, boron nitride, and cobalt nitrate (Bahremandi-Toloue et al., 2022).

4.2.2.1 Zirconia

Zirconia, also called zirconium oxide with a chemical formula of ZrO_2, is a class of ceramics with a density of about 5.68 g/cm^3 and a melting point of 2,715°C and is usually white in color. It has good mechanical strength, biocompatibility, and very high thermal, wear, and corrosion resistance (Reference). Its favorable mechanical properties such as high fracture toughness and hardness, adding to its white color, make it ideal for bioceramic scaffolds for bone tissue engineering. It is found in its natural state but can be chemically formed from zircon. Apart from bone engineering, it is used in electronic, welding, and friction applications.

4.2.2.2 Carbon

Carbon is a very good material for bone engineering due to its similarity to the bone's mechanical properties and compatibility with the bone and other tissues. Carbon exists in various forms and hardly undergoes fatigue which makes it good for bone and dental applications, but its brittleness and low tensile strength make its use limited in its uncombined state.

4.2.2.3 Alumina

Alumina (Al_2O_3) is one of the bioinert ceramics known to have high friction and wear resistance and relatively high hardness. It also possesses a smooth surface which is one of the reasons for its high abrasion resistance. It is known to possess surface energy with a hexagonal structure. Research shows that alumina ceramics have no cytotoxicity and as such are nontoxic to the human body when used as implants in bone marrows.

4.2.3 Bioactive Ceramics

This group of ceramics interacts with the skeletal structure and surrounding tissues in the human body. They interact positively with living tissues and chemically bond with the bone (Heimann, 2002). This is mostly done through the ion release from the ceramics and the body fluids which results in the formation of biologically active carbonates like minerals in the human bone. Some examples are HA, bioglass (BG), silicates, and apatite–wollastonite (A-W) glass–ceramic.

4.2.3.1 Hydroxyapatite

The human bone is about 65% of carbonated hydroxyapatite; therefore, HA is one of the most important bioceramic scaffolds used for bone repair. HA ($Ca_{10}(PO_4)_6(OH)_2$) is known to facilitate the formation of specialized cell types in and around the bone tissues. It also has good bone tissue integration, as its microporosity supports capillary ingrowth and the breakdown and assimilation of the old bone for the entire bone growth process. This is possible due to its hexagon symmetry (Cox et al., 2015). HA has a rough surface and large interconnection due to its micropore size which enables easy integration of the scaffolds in vivo. As such, it is one of the major bioceramic materials for bone tissue application (Jazayeri et al., 2018). Due to its similar chemical composition to that of the human bone and its biological properties, such as adhesion and proliferation, it can also combine with other elements such as Al_2O_3, ZrO_2, and carbon fiber to improve its mechanical properties and also combine with natural polymers to achieve a better degradation rate and biocompatibility (Ma et al., 2018). Figure 4.3 shows the 3DP method used for the fabrication and the SEM image of HA bioceramic scaffolds fabricated by the 3DP process (Kim et al., 2022).

4.2.3.2 Apatite–Wollastonite (A–W) Glass–Ceramic

AW glass ceramics are a group of bioceramics used for repairing and filling bone defects. AW glass ceramics were found to be used as a prosthetic application and

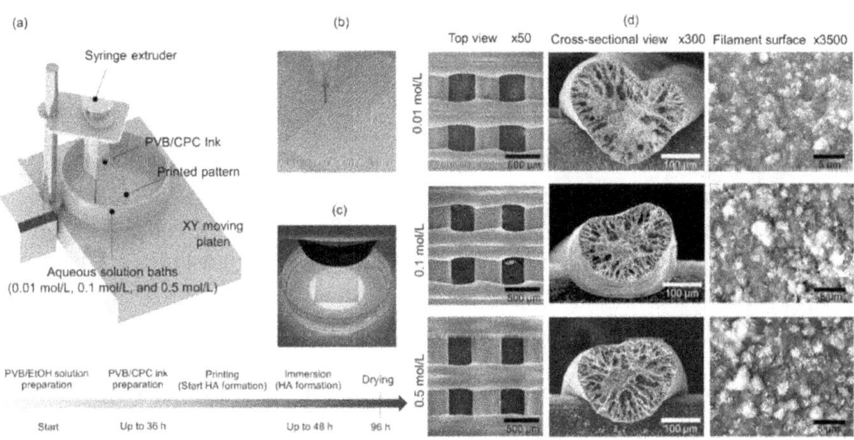

FIGURE 4.3 Schematic diagram of the 3D printing process and fabricated hydroxyapatite scaffolds (Kim et al., 2022).

bone substitute in cemented hip arthroplasty. The AW glass possesses good mechanical properties, interconnected porosity, and the ability to bond chemically with the bone to facilitate bone growth (Da Li et al., 2009; Tatli et al., 2022). However, AW glass ceramics may suffer from lower fatigue life, although they may have higher strength (Kokubo et al., 1987). One of the major reasons for lower fatigue strength is that crystallization occurring on the AW glass bioceramics inside the human body causes an internal crack. As a result, many researchers investigated the effectiveness of various additives to improve both mechanical properties and bioactivity of AW glass bioceramics. For example, fluorine can be added to the AW glass ceramics to slow down crystallization, thus reducing or eliminating the internal cracks on the structural prosthetics (Calver et al., 2004). Zn addition has been found to improve the performance of AW glass bioceramics by increasing the pore size and decreasing the rate of apatite mineralization. However, the compressive strength was found to be reduced with the addition of Zn (Rattanachan et al., 2012). The addition of magnesium–zinc ferrite was found to slow down the apatite formation by forming a new chemical phase, where the magnetic property improved (Li et al., 2014). The bone growth performance of AW glass ceramics can be further improved by using a fibrin mixture (Ono et al., 1988).

Several different 3DP techniques are used to fabricate scaffolds made up of AW glass bioceramics, such as SLS, digital light processing, and powder-based 3D printing techniques. It was reported that although different processes used different techniques and different sizes of glass powders, all the 3D-printed scaffolds showed good mechanical strength and bioactivity (Elsayed et al., 2018). Other 3DP processes SLA, inkjet plotting, and dispense plotting were also reported to be in use for the fabrication of AW glass–ceramic scaffolds. AM of AW glass ceramics allows fabrication of complex and interconnected porous structured scaffolds that are difficult to be produced by traditional processes.

4.2.3.3 Silicates

Calcium silicate is another group of bioceramics widely used for bone tissue engineering applications, such as tissue repair or replacement, and tumor therapy. Calcium silicate scaffolds have found important applications in bone tissue repair because of excellent degradation rate and apatite mineralization, and better antibiosis and osteogenesis properties (Srinath et al., 2020; Yu et al., 2019). Calcium silicate was found to facilitate osteogenic differentiation of osteoblastic and endothelial cells, thus exhibiting growth of bone tissue. Calcium silicate bioceramics were also used in endodontic treatment (Song et al., 2020). Magnesium silicate bioceramics are explored nowadays to replace widely used calcium phosphate (CaP)-based bioceramics for bone cell growth and repair applications (Devi et al., 2019).

Although wollastonite- or calcium silicate-based bioceramics promote cell growth and show no cytotoxicity, they are found to be difficult to form in complex shapes. As a result, 3DP of calcium silicate bioceramics has become popular due to its capability of producing complex shapes with porosity to promote cell growth (Zocca et al., 2015). Lithium calcium silicate was found to be a new bioceramic that has both biodegradability and apatite mineralization ability, thus providing better performance in osteochondral interface reconstruction. The 3DP process enables

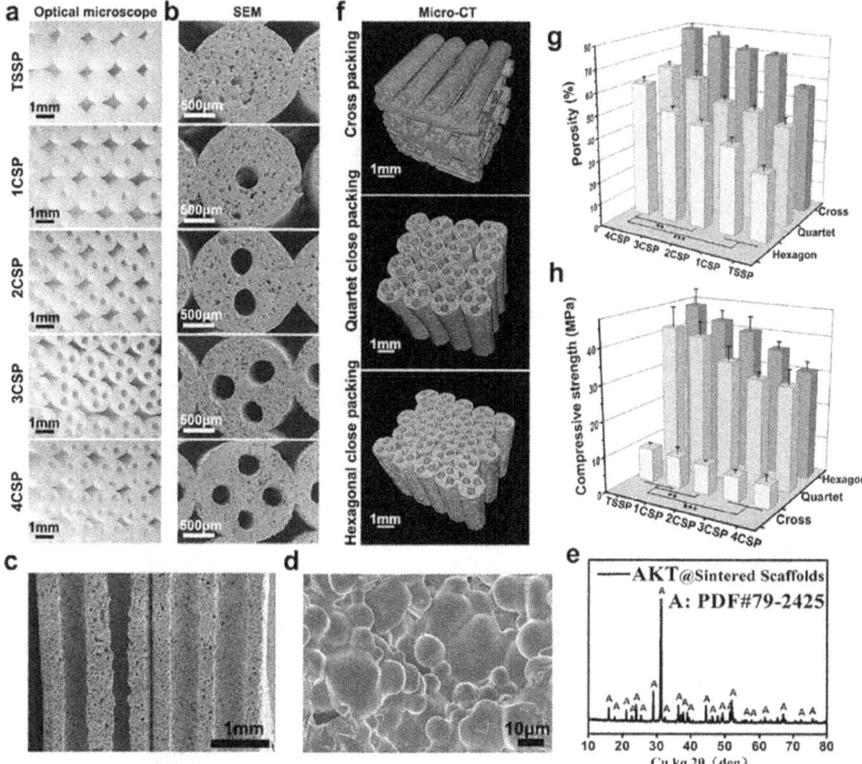

FIGURE 4.4 3D printing and characterization of calcium-magnesium-silicate ($Ca_2MgSi_2O_7$) bioceramic scaffold with a microstructure mimicking lotus root (Feng et al., 2017).

the fabrication of lithium calcium silicate scaffold using the sol-gel method, which otherwise was difficult to fabricate complex structures (Chen et al., 2019). Figure 4.4 shows characterization of surface topography, porosity, and compressive strength of the 3D-printed calcium-magnesium-silicate bioceramics. It can be seen that the compressive strength of silicate bioceramics can vary significantly with different 3DP parameters.

4.2.4 Biodegradable Ceramics

These are bioceramics that dissolve gradually inside the human body and are slowly being replaced by bones or other tissues. Examples of this category are tricalcium phosphate (TCP), calcium sulfate, and calcium oxide (Bahremandi-Toloue et al., 2022).

4.2.4.1 Calcium Phosphate

Calcium phosphate (CaP) in its natural state without the addition of other elements has limited bone tissue engineering application due to its brittle nature and its ionic bond and poor mechanical property. As a result, medical application is limited to

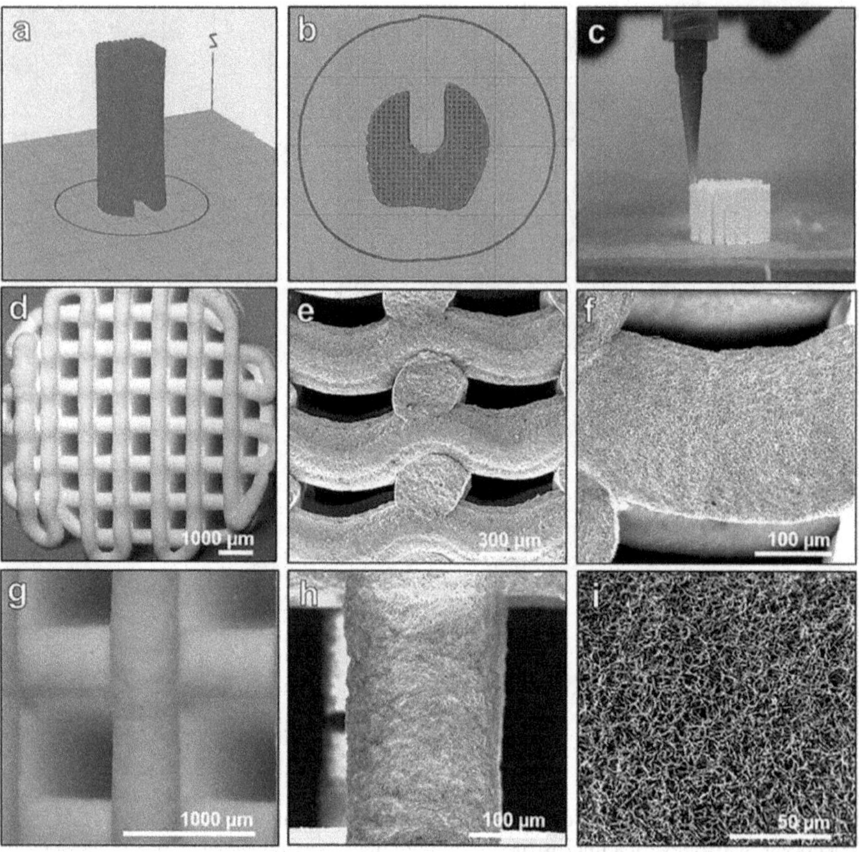

FIGURE 4.5 The 3D printing process of fabrication and images of printed calcium phosphate bioceramic scaffolds (Vidal et al., 2020).

non-load-bearing implants, like implants for ear surgery, coating on dental implants, filling of bone defects in the skeletal structure and oral cavity, etc. To enhance the scaffolds, CaP is enhanced through some special formations (Ma et al., 2018). Figure 4.5 shows an example of 3DP process used for the fabrication of CaP bioceramic scaffolds. SEM images of the scaffold shown in Figure 4.5 demonstrate the structure and surface morphological characteristics of the fabricated CaP scaffold.

4.2.4.2 Calcium Sulfate

Calcium sulfate is an absorbable bioceramic most commonly used for bone healing. As this is an absorbable biomaterial, it can be mixed with an antibiotic and be used for delivering antibiotics locally (Ene et al., 2021). It is also used as bone void fillers and is used as a bone graft substitute for a bone cyst. It has a higher rate of biodegradation and has good biocompatibility and limited complications inside the body and is capable of complete biodegradation. Calcium sulfate antibiotic beads are used inside the body for infection prevention (Boyan and Schwartz, 2017).

It has demonstrated excellent performance as a drug delivery material (Ene et al., 2021). However, it is difficult to use calcium sulfate bioceramics for structural applications due to its higher rate of biodegradability.

3D-printed calcium sulfate biocomposites have excellent porosity and the ability to mineralize apatite and provide excellent performance in cell adhesion and proliferation that can promote bone generation (Qi et al., 2017). The degradation rate of 3D-printed calcium sulfate scaffold can be controlled to match that with the growth rate of bone by engineering the structure and composition of calcium sulfate bioceramics (Zhang et al., 2019c).

4.3 IMPORTANT CHARACTERISTICS OF BIOCERAMIC SCAFFOLDS

4.3.1 Porosity

Bioceramic scaffolds must be printed with some amount of porosity and interconnectivity because the bone's cortical and trabecular structures are designed to allow for the transportation of blood, nutrients, and osteogenic cells throughout the skeletal structure. However, the porosity should be of similar size and specification to that of the human bone; this is so as not to have too large pores in the scaffolds which could also affect the mechanical strength and osteogenesis of the bioceramics (Jazayeri et al., 2018). The porosity inside the human bone can vary from macroscale to micro- or even nanoscale. As a result, the fabricated scaffolds should also consider various levels of porosity to match those of bones. Figure 4.6 shows how the microstructure and porosity levels can vary from macroscale to nanoscale on the bioceramic scaffolds fabricated by the 3DP process (Ma et al., 2018).

4.3.2 Mechanical Properties

The skeletal system is the weight-carrying part of the body, and scaffolds should possess mechanical properties like that of the bone. Bioceramics generally have high compressive strength; however, the compressive strength of bioceramic scaffolds can be further enhanced either with other materials or the addition of fillers while designing 3DP of the scaffolds to meet the required specification (Jazayeri et al., 2018).

4.3.3 Biodegradability

As the bone tissue regenerates, the bioceramic scaffold begins to degrade; these processes must be studied and controlled for uniformity. The ability of bioceramics to undergo biodegradation makes them applicable for bone tissue engineering. The rate at which bioceramic scaffolds undergo biodegradation is determined by the material's bioreactivity and the composition it is made up of. Biodegradation is very important as we don't want to leave the scaffold in the body forever or surgically remove it (Jazayeri et al., 2018).

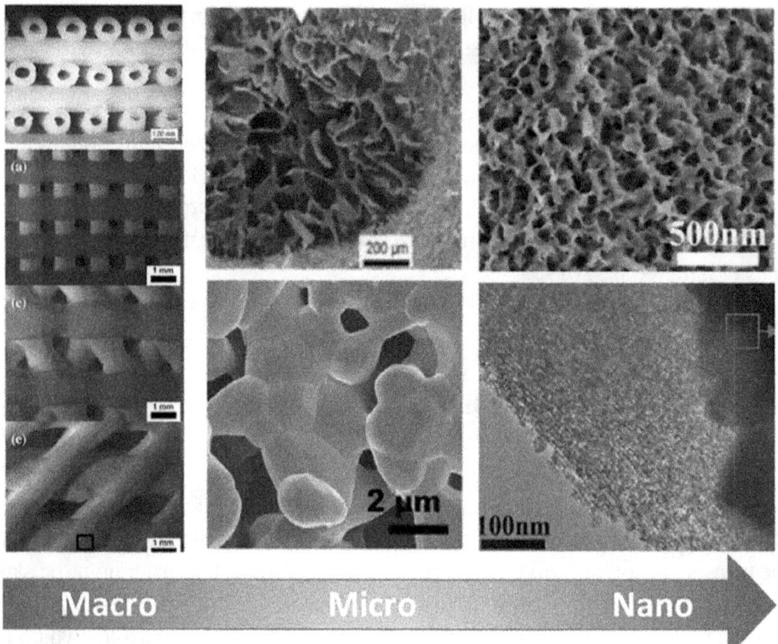

FIGURE 4.6 Range of scales for microstructures and porosity level for 3D-printed bioceramics (Ma et al., 2018).

4.3.4 Biocompatibility

For any bioceramic scaffolds to work, there should exist a level of physical, mechanical, or chemical compatibility within the parts. It must not have any negative interaction with the body's living tissue, and it must be safe and nontoxic (Heimann, 2002). Bioceramic scaffolds for bone and tissue engineering are integrated into the human body; they must possess the ability to initiate and facilitate molecular and cellular bioactivities with the body tissues and bones, such as blood vessel formation, osteoinductivity, and proliferation (Jazayeri et al., 2018).

4.4 3D-Printing Techniques of Bioceramics Scaffolds

It was said that Professor Sachs of the Massachusetts Institute of Technology (MIT) first proposed 3DP. 3DP can be classified into the fabrication method, which are majorly SLA, binder based, FDM, DLP, SLS, and so many others. The most used 3DP techniques for bioceramic scaffold printing are SLA, DIW, 3DP, and SLS which are grouped either as powered-based (3DP and SLS)—here the bioceramics are dispersed on a printing platform and bonded with a binder—or lasers or slurry-based (DIW and SLA) where very smoothly bioceramics are prepared in a liquid form and printed as ink or paste (Lin et al., 2019). Other 3DP techniques for bioceramics include BJ, DLP, FDM, material jetting (MJ), etc.

4.4.1 Selective Laser Sintering (SLS)

SLS was invented by Deckard and Beaman in 1986 at the University of Texas at Austin. In SLS, a high laser is used to coalesce the fine particles of the powder together; this is done by scanning through the area on the powder bed surface. This is done layer by layer until the desired programed shape is achieved. The printing can be direct or indirect. In the direct method, the bioceramic powder does not totally melt to retain the dimension and shape. Even at a high laser power to initiate sintering, this process is difficult to attain due to the high melting temperature of ceramics. In the indirect SLS technique, bioceramics are coated with an organic polymer that gives away after undergoing the laser; this is done for the particles to bind together; and thereafter, it is sintered to form the desired scaffold at a high temperature (Lin et al., 2019). Figure 4.7 presents the step-by-step SLS process of printing bioceramic scaffolds and fabricated scaffolds (Lin et al., 2019).

4.4.2 Robocasting/Direct Ink Writing (DIW)

This technique was developed in 1997 by Cesarano, where the ceramic scaffold is printed from the extrusion of ink through a controlled nozzle. Therefore, preparing a ceramic ink with a good amount of ceramic powder that could flow through a small nozzle and still retain its shape and properties is the challenge faced in robocasting

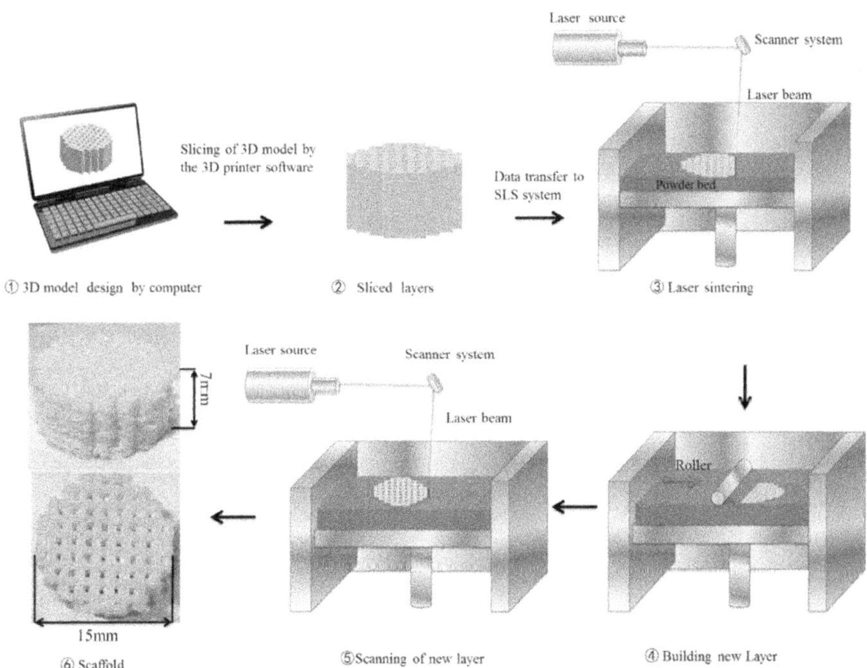

FIGURE 4.7 Schematic of step-by-step SLS 3D printing process and fabricated bioceramic scaffold (Lin et al., 2019).

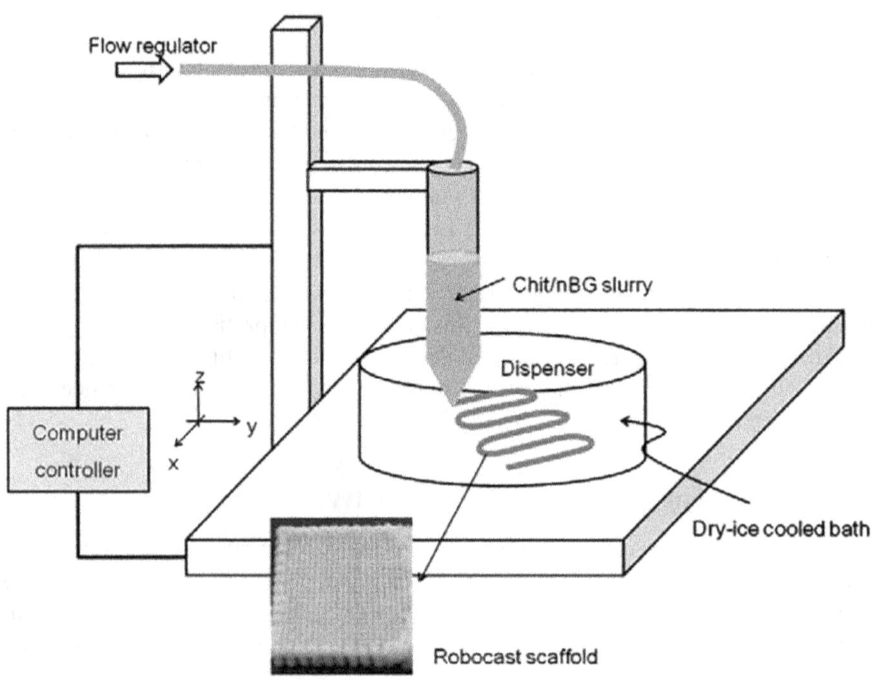

FIGURE 4.8 Schematic diagram of robocasting process (Dorj et al., 2012).

(Lin et al., 2019). Robocasting 3DP technique involves the formation of 3D-printed parts or scaffolds through layer-by-layer deposit of material through a nozzle. It is said to have low cost and could be used for different materials as well as in the printing of various shapes of 3D parts as compared to other 3DP techniques (Hou et al., 2004).

Robocasting 3DP is accompanied by a high temperature, which can alter the desired mechanical properties of the scaffold as in the case of bioceramics with CaP composite (Paterlini et al., 2021). An example of this DIW process is the BCN3D+ model produced by BCN3D technologies (Barcelona, Spain). It could use a syringe in place of a filament nozzle. A cogged plunger was used to allow for easy movement along the z-axis. The support platform could be as much as 252 × 200 × 200 mm. The driving software is Slic3r, which supports STL files (Paterlini et al., 2021). Figure 4.8 shows the schematic presentation of DIW process using robocasting technique (Dorj et al., 2012).

4.4.3 Stereolithography (SLA)

Stereolithography, a Greek word, stereo-solid and lithography was invented in the early 1980s by Hideo Kodama but was patented by Chuck Hull in 1984. It can be used to print scaffolds with very good dimensional accuracy, different shapes, and versatility. However, SLA needs a temporary supporting material to make the printing possible. The most challenging is the removal of the supporting material, which

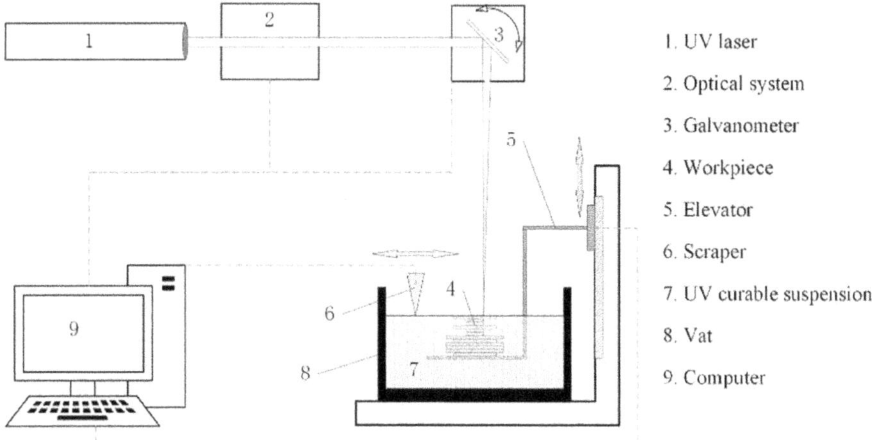

FIGURE 4.9 Schematic diagram of SLA (Wang et al., 2019).

if not properly removed could cause deformation to the shape and dimension of the ceramic scaffold and could initiate cracks in the scaffold (Griffith and Halloran, 1994). SLA technique is used in printing thin layers of parts starting from the bottom to the top until the object is completely printed. It is one of the commonly used methods of 3DP techniques due to its quality surface finishing and easy-to-print complex shapes (Lin et al., 2019).

Figure 4.9 shows the schematic diagram of the SLA process indicating different components of the SLA 3DP system (Wang et al., 2019). In SLA, the ultraviolet laser is used to solidify photosensitive resin or ceramic particles with a very fine size, which is done layer by layer. Dispersant agents are used to stabilize and prevent the ceramic particles from agglomeration. A smaller ceramic particle size is recommended for the SLA technique because the printing time for ceramic particles compared to resins is longer. The higher printing time for ceramic particles is due to the diffraction of the UV light by the ceramic particles, which results in low resolution and cure depth (Griffith and Halloran, 1994). Uniform distribution of the ceramic particle is required all through the printing process because the particles are inert to UV light. To eliminate the formed organic particles, the scaffold is calcined and thereafter sintered to bind the particles together (Pfaffinger et al., 2015).

4.4.4 Binder Jetting (BJ)

BJ was invented at MIT in 1993, and Z Corporation got a license for it from MIT in 1995 (Ziaee and Crane, 2019). The BJ process sprays binding agents in a liquid form on powders to produce 3D objects on a layer on a powder bed. BJ 3DP technique is usually preferred when it comes to printing brittle materials, like ceramics. It is ideal for ceramics due to its ability to form scaffolds at a high temperature. BJ has two stages for fabrication before sintering of the scaffold is done to improve the bioceramic properties; they are the spreading of thin powder and inkjet deposition (Oropeza and Hart, 2021).

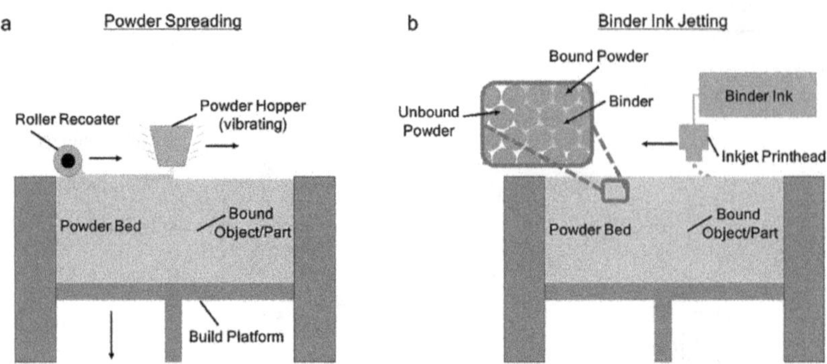

FIGURE 4.10 Schematic of binder jetting fabrication processes (Oropeza and Hart, 2021).

For BJ of bioceramics, the ceramic powder is spread with a roller at a desired thickness on a printing bed. The solution is placed on the printing bed using a spray printhead as designed with CAD, which controls the movement, speed, and directions. Before it is made into a solid form by heating, the particles and the solution are bonded together. At the completion of each printing layer, the process is done again for the next layer until the whole part is completed (Bose et al., 2013). Figure 4.10 shows the schematic diagram of the BJ 3DP process to make bioceramic scaffolds (Oropeza and Hart, 2021).

Before using a BJ for bioceramic scaffold printings, several factors must be considered, such as powder–binder wettability, the paste viscosity and concentration, how fine the powder is, and the morphology and surface texture of the powder particles. The particles, if not very fine, could cause defects such as cracks while removing the residues before sintering. The sintering process which is done after printing is essential to achieve the desired mechanical strength; as the scaffold is heated, there is an amount of shrinkage in the scaffold volume depending on the bioceramic material and the printing method. On the other hand, very fine bioceramic particles form the paste and the quality of the print and the shrinkage percentage are minimal. The fine particle leads to a smoother surface and fine edge finishing. Fine particle size also facilitates smooth and easy flow of the ceramic paste onto the printing bed. But when the particle sizes are too fine, it poses a problem of non-proper integration with the binder (Lin et al., 2019).

There are typically two types of binders: acid based and organic based. In acid based, such as citric or phosphoric acids, sintering is not necessary, but the scaffolds are brittle. However, they can have better mechanical properties by undergoing further post-processing activities. This technique is mainly used for CaP powders (Lin et al., 2019). BJ has some outstanding advantages that it can be used to fabricate large sizes, low cost, easy-to-design scaffolds with varied pore sizes, interconnectivity, pore geometry, and shape.

4.4.5 Digital Light Processing (DLP)

Digital light processing (DLP) technology is a fast-rising 3DP method in fabricating ceramic scaffolds due to its ability to develop complex shapes and precision in

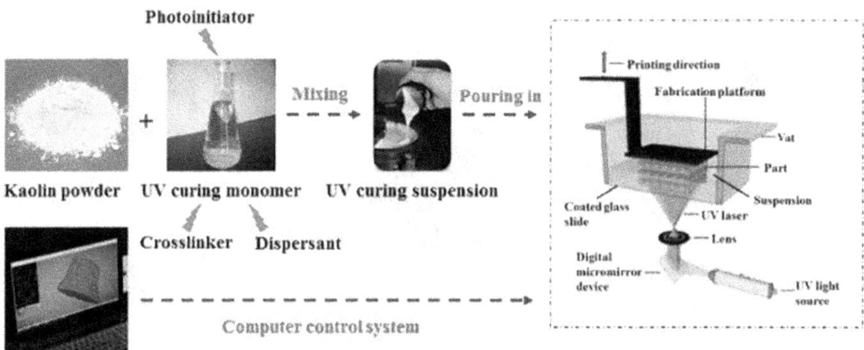

FIGURE 4.11 DLP schematic diagram (Li et al., 2020).

fabrication. The DLP technology has found application in oral and dental medicine (Li et al., 2020). It works on the principle of light mask projecting using a micromirror component. The scaffolds are built layer by layer by exposing the resin or ceramic particles to UV light. This is done by first coating the printing bed with the resin and then creating an oxygen-permeable layer on the printing platform. This process is done to easily remove the cured resin and then the object is solidified from the part of the liquid resin that is exposed to the UV light. The platform moves up with every completed layer until the object is completely printed and then the parts and its build support are removed. Excess resin may be formed on the part surfaces and edges, but this can be easily removed without causing damage to the parts and the scaffold is calcined at high temperature to improve the mechanical characterization (Wiese et al., 2021). An example of the DLP process for the fabrication of ceramic parts is shown in Figure 4.11 (Li et al., 2020).

4.4.6 Fused Deposition Modeling (FDM)

FDM process is widely used for household, research, and industrial applications, as it is easy to use and cost effective. It forms 3D objects by layer-by-layer melt-extrusion of filaments to form the designed 3D parts. A 3D design is created using any of the CAD software and then the printing is executed layer by layer until the model is formed.

FDM 3DP technique has been in use for over 20 years. It was invented by Scott Crump in the 1980s (Palermo, 2013) and it is suitable for mass-producing 3D parts due to its low cost of production and the production time (Tagliaferri et al., 2019). The quality of part production by FDM is dependent on the prototype CAD design, then the slicing into layers and thereafter communication to the printer, the type of printing material used, printing orientation, printing speed, infill density, extrusion rate, stacking sequences, bed and extrusion temperatures, the design shape, and geometry (Ansari and Kamil, 2021).

The FDM process begins with the design of the part in a computer using the CAD software as .stl file and then the slicing software converts it into a G-code file for the

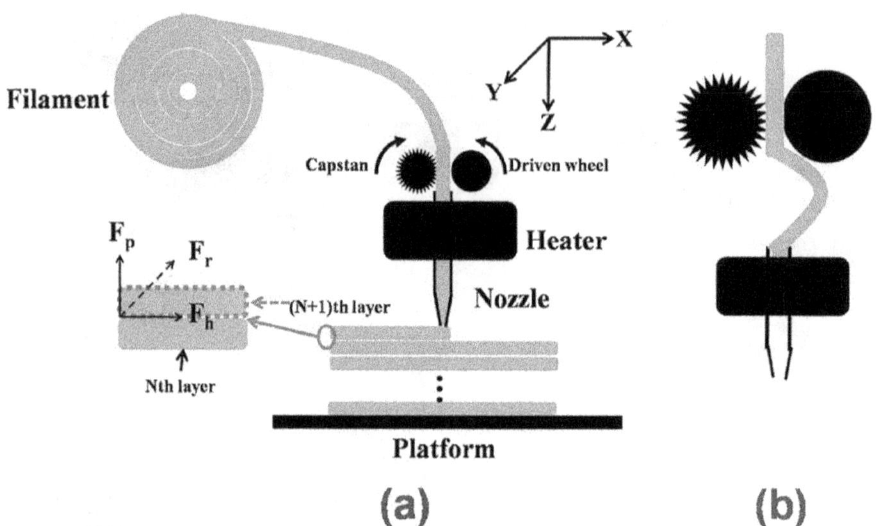

FIGURE 4.12 Schematic diagram of FDM 3D printer (Zhang et al., 2020).

printer to recognize its specifications. The material for print is mostly in the form of a filament wire or coil fed into a temperature nozzle for heating. The nozzle brings the temperature of the filament material near or over its melting point and extrudes the material layer by layer on the printing bed through the nozzle (Cano-Vicent et al., 2021). The printing bed has its own controlled temperature; this layer-by-layer print is continued until the part is fully formed. Figure 4.12 shows the schematic diagram of the FDM process demonstrating the mechanism of the printing process (Zhang et al., 2020).

FDM 3DP technology has applications across many industries. Wide research has been done on various filament materials that could be used in FDM printing. Some of these filaments are polylactic acid with wood, polycarbonate, polyethylene terephthalate, and cellulose materials, among others. 3D-printed cellulose parts have found significant applications in biomedical engineering such as scaffolds for tissue engineering, bone models, prosthesis, dentistry, and implants. FDM process has the capability of printing bioceramics of various materials by creating filaments of the target ceramic materials using a binder, printing the green part, and then sintering the bioceramic part. Figure 4.13 shows how the FDM-printed bioceramic scaffolds are used in both in vitro and in vivo bone cell engineering applications (Cano-Vicent et al., 2021).

4.4.7 Laminated Object Manufacturing (LOM)

LOM was developed by Helisys Inc. (now known as Cubic Technologies), in the United States (Gupta et al., 2020), but according to B. Dermeik and N. Travitzky, LOM began with the patent of DiMatteo in 1974. LOM is a sheet lamination process to produce 3D parts. It is said to be cost effective and can be processed with a high

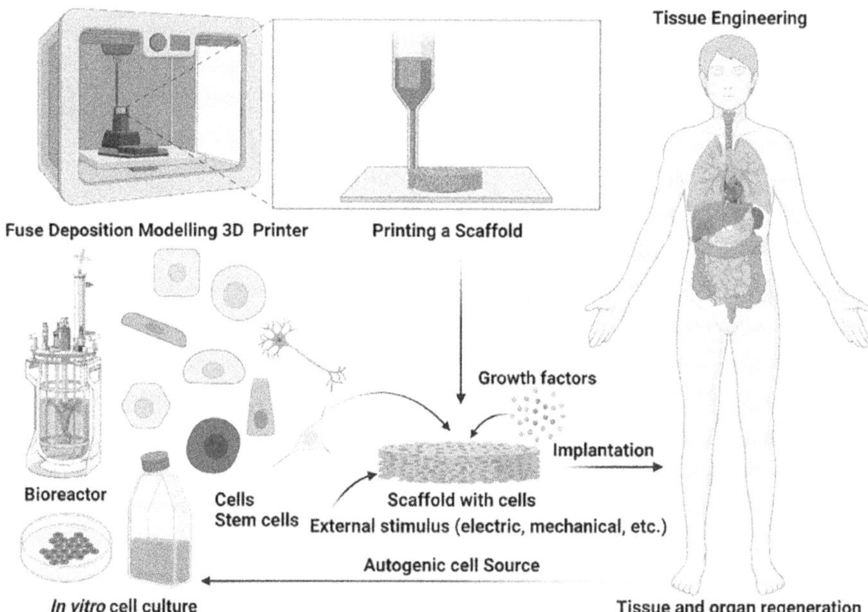

FIGURE 4.13 FDM-printed 3D scaffold for tissue engineering (Cano-Vicent et al., 2021).

production velocity. LOM-printed bioceramic scaffolds are characterized to have high temperature resistance and good strength-to-weight ratio. In LOM, layers can be cut, then laminated or the layers could be first laminated before cutting; many prefer the first as it reduces errors and wastage of materials (Dermeik and Travitzky, 2020).

LOM is known to use a wide range of different forms of materials such as fiberglass and ceramic fillers. It is inexpensive and suitable for large-scale production of 3D parts and can produce complex-shaped objects. It is also nontoxic which makes it ideal for the printing of bioceramic scaffolds and parts (Tofail et al., 2018). Just like every other AM technique, the LOM 3DP process starts with the CAD design and then proceeds to the slicer which converts it into G-code. To reduce wastage of material, manpower, and time, feed, cut, press, and adhesive spraying is recommended. In this process, a new layer is added in the feed process, and then the laser path controller cuts the new layer; this layer thickness is usually 0.1 mm. Then, after the cutting process, the material is pressed with a pressing head; thereafter, with a nozzle, the adhesive spraying process is done (Chiu et al., 2003). Figure 4.14 demonstrates the working mechanism of the LOM-based 3DP of bioceramic scaffolds (Dermeik and Travitzky, 2020).

4.4.8 Material Jetting (MJ)

Unlike some 3DP techniques like BJ, DLP, FDM, SLS, and SLA, MJ is one 3DP technology that much research has not been done on. MJ-printed parts have good mechanical properties (Bass et al., 2016), high dimensional accuracy (Yap et al., 2017), and better surface finishing (Zhang et al., 2019).

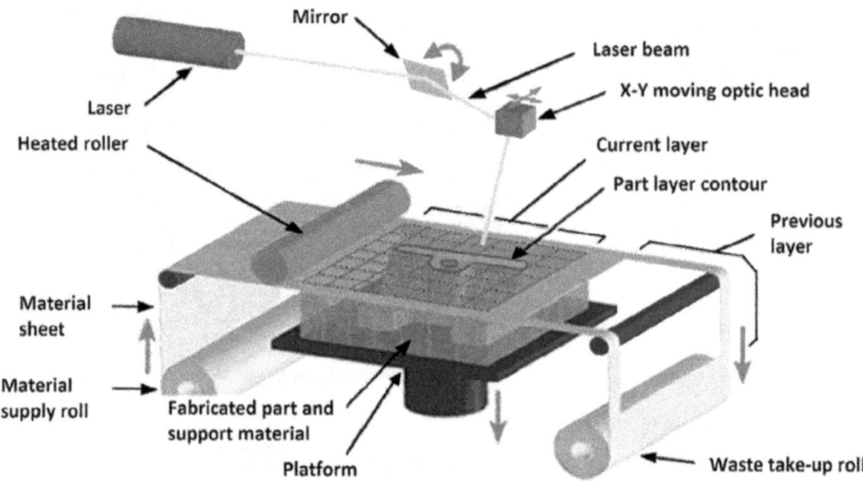

FIGURE 4.14 Schematic diagram of LOM 3D printing technology (Dermeik and Travitzky, 2020).

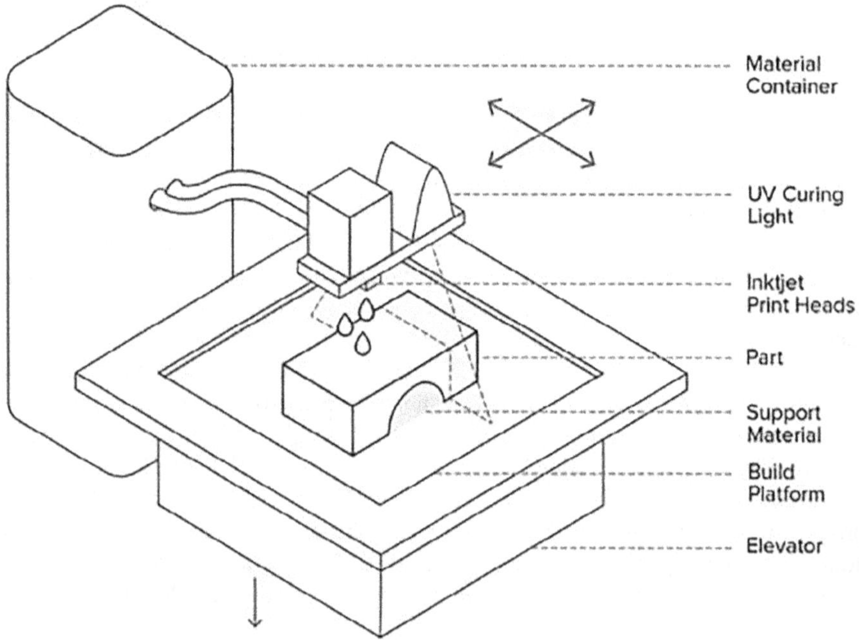

FIGURE 4.15 Schematic of material jetting printer (Tyagi et al., 2021).

The schematic diagram of the MJ process of fabricating bioceramic scaffolds is shown in Figure 4.15 (Tyagi et al., 2021). For MJ process, the desired CAD design is converted into .stl file, then a G-code, and then transferred to the printer. The material to be printed is then heated up to an adequate temperature. With the print head movement, little droplets of the material are deposited on the printing bed and the UV light solidifies the droplets in a layer, then the process is repeated until the parts are fully formed. MJ printing process is known to be fast compared to other printing technologies, thereby printing multiple layers in a short time, and it does not necessarily require sintering of the printed object (Tyagi et al., 2021).

4.4.9 Selective Laser Melting (SLM)

SLM is one of the emerging 3DP technologies used mostly in metal matrix composite. However, research shows that the SLM process is also an excellent option for producing bioceramic and biomaterial scaffolds. It is made up of high-energy laser beam, which is used to melt the material into a molten state where it is used to produce the 3D part layer by layer (Zhang et al., 2019). It is an excellent choice for producing 3D objects with complex geometry, as it eliminates the traditional way of casting with molds. The printed part dimensions are accurate, thus eliminating the need for surface filing and machining processes. It also eliminates the welding of parts in metallic composites. Generally, SLM technology shortens the production cycle and time.

SLM uses models designed by a CAD program. When the file is fed to the SLM printer, high laser energy is applied to the powder beds, the desired cross-sectional area is scanned by the laser, melted, and sintered, and the particles are bonded together. Then, the powder is spread again on the top over the just-scanned layer to form a new layer; this process is repeated until the 3D object is formed and bonded together (Olakanmi et al., 2015). Figure 4.16 demonstrates schematically how the SLM process can be used to fabricate bioceramic scaffolds for bone cell engineering applications (Gunasekaran et al., 2021).

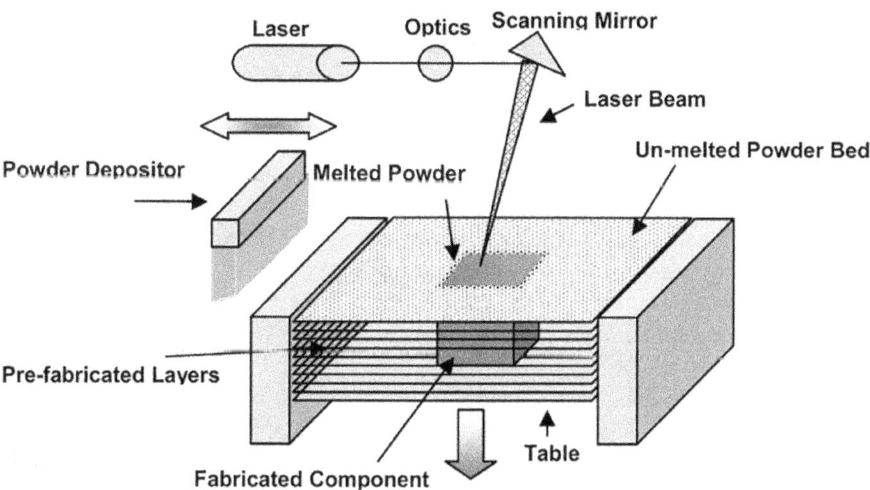

FIGURE 4.16 Schematic of SLM printing process (Gunasekaran et al., 2021).

4.5 CURRENT RESEARCH ON 3D PRINTING OF BIOCERAMICS

Kaur et al. in their work used DLP 3DP technique to fabricate superhydrophobic scaffolds. Autodesk inventor computer software was used to design the 3D parts, which was converted to .stl file and transferred to the DLP printer (Freeform Plus 39, Asiga) which is an inverted printer with a digital mirror device. The printing ink had no hydrophobic fumed silica and fluorinated acrylates; it was a mixture of dipropylene glycol diacrylate and trimethylolpropane triacrylate in the ratio of 95%–5% (w/w), respectively. To allow for better printing resolution, 1.5 wt% of photoinitiator and 0.001 wt% of methyl red dye were added to the mixture. To reduce the effect of oxygen inhibition, a UV light source was used to polymerize the microstructure of the parts layer by layer. The printed parts have a thickness of 25 μm, printed with 26 mW/cm^2 light intensity (Kaur et al., 2020).

Brunello et al. used 3DP, SLA, FDM, and SLS to fabricate powder-based bone tissue engineering scaffolds that could possibly improve bone cell and tissue development. Some examples of these powder-based materials are HA, TCP, ZP113, and calcium sulfate-based powder (zp150). They also used some binders in the 3DP process such as cyanoacrylate, pyrrolidone, and phosphoric acid. They finally stated that CaP bioceramic shows promising properties that are suitable for bone and tissue engineering (Brunello et al., 2016).

Charbonnier et al. fabricated porous HA bioceramics by using a 3DP technology. This process includes impregnation of 3D-printed wax mold, slurry feature optimization, and deposition of HA powder that was prepared by an aqueous precipitation technique to develop a robust bioceramic wax mold. They examined the optimization of slurry-based HA and found out that the slurry drying rate determines the quality of the bioceramic scaffolds. Using this process to develop bioceramic scaffolds could advance bone tissue regeneration (Charbonnier et al., 2016).

Zhong et al. aimed at fabricating high-resolution bioceramic scaffolds of HA paste through a low-temperature extrusion process of the 3DP technique. The characterization approach focused on the extrusion process of bioceramics for 3DP of bone tissue engineering scaffolds. The results indicated that bioceramic scaffolds fabricated by low-temperature extraction of a 3DP show uniform microstructure property that is suitable as a bioceramic scaffold for bone tissue engineering (Zhong et al., 2017).

Mariani et al. used BJ, a powder-based 3D printing technology, to produce alumina bioceramic scaffold. Alumina (Al_2O_3) is a widely used bioceramic due to its wear resistance and biocompatibility and is not toxic to the human body. The alumina powder was dried for 5 hours at a temperature of 200°C and later mixed for 5 minutes. They used an Innovent Plus printer by ExOne to print the scaffold. A flexural test was carried out on the scaffold via a three-point bending test. Results produced by the BJ technique possess the potential for a wide range of applications (Mariani et al., 2021).

Bouchart et al. in their study aimed at providing a solution to bone infection using a 3D-printed bioceramic scaffold. The bone infection was found to slow down bone healing after bone surgery. SLA 3DP method was used for printing the bioceramic scaffold. The material used was diammonium phosphate solution $(NH_4)_2HPO_4$ and calcium nitrate solution $Ca(NO_3)_2.4H_2O$. Calcium phosphate $Ca_3(PO_4)_2$ powder was prepared from the precipitation of $(NH_4)_2HPO_4$ and $Ca(NO_3)_2.4H_2O$ at a temperature of 35°C for 20 hours. The filtered particle was dried at 80°C and calcined at 850°C. Thereafter, the powder was ground for 3 hours using a ball milling machine. A ceramic surface structure analysis was carried out on the sample using electron microscopy, which showed a regular-spaced parallel streak in the printing layers which can be easily used as a bioceramic bone scaffold. Remus and lambda phage coating was determined on the ceramic, and the phages were removed from the pellets by washing, accounting for better interaction between the material layers. Test to ascertain the toxicity of the material was done by pre-osteoblast cell growth proliferation for 2 days in an incubator at 37°C. This was done using various samples of phage coating or without coating. The cells were removed from the pellets at different intervals of incubation. The result showed that the osteoblast cells are not affected by ceramics coated with the Remus phage. It showed that bioceramics coated with the Remus phage were very efficient in fighting against *Staphylococcus aureus* bone infection and could protect the osteoblast cells as the phages were able to prevent the rupture of the cells due to the actions of the *S. aureus* infection (Bouchart et al., 2020).

Tonelli et al. investigated the properties of magnesium silicate bioceramic cement paste for possible use as a scaffold for bone tissue engineering using 3DP technology. The primary materials used are magnesium nitrate and sodium silicate.

Cement pastes were made by hydration of MgO and SiO_2 at room temperature at a composition of 1–1 and 2–1 of MgO/SiO_2 composition ratio, which was cast into molds of 10 mm diameter or printed with an engine standard resolution printer (Hyrel) using a dimension of $20 \times 20 \times 3$ mm and $10 \times 10 \times 10$ mm. The 3DP method was the injection method. The bioceramic was printed at 8 mm/s with a layer thickness of 1 mm and an infill ratio of 80%. The materials, both molded and printed, were then calcined at 1,000°C in a muffle (Nabertherm) for 2 hours. Some of the tests and characterization done were X-ray diffraction, injectability test, confocal Roman microscopy, field emission-scanning electron microscopy, atomic force microscopy, gas porosimetry, X-ray micro-computed tomography, and compression strength test. The characterization results showed that preparing the sample at a settling temperature of 37°C resulted in shortening the preparation time over that of room temperature (25°C). A composition ratio of 2–1 MgO/SiO_2 gave a high brucite content. For the confocal Roman microscopy, both the molded and 3D samples showed a similar chemical surface composition. The 3D-printed sample had better and homogeneous layer bonding (Tonelli et al., 2021).

Using a homemade 3D printer, Chang et al. aimed at producing a bioceramic bone scaffold with good mechanical strength for skeletal support. The main materials used are SiO_2 powder and SiO_2 sol. The average SiO_2 powder particle size was 23 μm and SiO_2 sol had a 40% solid content and nanoparticle diameter of 50 nm. Other materials are $CaCO_3$, which served as an additive filler. SiO_2 powder and SiO_2 sol were mixed at a ratio of 20/80, and $CaCO_3$ was then added. This was then printed with the homemade 3D printer; the sample has a dimension of 4.5 mm $\times$ 4.5 mm $\times$ 1 mm. It was then calcined for 2 hours at a temperature between 900°C and 1,100°C to form a bioceramic scaffold. The homemade 3D printer produced parts using thermal energy from a CO_2 laser. A laser scanner, scraper, feeder, and flat working platform for the printed samples were various parts of the in-house built 3D printer. Mechanical property test such as compressive strength test and biocompatibility test such as cytotoxicity test, surface topography test, and cell affinity test were carried out. The 5% composition of $CaCO_3$ sample showed a better compressive strength of 45 MPa with a porosity of 34%; the sample showed no presence of cytotoxicity, with a cell affinity of MG-63 osteoblast-like cell and had a better surface topography. It was found that calcined temperature influences the compressive strength of the scaffold. From the results, bioceramic scaffold of SiO_2 and $CaCO_3$ is a good bone scaffold for bone and tissue engineering (Chang et al., 2015).

Ma et al. focused on the investigation of properties of BG-reinforced β-TCP scaffolds for potential bone tissue engineering. They used calcium nitrate solution $Ca(NO_3)_2.4H_2O$ and diammonium phosphate solution $(NH_4)_2HPO_4$, sodium carbonate, calcium carbonate, magnesium carbonate, and hydroxide pentahydrate as materials for making bioceramics. Quantities of 52.9 g of $(NH_4)_2HPO_4$ and 141.7 g of $Ca(NO_3)_2.4H_2O$ were dissolved in 800 and 1,200 mL of deionized water respectively, with the addition of calcium solution to the phosphate solution and stirred for about 4 hours. It was left for 24 hours and filtered and dried at 110°C and heated at 10°C/min to form the powder, which was then calcined for 2 hours at 800°C. Also, for BG, deionized water was mixed with $(MgCO_3)_4.Mg(OH)_2.5H_2O$, $NH_4H_2PO_4$, Na_2CO_3, and $CaCO_3$ and then stirred while heating to remove NH_3. It was melted at 900°C and then quenched in water to obtain a 2O–CaO–MgO–P_2O_5 bioactive glass. Ball milling was used to mix β-TCP and BG contents at different wt% and then dried for 24 hours at 60°C. After mixing β-TCP/BG powders, sodium alginate suspension, and pluronics F-127 solution in deionized water, the inks were poured into the printing tubes of a homemade 3D printer to produce the BG-reinforced bioceramic scaffold. A compressive strength test was carried out on the scaffold using a universal testing machine. The morphology, internal micropore structure, and interconnected pore structure were evaluated using an SEM, degradation of bioceramic property was observed using Tris-HC1, and the cell viability was also evaluated using MTT (3-[4,5-dimethylthiazol-2-yl]-2,5 diphenyl tetrazolium bromide) assay. The results show that the 3D-printed scaffold exhibited a compressive strength of 8.34 MPa and an elastic modulus of 208.5 MPa. There was no visible cytotoxicity in the scaffold; therefore, the scaffold will cause no harm to the bone and cells. From the study in this paper, Na_2O–CaO–MgO–P_2O_5 BG-reinforced beta-TCP ceramic scaffold has great potential in bone engineering following the compressive strength of the scaffold, its degradability rate, and its non-cytotoxicity nature to cells (Ma et al., 2019).

He et al. investigated the feasibility of 3DP of Mg-substitute wollastonite-reinforcing diopside porous bioceramics with enhanced mechanical and biological performance. The authors evaluated the properties of printed CSM10-reinforced diopside porous scaffold for potential clinical use as a bioceramic rod in osteonecrosis of the femoral head treatment. Femoral head osteonecrosis is a medical condition where there is inadequate blood supply of the subchondral bone in the femoral head. It can be treated surgically with the insertion of bio-artificial rods to the affected bone. The materials used were wollastonite powder ($CaSiO_3$) (with 10% Mg, denoted as CSM10) and diopside ($MgCaSi_2O_6$). Using ball milling, diopside and CSM10 were mixed in ethanol for about 30 minutes and dried at 90°C for about 12 hours. The obtained powder was mixed in a polyvinyl alcohol solution and this was printed using a homemade 3D printer to prepare the scaffold. Dynamic light scattering was used to investigate the particle size distribution while SEM was used to observe the morphology of the scaffold. The porosity of the scaffold was also observed using the Archimedes method in distilled water. Degradation and mechanical behavior of the scaffold were evaluated using an Instron testing machine. From the results, an increase in CSM10 improved the inorganic CaP salt biochemical processes and ability in the scaffold pores. The scaffold showed a flexural strength of about 30 MPa and compressive strength of 37 MPa. The scaffold also showed a slow decay rate in the Tris buffer solution. The findings showed that wollastonite powder and diopside 3D-printed scaffold could have great potentials in the osteonecrosis of the femoral head treatment (He et al., 2016).

Zhang et al. aimed at fabricating 3D-printed β-TCP bioceramic scaffold for tissue engineering. They used the injection 3DP technique to fabricate the scaffolds and allowed them to dry at room temperature for 24 hours and calcined for 3 hours. Graphite oxide (GO), anhydrous iron (III) chloride, polyvinyl alcohol, sodium hydroxide, and beta- TCP powder were the materials used. For β-TCP–Fe–GO scaffolds, 0.30 g of Fe_3O_4/GO nanocomposites was mixed homogeneously in 20 mL deionized water for 5 minutes. Then, 0.5 g of GO was mixed in 500 mL deionized water for 4 hours and β-TCP scaffolds were added to the mixture for 10 minutes and allowed to dry for 30 minutes. To cover the Fe_3O_4 particles, the scaffolds were soaked in the GO–water mixture once more, thereby forming a sandwich layer of GO–Fe_3O_4–GO. Using optical microscopy, the morphology of the scaffold was observed, which showed that the scaffold has spherical morphology with a uniform 250 nm pore size. The surface microstructure was studied using an SEM. A computer was used to evaluate the magnetothermal property of the scaffold, which showed a positive result as the tumor cells died almost completely. The magnetic property of the scaffold was produced primarily by Fe_3O_4. From the results, it could be deduced that β-TCP–Fe–GO scaffolds possess a magnetic field that could kill the tumor cell, which showed that it would be an effective scaffold for bone tumor therapy and bone regeneration (Zhang et al., 2016).

4.6 APPLICATIONS OF 3D-PRINTED BIOCERAMICS

In recent years, 3D-printed bioceramics have found exponential application in various aspects of medical engineering due to their mechanical properties, biocompatibility, and nontoxic nature. Their application could be seen in implants, bio-models, biotools, biocompatible tissues, antibacterial drugs, and drug manufacturing (Andres-Cano et al., 2021). Bioceramic scaffolds have found applications in almost every part of the human body ranging from skull to the ankle. Some of the applications are implants for cranial bones, tooth root, temporomandibular bone, and shoulder, elbow, hip, knee, and ankle joints. Figure 4.17 shows how bioceramic implants can be used in various parts of the human body (Shekhawat et al., 2021). Some of the specific examples of applications of different bioceramic materials in real life are shown in Figure 4.18 (Stanford Advanced Materials). A brief description of application examples of bioceramics, as reported in the literature, is presented in this section.

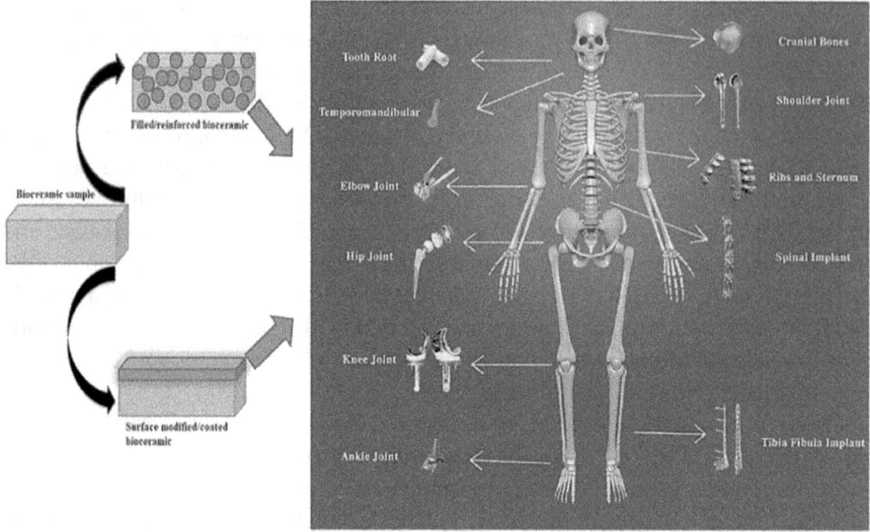

FIGURE 4.17 Application areas of 3D-printed bioceramics in various parts of the human body (Shekhawat et al., 2021).

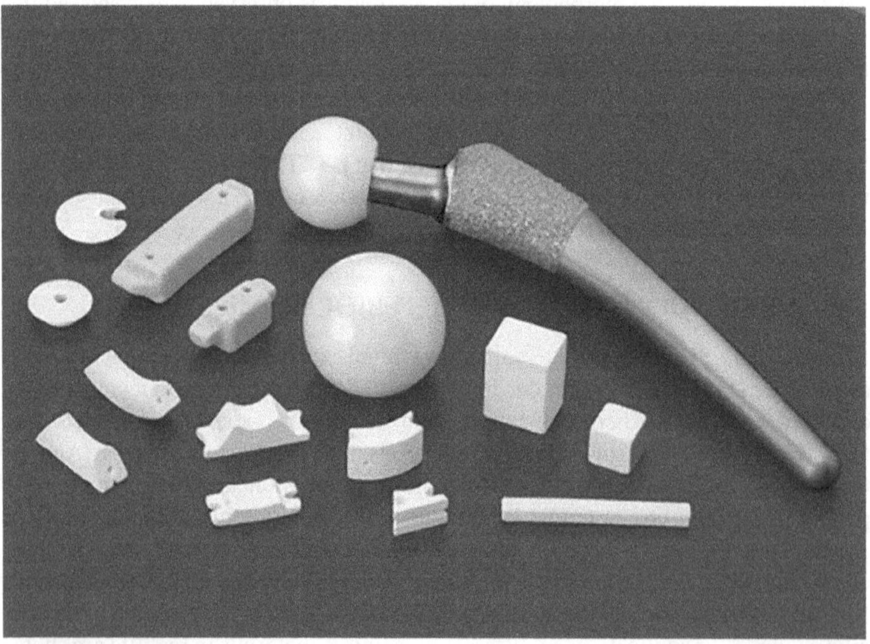

FIGURE 4.18 Examples of real-life bioceramic parts used as implants inside the human body. (Photo Courtesy: Stanford Advanced Materials.)

- Carbonated HA-reinforced PLA matrix is suitable as a replacement for bone tissues (Oladapo et al., 2019). For the enhancement of antibacterial and cells in bone tissue (Touri et al., 2019).
- Al_2O_3, ZrO_2, Y-TZP, Mg-PSZ, and Zircalon are known to be excellent for endosseous and femoral head implants, and fractured tooth, hip, elbow, shoulder, knee, and wrist replacements.
- CaP is used for bone grafting and tissue replacement. HA is also used as a filler for bone defects and bone grafting (Shekhawat et al., 2021).
- The antibacterial forsterite (Mg_2SiO_4) scaffold was reported to be a promising bioceramic for load-bearing applications. Forsterite bioceramics possess good mechanical strength and biodegradability properties, and they are good as antibacterial bioceramics against bacterial pathogens (Choudhary et al., 2018).
- 3D-printed diopside ($CaMgSi_2O_6$) bioceramic scaffolds could be used as a synthetic bone substitute and implant in orthopedics (Palakurthy et al., 2021).
- 3D-printed HA scaffolds have both interconnected and microporosity which supports the capillary ingrowth of new cells as well as replacement of the old bone (Cox et al., 2015). HA has become a major bioceramic material for bone tissue application (Jazayeri et al., 2018).
- 3D-printed AW glass–ceramic scaffolds are used as a prosthetic application due to their high compressive strength and load-bearing capacity. The 3D-printed AW glass scaffolds also have interconnected porosity and support bone growth and are used as a bone substitute in cemented hip arthroplasty (Da Li et al., 2009; Tatli et al., 2022).
- 3D-printed calcium silicate scaffolds were found applications in endodontic treatment (Song et al., 2020), whereas magnesium silicate was found applications in bone repair (Devi et al., 2019). 3D-printed lithium calcium silicate is another reported bioceramic that was used in osteochondral interface reconstruction (Chen et al., 2019).
- 3D-printed calcium sulfate scaffold is used to fill voids inside the bone and is also a medium for delivering antibiotics inside the body, mostly because of its controlled absorbability and biodegradability.

4.7 CURRENT CHALLENGES AND FUTURE RESEARCH DIRECTIONS

3D printing has some great advantages over other processes for making bioceramic scaffolds, such as fabricating complex structures which are difficult to be fabricated otherwise, and control over the density and porosity of the part. Some of the general challenges in 3DP of bioceramics are comparatively lower accuracy and resolution and poor surface finish and mechanical properties. These challenges may be more experienced in 3DP processes that are associated with melting and solidification processes (Kolan et al., 2012; Lin et al., 2019). An additional benefit of 3DP is that two or more bioceramic materials can be added to the scaffold. Even different types of materials can be mixed together as well, such as polymer and ceramics, bioceramics, medicine, etc., thus allowing multiple functions of scaffolds (structural or load bearing, bone repair, bone growth, controlled drug delivery, etc.) (Lin et al., 2019). However, this also brings additional challenges in the uniform processing of two materials and controlling of different phases and chemical structures formed by those. However, there are future research opportunities in revealing the science behind multi-material processing.

Larger powder particles are easier to spread and the binder can penetrate through them easily in the BJ-based 3D printing of bioceramics, but they cause low part density and also create instability during melt pool formation. Many factors such as particle size, powder flowability, binder material, amount of binder, wettability of powder in binder liquid, etc., all play a role in BJ (Hwa et al., 2017). Future research should focus on data analytics and multivariable optimization of process parameters and make the parameter selection as a data-driven process to improve the performance of BJ 3DP process.

Ceramic particles are known to cause scattering on the UV lights, thus providing poor resolution and increased printing time (Griffith and Halloran, 1994) in the SLA process. Agglomeration of ceramic particles is another challenge during the SLA process due to the instability of the slurry. Agglomeration and formation of small particles after laser melting are also challenges for SLM- or SLS-based 3D printing of ceramics (Wang et al., 2015). Four-dimensional (4D) printing by adding ultrasonic vibration has been investigated by several researchers to overcome some of those challenges with powder agglomeration. Future research should explore more in this area and should come up with innovative process development to get rid of the challenges and improve printing performance of the above-mentioned 3DP processes.

SLS of ceramics is itself challenging due to high melting point of ceramics. The current process mechanism causes some dimensional accuracy issues because of the uneven melt pool. In addition, scattering of laser light by the ceramics, high energy consumption, and slow cooling rates make the process cost-ineffective (Kolan et al., 2012). These challenges open up opportunities for future research.

Load-bearing capability can be a challenge of 3D-printed bioceramics. The internal microstructure obtained by the 3DP process and the solidification mechanism used in various bioceramics are responsible for a low load-bearing capacity. Not all bioceramics have good load-bearing capacity. Several bioceramics have been found to be suffered from fatigue failure due to internal cracks generated during solidification. Therefore, material aspects of the 3DP of bioceramics remain a major area of improvement for future researchers working in this area.

Post-implantation performance of 3D-printed bioceramic scaffolds remains an area of investigation. The supply of nutrition to the newly placed bioceramic scaffolds is important to ensure cell growth around the bone. This is especially challenging when replacing a bone or filling in large voids in bone.

Post-processing of bioceramics has become an effective mechanism to improve surface finish, dimensional accuracy, and functionality of scaffolds. Ceramics can be post-processed or machined by various conventional and nonconventional processes, such as electrical discharge machining, electrochemical machining, electrochemical discharge machining, laser processing, etc. (Rashid et al., 2019; Bilal et al., 2021). Future research should focus on exploring the possibility of using sustainable machining strategies (i.e., cryogenic machining) for post-processing of 3D-printed bioceramics (Cococcetta et al., 2021). At the same time, existing nonconventional machining or post-processing should be explored for 3D-printed ceramics, as most of the research on nonconventional machining of ceramics is done on traditionally manufactured ceramic materials (Bilal et al., 2018).

Finally, future research should focus on the design and development of new structures of scaffolds using bioceramics considering the structure of the bones, cell adhesion, and proliferation, bioactivity, and load-bearing capacity. Having similar human bone macro- and microstructures in scaffolds will help in the improved performance of bioceramic implants inside the human body. Major research focuses for future researchers working in this area will be designing bioinspired scaffolds by mimicking various structures of bones and cells inside the human body. Because of the ability of 3D printers to fabricate complex structures, bioinspired scaffolds with structures similar to human bone can be achieved with much ease compared to other processes of making bioceramic implants. Figure 4.19 shows various designs of unit cells and scaffolds made up of bioceramic materials that have been already reported

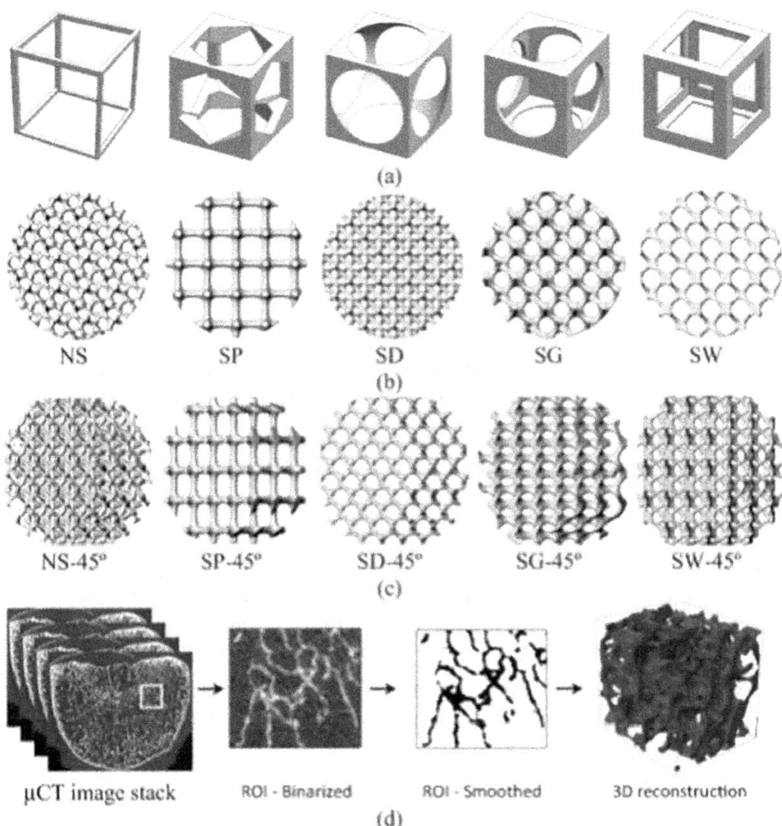

FIGURE 4.19 Various designs of 3D-printed scaffolds used for printing bioceramic composites for bone tissue engineering applications (Gomez et al., 2016).

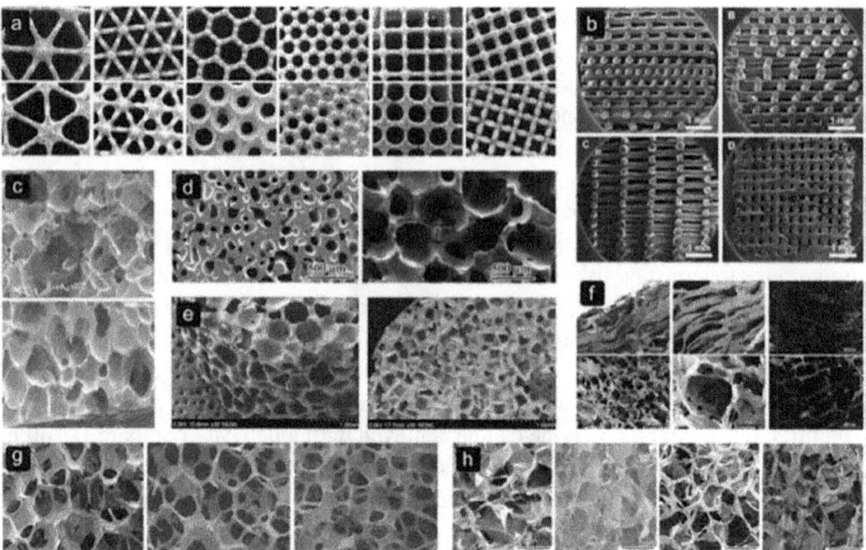

FIGURE 4.20 Examples of different pore sizes, shapes, and biomaterials used in 3D printing of scaffolds (Abdulghani and Mitchell, 2019).

by researchers (Gomez et al., 2016). Figure 4.20 shows images of actual bioceramic scaffolds with various structures and porosity demonstrating the capability and range of the 3DP process for processing bioceramic composites (Abdulghani and Mitchell, 2019).

4.8 SUMMARY

This review highlighted some of the advantages of 3DP over the traditional method of manufacturing bioceramic scaffolds, showing that 3D-printed scaffolds have smoother and better surface finishing, better interconnectivity between the layers, good microstructure, better porosity, and enhanced mechanical strength when compared to the traditional method. The study outlined various 3DP technologies that could be used depending on the material, shape, and desired properties of the scaffold. 3D-printed bioceramics have shown that they are nontoxic to the human body, similar in chemical composition to the human bone, compatible and integrate with the human tissue, and could be used as a load-bearing replacement in the body as implants.

3D-printed bioceramic scaffold is an excellent biomaterial suitable for dental and tooth replacement, pharmaceutical application, and drug manufacturing; it could be used in surgical implants, for the manufacturing of bio-models and bio-tools, and as an antibacterial against pathogens in the body.

REFERENCES

Abdulghani, S., Mitchell, G.R., 2019. Biomaterials for in situ tissue regeneration: A review, *Biomolecules* 9(11), 750.

Andres-Cano, P., Calvo-Haro, J., Fillat-Goma, F., Andres-Cano, I., Perez-Mananes, R., 2021. Role of the orthopaedic surgeon in 3D printing: Current applications and legal issues for a personalized medicine, *Revista Espanola de cirugia ortopedica y traumatologia* 65, 138–151.

Ansari, A., Kamil, M., 2021. Effect of print speed and extrusion temperature on properties of 3D printed PLA using fused deposition modeling process, *Materials Today: Proceedings*, 5462–5468.

Bass, L., Meisel, N., Williams, C., 2016. Exploring variability of orientation and aging effects in material properties of multi-material jetting pars, *Rapid Prototyping Journal* 22(5), 826–834.

Bilal, A., Jahan, M.P., Talamona, D., Perveen, A., 2018. Electro-discharge machining of ceramics: A review, *Micromachines* 10(1), 10.

Bilal, A., Perveen, A., Talamona, D., Jahan, M.P., 2021. Understanding material removal mechanism and effects of machining parameters during EDM of zirconia-toughened alumina ceramic, *Micromachines* 12(1), 67.

Bose, S., Vahabzadeh, S., Bandyopadhyay, A., 2013. Bone tissue engineering using 3D printing, *Material Today* 16, 496–504.

Bouchart, F., Vidal, O., Lacroix, J., Spriet, C., Chamary, S., Brutel, A., Hornez, J., 2020. 3D printed bioceramic for phage therapy against bone nosocomial infection, *Materials Science & Engineering C* 111, 110840.

Boyan, B.D., Schwartz, Z., 2017. *Comprehensive Biomaterials II.* Elsevier, ISBN 978-0-08-100692-4.

Brunello, G., Sivolella, S., Meneghello, R., Ferroni, L., Gardin, C., Piattelli, A., Zavan, B., Bressan, E., 2016. Power-based 3D printing for bone tissue engineering, *Biotechnology Advances* 34, 740–753.

Calver, A; Hill, R G; Stamboulis, A., 2004. Influence of fluorine content on the crystallization behavior of apatite-wollastonite glass-ceramics, *Journal of Materials Science* 39(7), 2601–2603.

Cano-Vicent, A., Tambuwala, M., Hassan, S., Barh, D., Aljabali, A., Birkett, M., Arjunan, A., Serrano-Aroca, A., 2021. Fuse deposition modelling: Current status, methodology, applications and future prospects, *Additive Manufacturing* 47, 102378.

Chang, C., Lin, C., Liu, F., Chen, M., Lin, C. P., Ho, H., Liao, Y., 2015, 3D printing bioceramic porous scaffolds with good mechanical property and cell affinity, *PLoS One* 10(11), e0143713.

Charbonnier, B., Laurent, C., Marchat, D., 2016. Porous hydroxyapatite bioceramics produced by impregnation of 3D-printed wax mold: Slurry feature optimization, *Journal of the European Ceramic Society* 36, 4269–4279.

Chen, L., Deng, C., Li, J., Yao, Q., Chang, J., Wang, L., Wu, C., 2019. 3D printing of a lithium-calcium-silicate crystal bioscaffold with dual bioactivities for osteochondral interface reconstruction, *Biomaterials* 196, 138–150.

Chiu, Y., Liao, Y., Hou, C., 2003. Automatic fabrication for bridged laminated object manufacturing (LOM) process, *Journal of Materials Processing Technology* 140, 179–184.

Choudhary, R., Chatterjee, A., Venkatraman, S., Koppala, S., Abraham, J., Swamiappan, S., 2018. Antibacterial forsterite ($Mg2SiO_4$) scaffold: A promising bioceramic for load bearing applications, *Bioactive Materials* 3, 218–224.

Cococcetta, N., Jahan, M.P., Schoop, J., Ma, J., Pearl, D., Hassan, M., 2021. Post-processing of 3D printed thermoplastic CFRP composites using cryogenic machining, *Journal of Manufacturing Processes* 68, 332–346.

Cox, S., Thornby, J., Gibbons, G., Williams, M., Mallick, K., 2015. 3D printing of porous hydroxyapatite scaffolds intended for use in bone tissue engineering applications, *Materials Science and Engineering C* 47, 237–247.

Da Li, G., Zhou, D. L., Pan, T. H., Chen, G. S., Lin, Y., Mao, M., Yan, G., 2009. Effect of Mn-Zn ferrite on apatite-wollastonite glass-ceramic (A-W GC), *Biomedical Materials* 4(4), 045001.

Dermeik, B., Travitzky, N., 2020. Laminated object manufacturing of ceramic-based materials, *Advanced Engineering Materials* 22, 2000256.

Devi, K.B., Nandi, S.M., Roy, M., 2019. Magnesium silicate bioceramics for bone regeneration: A review, *Journal of the Indian Institute of Science* 99, 261–288.

Dorj, B., Park, J., Kim, H., 2012. Robocasting chitosan/nanobioactive glass dual-pore structured scaffolds for bone engineering, *Materials Letters* 73, 119–122.

E Bahremandi-Toloue, Z Mohammadalizadeh, S Mukherjee, S Karbasi, 2022. Incorporation of inorganic bioceramics into electrospun scaffolds for tissue engineering applications: A review, *Ceramics International* 48 (7), 8803–8837.

Elsayed, H., Zocca, A., Schmidt, J., Günster, J., Colombo, P., Enrico Bernardo, E., 2018. Bioactive glass-ceramic scaffolds by additive manufacturing and sinter-crystallization of fine glass powders, *Journal of Materials Research* 33(14): Focus Issue: 3D Printing of Biomaterials, 1960–1971.

Ene, R., Nica, M., Ene, D., Cursaru, A., Cirstoiu, C., 2021. Review of calcium-sulphate-based ceramics and synthetic bone substitutes used for antibiotic delivery in PJI and osteomyelitis treatment, *EFORT Open Reviews* 6(5), 297–304.

Feng, C., Zhang, W., Deng, C., Li, G., Chang, J., Zhang, Z., Jiang, X., Wu, C., 2017. 3D printing of lotus root-like biomimetic materials for cell delivery and tissue regeneration, *Advanced Science* 4, 1700401.

Gao, C., Peng, S., Feng, P., Shuai, C., 2017. Bone biomaterials and interactions with stem cells, *Bone Research* 5, Article number: 17059.

Gómez, S., Vlad, M.D., López, J., Fernández, E., 2016. Design and properties of 3D scaffolds for bone tissue engineering, *Acta Biomaterialia* 42, 341–350.

Griffith, M., Halloran, J., 1994. *Ultraviolet Curable Ceramic Suspensions for Stereolithography of Ceramics*. ASME, New York, NY, Volume 68-2, pp. 529–534.

Gunasekaran, J., Sevvel, P., Solomon, I., 2021. Metallic materials fabrication by selective laser melting: A review, *Materials Today: Proceedings* 37, 252–256.

Gupta, R., Dalakoti, M., Narasimhulu, A., 2020. A critical review of process parameters in laminated object manufacturing process. In: Singh, I., Bajpai, P., Panwar, K. (eds) *Advances in Materials Engineering and Manufacturing Processes. Lecture Notes on Multidisciplinary Industrial Engineering*. Springer, Singapore. https://doi.org/10.1007/978-981-15-4331-9_3

He, D., Zhuang, C., Xu, S., Ke, X., Yang, X., Zhang, L., Yang, G., Chen, X. Mou, A. Liu, Z. Gou, 2016. 3D printing of Mg-substitute wollastonite reinforcing diopside porous bioceramics with enhanced mechanical and biological performances, *Bioactive Materials* 1, 85–92.

Heimann, H. B., 2002. Material science of crystalline bioceramics: A review of basic properties and applications, *CMU* 1(1), 23–46.

Hou, Q., De Bank, P., Shakesheff, A., 2004. Injectable scaffolds for tissue regeneration, *Chemical* 14, 1915–1923.

Hwa, L.C., Rajoo, S., Noor, A.M., Ahmad, N., Uday, M.B., 2017. Recent advances in 3D printing of porous ceramics: A review, *Current Opinion in Solid State and Materials Science* 21, 323–347.

iO3DPRINT, n.d. A comprehensive list of 3D printer manufacturers, Available at https://io3d-print.com/3d-printer-manufacturers/ (last accesssed on September 8, 2022).

Jazayeri, H., Rodriguez-Romero, M., Razavi, M., Tahriri, M., Ganjawalla, K., Rasoulianboroujeni, M., Malekoshoaraie, M., Khoshroo, K., Tayebi, L., 2018. The cross-disciplinary emergence of 3d printed bioceramic scaffolds in orthopedic bioengineering, *Ceramics International* 44, 1–9.

Kaur, G., Marmur, A., Magdassi, S., 2020. Fabrication of superhydrophobic 3D object by digital light processing, *Additive Manufacturing* 36, 101669.

Kim, Y., Lee, E.J., Kotula, A.P., Takagi, S., Chow, L., Alimperti, S., 2022. Engineering 3D printed scaffolds with tunable hydroxyapatite, *Journal of Functional Biomaterials* 13(2), 34.

Kokubo, T., Ito, S., Shigematsu, M., Sanka, S., Yamamuro, T., 1987. Fatigue and life-time of bioactive glass-ceramic A-W containing apatite and wollastonite, *Journal of Materials Science* 22, 4067–4070.

Kolan, K.C.R., Leu, M.C., Hilmas, G.E., Velez, M., 2012. Effect of material, process parameters, and simulated body fluids on mechanical properties of 13-93 bioactive glass porous constructs made by selective laser sintering, *Journal of the Mechanical Behavior of Biomedical Materials* 13, 104–124.

Li, F., Ji, X., Wu, Z., Qi, C., Lai, J., Xian, Q., Sun, B., 2020. Digital light processing 3D printing of ceramic shell for precision casting, *Materials Letters* 276, 128037.

Li, H., Xue, K., Kong, N., Liu, K., Chang, J., 2014. Silicate bioceramics enhanced vascularization and osteogenesis through stimulating interactions between endothelia cells and bone marrow stromal cells, *Materials* 35(12), 3803–3818.

Lin, K., Sheikh, R., Romanazzo, S., Roohani, I., 2019. 3D printing of bioceramic scaffolds-barriers to the clinical translation: From promise to reality, and future perspectives, *Materials* 12(17), 2660.

Ma, Y., Dai, H., Huang, X., Long, Y., 2019. 3D printing of bioglass-reinforced β-TCP porous bioceramic scaffolds, *Journal of Materials Science* 54, 10437–10446.

Ma, H., Feng, C., Chang, J., Wu, C., 2018. 3D-printed bioceramic scaffolds: From bone tissue engineering to tumor therapy, *Acta Biomaterrialia* 79, 37–59.

Mariani, M., Beltrami, R., Brusa, P., Galassi, C., Ardito, R., Lecis, N., 2021. 3D printing of fine alumina powders by binder jetting, *Journal of the European Ceramic Society* 41, 5307–5315.

Oladapo, B., Zahedi, S., Adeoye, A., 2019. 3D printing of bone scaffolds with hybrid biomaterials, *Composites Part B* 158, 428–436

Olakanmi, E., Cochrane, R., Dalgarno, K., 2015. A review on selective laser sintering/melting of aluminum alloy powders: Processing, microstructure, and properties, *Progress in Materials Science* 74, 401–477.

Ono, K., Yamamuro, T., Nakamura, T., Kakutani, Y., Kitsugi, T., Hyakuna, K., Kokubo, T., Oka, M., Kotoura, Y., 1988. Apatite-wollastonite containing glass ceramic-fibrin mixture as a bone defect filler, *Journal of Biomedical Materials Research* 22(10), 869–885.

Oropeza, D., Hart, A., 2021, Reactive binder jet additive manufacturing for microstructural control and dimensional stability of ceramic materials, *Additive Manufacturing* 48, Part B, 102448.

Palakurthy, S., Azeem, A., Reddy, V., Padala, C., Manavathi, B., Rayavarapu, P., 2021. A novel cost-effective approach to fabricate diopside bioceramics: A promising ceramic for orthopedic applications, *Advanced Powder Technology* 32, 875–884.

Palermo, E., 2013. Fused deposition modelling: Most common 3D printing method, *Live Science*, 39810, livescience.com/39810-fused-depostion-modeling.html

Paterlini, A., Le grill, S., Brouillet, F., Combes, G., Grossin, D., Bertrand, G., 2021. Robocasting of self-setting bioceramics: From paste formulation to 3D part characteristics. *Open Ceramics* 5, 100070.

Pfaffinger, M., Mitteramskogler, G., Gmeiner, R., Stampfl, J., 2015. Thermal debinding of ceramic-filled photopolymers, *Materials Science Forum* 825–826, 75–81.

Qi, X., Pei, P., Zhu, M., Du, X., Xin, C., Zhao, S., Li, X., Zhu, Y., 2017. Three dimensional printing of calcium sulfate and mesoporous bioactive glass scaffolds for improving bone regeneration in vitro and in vivo, *Scientific Reports* 7, Article number: 42556.

Rashid, A., Bilal, A., Liu, C., Jahan, M.P., Talamona, D., Perveen, A., 2019. Effect of conductive coatings on micro-electro-discharge machinability of aluminum nitride ceramic using on-machine-fabricated microelectrodes, *Materials* 12(20), 3316.

Rattanachan, S.T., Srakaew, N.L., Pethnin, R., Suppakarn, P., 2012. Effect of Zn addition on sol-gel derived apatite/wollastonite glass-ceramics scaffolds, *Journal of Metals, Materials and Minerals* 22(2), 61–65.

Shekhawat, D., Singh, A., Banerjee, M., Singh, T., Patnaik, A., 2021. Bioceramic composite for orthopaedic applications: A comprehensive review of mechanical, biological and microstructural properties, *Ceramics International* 47, 3013–3030.

Song, W., Sun, W., Chen, L., Yuan, Z., 2020. In vivo biocompatibility and bioactivity of calcium silicate-based bioceramics in endodontics, *Frontiers in Bioengineering and Biotechnology* 8, 580954.

Srinath, P., Azeem, P.A., Reddy, K.V., 2020. Review on calcium silicate-based bioceramics in bone tissue engineering 17(5), 2450–2464.

Stanford Advanced Materials. (n.d.). Bioceramics are entering our bodies https://www.samaterials.com/content/bioceramics-are-entering-our-bodies.html (last accessed on September 9, 2022).

Tagliaferri, V., Trovalusci, F., Guarino, S., Venettaci, S., 2019. Environmental and economic analysis of FDM, SLS and MJF additive manufacturing technologies, *Materials* 12, 4161.

Tatli, Z., Bretcanu, O., Çalışkan, F., Dalgarno, K., 2022. Fabrication of porous apatite-wollastonite glass ceramics using a two steps sintering process, *Materials Today Communications* 30, 103216.

Tofail, S., Koumoulos, E., Bandyopdhyay, A., Bose, S., O'Donoghue, L., Charitidis, C., 2018. Additive manufacturing: Scientific and technological challenges, market uptake and opportunities, *Materials Today* 21, 22.

Tonelli, M., Faralli, A., Ridi, F., Bonini, M., 2021. 3D printable magnesium-based cements towards the preparation of bioceramics, *Journal of Colloid and Interface Science* 598, 24–35.

Touri, M., Moztarzadeh, F., Osman, N., Dehghan, M., Mozafari, M., 2019. Optimization and biological activities of bioceramic robocast scaffolds provided with an oxygen-releasing coating for bone tissue engineering application, *Ceramics International* 45, 805–816.

Tyagi, S., Yadav, A., Deshmukh, S., 2021. Review on mechanical characterization of 3D printed parts created using material jetting process, *Materials Today Proceedings* 51, Part 1, 1012–1016

Vidal, L., Kampleitner, C., Krissian, S., Brennan, M.A., Hoffmann, O., Raymond, Y., Maazouz, Y., Ginebra, M.P., Rosset, P., Layrolle, P., 2020. Regeneration of segmental defects in metatarsus of sheep with vascularized and customized 3D-printed calcium phosphate scaffolds, *Scientific Reports* 10, Article number: 7068.

Wang, Z., Huang, C., Wang, J., Zou, B., 2019. Development of a novel aqueous hydroxyapatite suspension for stereolithography applied to bone tissue engineering, *Ceramics International* 45, 3902–3909.

Wang, J., Yang, M., Zhang, Y., 2015. A nonequilibrium thermal model for direct metal laser sintering, *Numerical Heat Transfer; Part A: Applications* 67, 249–267.

Wiese, M., Kwauka, A., Thiede, S., Harrmann, C., 2021. Economic assessment for additive manufacturing of automotive end-use parts through digital light processing (DLP), *CIRP Journal of Manufacturing Science and Technology* 35, 268–280.

Yap, Y., Wang, C., Sing, S., Dikshit, V., Yeong, W., Wei, J., 2017. Material jetting additive manufacturing: An experimental study using designed metrological benchmarks, *Precision Engineering* 50, 275–285.

Yu, Q., Chang, J., Wu, C., 2019. Silicate bioceramics: From soft tissue regeneration to tumor therapy, *Journal of Materials Chemistry B* 7, 5449–5460.

Ziaee, M., Crane, N., 2019. Binder jetting: A review of process, materials and methods, *Additive Manufacturing* 28, 781–801.

Zhang, X., Fan, W., Liu, T., 2020. Fused deposition modeling 3D printing of polyamide-based composites and its applications, *Composites Communications* 21, 100413.

Zhang, F., Saleh, E., Vaithilingam, J., Li, Y., Tuck, C., Hague, R., Wildman, R., He, Y., 2019a. Reactive material jetting of polyimide insulators for complex circuit board design, *Additive Manufacturing* 25, 477–484.

Zhang, J., Song, B., Wei, Q., Bourell, D., Shi, Y., 2019b. A review of selective laser melting of aluminum alloys: Processing, microstructure, property and developing trends, *Journal of Materials Science & Technology* 35, 270–284.

Zhang, B., Sun, H., Wu, L., Ma, L., Xing, F., Kong, Q., Fan, Y., Zhou, C., Zhang, X., 2019c. 3D printing of calcium phosphate bioceramic with tailored biodegradation rate for skull bone tissue reconstruction, *Bio-Design and Manufacturing* 2, 161–171.

Zhang, Y., Zhai, D., Xu, M., Yao, Q., Chang, J., Wu, C., 2016. 3D-printed bioceramic scaffolds with a $fe_3o_{4/}$graphene oxide nanocomposite interface for hyperthermia therapy of bone tumor cells, *Journal of Materials Chemistry B* 4, 2874.

Zhong, G., Vaezi, M., Liu, P., Yang, S., 2017. Characterization approach on the extrusion process of bioceramics for the 3D printing of bone tissue engineering scaffolds, *Ceramics International* 43, 13860–13868.

Zocca, A., Elsayed, H., Bernardo, E., Gomes, C. M., Lopez-Heredia, M. A., Knabe, C., Colombo, P., Günster, J., 2015. 3D-printed silicate porous bioceramics using a non-sacrificial preceramic polymer binder, *Biofabrication* 22(2), 025008.

Section B

Properties and Processing

5 Structural, Chemical, Electrical, Thermal, and Mechanical Properties of Bioceramics

Md Enamul Hoque, Samira Islam Shaily, and Asif Mahmud Rayhan
Military Institute of Science and Technology

5.1 INTRODUCTION

The term "bioceramics" has been assigned numerous definitions. As the definition evolved, so did the theoretical underpinnings and the anticipated performance of natural systems, it evolved along with age (Bongio et al., 2010). The creation of biomaterials, especially a variety of ceramics for bone healing and reconstruction, has made significant strides in recent decades (Kinnari et al., 2009; Muster, 1992). Dental and bone implantation made of ceramic are increasingly widely employed in the medical industry. Surgical instruments are frequently utilized. Bioceramic materials are frequently used to coat joint replacements to lessen wear and inflammatory reaction (Kassinger, 2003). Bioceramics must be durable since body implants must last longer than a person's lifetime. This substance is corrosion-resistant and extremely durable. Due to its extraordinary intrinsic hardness, it exhibits amazing mechanical qualities comparable to any technological ceramic. Durable bioceramic parts can be made using alumina- and zirconia-based materials like ZTA (zirconia-toughened alumina) and ATZ (alumina-toughened zirconia) (*Bioceramics, What Do We Talk About?*, 2022). Bioceramics are created in several ways and stages, and they have a wide range of uses in human recovery. Bioceramics are employed in biomedical applications as bulk materials with specified shapes that are called transplants, replacements, or prosthetic devices. The replacement and regeneration of aging and decaying joints with bioceramics that can be transferred to the new matured bone without temporary loss of assistance structure present two significant challenges for its medical application (Ducheyne et al., 2017; Library, 2006; Ratner et al., 2013). Research in the broad topic of bioceramics is moving quite quickly as well. In light of this, it is challenging to thoroughly review the most recent data on bioceramics. Since they are made from biodegradable natural fiber, biocomposite materials have become much more appealing (Hoque et al., 2021). The primary function of bioceramics is the regeneration of damaged tissue orthopedic system components.

DOI: 10.1201/9781003258353-7

The sort of bioceramic/tissue attachment desired will determine which bioceramic is most suited for a given application.

To govern interactions with elements of living systems, a biomaterial is a substance that has been created to assume the shape and is employed alone or as a component of a convoluted process to carry out any treatment or examination method. Ceramics, polymeric materials, and metal alloys are the three broad categories of biomaterials. The three categories of bioceramics are bioinert ceramics, bioactive ceramics, and biodegradable ceramics. Ceramics used to transport medications or to trigger tissue responses for bonding, like calcium phosphate and ceramic glass, are typically reactive. The majority of other ceramics, such as zirconia, alumina, sialon, and cermets, are bioinert (Park, 2009). Based on the use, bioceramics can directly make contact with the local tissue, either promoting tissue growth or, in the case of bioactive ceramics, triggering fresh tissue repair. As in the instance of bioinert ceramics, it can also remain latent at the site of application and serve as a biomechanical element. A ceramic substance must possess the necessary physical, biological, and mechanical qualities to serve as a biomaterial. The fundamental difficulty in using ceramics as permanent implants within the body is replacing deteriorated, aged bone having something that will last the patient's remaining years. A stable interaction with live tissue must form for a bioceramic to survive (*Bioceramics – An Overview | ScienceDirect Topics*, 2022).

5.2 PROPERTIES OF BIOCERAMICS

Bioceramics are an important subset of biomaterials that are primarily used in the healthcare industry. Dentistry cannot think without bioceramics. To be suitable for use in medicine, the majority of bioceramic products need to have the following four qualities: biocompatibility, durability, aesthetic properties, and radiopacity. The materials employed in the production of bioceramics are chemically inert and antimicrobial, which contributes to their biocompatibility. Biomaterial development and use should be constrained by morally responsible practices (Kumar et al., 2021). Zirconia is also one of the best materials for bioceramics because it is not hazardous to cells. For people to continue to benefit, bioceramic components must be durable. Table 5.1 gives some common characteristics of various glass and ceramics. Bioceramics used in hip replacements, for instance, must be able to withstand wear and tear from repeated motion. Aesthetic components that resemble natural teeth are required for dental applications of bioceramics. Lastly, the dentistry sector demands that bioceramics be visible on radiograph (*Bioceramics*, 2022). Synthetic materials known as biomaterials have the potential to be utilized for the replacement or repair of biological tissues with damage. They can be used, for example, to create organ replacements that serve a variety of functions, such as creating therapeutic tools, organ transplants, organ additions, etc. (Golieskardi et al., 2020).

A ceramic substance is an inorganic, nonmetallic substance that is frequently crystallized oxide, nitride, or carbide. Ceramics are robust, stiff, and rigid under compression but insufficient when under stress or shear. Due to its amorphous (noncrystalline) nature, glass is frequently not regarded as a ceramic. However, there are multiple processes involved in the ceramic manufacturing process while making

TABLE 5.1
Some Properties of Ceramics and Glasses (Jones & Ashby, 2012)

Components	Specific Gravity (g/cm^2)	Modulus of Young (Gpa)	Strength under Compression (Mpa)	Rupture Modulus (Mpa)	Exponent Weibull (m)	Exponential Time (n)
Glass ceramic (bioglass)	~2.8	~100	~1,000	~100	~10	10
Alumina (Al_2O_3)	3.9	380	3,000	300–400	10	10
Hydroxyapatite $Ca_{10}(PO_4)_6(OH)_2$	3.15	100~200	~1,000	<100	~10	10
Silicon carbide	3.2	410	2,000	200–500	10	40
Silicon nitride	3.2	310	1,200	300–850	–	40
Concrete	2.4	30–50	50	7	12	40

glass, and the mechanical attributes of glass are similar to those of ceramic materials. Thus, biocompatible bioactive ceramic glasses are made.

Any substance put into the body of a person to stay a protracted period must be acceptable to the organism. Therefore, it is essential to recognize how the structural, chemical, electrical, thermal, and mechanical properties of bioceramics are capable of producing useful bioceramics. For example, to comprehend how HAp functions inside a hierarchical structure organization, it is vital to comprehend the structural relationships among the several stages. Moreover, as a composite material, CaP gives bone its mechanical toughness, stiffness, and outstanding resilience to compression, while collagen gives it its elasticity and resistance to stress.

5.3 STRUCTURAL PROPERTIES

Traditional mechanical engineering ideas, which primarily underlie the fracture strength, dependability, and tribological response of biomaterials, severely restrict the choice of bioceramics for artificial joints (Pezzotti, 2014). In Figure 5.1, a ceramic femoral head hip prosthesis is shown which is attached to its plastic acetabular cup. An illustration of the adherence of collagen fibers that come before osseointegration is provided by the chemical deposition of biomaterials like the oxide layer of chondroitin sulfate. Implant and tissue will join together tightly. Occasionally, intermingled tissue is forming at this point, producing a fibrous connective capsule that prevents osteointegration and leads to implant loosening. That is why bioceramics have the following structural properties: strong melting temperature, good wear durability, excellent resistance to corrosion and oxidation attack, high resilience to weather, high stiffness, high Young's modulus, low ductility, and so on.

5.3.1 Atomic Bond, Structure, and Arrangement

Bioceramics are created in several ways and stages, and they serve a broad range of purposes in the healing of the body system. Biomaterials have a wide range of

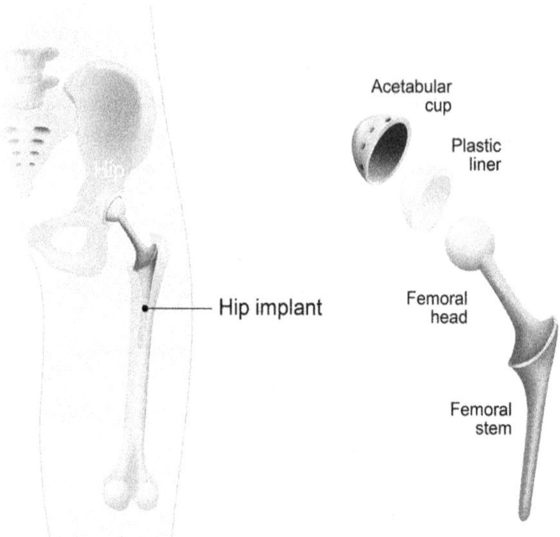

FIGURE 5.1 Ceramic femoral head hip prosthesis is attached to its plastic acetabular cup (*Bioceramics | Britannica*, 2022).

functions within the body system, including encasing to produce tissues using cells and sending data to tissues to regulate their activity (Biswas et al., 2022). Ionization is the process of transferring electrons between (neutral) atoms, such as nonmetallic chlorine and metallic sodium. The powerful attraction between the ions, both positive and negative, allows sodium and chlorine to form an ionic compound. Positively and negatively charged ions differ significantly in size as a result of the gain and loss of valence electrons.

$$\mathrm{Na} \leftrightarrow \mathrm{Na}^{+} + e^{-}$$

$$\mathrm{Cl} + e^{-} \leftrightarrow \mathrm{Cl}^{-}$$

According to their structural compounds, ceramics can be categorized, with AmXn serving as an instance (A stands for metal, X for a nonmetal element, and *m* and *n* are integers.). Table 5.2 gives some selected AmXn structure. The most basic illustration of this system is the AX structure ($m = n = 1$), which has three variants. The cubic structure of CsCl is straightforward. Comparing this to the body-centered cubic (bcc) structure, the central unit cell (body) would be occupied by the same ion or atom. The CsCl structure of CsBr and CsI is the same.

The ions' respective sizes account for the discrepancy between these structures. Tetragonal or octagonal gaps can hold positive ions produced by larger negative ions when the relative sizes of the ions are dramatically different from one another. A2X3 ($m \neq n \neq 1$) is the type of structure found in aluminum and chromium oxide. The positive ions (Al^{3+}, Cr^{3+}) fill up two-thirds of the octahedral sites, leaving one-third free,

TABLE 5.2
Some AmXn Structures of Bioceramics (Park, 2009)

Prototype Complex rA/RX	Lattice of A Other Complex	CN of A Sites	Available Sites Occupied	Minimal Value	
CsCl	Simple cubic	8	All	0.732	Csl
ZnS	fcc	4	12	0.225	CdS, AlP
Al_2O_3	hcp	6	2/3	0.414	Cr_2O_3

forming a tightly packed hexagonal (hcp) arrangement, the remaining three-quarters are filled with oxygen ion molecules (Park, 2009).

Ionic and covalent bonds both exist in ceramics. Covalent bonds are present in a variety of ceramic formations, including SiC, BN, and diamond. Each pair of nearest neighbors in these ceramics is designed so that they share a pair of electrons to create a chemical connection. The tetrahedron of silica (SiO_2), which is connected to form a network with two or three dimensions, serves as the basis for silicate glasses. The microstructure of ceramics determines their physical characteristics, which may be described by the quantity and types of phases that are present, the size, form, and alignment of each phase, as well as the proportions of each. Porosity is the most crucial characteristic of brittle materials like ceramics and glassware. Due to the significant increase and intensification of stress, brittle substances' physical qualities are greatly influenced by pore volume. Al_2O_3, SiO_2, hydroxyapatite (HA), and brushite, four different forms of bioceramic films, were examined to see if they could be used as orthodontic brackets and to improve their structural and mechanic characteristics (Cho et al., 2019). The formation rates of each layer were evaluated, and variations in the beginning particle-derived bioceramic films' crystallite sizes were investigated.

5.3.2 Structural Symptoms of Hydroxyapatite [$Ca_{10}(PO_4)_6(OH)_2$]

Hydroxyapatite (HAP), shown in Figure 5.2, is thought to be employed to create structural matrices in the mineral phases of enamel, dentin, and bone. The primary elemental inorganic hard tissues, including bones and teeth, are calcium phosphate, namely, hydroxyapatite (HA) (Bang et al., 2015; Märten et al., 2010; Ooi ct al., 2007; Ramesh et al., 2018; Sheikh et al., 2015; Sopyan et al., 2007). Due to a calcium-rich environment, phosphate, and hydroxyl ions in HAP, which are substances that exist in the human system spontaneously, there is no installation damage (Anita Lett et al., 2021). The majority of naturally occurring hydroxyapatite is derived from biological wastes or sources, such as mammalian bone, where the synthesis of HA typically entails producing 0.5 mol of $Ca(NO_3)_2.4H_2O$ and 0.3 mol of $NH_4H_2PO_4$ in line having a Ca/P ratio of 1.67. In a glass beaker, the two precursors were initially placed and individually dissolved in deionized water while being stirred and heated to 90°C on a magnetic hotplate. The calcium solution was then stirred at 90°C while adding drops of the phosphate solution. Then came the titration procedure. The mixture's pH was kept at roughly 10 ± 0.5 by mixing ammonia solution into the mixture.

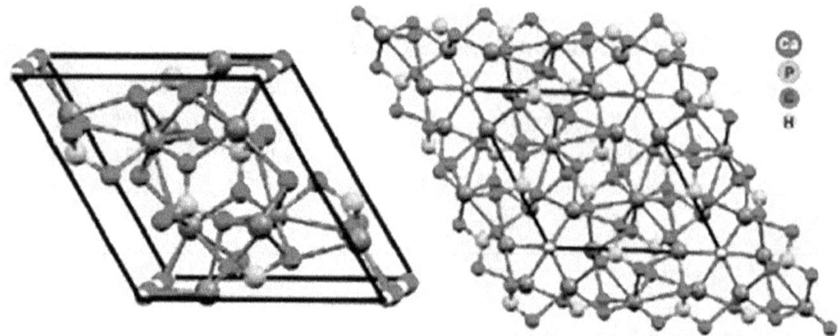

FIGURE 5.2 The Hap's crystal structure (left) and the plane-based projection of the Hap structure 001 (right) (Gomes et al., 2019).

Following titration, the combination was vigorously agitated until at least 3 more hours before being left to mature at room temperature for the night (Wahid et al., 2015). A Rigaku diffractometer was used to perform an XRD examination of the samples at three distinct degrees of Zn doping.

Analyzing phases with XRD of the undoped and Zn-doped hydroxyapatite powders as-synthesized revealed conventional matching diffraction patterns in the reference identical file for hydroxyapatite (ICDD No.9-432). Given that no secondary phase was found, HA only exists as a single phase (Mardziah et al., 2020).

A process known as contact osteogenesis is what creates the link between the bone matrix and an implant made of a metallic titanium alloy. Many times, in clinical practice, a bioactive hydroxyapatite layer is supplied to enable bonding osteogeneses that can convey these tensile and shear pressures through "bony in-growth". Here given some examples in Table 5.3 of attempting to take HAP out of mammalian origins.

TABLE 5.3
Some Structural Characteristics of Hap out of Mammalian Origins

Origin	Extraction Technique	Ca/P Ratio	Crystallization Epochs	Species Size	Shape	Reference
Bovine bone	Alkaline Hydrolysis	1.86	Hap	Nanosize	Nanoflakes	Barakat et al. (2009a)
	Calcination	1.5	Hap, β-TCP (850°C)	420–500 μm	Nanoflakes	Sun et al. (2017)
Camel bone	Calcination	1.66	Hap (1,000°C)	79–0.9 nm	Irregular	Jaber et al. (2018a)
Horse bone	Alkaline heat					
Pig bone	treatment + calcination	1.709 (800°C)	Hap (800°C)	70–180 nm (800°C) & 200–700 nm (1,200°C)	Irregular	Janus et al. (2008)

There is growing proof that a 150-m long-lasting bioactive hydroxyapatite covering will affect the surface chemistry of the implant material by attracting noncollagenous proteins like silylated g proteins, osteocalcin, osteonectin, and proteoglycans to the interaction of the implant substance. The benefits of biodegradable films involve (i) preventing the formation of a connective tissue capsule consisting of fibrous tissue surrounding the implantation; (ii) rapid rates of adhesion in the presence of bone; (iii) bonding osteogenesis, which establishes a consistent, robust, and able to transfer compressive, tensile, and shear forces interaction between transplant and tissue; (iv) increased recovery speed compared to implant materials without a biocompatible covering; and (v) the apparent risk of a cytotoxic reaction is lessened due to a lower discharge to the surrounding tissue of titanium ions (Heimann, 2002).

5.4 CHEMICAL PROPERTIES OF BIOCERAMICS

Numerous bioceramic materials have outstanding compressive strength, chemical resistance, and wear resistance. However, certain bioceramic materials have a bioactive response that is inert. Modern medicine is heavily reliant on biotechnologies, the advancement of which is governed by the caliber of cutting-edge biomaterials. The physicochemical and mechanical properties of biomaterials, including chemical inertness, thermal properties, radiation resistance, structural strength, flexibility, and microstructural diversity, are the subject of research in the area of bioactive molecules.

5.4.1 Inert Ceramic Materials: Oxide Bioceramics

No substance inserted into living tissue can be regarded as entirely inert. While metals and plastics are susceptible to corrosion and degradation in biological systems, oxide bioceramics are not, by their very nature, susceptible to these problems. Ceramics are often stable inside live bodies and offer a high level of acceptability due to their proximity to surrounding muscles despite having molecular structures that are entirely different than those of living tissues (Ichikawa et al., 1992). Alumina (Al_2O_3) and zirconium dioxide (ZrO_2, zirconia) are inert oxide ceramics that are used in bioceramics. In contrast to bioactive ceramics, inert ceramics do not attach to bone in the same way. Other oxides help stable zirconia bioceramics to some extent, such as the oxides of yttrium, calcium, or magnesium, whereas alumina bioceramics are in the pure form of aluminum oxide (Li & Hastings, 2016).

Alveolar ridge restoration (Hammer et al., 1973), maxillary reconstruction, ossicular bone replacements (Grote, 1984), ophthalmology (Polack & Heimke, 1980), knee arthroplasty (Li & Hastings, 1998), bone screws, and other uses as oral biomaterials, including such dental crown base, post, bracket, and inset (Andersson & Odén, 1993; Kittipibul & Godfrey, 1995), are all possible using alumina and zirconia ceramics. Alumina resists corrosion and is chemically stable. It is hardly soluble in strong acids and alkalies and dissolvable. Therefore, at a normal pH of 7.4, there is essentially no discharge of ions from the alumina. Zirconia ceramics are classified as polymorphic. They manifest three established polymorphs, including the tetragonal, cubic, and monoclinic phases, as well as the high-pressure orthorhombic variant. The tetragonal

TABLE 5.4
Engineering Psychochemical Characteristics of Zirconia and Alumina (Li & Hastings, 2016)

Property	Al_2O_3	ZrO_2
Melting point (°C)	2,040	2,680
Hardness (GPa)	22	12.2
Purity (%)	>99.5	>99.5
Porosity (%)	0	0
Microhardness (GPa)	23	–
Resistance to wear (mm^3/h)	0.01	–
Oxidation resistance (mg/m^2d)	<0.1	–

condition is secure upward to 2,370°C and changes from the monoclinic structure at about 1,170°C, while a cubed phase remains up to 2,680°C, the melting temperature. The progress of using ceramic materials in biomedical applications relies heavily on the physicochemical features of these materials, as is well recognized. Here in Table 5.4 given some physiochemical properties of inert ceramics.

Alumina is known to have exceptional hardness and abrasion-resistant qualities (Banijamali et al., 2012; Liu et al., 2003). The remarkable wear and friction qualities of alumina are facilitated by its smoothness and surface energy (Banijamali & Ebadzadeh, 2016; Portilla & Halik, 2014). Alumina, which has been demonstrated to be the sole thermochemical stable form, is an excellent option for skeletal reconstruction due to its mechanical qualities and chemical stability (Thamaraiselvi & Rajeswari, 2004). By heating aluminum's hydrates, aluminum oxide is created. All other structures are permanently changed to hexagonal corundum, an arrangement of oxygen ions, at temperatures exceeding 1,200°C. Therefore, the only stable structure over 1,200°C is alpha-alumina. The friction stress between the alumina femoral heads and the acetabular cup is extremely low, as shown in Figure 5.3, which results in a small amount of wear. This is due to the chemical stability, excellent surface finish, and exact dimensions. In the lab, ceramic heads with UHMWPE cups have higher wear resistance than metal, and UHMWPE increased from 1.3 to 34 times and the clinical setting between three and four times (Murphy et al., 2016). There were no alumina wear particles in the recovered ceramic/UHMWPE, but there were UHMWPE wear particles in the recovered surrounding tissues that ranged in size from microns to millimeters (Li & Hastings, 1998). Ceramic balls in a ceramic:UHMWPE combination have essentially no breakage.

Zirconia ceramics shatter because the volume shift brought on by phase transformation is sufficient to push them past their elastic and fracture limits. Zirconia must therefore be combined with CaO, MgO, and Yttria (Y_2O_3) that are examples of added chemicals to stable the substance in either the tetragonal or the cubic state. Zirconia ceramics have such a low microhardness and a high density due to the presence of heavy zirconium ions. The efficiency of zirconia materials even above alumina ceramics in terms of reduced shrinking and wearing, as well as the excellent wear

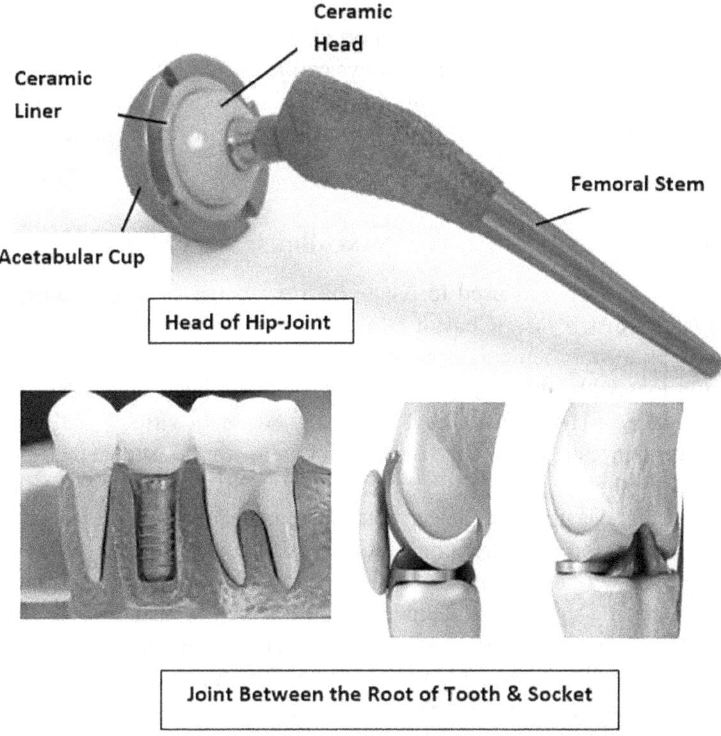

FIGURE 5.3 Inert ceramics material for medical application (Resan, 2018).

resistance of zirconia versus UHMWPE. A pin-on-disc wear experiment and a hip simulator both revealed a significantly lower rate of wear for metal ball caps versus zirconia ball heads (Derbyshire et al., 1994). Radiation and deterioration are two negative aspects of the applicability of zirconia in bioceramics. Despite no appreciable decrease in structural rigidity, the zirconia balls' surface deterioration caused by the phase formation during stress appears to be an issue (Cales et al., 1994; Shimizu et al., 1993).

5.5 ELECTRICAL PROPERTIES OF BIOCERAMICS

Bioceramics are such a unique category of modern ceramics that are largely used in the healthcare sector for a variety of purposes. The features required for various medicinal uses include dimensional stability, antibacterial, anticorrosive, and total chemical inertness in finished bioceramic products. Ceramics are typically inorganic substances with covalent and ionic bonds. As a result, they are relatively hard, have sharp melting temperatures, and have limited electricity and heat conduction. Glass ceramics, bioactive glass, zirconia, alumina, hydroxyapatite, and resorbable calcium phosphates are a few examples of materials that fall under the category of bioceramics.

Electrical potentials have been linked to indicators in the bone remodeling cycle in mechanically loaded bone. There has reportedly been an emerging interest in using this phenomenon to create electrically active ceramics that can be implanted in bone grafts and perhaps enhance biological responses. Due to HA's polarization, hydroxyapatite has an excellent electrical potential, which benefits the ceramic's interfacial reactions (Bowen et al., 2010).

5.5.1 Electroactivity of Ag-HAp Nano-Bioceramics Materials

Moisture precipitation was used to create hydroxyapatite (HAp) bioceramics with a nanocrystalline structure (Chavan et al., 2010). Additionally, the HAp structure was altered by a room-temperature ion-exchange technique that switched some of the calcium ions with silver ions (Ag^+). Using synthetic body fluid (SBF) in a static state, HAp and Ag-HAp scaffoldings were studied for in vitro bioactivity. For in-depth dielectric investigations, Ag-HAp scaffolding with a constant silver proportion (0.005 M) was submerged in SBF for varying lengths of time. For scaffolds that had already been constructed, dielectric measurements of the scaffolds' impedance, dissipation factor, and dielectric constant were made using a Quad Tech LCR 7600 meter at ambient temperature.

Here, Figure 5.4a shows the fluctuation of the dielectric constant for Ag-HAp scaffolding with a frequency change of an induced concentration-dependent AC field. The graphs are evaluated to those for Ca-HAp. Figure 5.4b and c demonstrates the effect of incubating on Ag-dielectric HAp's constancy depending on how often an AC field is applied. All of the scaffolds exhibit a consistent pattern of falling dielectric constant with rising frequency. After adding silver ions to the HAp structure, the Ca-HAp relative dielectric is significantly altered. Figure 5.4c shows that the dielectric characteristics for Ag-HAp scaffold are far less than those for Ca-HAp. The dielectric constant is seen to decrease as the silver ion level rises. Considering this concentration of silver toward others, the greater silver ion concentration (0.025 M) is shown to have the greatest dielectric constant (roughly close to 8). The study unambiguously shows that the dielectric strength reflects the concentration of ions. Before and after incubation, the dielectric characteristics of Ag-HAp demonstrate strong dielectric permittivity at a lower frequency that decreases with increasing frequency and reaches a constant rate for all specimens. Because hydroxyapatite is a bioceramic, the effect of dielectric properties is relating to autonomous polarization vibrating in an imposed shifting field, maximum levels of electronic properties at lower frequencies are clear. Electric dipoles, in particular those caused by hydroxyl ions, follow changes in the field at quite low alternating field rates and increase the dielectric properties at low frequency. As a result, the dielectric constant falls, and the capacitor's low characteristic impedance to the sinusoidal pulse minimizes power dissipation in the resistor.

Here, grain capacitance (C g), grain boundary capacitance (C Gb), grain boundary resistance (R Gb), and bulk resistance (R g) are listed, and their values are determined using the equations below.

$$\frac{C_g}{C_{gb}} = \frac{1}{\dfrac{R_g}{R_{gb}} * 2\Lambda f_c} \tag{5.1}$$

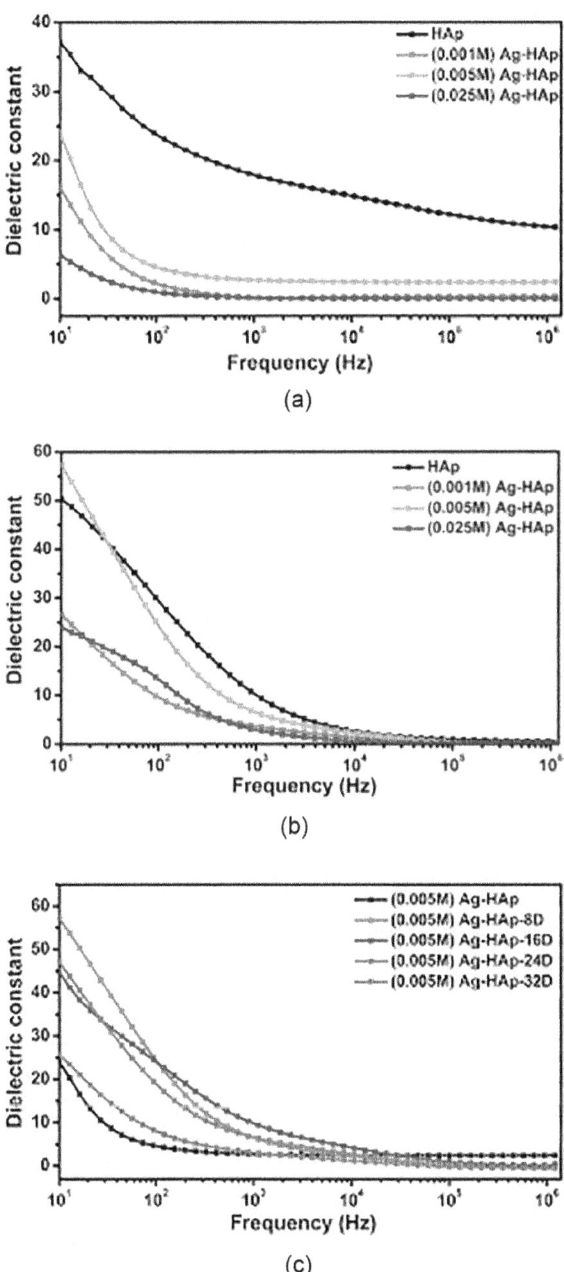

FIGURE 5.4 (a) Ag-HAp nano-ceramic scaffolds' dissipation factor as a versus frequency before incubation (Bahir et al., 2020); (b) Ag-HAp nano-ceramic scaffolds' dissipation factor as a versus frequency after incubation for fixed duration with a changeable silver ion content (Bahir et al., 2020); (c) Ag-HAp nano-ceramic scaffolds' dissipation factor as a versus frequency after incubation for changeable duration with a fixed silver ion content (Bahir et al., 2020).

Grain capacitance = C_g
Grain boundary capacitance = C_{gb}
Bulk resistance = R_g
Grain boundary resistance = R_{gb}
Frequency = f_c

Before and after being incubated in SBF, the Ca-HAp and Ag-HAp scaffolds' C_g and C_{gb} values, respectively, are given here in Tables 5.5 and 5.6.

Ag-HAp scaffolds incubated with SBF are monitored using cole-cole plots against incubation time and silver ion concentration (Bahir et al., 2020). Except for the scaffold with a silver ion content of 0.005 M, the high-frequency wave

TABLE 5.5
Values before Incubation in Simulated Body Fluid (SBF) (Bahir et al., 2020)

Before Incubation in Simulated Body Fluid (SBF)

Scaffolds	Resistant to Bulk (Rg) (KΩ)	Grain Growth Resistance (Rgb) (KΩ)	Bulk Capacitance (Cg)(nf)	Grain Growth Capacitance (Cgb) (nf)	Leisure Period (τ) (ms)
HAp	93.9	–	38	–	3.6
(0.001 M) Ag- HAp	69	253.6	24	76	1.7
(0.005 M) Ag- HAp	69.01	221.7	29	85	2
(0.025 M) Ag- HAp	43.2	346.5	39	43	1.7

TABLE 5.6
Values after Incubation in Simulated Body Fluid (SBF) for 8 Days (Bahir et al., 2020)

After Incubation in Simulated Body Fluid (SBF) for 8 Days

Scaffolds	Bulk Resistance (Rg) (KΩ)	Grain Boundary Resistance (Rgb) (KΩ)	Bulk Capacitance (Cg) (nf)	Grain Boundary Capacitance (Cgb) (nf)	Relaxation Time (τ) (ms)
Hap	95.2	–	10	–	1
(1.3 M) Ag- Hap	74.9	437.9	28	44	2.1
(0.005 M) Ag- Hap	87.9	609.1	31	31	2.7
(0.025 M) Ag- Hap	63.9	517.3	20	29	1.3

amplitude lines of the impedance range do not significantly alter with changes in ion concentration. The grain resistance effect on this scaffold has the lowest amplitude. Higher grain boundary impedance is produced by a scaffold with the highest silver content. A considerable low-frequency arcs' shifting amplitude and position is also seen after incubation. Ca-HAp scaffold exhibits the highest semicircle height for a fixed SBF treatment duration (0.005 M), followed by the Ag-HAp scaffold, coupled with the strongest grain boundary resistance. It is possible to see how the duration of the SBF treatment affects the grain boundary resistances and low-frequency arc amplitudes. For the Ag-HAp scaffold, the grain boundary resistance reduces when the incubation period is extended from 8 to 32 days (0.005 M). The growth of calcified fibroblasts on the scaffolding's surfaces may be the cause of the shift in the amplitude and location of low-frequency arcs in the impedance bands of the scaffolds under incubation. Gain boundary resistance is impacted by these modifications to the scaffolds following SBF treatment. These results imply that impedance spectroscopy may be a practical method for assessing the in vitro biocompatibility of scaffolds.

5.5.2 Electrical Impedance Measurement of Hydroxyapatite

Animal bones, including the bovine femur, can be used to make HA base materials. The choice of the bovine stems from its chemical similarity to that of pure bone (Vallet-Regí et al., 2008). The dielectric barrier discharge (DBD) plasma therapy approach is anticipated to be a promising approach to raising the strength in deformation of the HA scaffold. A crucial element that must be taken into account when deciding whether or not the HA scaffold is suitable for plasma DBD treatment is its impedance value. Impedance, denoted by the letter Z, is a unit of measurement for the resistance to electrical flow. Ohms are used to measuring it. Impedance is a type of electrical load that includes capacitive, inductive, and dissipative components. This is connected to resistance for dissipative, capacitive, and inductive systems, while it is associated with capacitance for dissipative systems. Each device's impedance is variable since it varies based on the material's nature and characteristics. HA scaffold, which contains pores and pore sizes that are challenging to control, is used in this investigation. Three separated HA scaffold thicknesses are presented here in Figure 5.5, 0.4 cm with pores ranging from 197.38 to 329.94 m, 0.5 cm with pores ranging from 163.30 to 523.15 m, and 0.6 cm with pores ranging from 144.71 to 515.45 m (Nurmanta et al., 2019).

Real and imaginary numbers are the two sorts of values that impedance typically exhibits. Resistance and capacitance, which are influenced by the material's impedances, have the potential to have an impact on imaginary numbers, whereas real numbers pertain to the material's resistance. Depending on how the chemical behaves, the value of this fake number could be either positive or negative. Because it is challenging to manage the pore diameters in HA scaffolds, the impedance results will also differ. Additionally, the capacitive value has a bigger impact than the inductive value. Because of the qualities of HAP, there are several uses for this chemical (Padmanabhan et al., 2020).

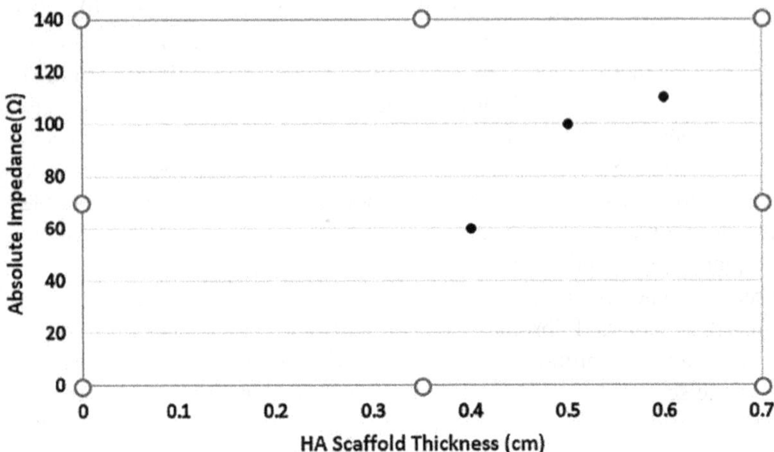

FIGURE 5.5 The relationship between thickness of the HA scaffold and the absolute impedance at the three distinct thicknesses of 0.4, 0.5, and 0.6 cm (Nurmanta et al., 2019).

5.6 THERMAL PROPERTIES OF BIOCERAMICS

The efficiency of the substance for implantation or for interacting with biological tissues can be greatly influenced by the thermal characteristics and conductivity of biomaterials. Temperature, pressure, and chemical composition are examples of thermodynamic variables that affect the phase. In theory, it is possible to mechanically separate the different phases of a substance. We will use fundamental thermodynamic concepts to comprehend some straightforward phase transitions happening in bioceramics and glassware. A good example is a zirconia (ZrO_2) (Park, 2009). The atomic structure and piling of a polymer material must be sufficient to transport energy dissipation from the electroactive elements to produce better thermally conductive polymer materials. This typically indicates that materials with a higher degree of amorphous perform well as thermal insulators and are also more prevalent in biologically produced materials (Thermal Conductivity of Biomaterials, 2022).

The relationship between free energy and enthalpy can be used to explain the phase shifts. Thermodynamic analysis, often known as "thermal mechanical analysis", has been used to study some dimensions caused by temperature variations (TMA) (Saleh et al., 2018).

The entropy is described by

$$dS = \frac{dq_{\text{rev}}}{T} = \left(\frac{C_p}{T}\right)dT \tag{5.2}$$

in which the specific heat at a pressure of constant

$$C_p = \left(\frac{\partial H}{\partial T}\right)_P = \left(\frac{dp}{dT}\right)_P \tag{5.3}$$

The following equation can be used to determine the linear expansion coefficient:

$$\alpha = \frac{1}{L_0} \cdot \frac{\Delta L}{\Delta T} \tag{5.4}$$

Coefficient of linear thermal expansion $= \alpha$

Change in length to temperature change $= \frac{\Delta L}{\Delta T}$

Original length $(m) = L_0$

5.6.1 Thermal Properties of Bioactive (SiO_2–CaO–Na_2O–P_2O_5) Glasses

The composition of SiO_2-CaO-Na_2O glasses and their derivatives was made using the melt-quench method with Na_2O as a P_2O_5 substitute. According to the results of the structural analysis, Si and P atoms are in clearly delineated tetrahedral groups with a bond spacing of 1.60 for both Si-O and P-O links, even though P has a larger average functional group than Si. There were tendentious changes in the behavior of the glass with increasing phosphate concentration (Fábián et al., 2020). Unfortunately, the limited load-bearing capacity and manufacturing challenges that are similar to all compositions prevent their widespread use in clinics. The creation of ceramic-glass composite materials using organic glass as a raw biomaterial either by thermal therapy or liquefy process is one way to solve these drawbacks (Lefebvre et al., 2007). Here given the diagram of the ternary system Na_2O-CaO-SiO_2 in Figure 5.6 illustrates the ratios of the glasses being examined after thermal treatment.

In linear-heating studies, a thermal study of the examined bioactive glasses was carried out. For each of the several glass compositions, the heating signs showed a glass change, solitary exothermic devitrification incident, and then multiple stages

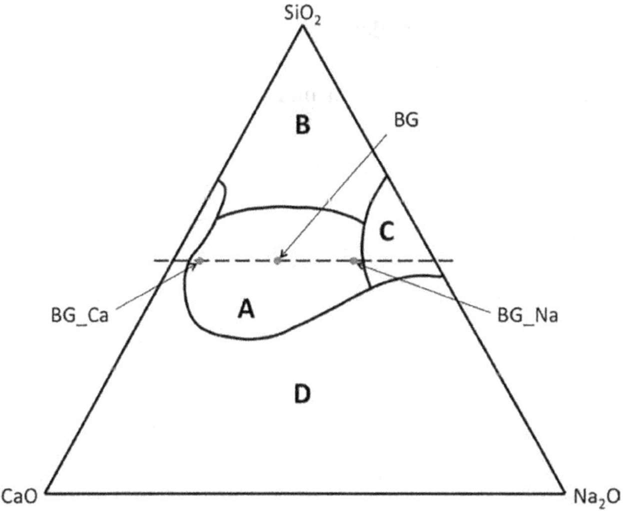

FIGURE 5.6 Diagram of the ternary system Na_2O-CaO-SiO_2 (Santoso et al., 2022).

TABLE 5.7
Characteristic Temperatures for the Different Bioactive Glass Compositions of Thermally Activated Processes (Fábián et al., 2020)

Specimen	T_g (°C)	T_x (°C)	T_m (°C)	TG Peak (°C)
Si45P0	516	662	1,215	503
Si45P1	527	689	1,192	467
Si45P3	540	754	1,209	418
Si45P5	550	690	1,183	410

of melting. Transition temperature (T_g), exothermic devitrification event peak temperature (T_x), melting temperature (T_m), and weight loss rate peak temperature (T_G peak) are examples of the appropriate temperature for the various thermal processes (Fábián et al., 2020). In Table 5.7, typical temperatures for the various bioactive glass compositions' thermally activated processes are shown.

Because of the robust SiO_4 tetrahedral system, fusing silica glass solid crystals at temperatures as high as 1,200°C (Ojovan, 2009). It is generally known that even small amounts of alkali elements, particularly Na, effectively lower the temperature at which silica glass crystallizes by causing the SiO_4 tetrahedra to rearrange and break their network (Fanderlik, 2013). It is therefore not astonishing that the T_g and T_x values for the Ca and Na-containing Si45P0 specimen are significantly lower than for fused silica. Low-concentration phosphate addition to the glass causes a gradual rise in T_g and T_x since P is likewise a network-forming component. Phosphate and liquid silica are miscible at elevated temperatures, around cc. 1,000°C. This improved miscibility presumably continues to exist in the glassy condition as well. The steadily rising T_g shows that the phosphorus glass structure's cohesiveness is improving.

Figure 5.7 shows the results of tests using linear heating to perform a thermal analysis on the bioactive glasses under study. For all the various components of glass, Figure 5.7a depicts the thermal signals showing a transform from glass, a unique exothermic display event, tracked by a multistage melting. The curves of weight

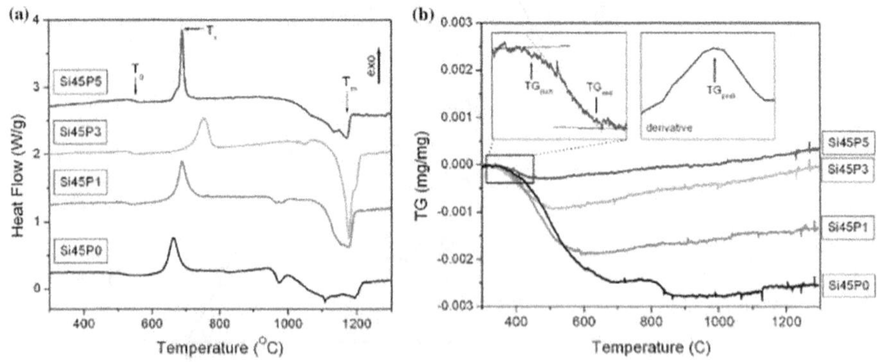

FIGURE 5.7 Thermal analysis of different bioactive glasses (Fábián et al., 2020).

reduction discovered during the identical linear thermal DTA scans are shown in Figure 5.7b. Thermal examination of the glasses shows the exothermic devitrification event (T_x), melting (T_m), and glass transition (T_g), as well as the weight loss lines moved to a typical source at 300°C. The Si45P5 sample's enlarged TG(T) curve well with TG_{start} and TG_{end} temperatures is shown in the insets, along with the associated normalized loss weight rate line with the TG_{peak} temperature (Fábián et al., 2020).

Given that phosphate glasses normally have lower melting points than silica glass, this behavior is unexpected. The T_x transition temperature up to a 3% PO_4 component exhibits the same pattern. Phase separation is recommended since, above this intensity, the phosphorus component becomes saturated. The decline of T_x for the Si45P5 sample indicates the deposition in the subcooled liquid state. The overall temperature changes during the weight loss procedure and weight loss are both decreased by the elevation in phosphate content. We predict that melting of the PO_4-containing surroundings will cause growth in the glass structure because of the substantial melting point gap seen between silica and phosphate stages (Fulchiron et al., 2015), which shows that the PO_4-containing configuration is less stable at high temperatures. The weight loss temperature is lowered at increased phosphate concentrations as a result of this process. All of the samples contain the same phase, hexagon sodium calcium silicate ($Na_6Ca_3Si_6O_{18}$). Above 600°C, this phase was also found in the thermal degradation of bioglass. Additionally, crystalline structure beta sodium calcium phosphate silicate ($Na_2Ca_4(PO_4)_2$ SiO_4) is present in sample Si45P5, just like in bioglass. This later phase vanishes in sample Si45P3, which has less phosphorus, leaving just the hexagonal phase. In sample Si45P1, bands of cubed sodium calcium silicate ($Na_{15.6}Ca_{3.84}(Si_{12}O_{36})$) appear as the phosphorus level continues to decline. Similar to Si45P1, sample Si45P0's diffractogram exhibits the same possibility, with the traces of the cubic structure and the undetermined lines being stronger than the traces of the hexagonal structure (Lefebvre et al., 2007).

5.6.2 Thermal Conductivity of Ion-Substituted Hydroxyapatite Bioceramics

One of the most popular bioceramics is calcium phosphate, which contains the mineral hydroxyapatite (HAP), which is similar to the minerals found in the bone and teeth (Padmanabhan et al., 2019). Due to similarities concerning the mineral component of bone tissue, hydroxyapatite is a substance that is frequently employed in the biomedical field. Because of its structure, which is remarkably similar to the mineral phase of hard connective tissue, hydroxyapatite (HAp) is among the most often utilized bioceramic materials in the treatment of many musculoskeletal system problems (Mohd Pu'ad et al., 2019). Additionally, HAp demonstrates bioactive activity, biocompatibility, nontoxicity, and osteoinduction capabilities, all of which are required for medical applications (Szcześ et al., 2017). Calcium phosphate biomaterials are ideal for such applications due to their bioactivity and bone regeneration qualities. Ca^{2+}, PO43, and OH have been partially replaced by other ions; calcium phosphates produce a large number of isomorphic materials due to their great thermal properties and the special flexibility of their fundamental structures (Legeros et al., 2017; Liu et al., 2013). By applying a constant thermal decomposition from

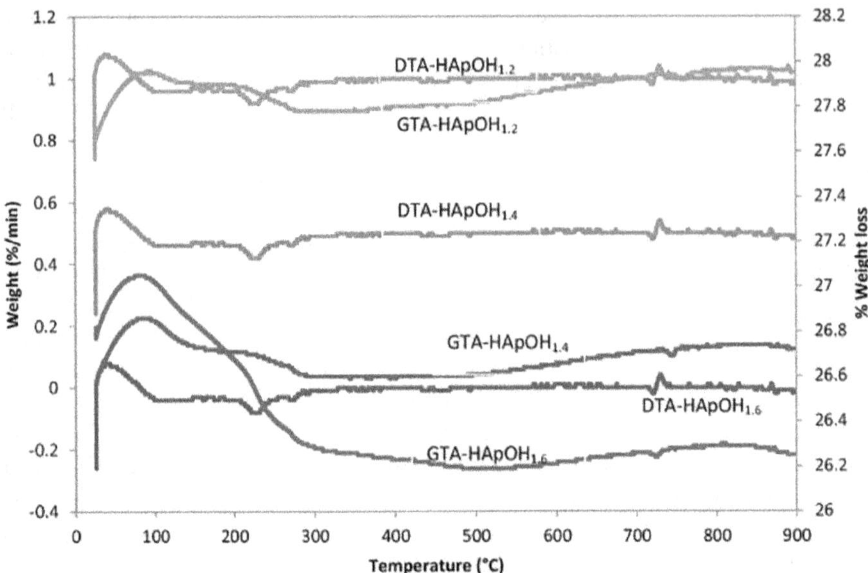

FIGURE 5.8 Ion-substituted HAp differential and thermal analysis curves (Hidouri et al., 2019).

ambient temperature to 900°C to the samples (OH group as HApOH2, HApOH1.6, HApOH1.4, and HApOH1.2) as-prepared, a divergent and thermal analysis was carried out. The results are shown in Figure 5.8. The products were identified by a consistent weight loss during the temperature increase, which was more pronounced with the compound HApOH1.2, according to the analysis of the TG thermograms (Hidouri et al., 2019).

HApOH1.6 and HApOH1.4 samples experienced a weight loss of around 2% at the heating rate of 50°C–200°C, whereas HApOH1.6 experienced a weight loss of about 4% due to the removal of solvent molecules. This occurs at the same time as an endothermic action at about 110°C. The removal of nitrates may be the cause of the next weight loss of about 2.5%, which began at a temperature of 250°C and concluded at 300°C. With these two occurrences, a third thermal effect was also detected and resulted from a novel phase that was later discovered. This effect was demonstrated by the occurrence of a minor exothermic peak at roughly 730°C. There was found a new stage in $Ca_3(PO_4)_2$ (β-TCP) at temperatures of about 730°C. When temperatures are high, β-TCP remains steady and its reflection energies grow. α-TCP and β-TCP, TTCP, and occasionally calcium oxide CaO are combined with HA (hydroxyapatite) and ACP (amorphous calcium phosphate), which make up the majority of the mixture. There was a chemical gradation in between the coating's inner and outer layers, with the inner layer having a larger ACP/HA ratio. Since the absorption bands of the various calcium phosphate phases differ, falling in order (Ducheyne et al., 1990; Johnsson & Nancollas, 1992; LeGeros, 1991; Osborn & Newesely, 1980):

$$\text{ACP} > \text{TTCP} >> \alpha - \text{TCP} >> \beta - \text{TCP} >> \text{HA}$$

The IR spectrum of specimens that were calcined at 1,150°C is shown in this figure. These HAp-specific spectra demonstrate that other than the removal of the nitrate bands, there was little to no change with rising temperature. Additional apexs of β-TCP were detected at 407, 427, and 964 cm[1], in addition to the previously mentioned maxima relative to multisubstituted Hap (Timchenko et al., 2017). Here, wavenumber spectrum reflects the biomolecules (proteins, lipids, carbohydrates, and nucleic acids) present in a sample in Figure 5.9.

Laser ceramic compact sintering rates are typically 200°C–300°C below the matching composition's melting point. Because there is a temperature point in which the rate of grain boundary rises exponentially, the choice of sintering temperature is influenced by the growth of ceramic grains (*Sintering Temperature – An Overview | ScienceDirect Topics*, 2022). The greater density was attained by boosting the sintering temperature. The initiation of the sintering process causes a dramatic increase in density when the sintering temperature goes up from 1,000°C to 1,050°C, and when the temperature continues to rise after that, the rate at which density grows is reduced.

This graph in Figure 5.10 shows how the relative density of sintering specimens' ion exchanges varies concerning sintering temperatures. It appears that each sample followed nearly the very same course in terms of relative density evolution as temperature-dependent. The densifying ratios of pure HAp, however, continued to be larger than those of ion-substituted ones. Densification proportions did not surpass 87% for temperatures between 900°C and 950°C, but they gradually increased as the temperature was raised, peaking at about 95% of theoretical density for ion-substituted samples at 1,150°C, while pure HAp achieved a high of 98% at 1,100°C.

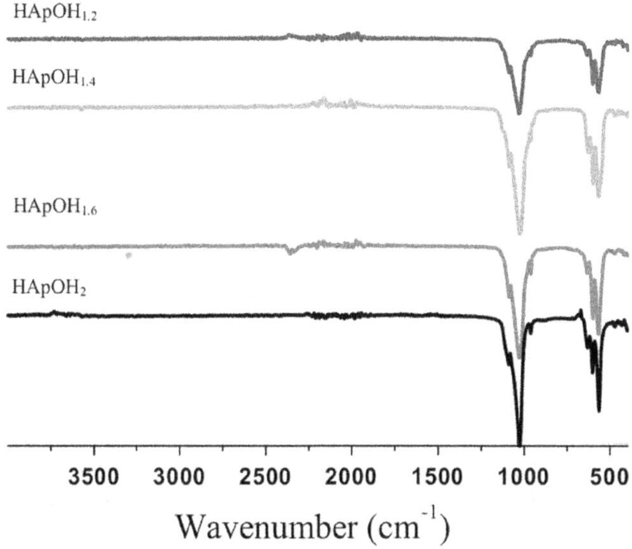

FIGURE 5.9 Ion-substituted HAp particles' FTIR spectra after being heated to 1,150°C (Hidouri et al., 2019).

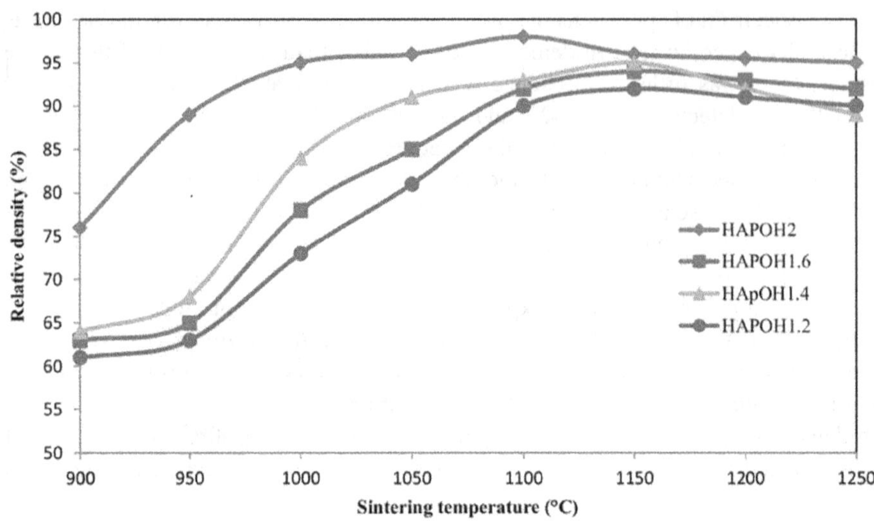

FIGURE 5.10 Increased density of purified and ion-substituted HAp as a result of sintering (Hidouri et al., 2019).

When temperatures rose above specified levels, the relative density marginally dropped. This behavior was consistent with what has been observed earlier with test specimens of hydroxyl and fluor apatite and magnesium (Gross & Rodrígueź-Lorenzo, 2004; Hidouri et al., 2009; Nsar et al., 2013; Senamaud et al., 1997). As a result, the interactivity in the present investigation was typically reduced when compared to neat HAp as a result of the contemporaneous cationic and anionic replacement in HAp.

The objective of this study is to highlight the impact of different ionic replacements of Mg^{2+}, Na^{+}, K^{+}, F, and Cl on the thermal features of HAp bioceramics. The nature of the synthetic chemicals was discovered to be nonstoichiometric. The subsequent stage of TCP is generated on account of the partial decomposition occurring at temperatures higher than 730°C, according to the thermal experiments. At 1,150°C, multisubstituted HAp sintered specimens achieve densifying levels of 95%.

5.7 MECHANICAL PROPERTIES OF BIOCERAMICS

Some of the most crucial characteristics of biomaterials that should be carefully examined and assessed prior to implantation are tensile strength, yield strength, elastic modulus, corrosion, creep, and hardness. The mechanical qualities are most important for applications involving hard tissue. For clinical use, bioactive implants must display strong interfacial interactions with both hard and soft tissues and match the mechanical properties of the host tissue (Chen et al., 2008; Day et al., 2004; Ducheyne & Qiu, 1999; Kim et al., 2007; Lu et al., 2003; Minardi et al., 2015; Shah et al., 2005).

A variety of studies have been conducted on mechanical properties of bioceramics and found that natural bone and bioinert ceramics have striking quantitative

TABLE 5.8
Some Mechanical Properties of Bioceramics and Bioactive Glasses (Park, 2009)

Mechanical Properties	Alumina (Al_2O_3)	Spinel ($MgAl_2O_4$)	Zirconia (ZrO_2), Cubic	Borosilicate Glass	Silicon Nitride (Si_3N_4)
Elastic modulus E (GPa)	390	284	152	62	304
Shear modulus G (GPa)	154	–	58	26	123
Poisson's ratio, ν	0.27	–	0.32	0.20	0.24
Density (g/m³)	3.9	3.55	5.56	2.23	3.44
Specific modulus (GPa.cm³/g)	100	80	27	28	88

differences in their mechanical characteristics. Here given some mechanical properties of bioceramics and bioactive glasses in the Table 5.8. These variations result in significant gradients in the elastic modulus (Young's modulus), which cause the so-called stress shielding. Inferring Young's modulus from the stress and strain curves' slope, the majority of brittle, homogeneous, and isotropic materials follow Hooke's rule, which is expressed as

$$\sigma = E\varepsilon \tag{5.5}$$

σ is the stress force per unit area of the cross-section (F/A_0),
ε is the strain,
E is known as the elastic modulus or Young's modulus.

The majority of studies characterize bone mineral crystals as having a plate-like structure and a wide variety of dimensions (Fratzl et al., 2004); the platelets' thickness varies between 1.5 and 9 nm, their length from 15 to 200 nm, and their width from 10 to 120 nm (Benezra et al., 1995; Betts et al., 1981; Danilchenko et al., 2002; Nyman et al., 2005; Panda et al., 2001).

A bioceramic and its appliances must, from a mechanical perspective, meet the following requirements: similar to natural tissue's Young's modulus, toughness, and stress-strain connection; high fracture hardness, low pressurized crack growth index, outstanding wear resistance, and low sensitivity to fatigue crack propagation; and all bioceramic appliances should go through a properly designed proof test to ensure their quality.

A general measuring device (Tinius Olsen, USA) was used to assess the flexural strength of composite material in the form of a rod with a 6,066 mm³ size using the traditional three-point bending method at a span of 40 mm and a cross-head speed of 0.5 mm/min (Ghosh & Sarkar, 2020). A testing procedure known as ASTM C1161 is used to evaluate the flexural strength of contemporary composite material. At room temperature, rectangular samples are loaded either in a four-point or a three-point configuration.

Flexural strength is a measurement of a beam's maximum bending strength. The specimen is only minimally subjected to the maximum stress in the three-point test configuration. Because of this, three-point flexural strengths are typically substantially higher than four-point flexural strengths. Four-point flexure testing is the preferred and suggested method for this test as well as for the majority of characterization conclusions (ASTM C1161 Ceramic Flexural Testing, 2022). The calculated flexural strength follows the formula:

$$\sigma_{\text{Flexural}} = \frac{3FL}{2BD^2} \tag{5.6}$$

5.7.1 Criteria of Alumina and Zirconia

Bioceramics have been designed specifically for use as medical and dental implants. For majority of the time, they are utilized to replace the body's hard tissues, including the bone and teeth. Alumina, zirconia, and the calcium phosphate substance hydroxyapatite are three common types of bioceramics (*Bioceramics*, 2022). With the chemical formula Al_2O_3, aluminum oxide is a combination of aluminum and oxygen. It is known as aluminum (III) oxide and is the most prevalent of numerous types of aluminum oxides. According to certain forms or uses, it may alternatively be referred to as aloxide, aloxite, or alundum in addition to the name alumina. On another, known as zirconia, this zirconium oxide is white and crystalline. The mineral baddeleyite is the most prevalent naturally occurring form of it and has a monoclinic crystalline structure. To be used as a gemstone and a diamond substitute, cubic zirconia, cubic structured zirconia with dopants, is produced in a variety of colors. Here in Table 5.9, given a comparison of mechanical properties between zirconia (Y-TZP), bone, alumina, and alumina ceramics (aluminia ISO 6474, aluminia new ISO norm).

While Y-TZP improved the mechanical performance than alumina considering fracture toughness, tensile and flexural strengths, and particularly, alumina is stiffer and has higher compressive strength. The ISO 6474 norm is used to contradict Müller and Greener (1970) (Hench & Wilson, 1993). The new ISO standard differs from the

TABLE 5.9
Comparison of Mechanical Properties Between Zirconia, Bone, Alumina, and Alumina Ceramics (Hench & Wilson, 1993; Kohn & Ducheyne, 2006; Mueller & Greener, 1970)

Mechanical Properties	Zirconia (Y-TZP)	Bone	Aluminia	Aluminia ISO 6474	Aluminia New ISO Norm
Density [g/cm^3]	6.08	1.7–2.0	3.98	>3.90	>3.94
E-modulus [GPa]	210	3–30	380–420	–	–
Compressive strength [MPa]	2,000	130–180	4,000–5,000	–	–
Tensile strength [MPa]	650	60–160	350	–	–
Flexural strength [MPa]	900	100	400–560	>400	>450
Fracture toughness [MNm$^{-3/2}$]	>9	2–12	4–6	–	–

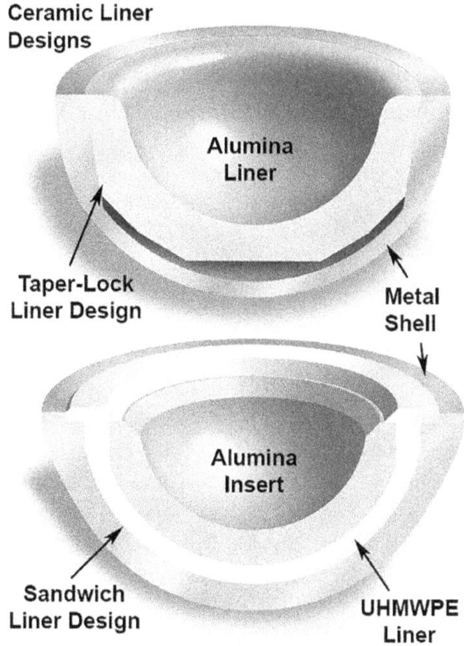

FIGURE 5.11 Two different metal-backed designs in acetabular bioceramic (alumina) components (*Chapter 2*, 2022).

previous standard in that a significantly smaller average grain size is specified along with a rise in flexural strength to over 450 MPa.

Zirconium silicate ($ZrSiO_4$), which generally is used to make zirconia, as is baddeleyite (monoclinic m-ZrO_2), and as a result, the processed material may contain trace amounts of uranium and thorium that have replaced the isovalent zirconium ion in the crystal lattice. This makes zirconia slightly radioactive. However, the addition of magnesium oxide along the alumina grain boundaries will produce a thin layer of spinel ($MgAl_2O_4$) on the surface that functions as a barrier to the movement of the grain boundaries necessary for the process of recrystallization. A ball-and-socket joint and the hip joint offer the stability required to support body weight while allowing for movement. High mechanical loads must be tolerated by the bioinert ceramics like alumina utilized in the femoral ball caps of hip endoprostheses (Figure 5.11) or zirconia. Zirconia femoral balls come in three standard diameter sizes (22, 28, and 32 mm) with laser markings to clearly identify the specific component.

5.7.2 Bioactive Glass-Ceramics Component

Ceramic glasses demonstrated bioactivity as well as a respectably high mechanical strength that only gradually deteriorated even when subjected to load-bearing circumstances in the body. Bioceramics, which are solid, impermeable, and hard materials, include bioactive glasses. They are mainly made of silicon dioxide (or silicate),

TABLE 5.10
Mechanical Properties of Bioactive Glasses, Ceramics, and Human Bones (Crovace et al., 2016; Duminis et al., 2017; Parnell, 2018)

		Mechanical Properties				
Materials	Structure	Compressive Modulus (GPa)	Compressive Strength (MPa)	Fracture Toughness (MPa m$^{1/2}$)	Bending Strength (MPa)	Vickers Hardness (MPa)
HA	Ceramic	35–120	100–150	0.8–1.2	60–120	90–140
Bioglass* 45S5	Glass	60	–	0.6	40	–
Bioglass* 52S4.6	Glass	60	–	–	40	–
Cerabone* AW	Glass ceramic	120	1,080	2	215	680
Ceravital*	Glass ceramic	100–160	500	–	100–150	–
Bioverit* I	Glass ceramic	70–90	500	1.2–2.1	140–180	–
Bioverit* II	Glass ceramic	70	450	1.2–1.8	90–140	–
Bioverit* III	Glass ceramic	45	–	0.6	60–90	–
Trabecular bone	–	0.05–0.6	1.5–7.5	0.1–0.8	10–20	40–60
Cortical bone	–	7–30	100–135	2–12	50–150	60–75

HA, hydroxyapatite; AW, apatite wollastonite.

which serves as the primary component, together with three additional fundamental elements: sodium dioxide, calcium oxide, and phosphorous. Composite materials are made of bioactive glass fibers, particularly biodegradable composite materials. The primary elements of bioactive glasses are SiO_2, CaO, and P_2O_5, which allow for the formation of a coating of hydroxyapatite (HA) on the surface of living organisms and facilitate their interaction with bones (Sharifianjazi et al., 2020; Wu et al., 2020; Zhang & Le, 2020). Some mechanical properties of bioactive glasses are given here in Table 5.10, with ceramics and human bones.

Hench et al. originally created bioactive glasses in 1969, which are a class of reactive substances capable of adhering to calcified bone tissue in a natural environment (*Bioactive Glass – An Overview | ScienceDirect Topics*, 2022). Even though bioactive glasses and ceramics have remarkable bioactive qualities, insufficient mechanical capacity and fracture toughness are their main drawbacks. Brittleness in glasses is a result of the three-dimensional network structure's lack of plastic deformation.

By altering the molar or particular volume of layers, glass ceramics can be surface-crystallized. The interior of glass or ceramic tends to shrink but is constrained by the stiff surface layer that is cooled by quenching, putting the surface in compression and the interior in tension. Rapid cooling (quenching) can have similar consequences. Making surface layers compressive in relation to the interior of a ceramic material might increase its resistance to tensile stress because tensile force cannot take control unless applied force is greater than compressive force. Ion exchange, quenching, and surface crystallization all have the potential to cause surface compression.

When ion-exchange processes occur at the contact between the biologic fluids and transplant surfaces mostly pertaining to the lingering amorphous phase, the fundamental issue with utilizing glasses including some crystalline is the accompanying loss in bioactivity.

Hardening melt-derived bioactive glass at three different temperatures produced glass ceramics—750°C, 800°C, and 850°C—with a mole percent composition of $33SiO_2$–21CaO, $32.5Na_2O_{12}P_2O_5$, and 1.5MgO (Mirza et al., 2017). As would be expected, as the sintering temperature is raised, the structure gets denser, porosity goes down, and crystallinity goes up, giving it better mechanical qualities (Mirza et al., 2017). By melting and quenching, bioglass doped with NiO was created. Between 0.4 and 1.65 mol% of NiO was present (Vyas et al., 2016). At temperatures between 700°C and 1,000°C, multistep sintering of AW and monoclinic ZrO_2 ($mZrO_2$) or 8 wt% Y_2O_3, a little bit stable tetragonal ZrO_2 ($tZrO_2$) mixtures produced AW bioglass doped with ZrO_2 (Li et al., 2014). By using $tZrO_2$-AW glass ceramics with a $tZrO_2$ content of 5%, the highest values for yield stress and fracturing resistance were attained ($5tZrO_2$-AWglass-ceramic). The fracture toughness rose from 0.92 0.02 $MPam^{1/2}$ (for pure AW glass ceramic) to 1.67 0.01 MPa•$m^{1/2}$ by adding 5 wt% of ZrO_2 (for $5tZrO_2$-AW glass ceramic).

5.7.3 Mechanical Performance of Silicate Bioceramics

Wollastonite, a silicate-based ceramic (Figure 5.12), was found to produce osteoclast hydroxyapatite (HA) coating on its surface both in vitro and in vivo, making them closely attached to living bone and demonstrating good bioactivity (De Aza et al., 2004; Ohura et al., 1991). Based on silicate bioceramics' innate capacity to heal tissue, bioactive ions are directly introduced into silicate bioceramics to give

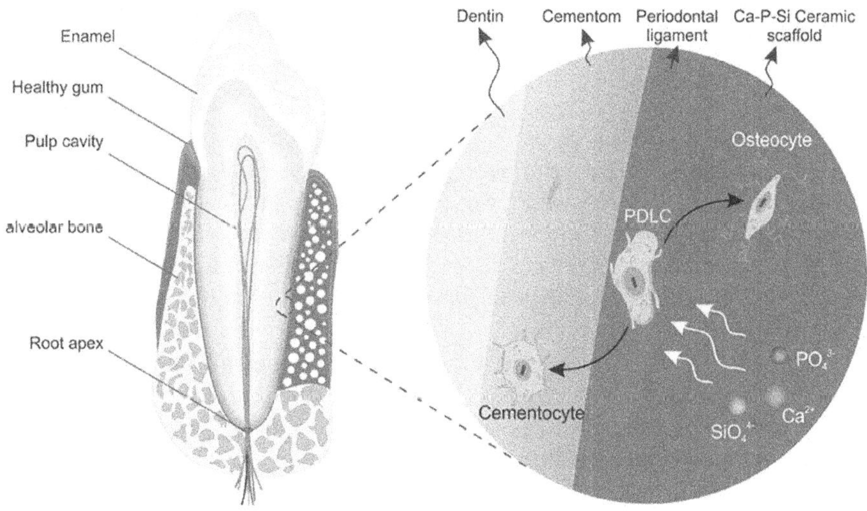

FIGURE 5.12 Silicate-containing calcium phosphate bioceramics in human body (Sepantafar et al., 2018).

them additional biological qualities. As a result, the range of applications for multifunctional silicate bioceramics is greatly increased. The past few years have seen a lot of interest in silicate bioceramics. Here in Table 5.11, given some names of silicate bioceramics and ions that release from specific silicate, ceramics composition, mechanical properties, apatite mineralization, and dissolution behaviors. The major justification for this is that silicate bioceramics may effectively promote tissue cell proliferation, differentiation, and gene expression as well as bone tissue regeneration through the production of Si-containing ionic compounds (Wu & Chang, 2013a). Many studies have discovered that silicate biomaterials and their combinations can improve tissue regeneration by releasing SiO44- ions, which can induce vascularization both in vitro and in vivo. It is interesting to note that SiO44- ions have also been shown to promote collagen synthesis and deposition.

The active mineralization sites in the joints of the human body are thought to include silicon (Si), one of the essential trace elements (Carlisle, 1970). Si has been discovered to play a direct role in the calcification of bone during the bone-growing process. Nearly 100 and 200–550 ppm quantities of silicon are found in bone and matrix proteins, respectively (Schwarz, 1973). Calcium silicate bioceramics have more varied chemical compositions than traditional phosphate-based bioceramics like HA and TCP, which allows them to have physicochemical attributes like bioactive activity, mechanical properties, and degradation mechanisms that are more precisely controlled. Maximum bending strength and fracture toughness for HA are 120 MPa and 1.2 $MPam^{1/2}$, respectively. As seen in the table, several silicate bioceramics, such as akermanite, bredigite, and hardystonite, among others, have fracture toughness that is typically higher than that of hydroxyapatite. Most silicate ceramic monoliths have bending strengths and elastic moduli that are comparable to those of human cortical bone. As shown in the table, calcium silicate ceramic materials also have a unique capacity for apatite mineralization. The ability of apatite to mineralize is described by the chemical makeup and dissolution of bioceramics. Usually, calcium silicate biomaterials with high Ca concentrations showed better apatite mineralization. The ability of calcium silicate ceramics to mineralize apatite is dramatically reduced whenever additions include Mg, Zn, and Zr trace metals. Forsterite Mg_2SiO_4 (M2S) composite materials cannot be employed for applications requiring weight bearing despite having superior mechanical qualities as calcium phosphate ceramic materials and lesser tensile stability than cortical bone. Despite having strong biocompatibility, M2S degrades slowly and has a weak capacity to generate apatite (Ni et al., 2008). To develop nanomaterials based on a bioactive approach that will help ceramics, research was concentrated in this area (Kaur et al., 2019).

A chemical bond is created between organic and inorganic phases of molecularly based organic-inorganic composite materials, either regardless of whether coupling agents are used. To create a covalent link with the inorganic phase, a bonding reagent can provide the polymer with a function. Both organic and inorganic portions are covalently bonded, giving them greater mechanical characteristics than the sum of their parts. Poly(dimethyl siloxane, or PDMS, $Si(OC_2H_6)n$) is a polymer that serves as the foundation for organic-inorganic composite biomaterials generated via sol-gel. The silica matrix and PDMS are joined via covalent bonds created

TABLE 5.11
Some Names of Silicate Bioceramics, Ions That Release from Specific Silicate, Ceramics Composition, Mechanical Properties, Apatite Mineralization, and Dissolution Behaviors

			Mechanical Behaviors of Silicate Bioceramics						
Name of Ceramics	Ions That Release from Silicate	Composition	Bending Strength (MPa)	Elastic Modulus (GPa)	Compressive Strength (MPa)	Fracture Toughness ($MPam^{1/2}$)	Apatite Mineralization	Dissolution Behavior	Reference
Tricalcium silicate	SiO_4^{4-}, Ca^{2+}	$CaSiO_3$	93.4	36.7	–	1.93	Excellent	Rapid	De Aza et al. (2001b), Wu and Chang (2013b)
Magnesium silicate	SiO_4^{4-}, Mg^{2+}	$MgSiO_3$	32	8.5	–	–	Poor	Very slow	Tavangarian and Emadi (2011b), Xu et al. (2008b)
Akermanite	SiO_4^{4-}, Ca^{2+}, Mg^{2+}	$Ca_2MgSi_2O_7$	–	–	0.53–1.13	0.63–1.72	Good	Moderate	Han et al. (2014), Wu and Chang (2006)
Bredigite	SiO_4^{4}, Ca^{2+}, Mg^{2+}	$Ca_7MgSi_4O_{16}$	156	43	0.233	1.57	Excellent	Rapid	Wu and Chang (2007)
Hardystone	SiO_4^{4-}, Ca^{2+}, Zn^{2+}	$(Sr,Ca)SiO_3$, $Ca_2ZnSi_2O_7$	136	37	–	1.37	Poor	Very slow	Jin et al. (2011)
Strontium	SiO_4^{4-}, Sr^{2+}, Zn^{2+}	$Sr_2ZnSi_2O_7$	–	–	–	–	Poor	Very slow	Ramaswamy et al. (2009b), Zreiqat et al. (2010b)
Nagelschmidtite	SiO_4^{4-}, Ca^{2+}	$Ca_7Si_2P_2O_{16}$	–	–	–	–	Excellent	Rapid	Lin et al. (2005b), Lu et al. (2012)

by PDMS-derived hybrids. These combinations are not biologically active, though, unless the network contains Ca^{2+} ions (Allo et al., 2012). Biocompatible hybrid materials with low Young's modulus can be created when flexible organic components are closely coupled with bioactive glasses comprising silica and titania at the nanoscale. Poly(glycerol sebacate) elastomer is among specially formulated polymers for regenerating flexible tissue. PGS's limited bioactivity, hydrophilicity, and structural rigidity (0.3–1.5 MPa) prevent it from being used to regenerate bone, cartilage, or tissues (Zhao et al., 2015). The Young's modulus improved from 1.75 0.15 MPa (solo PGS) to 34.52 4.73 MPa, while the tensile strength improved noticeably from 1.13 0.1 MPa for pure PGS to 4 0.84 MPa for PGS-SC (PGS-SC). After being submerged in SBF, PGS-SC becomes bioactive and is not harmful to MC3T3 osteoblasts (Zhao et al., 2015).

Bioinert materials are those that do not react or interact with biological tissue when they are introduced. Al_2O_3 and ZrO_2 are bioinert ceramics that are used for the artificial hip joint and acetabular cupping because of their exceptional mechanical strength and a high degree of durability (Rahaman et al., 2007). This is the major justification for why calcium phosphate bioceramics have been proposed to repair osseous malformations with bone, and it shows a higher biochemical and anisotropy resemblance with the mineralization of bone (Dorozhkin, 2016). Since hydroxyapatite is often nonresorbable or only very slowly biodegradable and biocompatible, unless it is manufactured in the form of nanoparticles, its mechanical strength is gaining significance for bone replacement. It also does not tend to weaken with time (Chen et al., 2012). According to sample collecting data, materials with the strongest HA content (75.96 wt%) have the finest four-point bending properties, while those with the weakest HA content (3.28 wt%) have the least (Hsu et al., 2009). Despite being the most widely utilized osteoconductive material for bone tissue creation, hydroxyapatite has a lower compressive modulus and a higher fracture toughness than a human cortical skeleton. Premature failures and structural difficulties between implants and native tissue may result from these variances.

5.8 CONCLUSIONS AND FUTURE DIRECTIONS

Over the past 30–40 years, significant advancements have been made in the fields of bioactive ceramics, hydroxyapatite, glasses, and glass ceramics. Currently, bioceramics are applied in several ways throughout the human body. They can be characterized by either "bioinert" or "bioactive" bioceramics depending on the kind of bioceramics utilized and how they engage with the living tissues, both biomaterials can serve a variety of purposes inside a body system (Best et al., 2008). There is still room for significant progress to be achieved in the administration of endodontic sealers, bone regeneration, and bonded osteogenesis due to the introduction of novel doping calcium silicate hydroxyapatite bioceramics. It has begun to be studied how the Ca-to-P atomic ratio, nano-CaP, biphasic and triphasic formulations, composites, and functionally graded materials affect the structure, properties, and applications of the various phases of HAP. The biological system still requires more understanding. The precise mechanism by which bone minerals and tissue are joined together is still in question. Although there is no discernible loss of structural rigidity when zirconia is employed as bioceramics, radiation and deterioration are two potential downsides. The deterioration of the zirconia pellets' surface brought on by phase formation under stress seems to be a problem. However, a lot of bioceramics are utilized in medical science due to their exceptional mending capabilities. Improvements in bioactive ceramics' mechanical behavior, gene activation, mechanical stability, the capacity to distribute biological agents, the development of smart materials that combine sensing and excellent biocompatibility, and the improvement of biomaterials composites are some of the main challenges in future advances that are listed with recent developments. It is necessary to overcome several obstacles first. We will be able to create better bioceramics in the future if we can fully comprehend how certain ions, tissues, and bones react with bioceramics and the signals they trigger.

REFERENCES

Allo, B. A., Costa, D. O., Dixon, S. J., Mequanint, K., & Rizkalla, A. S. (2012). Bioactive and biodegradable nanocomposites and hybrid biomaterials for bone regeneration. *Journal of Functional Biomaterials*, *3*(2), 432–463. https://doi.org/10.3390/jfb3020432

Andersson, M., & Odén, A. (1993). A new all-ceramic crown. A dense-sintered, high-purity alumina coping with porcelain. *Acta Odontologica Scandinavica*, *51*(1), 59–64. https://doi.org/10.3109/00016359309041149

Anita Lett, J., Sagadevan, S., Fatimah, I., Hoque, M. E., Lokanathan, Y., Léonard, E., Alshahateet, S. F., Schirhagl, R., & Oh, W. C. (2021). Recent advances in natural polymer-based hydroxyapatite scaffolds: Properties and applications. *European Polymer Journal*, *148*, 110360. https://doi.org/10.1016/j.eurpolymj.2021.110360

ASTM C1161 Ceramic Flexural Testing. (2022, July 20). *ADMET*. https://www.admet.com/testing-applications/testing-standards/astm-c1161-ceramic-flexural-testing/

Bahir, M. M., Khairnar, R. S., & Mahabole, M. P. (2020). Electrical properties of newly calcified tissues on the surface of silver ion administrated hydroxyapatite scaffolds. *Journal of Biomaterials and Nanobiotechnology*, *11*(2), 83–100. https://doi.org/10.4236/jbnb.2020.112006

Bang, L. T., Ramesh, S., Purbolaksono, J., Long, B. D., Chandran, H., Ramesh, S., & Othman, R. (2015). Development of a bone substitute material based on alpha-tricalcium phosphate scaffold coated with carbonate apatite/poly-epsilon-caprolactone. *Biomedical Materials*, *10*(4), 045011. https://doi.org/10.1088/1748-6041/10/4/045011

Banijamali, S., & Ebadzadeh, T. (2016). Glass-forming ability, sinter-crystallization behavior and microwave dielectric properties of MgO-B2O3-(Al_2O_3)-SiO2 glass-ceramics. *Journal of Non-Crystalline Solids*, *441*, 34–41. https://doi.org/10.1016/j.jnoncrysol.2016.03.014

Banijamali, S., Yekta, B. E., & Aghaei, A. R. (2012). The effect of ionic and metallic silver on the crystalline phases developed in CaO-Al_2O_3-TiO2-P2O5 glasses. *Journal of Non-Crystalline Solids*, *358*(2), 303–309. https://doi.org/10.1016/j.jnoncrysol.2011.09.026

Barakat, N. A. M., Khil, M. S., Omran, A. M., Sheikh, F. A., & Kim, H. Y. (2009). Extraction of pure natural hydroxyapatite from the bovine bones bio waste by three different methods. *Journal of Materials Processing Technology*, *209*(7), 3408–3415. https://doi.org/10.1016/j.jmatprotec.2008.07.040

Benezra, V., Spector, M., & Hobbs, L. W. (1995). Characterization of mineral deposits on plasma-sprayed HA-coated Ti-6Al-4V. *MRS Online Proceedings Library (OPL)*, *414*, 165. https://doi.org/10.1557/PROC-414-165

Best, S., Porter, A., Thian, E. S., & Huang, J. (2008). Bioceramics: Past, present and for the future. *Journal of the European Ceramic Society*, *28*, 1319–1327. https://doi.org/10.1016/j.jeurceramsoc.2007.12.001

Betts, F., Blumenthal, N. C., & Posner, A. S. (1981). Bone mineralization. *Journal of Crystal Growth*, *53*(1), 63–73. https://doi.org/10.1016/0022-0248(81)90056-7

Bioactive Glass – An Overview | ScienceDirect Topics. (2022, July 20). https://www.sciencedirect.com/topics/pharmacology-toxicology-and-pharmaceutical-science/bioactive-glass

Bioceramics. (2022, July 20). Science learning hub. https://www.sciencelearn.org.nz/resources/1776-bioceramics

Bioceramics – An Overview | ScienceDirect Topics. (2022, July 26). https://www.sciencedirect.com/topics/medicine-and-dentistry/bioceramics

Bioceramics | Britannica. (2022, July 26). https://www.britannica.com/technology/bioceramics

Bioceramics, What Do We Talk About? – Baikowski(r). (2022, July 26). https://www.baikowski.com/en/page-bioceramics/

Biswas, M. C., Jony, B., Nandy, P. K., Chowdhury, R. A., Halder, S., Kumar, D., Ramakrishna, S., Hassan, M., Ahsan, M. A., Hoque, M. E., & Imam, M. A. (2022). Recent advancement of biopolymers and their potential biomedical applications. *Journal of Polymers and the Environment*, *30*(1), 51–74. https://doi.org/10.1007/s10924-021-02199-y

Bongio, M., van den Beucken, J. J. J. P., Leeuwenburgh, S. C. G., & Jansen, J. A. (2010). Development of bone substitute materials: From 'biocompatible' to 'instructive'. *Journal of Materials Chemistry*, *20*(40), 8747. https://doi.org/10.1039/c0jm00795a

Bowen, C., Turner, I., & Dent, A. (2010). Electrically active bioceramics: A review of interfacial responses. *Annals of Biomedical Engineering*, *38*, 2079–2092. https://doi.org/10.1007/s10439-010-9977-6

Cales, B., Stefani, Y., & Lilley, E. (1994). Long-term in vivo and in vivo aging of a zirconia ceramic used in orthopaedy. *Journal of Biomedical Materials Research*, *28*(5), 619–624. https://doi.org/10.1002/jbm.820280512

Carlisle, E. M. (1970). Silicon: A possible factor in bone calcification. *Science*. https://doi.org/10.1126/science.167.3916.279

Chapter 2. (2022, July 20). https://biomed.drexel.edu/orthoceramics/Encyclopedia/chapter2.html

Chavan, P. N., Bahir, M. M., Mene, R. U., Mahabole, M. P., & Khairnar, R. S. (2010). Study of nanobiomaterial hydroxyapatite in simulated body fluid: Formation and growth of apatite. *Materials Science and Engineering: B*, *168*(1), 224–230. https://doi.org/10.1016/j.mseb.2009.11.012

Chen, F., Zhu, Y., Wu, J., Huang, P., & Cui, D. (2012). Nanostructured calcium phosphates: Preparation and their application in biomedicine. *Nano Biomedicine and Engineering*, *4*(1), 41–49. https://doi.org/10.5101/nbe.v4i1.p41-49

Chen, Q.-Z., Harding, S. E., Ali, N. N., Lyon, A. R., & Boccaccini, A. R. (2008). Biomaterials in cardiac tissue engineering: Ten years of research survey. *Materials Science and Engineering: R: Reports*, *59*(1), 1–37. https://doi.org/10.1016/j.mser.2007.08.001

Cho, M.-Y., Lee, D.-W., Kim, I.-S., Kim, W.-J., Koo, S.-M., Lee, D., Kim, Y.-H., & Oh, J.-M. (2019). Evaluation of structural and mechanical properties of aerosol-deposited bioceramic films for orthodontic brackets. *Ceramics International*, *45*(6), 6702–6711. https://doi.org/10.1016/j.ceramint.2018.12.159

Crovace, M. C., Souza, M. T., Chinaglia, C. R., Peitl, O., & Zanotto, E. D. (2016). Biosilicate(r) – A multipurpose, highly bioactive glass-ceramic. In vitro, in vivo and clinical trials. *Journal of Non-crystalline Solids*, *432*, 90–110. https://doi.org/10.1016/j.jnoncrysol.2015.03.022

Danilchenko, S. N., Kukharenko, O. G., Moseke, C., Protsenko, I. Yu., Sukhodub, L. F., & Sulkio-Cleff, B. (2002). Determination of the bone mineral crystallite size and lattice strain from diffraction line broadening. *Crystal Research and Technology*, *37*(11), 1234–1240. https://doi.org/10.1002/1521-4079(200211)37:11<1234::AID-CRAT1234>3.0.CO;2–X

Day, R. M., Boccaccini, A. R., Shurey, S., Roether, J. A., Forbes, A., Hench, L. L., & Gabe, S. M. (2004). Assessment of polyglycolic acid mesh and bioactive glass for soft-tissue engineering scaffolds. *Biomaterials*, *25*(27), 5857–5866. https://doi.org/10.1016/j.biomaterials.2004.01.043

De Aza, P. N., Fernández-Pradas, J. M., & Serra, P. (2004). In vitro bioactivity of laser ablation pseudowollastonite coating. *Biomaterials*, *25*(11), 1983–1990. https://doi.org/10.1016/j.biomaterials.2003.08.036

De Aza, P. N., Luklinska, Z. B., Anseau, M. R., Guitian, F., & De Aza, S. (2001). Transmission electron microscopy of the interface between bone and pseudowollastonite implant. *Journal of Microscopy*, *201*(1), 33–43. https://doi.org/10.1046/j.1365-2818.2001.00779.x

Derbyshire, B., Fisher, J., Dowson, D., Hardaker, C., & Brummitt, K. (1994). Comparative study of the wear of UHMWPE with zirconia ceramic and stainless steel femoral heads in artificial hip joints. *Medical Engineering & Physics*, *16*(3), 229–236. https://doi.org/10.1016/1350-4533(94)90042-6

Dorozhkin, S. V. (2016). Calcium orthophosphates (CaPO4): Occurrence and properties. *Progress in Biomaterials*, *5*(1), 9–70. https://doi.org/10.1007/s40204-015-0045-z

Ducheyne, P., Grainger, D. W., Healy, K. E., Hutmacher, D. W., & Kirkpatrick, C. J. (2017). *Comprehensive Biomaterials II*. Elsevier.

Ducheyne, P., & Qiu, Q. (1999). Bioactive ceramics: The effect of surface reactivity on bone formation and bone cell function. *Biomaterials*, *20*(23), 2287–2303. https://doi.org/10.1016/S0142-9612(99)00181-7

Ducheyne, P., Radin, S., Heughebaert, M., & Heughebaert, J. C. (1990). Calcium phosphate ceramic coatings on porous titanium: Effect of structure and composition on electrophoretic deposition, vacuum sintering and in vitro dissolution. *Biomaterials*, *11*(4), 244–254. https://doi.org/10.1016/0142-9612(90)90005-b

Duminis, T., Shahid, S., & Hill, R. G. (2017). Apatite glass-ceramics: A review. *Frontiers in Materials*, *3*. https://www.frontiersin.org/articles/10.3389/fmats.2016.00059

Fábián, M., Kovács, Zs., Lábár, J. L., Sulyok, A., Horváth, Z. E., Székács, I., & Kovács Kis, V. (2020). Network structure and thermal properties of bioactive (SiO2-CaO-Na2O-P2O5) glasses. *Journal of Materials Science*, *55*(6), 2303–2320. https://doi.org/10.1007/s10853-019-04206-z

Fanderlik, I. (2013). *Silica Glass and Its Application*. Elsevier.

Fratzl, P., Gupta, H. S., Paschalis, E. P., & Roschger, P. (2004). Structure and mechanical quality of the collagen-mineral nano-composite in bone. *Journal of Materials Chemistry*, *14*(14), 2115–2123. https://doi.org/10.1039/B402005G

Fulchiron, R., Belyamani, I., Otaigbe, J. U., & Bounor-Legaré, V. (2015). A simple method for tuning the glass transition process in inorganic phosphate glasses. *Scientific Reports*, *5*(1), 8369. https://doi.org/10.1038/srep08369

Ghosh, R., & Sarkar, R. (2020). Comparative analysis of novel calcium phosphate based machinable bioceramic composites. *Transactions of the Indian Ceramic Society*, *79*(3), 131–138. https://doi.org/10.1080/0371750X.2020.1773931

Golieskardi, M., Satgunam, M., Ragurajan, D., Hoque, M. E., & Ng, A. M. H. (2020). Microstructural, tribological, and degradation properties of Al_2O_3- and CeO2-doped 3 mol.% yttria-stabilized zirconia bioceramic for biomedical applications. *Journal of Materials Engineering and Performance*, *29*(5), 2890–2897. https://doi.org/10.1007/s11665-020-04829-3

Gomes, D. S., Santos, A. M. C., Neves, G. A., & Menezes, R. R. (2019). A brief review on hydroxyapatite production and use in biomedicine. *Cerâmica*, *65*(374), 282–302. https://doi.org/10.1590/0366-69132019653742706

Gross, K. A., & Rodríguez-Lorenzo, L. M. (2004). Sintered hydroxyfluorapatites. Part I: Sintering ability of precipitated solid solution powders. *Biomaterials*, *25*(7), 1375–1384. https://doi.org/10.1016/S0142-9612(03)00565-9

Grote, J. J. (1984). *Biomaterials in Otology*. Springer, Dordrecht. https://doi.org/10.1007/978-94-009-6756-4; https://digital.bibliotecaorl.org.br/handle/forl/342

Hammer, W. B., Topazian, R. G., Mc Kinney, R. V., & Hulbert, S. F. (1973). Alveolar ridge augmentation with ceramics. *Journal of Dental Research*, *52*(2), 356–361. https://doi.org/10.1177/00220345730520022801

Han, Z., Feng, P., Gao, C., Shen, Y., Shuai, C., & Peng, S. (2014). Microstructure, mechanical properties and in vitro bioactivity of akermanite scaffolds fabricated by laser sintering. *Bio-medical Materials and Engineering*, *24*(6), 2073–2080. https://doi.org/10.3233/BME-141017

Heimann, R. B. (2002). *Materials Science of Crystalline Bioceramics: A Review of Basic Properties and Applications*, *1*, 24.

Hench, L. L., & Wilson, J. (1993). *An Introduction to Bioceramics*. World Scientific.

Hidouri, M., Boughzala, K., Lecompte, J. P., & Bouzouita, K. (2009). Sintering and mechanical properties of magnesium-containing fluorapatite. *Comptes Rendus Physique*, *10*(2), 242–248. https://doi.org/10.1016/j.crhy.2009.04.001

Hidouri, M., Dorozhkin, S. V., & Albeladi, N. (2019). Thermal behavior, sintering and mechanical characterization of multiple ion-substituted hydroxyapatite bioceramics. *Journal of Inorganic and Organometallic Polymers and Materials*, *29*(1), 87–100. https://doi.org/10.1007/s10904-018-0969-6

Hoque, M. E., Rayhan, A. M., & Shaily, S. I. (2021). Natural fiber-based green composites: Processing, properties and biomedical applications. *Applied Science and Engineering Progress*, *14*(4), 689–718. https://doi.org/10.14416/j.asep.2021.09.005

Hsu, Y. H., Turner, I. G., & Miles, A. W. (2009). Mechanical properties of three different compositions of calcium phosphate bioceramic following immersion in Ringer's solution and distilled water. *Journal of Materials Science: Materials in Medicine*, *20*(12), 2367. https://doi.org/10.1007/s10856-009-3809-y

Ichikawa, Y., Akagawa, Y., Nikai, H., & Tsuru, H. (1992). Tissue compatibility and stability of a new zirconia ceramic in vivo. *The Journal of Prosthetic Dentistry*, *68*(2), 322–326. https://doi.org/10.1016/0022-3913(92)90338-B

Jaber, H. L., Hammood, A. S., & Parvin, N. (2018). Synthesis and characterization of hydroxyapatite powder from natural Camelus bone. *Journal of the Australian Ceramic Society*, *54*(1), 1–10. https://doi.org/10.1007/s41779-017-0120-0

Janus, A. M., Faryna, M., Haberko, K., Rakowska, A., & Panz, T. (2008). Chemical and microstructural characterization of natural hydroxyapatite derived from pig bones. *Microchimica Acta*, *161*(3), 349–353. https://doi.org/10.1007/s00604-007-0864-2

Jin, X., Chang, J., Zhai, W., & Lin, K. (2011). Preparation and characterization of clinoenstatite bioceramics. *Journal of the American Ceramic Society*, *94*(1), 66–70. https://doi.org/10.1111/j.1551-2916.2010.04032.x

Johnsson, M. S., & Nancollas, G. H. (1992). The role of brushite and octacalcium phosphate in apatite formation. *Critical Reviews in Oral Biology and Medicine: An Official Publication of the American Association of Oral Biologists*, *3*(1–2), 61–82. https://doi.org/10.1177/10454411920030010601

Jones, D. R. H., & Ashby, M. F. (2012). *Engineering Materials 2: An Introduction to Microstructures and Processing*. Butterworth-Heinemann.

Kassinger, R. (2003). *Ceramics: From Magic Pots to Man-Made Bones*. Twenty-First Century Books.

Kaur, G., Kumar, V., Baino, F., Mauro, J. C., Pickrell, G., Evans, I., & Bretcanu, O. (2019). Mechanical properties of bioactive glasses, ceramics, glass-ceramics and composites: State-of-the-art review and future challenges. *Materials Science and Engineering: C*, *104*, 109895. https://doi.org/10.1016/j.msec.2019.109895

Kim, S.-S., Ahn, K.-M., Park, M. S., Lee, J.-H., Choi, C. Y., & Kim, B.-S. (2007). A poly(lactide-co-glycolide)/hydroxyapatite composite scaffold with enhanced osteoconductivity. *Journal of Biomedical Materials Research Part A*, *80A*(1), 206–215. https://doi.org/10.1002/jbm.a.30836

Kinnari, T. J., Esteban, J., Gomez-Barrena, E., Zamora, N., Fernandez-Roblas, R., Nieto, A., Doadrio, J. C., López-Noriega, A., Ruiz-Hernández, E., Arcos, D., & Vallet-Regí, M. (2009). Bacterial adherence to SiO2-based multifunctional bioceramics. *Journal of Biomedical Materials Research. Part A*, *89*(1), 215–223. https://doi.org/10.1002/jbm.a.31943

Kittipibul, P., & Godfrey, K. (1995). In vitro shearing force testing of the Australian zirconia-based ceramic Begg bracket. *American Journal of Orthodontics and Dentofacial Orthopedics*, *108*(3), 308–315. https://doi.org/10.1016/s0889-5406(95)70026-9

Kohn, D. H., & Ducheyne, P. (2006). Materials for bone and joint replacement. In *Materials Science and Technology*. John Wiley & Sons, Ltd. https://doi.org/10.1002/9783527603978.mst0161

Kumar, R., Kumar, P. S., Manju, R., & Hoque, M. E. (2021). 19: Ethical issues of biocomposites. In M. E. Hoque, A. Sharif, & M. Jawaid (Eds.), *Green Biocomposites for Biomedical Engineering* (pp. 409–420). Woodhead Publishing. https://doi.org/10.1016/B978-0-12-821553-1.00019-3

Lefebvre, L., Chevalier, J., Gremillard, L., Zenati, R., Thollet, G., Bernache-Assolant, D., & Govin, A. (2007). Structural transformations of bioactive glass 45S5 with thermal treatments. *Acta Materialia*, *55*(10), 3305–3313. https://doi.org/10.1016/j.actamat.2007.01.029

LeGeros, R. Z. (1991). Calcium phosphates in oral biology and medicine. *Monographs in Oral Science*, *15*, 1–201.

Legeros, R. Z., Brown, P. W., Glimcher, M. J., Constantz, B., Khan, S. R., Roy, D. M., Smith, D. K., Yoshimura, M., Suda, H., Nancollas, G. H., Zhang, J., Hackett, R. L., Tomazic, B. B., Hanlon, J. G., Spector, M., Chow, L. C., Takagi, S., Ishikawa, K., Markovic, M., … Adair, J. H. (2017). *Hydroxyapatite and Related Materials*. CRC Press. https://doi.org/10.1201/9780203751367

Li, H. C., Wang, D. G., Meng, X. G., & Chen, C. Z. (2014). Effect of ZrO2 additions on the crystallization, mechanical and biological properties of MgO-CaO-SiO2-P2O5-CaF2 bioactive glass-ceramics. *Colloids and Surfaces B: Biointerfaces*, *118*, 226–233. https://doi.org/10.1016/j.colsurfb.2014.03.055

Li, J., & Hastings, G. W. (1998). Oxide bioceramics: Inert ceramic materials in medicine and dentistry. In J. Black & G. Hastings (Eds.), *Handbook of Biomaterial Properties* (pp. 340–354). Springer US. https://doi.org/10.1007/978-1-4615-5801-9_21

Li, J., & Hastings, G. W. (2016). Chapter 5: Oxide bioceramics: Inert ceramic materials in medicine and dentistry. In W. Murphy, J. Black, & G. Hastings (Eds.), *Handbook of Biomaterial Properties* (pp. 339–352). Springer. https://doi.org/10.1007/978-1-4939-3305-1_21

Library. (2006). *MRS Bulletin*, *31*(1), 58–60. https://doi.org/10.1557/mrs2006.17

Lin, K., Zhai, W., Ni, S., Chang, J., Zeng, Y., & Qian, W. (2005). Study of the mechanical property and in vitro biocompatibility of CaSiO3 ceramics. *Ceramics International*, *31*(2), 323–326. https://doi.org/10.1016/j.ceramint.2004.05.023

Liu, Q., Huang, S., Matinlinna, J. P., Chen, Z., & Pan, H. (2013). Insight into biological apatite: Physiochemical properties and preparation approaches. *BioMed Research International*, *2013*, e929748. https://doi.org/10.1155/2013/929748

Liu, Y., Fischer, T. E., & Dent, A. (2003). Comparison of HVOF and plasma-sprayed alumina/titania coatings-Microstructure, mechanical properties and abrasion behavior. *Surface and Coatings Technology*, *167*(1), 68–76. https://doi.org/10.1016/S0257-8972(02)00890-3

Lu, H. H., El-Amin, S. F., Scott, K. D., & Laurencin, C. T. (2003). Three-dimensional, bioactive, biodegradable, polymer-bioactive glass composite scaffolds with improved mechanical properties support collagen synthesis and mineralization of human osteoblast-like cells in vitro. *Journal of Biomedical Materials Research Part A*, *64A*(3), 465–474. https://doi.org/10.1002/jbm.a.10399

Lu, W., Duan, W., Guo, Y., & Ning, C. (2012). Mechanical properties and in vitro bioactivity of Ca5(PO4)2SiO4 bioceramic. *Journal of Biomaterials Applications*, *26*(6), 637–650. https://doi.org/10.1177/0885328210383599

Mardziah, C. M., Ramesh, S., Abdul Wahid, M. F., Chandran, H., Sidhu, A., Krishnasamy, S., & Purbolaksono, J. (2020). Effect of zinc ions on the structural characteristics of hydroxyapatite bioceramics. *Ceramics International*, *46*(9), 13945–13952. https://doi.org/10.1016/j.ceramint.2020.02.192

Märten, A., Fratzl, P., Paris, O., & Zaslansky, P. (2010). On the mineral in collagen of human crown dentine. *Biomaterials*, *31*(20), 5479–5490. https://doi.org/10.1016/j.biomaterials.2010.03.030

Minardi, S., Corradetti, B., Taraballi, F., Sandri, M., Van Eps, J., Cabrera, F. J., Weiner, B. K., Tampieri, A., & Tasciotti, E. (2015). Evaluation of the osteoinductive potential of a bio-inspired scaffold mimicking the osteogenic niche for bone augmentation. *Biomaterials*, *62*, 128–137. https://doi.org/10.1016/j.biomaterials.2015.05.011

Mirza, A., Riaz, M., Zia, R., Hussain, T., & Bashir, F. (2017). Effect of temperature on mechanical and bioactive properties of glass-ceramics. *Journal of Alloys and Compounds*, *726*, 348–351. https://doi.org/10.1016/j.jallcom.2017.07.285

Mohd Pu'ad, N. A. S., Koshy, P., Abdullah, H. Z., Idris, M. I., & Lee, T. C. (2019). Syntheses of hydroxyapatite from natural sources. *Heliyon*, *5*(5), e01588. https://doi.org/10.1016/j.heliyon.2019.e01588

Mueller, H. J., & Greener, E. H. (1970). Polarization studies of surgical materials in ringer's solution. *Journal of Biomedical Materials Research*, *4*(1), 29–41. https://doi.org/10.1002/jbm.820040105

Murphy, W., Black, J., & Hastings, G. (2016). *Handbook of Biomaterial Properties*. Springer.

Muster, D. (1992). *Biomaterials: Hard Tissue Repair and Replacement*. North-Holland.

Ni, S., Chang, J., & Chou, L. (2008). In vitro studies of novel CaO-SiO2-MgO system composite bioceramics. *Journal of Materials Science: Materials in Medicine*, *19*(1), 359–367. https://doi.org/10.1007/s10856-007-3186-3

Nsar, S., Hassine, A., & Bouzouita, K. (2013). *Sintering and Mechanical Properties of Magnesium and Fluorine Co-Substituted Hydroxyapatites*. *2013*. https://doi.org/10.4236/jbnb.2013.41001

Nurmanta, D. A., Nuha, S. A., Kamasi, D. D., & Santjojo, D. J. D. H. (2019). Electrical impedance measurement of hydroxyapatite bovine scaffold based on thickness. *AIP Conference Proceedings*, *2202*(1), 020025. https://doi.org/10.1063/1.5141638

Nyman, J. S., Reyes, M., & Wang, X. (2005). Effect of ultrastructural changes on the toughness of bone. *Micron*, *36*(7), 566–582. https://doi.org/10.1016/j.micron.2005.07.004

Ohura, K., Nakamura, T., Yamamuro, T., Kokubo, T., Ebisawa, Y., Kotoura, Y., & Oka, M. (1991). Bone-bonding ability of P2O5-Free CaO · SiO2 glasses. *Journal of Biomedical Materials Research*, *25*(3), 357–365. https://doi.org/10.1002/jbm.820250307

Ojovan, M. I. (2009). Viscosity and glass transition in amorphous oxides. *Advances in Condensed Matter Physics*, *2008*, e817829. https://doi.org/10.1155/2008/817829

Ooi, C. Y., Hamdi, M., & Ramesh, S. (2007). Properties of hydroxyapatite produced by annealing of bovine bone. *Ceramics International*, *33*(7), 1171–1177. https://doi.org/10.1016/j.ceramint.2006.04.001

Osborn, J. F., & Newesely, H. (1980). The material science of calcium phosphate ceramics. *Biomaterials*, *1*(2), 108–111. https://doi.org/10.1016/0142-9612(80)90009-5

Padmanabhan, V. P., Kulandaivelu, R., Venkatachalam, V., Veerla, S. C., Mohammad, F., Al-Lohedan, H. A., Oh, W. C., Schirhagl, R., Obulapuram, P. K., Hoque, M. E., & Sagadevan, S. (2020). Influence of sonication on the physicochemical and biological characteristics of selenium-substituted hydroxyapatites. *New Journal of Chemistry*, *44*(40), 17453–17464. https://doi.org/10.1039/D0NJ03771K

Padmanabhan, V. P., Sankara Narayanan, N. T. S., Sagadevan, S., Hoque, M. E., & Kulandaivelu, R. (2019). Advanced lithium substituted hydroxyapatite nanoparticles for antimicrobial and hemolytic studies. *New Journal of Chemistry*, *43*(47), 18484–18494. https://doi.org/10.1039/C9NJ03735G

Panda, R. N., Hsieh, M.-F., Chung, R.-J., & Chin, T.-S. (2001). X-ray diffractometry and X-ray photoelectron spectroscopy investigations of nanocrytalline hydroxyapatite synthesized by a hydroxide gel technique. *Japanese Journal of Applied Physics*, *40*(8R), 5030. https://doi.org/10.1143/JJAP.40.5030

Park, J. (2009). *Bioceramics: Properties, Characterizations, and Applications.* Springer Science & Business Media.

Parnell, H. (2018). *The use of CES EduPack at all levels of Higher Education*, 15.

Pezzotti, G. (2014). Bioceramics for hip joints: The physical chemistry viewpoint. *Materials*, *7*(6), 4367–4410. https://doi.org/10.3390/ma7064367

Polack, F. M., & Heimke, G. (1980). Ceramic Keratoprostheses. *Ophthalmology*, *87*(7), 693–698. https://doi.org/10.1016/S0161-6420(80)35179-8

Portilla, L., & Halik, M. (2014). Smoothly tunable surface properties of aluminum oxide core-shell nanoparticles by a mixed-ligand approach. *ACS Applied Materials & Interfaces*, *6*(8), 5977–5982. https://doi.org/10.1021/am501155r

Rahaman, M. N., Yao, A., Bal, B. S., Garino, J. P., & Ries, M. D. (2007). Ceramics for prosthetic hip and knee joint replacement. *Journal of the American Ceramic Society*, *90*(7), 1965–1988. https://doi.org/10.1111/j.1551-2916.2007.01725.x

Ramaswamy, Y., Wu, C., Dunstan, C. R., Hewson, B., Eindorf, T., Anderson, G. I., & Zreiqat, H. (2009). Sphene ceramics for orthopedic coating applications: An in vitro and in vivo study. *Acta Biomaterialia*, *5*(8), 3192–3204. https://doi.org/10.1016/j.actbio.2009.04.028

Ramesh, S., Loo, Z. Z., Tan, C. Y., Chew, W. J. K., Ching, Y. C., Tarlochan, F., Chandran, H., Krishnasamy, S., Bang, L. T., & Sarhan, A. A. D. (2018). Characterization of biogenic hydroxyapatite derived from animal bones for biomedical applications. *Ceramics International*, *44*(9), 10525–10530. https://doi.org/10.1016/j.ceramint.2018.03.072

Ratner, B. D., Hoffman, A., Schoen, F. J., & Lemons, J. E. (2013). *Biomaterials Science: An Introduction to Materials: Third Edition* (p. 1555). https://doi.org/10.1016/B978-0-08-087780-8.00148-0

Resan, K. (2018). *Bio Ceramics MSc Lect. 4*. https://doi.org/10.13140/RG.2.2.18713.88163

Saleh, S. H., Hashim, I. H., Aadim, K., & Kareem, F. K. (2018). Improvement in mechanical properties and wear resistance of the nano composite PMMA / hydroxyapatite-zirconia (HA-ZrO2) by atmospheric plasma. *Eurasian Journal of Analytical Chemistry*. https://doi.org/10.29333/EJAC/99537

Santoso, I., Riihimäki, M., Sibarani, D., Taskinen, P., Hupa, L., Paek, M.-K., & Lindberg, D. (2022). Impact of recently discovered sodium calcium silicate solutions on the phase diagrams of relevance for glass-ceramics in the Na2O-CaO-SiO2 system. *Journal of the European Ceramic Society*, *42*(5), 2449–2463. https://doi.org/10.1016/j.jeurceramsoc.2022.01.010

Schwarz, K. (1973). A bound form of silicon in glycosaminoglycans and polyuronides. *Proceedings of the National Academy of Sciences of the United States of America*, *70*(5), 1608–1612.

Senamaud, N., Bernache-Assollant, D., Champion, E., Heughebaert, M., & Rey, C. (1997). Calcination and sintering of hydroxyfluorapatite powders. *Solid State Ionics*, *101–103*, 1357–1362. https://doi.org/10.1016/S0167-2738(97)00242-7

Sepantafar, M., Mohammadi, H., Maheronnaghsh, R., Tayebi, L., & Baharvand, H. (2018). Single phased silicate-containing calcium phosphate bioceramics: Promising biomaterials for periodontal repair. *Ceramics International*, *44*(10), 11003–11012. https://doi.org/10.1016/j.ceramint.2018.03.050

Shah, R., Sinanan, A. C. M., Knowles, J. C., Hunt, N. P., & Lewis, M. P. (2005). Craniofacial muscle engineering using a 3-dimensional phosphate glass fibre construct. *Biomaterials*, *26*(13), 1497–1505. https://doi.org/10.1016/j.biomaterials.2004.04.049

Sharifianjazi, F., Pakseresht, A. H., Asl, M. S., Esmaeilkhanian, A., Khoramabadi, H. N., Jang, H. W., & Shokouhimehr, M. (2020). Hydroxyapatite consolidated by zirconia: Applications for dental implant. *Journal of Composites and Compounds*, *2*(2), 26–34. https://doi.org/10.29252/jcc.2.1.4

Sheikh, Z., Sima, C., & Glogauer, M. (2015). Bone replacement materials and techniques used for achieving vertical alveolar bone augmentation. *Materials*, *8*(6), 2953–2993. https://doi.org/10.3390/ma8062953

Shimizu, K., Oka, M., Kumar, P., Kotoura, Y., Yamamuro, T., Makinouchi, K., & Nakamura, T. (1993). Time-dependent changes in the mechanical properties of zirconia ceramic. *Journal of Biomedical Materials Research*, *27*(6), 729–734. https://doi.org/10.1002/jbm.820270605

Sintering Temperature-An overview | ScienceDirect Topics. (2022, July 25). https://www.sciencedirect.com/topics/materials-science/sintering-temperature

Sopyan, I., Mel, M., Ramesh, S., & Khalid, K. A. (2007). Porous hydroxyapatite for artificial bone applications. *Science and Technology of Advanced Materials*, *8*(1–2), 116. https://doi.org/10.1016/j.stam.2006.11.017

Sun, R.-X., Lv, Y., Niu, Y.-R., Zhao, X.-H., Cao, D.-S., Tang, J., Sun, X.-C., & Chen, K.-Z. (2017). Physicochemical and biological properties of bovine-derived porous hydroxyapatite/collagen composite and its hydroxyapatite powders. *Ceramics International*, *43*(18), 16792–16798. https://doi.org/10.1016/j.ceramint.2017.09.075

Szcześ, A., Hołysz, L., & Chibowski, E. (2017). Synthesis of hydroxyapatite for biomedical applications. *Advances in Colloid and Interface Science*, *249*, 321–330. https://doi.org/10.1016/j.cis.2017.04.007

Tavangarian, F., & Emadi, R. (2011). Nanostructure effects on the bioactivity of forsterite bioceramic. *Materials Letters*, *65*(4), 740–743. https://doi.org/10.1016/j.matlet.2010.11.014

Thamaraiselvi, T. V., & Rajeswari, S. (2004). *Biological Evaluation of Bioceramic Materials – A Review*.

Thermal Conductivity of Biomaterials. (2022, July 22). *C-Therm Technologies Ltd.* https://ctherm.com/resources/webinars/measuring-the-thermal-conductivity-of-biomaterials/

Timchenko, P. E., Timchenko, E. V., Pisareva, E. V., Vlasov, M. Y., Red'kin, N. A., & Frolov, O. O. (2017). Spectral analysis of allogeneic hydroxyapatite powders. *Journal of Physics: Conference Series*, *784*, 012060. https://doi.org/10.1088/1742-6596/784/1/012060

Vallet-Regí, M., Navarrete, D. A., & Arcos, D. (2008). *Biomimetic Nanoceramics in Clinical Use: From Materials to Applications*. Royal Society of Chemistry.

Vyas, V. K., Kumar, A. S., Ali, A., Prasad, S., Srivastava, P., Mallick, S. P., Ershad, M., Singh, S. P., & Pyare, R. (2016). Assessment of nickel oxide substituted bioactive glass-ceramic on in vitro bioactivity and mechanical properties. *Boletín de La Sociedad Española de Cerámica y Vidrio*, *55*(6), 228–238. https://doi.org/10.1016/j.bsecv.2016.09.005

Wahid, M. F. A., Hyie, K. M., Mardziah, C. M., & Roselina, N. R. N. (2015). EFFECT OF SEVERAL CALCINATION TEMPERATURE ON DIFFERENT CONCENTRATION ZINC SUBSTITUTED CALCIUM PHOSPHATE CERAMICS. *Jurnal Teknologi*, *76*(10), Article 10. https://doi.org/10.11113/jt.v76.5802

Wu, C., & Chang, J. (2006). A novel akermanite bioceramic: Preparation and characteristics. *Journal of Biomaterials Applications*, *21*(2), 119–129. https://doi.org/10.1177/0885328206057953

Wu, C., & Chang, J. (2007). Synthesis and in vitro bioactivity of bredigite powders. *Journal of Biomaterials Applications*, *21*(3), 251–263. https://doi.org/10.1177/0885328206062360

Wu, C., & Chang, J. (2013b). A review of bioactive silicate ceramics. *Biomedical Materials*, *8*(3), 032001. https://doi.org/10.1088/1748-6041/8/3/032001

Wu, C.-T., & Chang, J. (2013a). Silicate bioceramics for bone tissue regeneration. *Journal of Inorganic Materials*, *28*, 29–39. https://doi.org/10.3724/SP.J.1077.2013.12241

Wu, P., Lü, L., Tang, S., Liu, C., Yue, H., Jiang, W., & Liang, B. (2020). The fouling properties of SiO2-CaO-P2O5 system in high-temperature rotary kiln phosphoric acid process. *Chinese Journal of Chemical Engineering*, *28*(7), 1824–1831. https://doi.org/10.1016/j.cjche.2020.01.005

Xu, S., Lin, K., Wang, Z., Chang, J., Wang, L., Lu, J., & Ning, C. (2008). Reconstruction of calvarial defect of rabbits using porous calcium silicate bioactive ceramics. *Biomaterials*, *29*(17), 2588–2596. https://doi.org/10.1016/j.biomaterials.2008.03.013

Zhang, K., & Le, Q. V. (2020). Bioactive glass coated zirconia for dental implants: A review. *Journal of Composites and Compounds*, *2*(2), 10–17. https://doi.org/10.29252/jcc.2.1.2

Zhao, X., Wu, Y., Du, Y., Chen, X., Lei, B., Xue, Y., & Ma, P. X. (2015). A highly bioactive and biodegradable poly(glycerol sebacate)-silica glass hybrid elastomer with tailored mechanical properties for bone tissue regeneration. *Journal of Materials Chemistry B*, *3*(16), 3222–3233. https://doi.org/10.1039/C4TB01693A

Zreiqat, H., Ramaswamy, Y., Wu, C., Paschalidis, A., Lu, Z., James, B., Birke, O., McDonald, M., Little, D., & Dunstan, C. R. (2010). The incorporation of strontium and zinc into a calcium-silicon ceramic for bone tissue engineering. *Biomaterials*, *31*(12), 3175–3184. https://doi.org/10.1016/j.biomaterials.2010.01.024

6 Biocompatibility and Biodegradability of Bioceramics

David Bahati, Meriame Bricha, and Khalil El Mabrouk
Euromed University of Fes (UEMF)

6.1 INTRODUCTION

Bioceramics form a group of biomaterials primarily used in medical applications, mainly in regenerative medicine. This group of biomaterials is composed of nonmetallic inorganic solid materials such as calcium phosphates (CaPs), bioactive glasses (BGs), and glass-ceramics (GCs). Among all bioactive materials, bioactive ceramics were the first artificial materials to form a firm bond with living tissues. Under this category, BGs were the first artificial material to reveal the peculiar properties of bonding with soft and hard native tissues. The field of hard tissue regeneration was initiated from the bonding ability of the bioactive ceramics with native tissues [1–4].

The discovery of bioactive glasses in 1969 originated following the U.S. Army material research conference held at Sagamore in New York City in 1967. The presentation of Larry Hench, a professor at Florida University, was about radiation-resistant of vanadium phosphate (V_2O_5-P_2O_5) glasses. Hench's talk about his discoveries took place before its official time in a bus ride with a U.S. colonel officer who had just returned from Vietnam War in 1967. Hench shared with the officer his scientific discovery of radiation-resistant material that could survive exposure to high-energy radiations. On the discussion, the colonel officer asked Hench if he could make a material that could sustain exposure in the human body without failure as an alternative to bioinert implant materials (plastic and metals) used during that time, which showed failure due to prolonged exposure in human body [5].

After the conference, Hench and his colleagues submitted a research proposal to the U.S. Army Medical Command to fund the project. The hypothesis of their project mainly focused on the biocompatibility of medical implants, relating the inorganic phase of natural human bone with their proposed new material. The hypothesis was as follows:

> The human body rejects metallic and synthetic polymeric materials by forming scar tissue because living tissues are not composed of such materials. However, bone contains a hydrated calcium phosphate component, hydroxyapatite (HA), and therefore if the material can form a HA layer *in vivo*, the body may not reject it.

 DOI: 10.1201/9781003258353-8

Hench used Na_2O-CaO-SiO_2 ceramic phase diagram to synthesize three compositions of ceramic glasses. The glass with composition (by weight) 45%SiO_2-24.5%Na_2O-24.5%CaO-6%P_2O_5 (currently known as Bioglass®) was selected for an *in vivo* experiment in a rat femoral implantation done by Dr. Greenlee at the Florida veteran administration hospital. After 6 weeks since implantation, Greenlee reported the implant ceramics to firmly bond to the rat femoral bone such that he could push, shove, and hit them without detaching [5].

The bond-forming ability between bioactive ceramics and living tissues is not an adequate decisive factor for their use in regenerative medicine as tissue implants. Among many other factors, the ceramic material must be biocompatible and, in addition, biodegradable. Biocompatibility of the ceramic material guarantees that the material becomes biologically body-friendly where it does not elicit any adverse response such as allergenic reactions, carcinogenic, cytotoxicity, immunogenicity, and formation of fibrous at the tissue-implant interface as a response to biological rejection. On the other hand, the biodegradability of ceramic material ensures that the degradation product at the physiological implant site results in nontoxic products that can be easily incorporated into the body's metabolism and secreted. Generally, based on their interactions with physiological body fluid and living tissues, ceramics used in medical applications can be categorized into three main groups (Figure 6.1), viz., bioinert, bioresorbable, and bioactive ceramics [6,7].

Bioinert ceramics refer to chemically and biologically inert materials when exposed under contact with physiological body fluid [8]. Bioinert ceramics belong to the first generation of biomaterials whose main function was to offer necessary mechanical support with minimum biological interactions when used in the human body as implants. Bioinert ceramics lack biological recognition from the body, leading to a thin fibrous layer at the tissue-implant interface to isolate the implant from the body [5,9]. Examples of such materials include alumina (Al_2O_3), zirconia (ZrO_2), and titania (TiO_2). These materials have been used to serve as permanent

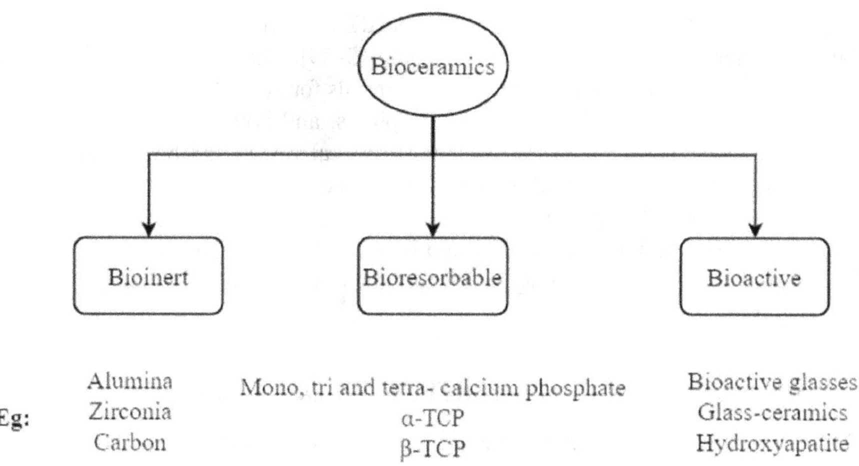

FIGURE 6.1 Classification of bioceramics based on *in vivo* interaction.

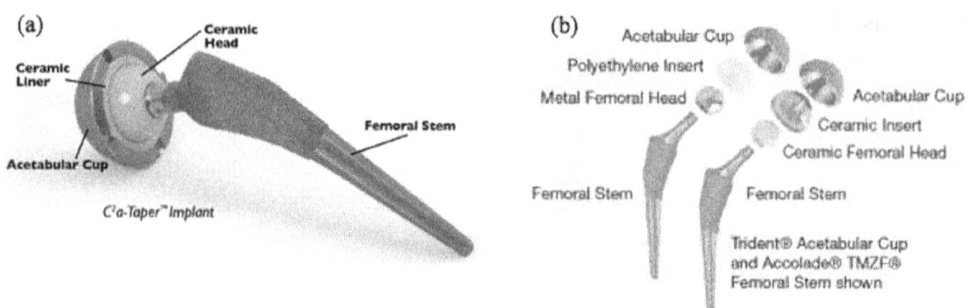

FIGURE 6.2 Keramos acetabular (a) and trident ceramic (b) systems for total hip arthroplasty. (Reproduced with permission from Ref. [27].)

tissue implants [10–12]. For example, ball heads and acetabular inlays used in total hip arthroplasty are made from bioinert ceramics, mostly alumina and zirconia (Figure 6.2).

Bioresorbable ceramics refer to the bioceramic materials that resorb *in vivo* with time and are finally entirely replaced by the living tissues [6]. Bioresorbable ceramics do not form stable interfacial contact with the native tissues [13]. These include ceramic materials such as calcium phosphates (e.g., mono-, tri-, and tetracalcium phosphates: $Ca(H_2PO_4)_2$, $Ca_3(PO_4)_2$, and $Ca_4(PO_4)_2O$), calcium sulfate dihydrate ($CaSO_4.2H_2O$), and calcium carbonates [14].

Bioresorbable ceramics spontaneously resorb *in vivo* and are finally replaced by the native tissues with their bioresorbable products metabolized and eventually eliminated from the body.

Bioactive ceramics refer to ceramic materials that are *in vivo* osteoinductive and can form interfacial bonding with living tissues. Examples of these materials include bioactive glasses, hydroxyapatite, and glass-ceramics. When implanted *in vivo*, bioactive ceramics induces specific biological and chemical interactions with physiological body fluid, leading to direct bond formation with living tissues.

Bioceramics have been used in various medical applications, such as replacing damaged knees, hips, ligaments, tendons, etc. [15–19]. They have been chiefly used in guided bone regeneration as osteogenic materials for alveolar ridge augmentation, pocket reduction in periodontitis, dental implants, and bone fillers [20]. Recently, the use of bioceramics has been extended to spinal fusion and bone space-filling after tumor removal [21–26]. Most bioresorbable ceramics are also bioactive. Their resorbabilities provoke specific biological and chemical interactions *in vivo*. This chapter presents the biocompatibility and biodegradability of bioceramics, emphasizing *in vitro* assays. Bioactivity of bioceramics in relation to biodegradability is also presented.

6.2 *IN VITRO* BIOCOMPATIBILITY OF BIOCERAMICS

During a lifetime, some human skeletal tissues such as the bones, teeth, and tendons can be damaged beyond their natural ability to repair. The tissue damage may be due to trauma, diseases, aging etc. The shortage of human-owned natural excess tissues that can save as implants (autograft), risks of infection transfer to a patient when a tissue implant is taken from a fellow human (allograft), immune rejection, disjoint and transfer of infections when implants are from animals (xenografts) have triggered researches focusing on alternative synthetic materials that could permanently or temporarily replace the damaged or diseased human tissues.

The coexistence ability between ceramic materials and human living tissues without eliciting any harm or undesirable response has been the main focus among many researchers linking material engineering with biomedical applications. Therefore, the overall interactions between synthetic ceramic materials and native living tissues without causing any adverse response are discussed under the context of the biocompatibility of the materials. Thus, biocompatibility of biomedical materials has been by default defined as an ability of the biomaterial to be nontoxic, nonimmunogenic, nonthrombogenic, noncarcinogenic, nonirritant, and many other side effect responses due to the introduction of the material into the host's physiological environment [28].

In 1986, the European Society for Biomaterials consensus conference was held with the major theme of acceptable definitions of various terms used in biomaterials [29]. From the conference, it was agreed that the biocompatibility of a biomaterial should be defined as "the ability of the material to perform with an appropriate host response in a specific situation" [30]. The definition mainly originated from positive observed biological and chemical interaction between the material and physiological environment, leading to favorable responses on both implant material and the host's biological systems. The definition from the consensus conference pinpointed the fact that the biocompatibility of material can only be evaluated based on a specific application. David Williams [28], a senior author in biomaterials in 2008, redefined biocompatibility as

> the ability of a biomaterial to perform its desired function with respect to a medical therapy without eliciting any undesirable local or systemic effect in the recipient or beneficial cellular or tissue response in that specific situation and optimizing the clinically relevant performance of that therapy.

As mentioned above, there has been increasing use and desire to use ceramic materials in the human body to heal, regenerate, augment, or replace damaged or diseased body tissues. The choice and test of these materials with reference to their biocompatibility are mainly based on material–host interaction that considers various interaction responses such as cytotoxicity, immunogenicity, mutagenicity, carcinogenicity, and genotoxicity. Some of these components of biocompatibility are measured in three successive stages, namely, *in vitro* test, *in vivo* test, and clinical test [31]. The basic concept of *in vitro* test of bioceramics is presented in the following sections. Usually, two regulatory standard bodies update procedures of biocompatibility test of a biomaterial in three stages from time to time based on novel advanced test techniques and avoiding preceding problems from the past version of the test protocol. The approved standard bodies currently are the international organization for standardization (ISO) and food and drug administration (FDA) [32].

6.3 BIOCOMPATIBILITY TESTS OF BIOCERAMICS

6.3.1 *In Vitro* Cytotoxicity Test of Bioceramics

In vitro biocompatibility serves as an initial screening test of the biomaterial before *in vivo* tests in complex biological systems. The ideal test of any biomaterial for medical application could be done on humans, but experimenting with the human is highly discouraged on legal and ethical grounds. The biocompatibility of a material *in vivo* can be assessed in two main perspectives: how native tissues respond to the foreign material and how the foreign material behaves *in vivo*. Thus, for any biomaterial to be approved for clinical applications, it must be sequentially passed through three test stages (Figure 6.3); *in vitro* test, *in vivo* test (test on animals), and clinical test (clinical trial of the biomaterial) on patients [33]. Good laboratory practices for the first two tests are very crucial to avoid any possible side effects during the clinical test [33]. It is worth noting that all *in vivo* tests must get approval from regulatory and ethical committees.

Usually, *in vitro* biocompatibility test focuses on evaluating cytotoxicity and genotoxicity on individual cells where it is intended to predict *in vivo* chemical and biological reactions between the body and the foreign material. The cytotoxicity test acts as an initial evaluation of the biocompatibility of a material. Under this consideration,

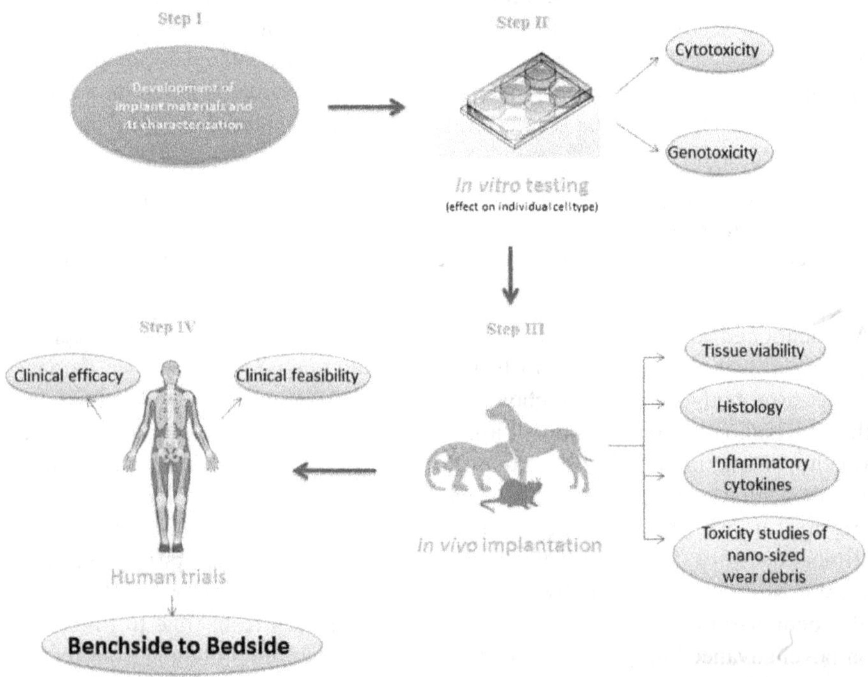

FIGURE 6.3 Steps toward approval of a biomaterial for clinical use. (Reproduced from Ref. [33] with permissions from the Royal Society of Chemistry.)

in vitro test is preferred as it is cheap, fast, highly reliable, and relatively simple compared to similar evaluation under *in vivo* test. However, the *in vitro* test results do not guarantee similar results under *in vivo* test due to complex physiological systems involved contrary to a single-cell system under *in vitro* test. Therefore, although *in vitro* test results do not perfectly predict *in vivo* test results, it is crucial to select the type of cells for *in vitro* cytotoxicity test based on the specific targeted application of the biomaterial.

There are several commonly available *in vitro* cytotoxicity assays such as Trypan blue, MTT, MTS, XTT, WST, LDH, NRU, GSH, Alama blue, etc. Each has its advantages and disadvantages that should be considered when choosing the assay protocol. The use of multiple assay techniques is preferably recommended for better screening before *in vivo* test of the biomaterial. Finally, the cytotoxicity is evaluated based on cell viability, morphology, migration, proliferation and differentiation [34–37].

6.3.1.1 *In Vitro* Cytotoxicity of Bioinert Ceramics

Bioinert ceramics are mainly employed in dental and prosthetic load-bearing applications due to their biological and chemical inertness and high mechanical strength [38]. Examples of such ceramics are aluminum oxide (Al_2O_3) and zirconium oxide (ZrO_2). The high chemical stability of alumina and zirconia is mainly attributed to their inherent high oxidative capability. These bioceramics find their popular use in orthopedics, mainly for joint replacement applications. Most published reports on *in vitro* biocompatibility tests of these materials reported significantly low cytotoxicity on various cell lines such as macrophages, fibroblasts, osteoblasts, and lymphocytes [39–42]. However, researchers reported *in vitro* study of alumina and zirconia of different forms such as powder, fibers, and dense ceramics to show dissimilar *in vitro* test results. The same material in a different form could produce different *in vitro* test results due to different surface-to-volume ratios influencing the interaction between the material and cells. For example, Li et al. [43] reported alumina and zirconia powder showing higher cytotoxicity on human fibroblasts than their corresponding denser ceramics.

Although bioinert ceramics do not biologically or chemically interact with the physiological systems and are well known for their high wear resistance, prolonged exposure in the physiological body fluid may lead to nano-sized particulate wear [33]. Cytotoxicity of zirconia wears debris of about 0.02 μm on the mouse, and Chinese hamster fibroblasts were reported by Standard [44]. It was observed a notable decrease in cell growth rate. However, the effect was insignificant and comparable to alumina and hydroxyapatite ceramics. In parallel, the study confirmed less than 5% cell death as a response to macrophages on zirconia and alumina commercially available fibers of sizes ranging 2–6 μm. The high-observed cell viability through a direct contact assay of alumina and zirconia confirms their high biocompatibility and bioinertness.

6.3.1.2 *In Vitro* Cytotoxicity of Bioresorbable Ceramics

The chemical similarity of the inorganic phase of human natural bone and calcium phosphate-based bioceramics such as tricalcium phosphates (TCPs) and hydroxyapatite (HA) makes the ceramic materials be mostly employed and more innovated for hard tissue engineering with the main focus being on the bones, joints, and teeth [45]. TCPs (α and $β\_Ca_3(PO_4)_2$) are bioresorbable ceramics materials that gradually resorb and are eventually replaced by the native tissues at the implantation sites [46]. β-TCP is well known for its excellent biocompatibility both *in vitro* and *in vivo* [47]. When used as an implant, β-TCP at the physiological site interacts with the surrounding tissues through ionic release of Ca^{2+} and PO_4^{3-} leading to the regulation of growth factors, cytokines, and osteoblast differentiation that promote bone healing at the defect site [48].

In vitro cytotoxicity of TCPs and most biomaterials is usually done by assessing cell viability and integrity as a response to direct contact with the materials or their extracts in a cell culture assay. As mentioned above, there are many assays, each with advantages and limitations. It is crucial to understand the limitations of the used cell assay before one draws final biocompatibility *in vitro* findings of the material. For instance, XTT is mainly based on metabolic deterioration tests that may be biased due to bacterial contamination leading to wrong evaluation of cell viability. On the other hand, crystal violet assay (CV) can mark dead cells that are still attached to the material, overestimating cell viability [49]. When higher *in vivo* bioresorbability is required, α-TCP are preferred due to their higher solubility compared to β-TCP [46]. Some studies have reported notable cytotoxicity of α-TCP, which could mainly be due to a remarkable decrease in pH of the surrounding fluid triggered by the formation of phosphoric acid as a result of hydrolysis of the α-TCP [50,51]. However, the remarkabe change of pH of the surrounding fluid *in vitro* could be stabilized through the presence of additives such as disodium hydrogen phosphate (Na_2HPO_4), citric acid ($C_6H_8O_7$), and tannic acid ($C_{76}H_{52}O_{46}$) [52].

6.3.1.3 *In Vitro* Cytotoxicity of Bioactive Ceramics

Bioactive ceramics also referred to as surface reactive ceramics interact with the native tissues at the physiological site through biophysical and biochemical interfacial reactions with the body fluid leading to the formation of a biologically active and thermodynamically stable at the pH of 7.4 hydroxyapatite layer. The formed hydroxyapatite layer is structurally and chemically similar to the inorganic phase of the natural bone. The formed hydroxyapatite layer is responsible for forming the firm interfacial bond between the implant material and the living tissues.

This biomaterials group includes bioceramics such as hydroxyapatite, glass-ceramics like apatite-wollastonite, certain compositions of silicate, phosphate, and borate bioactive glasses. These biomaterials are considered both bioresorbable and bioactive. *In vitro* biocompatibility assessment of these bioceramics is usually done by evaluating cell responses such as cell adhesion, colonization, and change of cell morphology due to their direct interaction with the biomaterial or its extracts.

Usually, cytotoxicity of bioactive ceramic materials is mainly due to the high rate of ionic exchange between the material and surrounding body fluid leading to a cytotoxic surface-modified material [53,54]. The *in vitro* cytotoxicity study on glass and glass-ceramic bioactive materials has shown the materials to be noncytotoxic at relatively low concentrations (e.g., ≤ 2 mg/mL) with the cytotoxicity increasing with increasing concentrations (e.g., ≥ 5 mg/mL) [53,55,56].

6.4 BIODEGRADABILITY OF BIOCERAMICS

Biodegradable ceramics have been used in decades as temporary tissue implants due to their osteoconductive and osteoinductive properties [57–59]. Matching biodegradation of bioceramics with the rate of regeneration of the living tissues is among the challenges in regenerative medicine [60,61]. In addition, the spontaneous degradation of the biomaterial *in vivo* compromises the mechanical integrity leading to inadequate mechanical support from the implant material before the target native tissue is completely restored. Controlled biodegradation of the bioceramics *in vivo* is essential for space creation in order for the native extracellular matrix to be deposited. *In vivo* biodegradation of bioceramics varies depending on the form, physiological site, composition, and processing of the ceramic material.

6.4.1 Biodegradation of Calcium Phosphates

Calcium phosphate-based (CaPs) ceramics are biodegradable and biocompatible due to their chemical and structural similarity with the inorganic phase of natural bone. CaPs are widely studied for their application in regenerative medicine as a temporary bone replacement due to their inherent resorbability and biocompatibility properties. These include tricalcium phosphate (TCP, $Ca(PO_4)_2$) [62–65], hydroxyapatite (HA, $Ca_{10}(PO_4)_6(OH)_2$) [66–69], and brushite ($CaHPO_4.2H_2O$) [70–72]. CaPs have been used in different forms as scaffolds, fillers, and coating agents.

Biodegradability of bioceramics at the implantation site is usually contributed by several factors mainly dominated by physicochemical dissolution (surface leaching) and cell-mediated (osteoclasts and phagocyte) resorption [73,74]. Degradation of the bioceramic material at the physiological site could also be due to physical nanoparticle wear caused by chemical attacks at the material's grain boundaries and fluid erosion [15].

Tricalcium phosphates (TCPs) appear in two different crystalline forms (α and β-TCP) with similar chemical structures ($Ca(PO_4)_2$). β-TCP shows a lower biodegradation rate compared to α-TCP but higher than HA [15,75,76]. Synthetic HA resulting from high-temperature processing is crystalline with lower biodegradation [70] than their amorphous counterparts [77]. Generally, high surface area, low crystallinity, crystal imperfection, and small grain sizes of calcium phosphate-based bioceramics promote a high biodegradation rate [15]. General mechanisms involved in the biodegradation of CaPs are summarized in Figure 6.4 [78].

6.4.1.1 Physicochemical Dissolution

Physicochemical dissolution of biodegradable ceramics such as calcium phosphates and hydroxyapatites at the implantation site is mainly due to the ionic exchange between the biomaterial and the surrounding fluid, usually leading to surface phase transformation of the material [79–82]. The physicochemical dissolution depends on the presence of soluble constituents in the ceramic material and the local pH [15]. The physicochemical ionic dissolution of CaPs occurs both *in vitro* and *in vivo*. The complex *in vivo* dissolution phenomena can be simplified in an *in vitro* assay by using simulated body fluid (SBF) and buffer solution. However, the *in vitro* assay

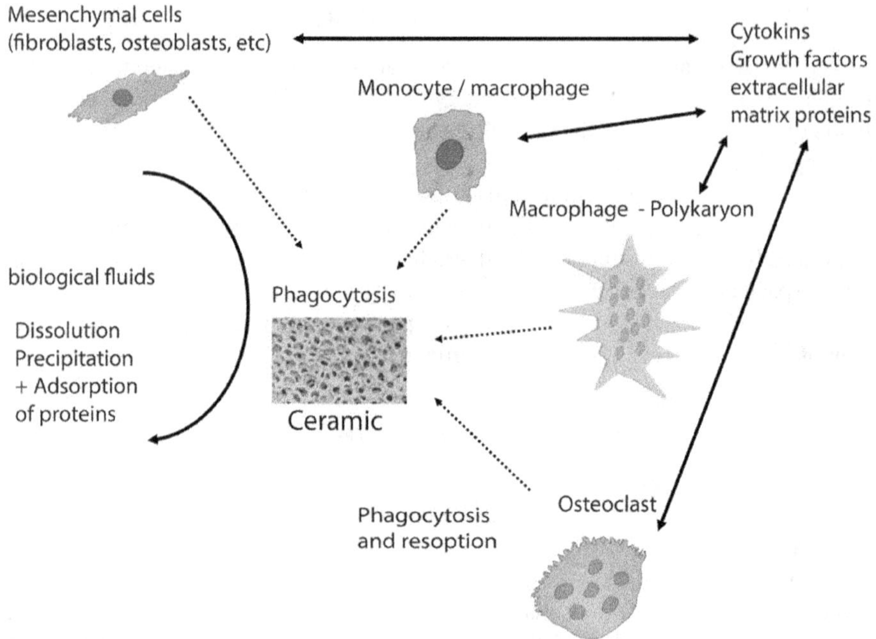

FIGURE 6.4 Cell-mediated and physicochemical degradation mechanisms of CaPs.

does not precisely mimic the complex *in vivo* ionic leaching process due to complex multicomponent involved processes. Physicochemical dissolution of CaPs can be controlled by incorporation of some metallic ions, varying material crystallinity, and varying mixing ratios between highly soluble and less soluble CaPs that result in the so-called biphasic calcium phosphates [83].

6.4.1.2 Cell-Mediated Degradation

Implantation of ceramic material (CaPs) induces recruitment of giant multinucleated cells, including osteoblasts at the physiological site. Since a long time ago, researchers have confirmed cellular degradation of ceramic implant materials. Osteoclasts were confirmed to resorb ceramics *in vivo* by changing the morphologies of lacunae in the ceramic implant, the observation that seemed to be similar to acidic-etched material [84,85]. The surface adhesion of mesenchymal cells and macrophages resorbs the foreign material through phagocytosis [86]. In contrast, giant cells (osteoclasts) resorb the material through the secretion of acid [87], such as lactic and citric acids [88], a phenomenon similar to the resorption in the bone remodeling process [89,90]. The mechanism under osteoclastic ceramic resorption is acidification of the extracellular matrix through proton pumping by osteoclast at the cell-material interface leading to reduction of local pH [78]. Under this consideration, the osteoclasts attach firmly to the surface of the foreign ceramic material and secret localized hydrogen ions (H^+), leading to a high acidic medium on the surface that leads to the dissolution of the material. Osteoclastic degradation is a crucial stage in the bone remodeling process.

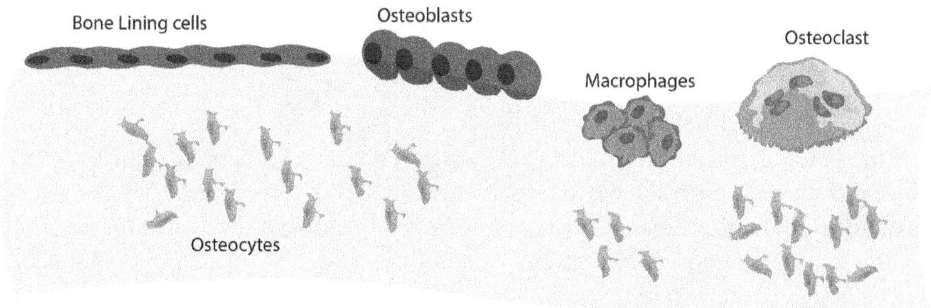

FIGURE 6.5 Mechanism of creeping substitution in bone remodeling.

New bones subsequently replace osteoclast-resorbed bone space through osteoblastic activity, leading to creeping substitution [91–93] (Figure 6.5). Degradation of implant biomaterials plays the same role of space creation for the native extracellular matrix to be deposited. The overall rate of material degradation should match the rate of native extracellular matrix deposition, and the material should be fully resorbed after the complete deposition of the native tissues.

The cellular degradation of CaPs ceramics depends on various factors such as intrinsic properties of the ceramic material, size of ceramic particulates [94], physiological implantation site, and the presence of various proteins and ions that make hormones, vitamins, cytokines etc. [78]. The ceramics degradation can also be due to the presence of other mesenchymal cells such as fibroblasts. It has been reported that doping bioceramics with metallic ions such as strontium, zinc, iron, sodium etc., could inhibit the osteoclastic activity [95–99], hence controlling biodegradation. Doping with magnesium, silicon, and strontium is preferred due to their ability to induce osteogenesis through stimulation of osteoblast proliferation with an added role of osteoclatic inhibition [98,100].

6.4.2 Biodegradation of Bioactive Glasses

6.4.2.1 Physicochemical Dissolution

In contact with simulated (SBF) or physiological body fluid, bioactive glasses release cations (Ca^{2+}, Na^{+}, etc.) to the surrounding fluid while in parallel recruiting proton (H^{+}) or hydronium ions (H_3O^{+}) from it. The ionic exchange between the bioactive glass and the surrounding fluid increases the pH that finally breaks the materials glass network (e.g., Si-O-Si bond in silicate-based glasses). The ionic dissolution of the bioactive glasses is an initial stage of glass surface reaction toward bioactivity, leading to a firm bond between the glass material and native tissues. The dissolution rate of bioactive glasses depends on the glass compositions, glass particle size, type, the pH of the dissolution media, etc. [101]. Generally, the ionic dissolution rate of borate-based glasses is higher than that of phosphate and silicate-based glasses [102–104]. Under this comparison, silicate-based bioactive glasses show the lowest dissolution rate.

Bioactive glasses can be used for medical applications in different forms such as particulate, putty, and bulky sintered scaffold blocks. However, the dynamic fluid environment at the implantation site can degrade the implant glass material via fluid erosion, leading to the decline of the implant's mechanical function.

6.4.2.2 Cell-Mediated Degradation

Cell-mediated degradation of bioactive glasses is not widely reported. However, bioactive glasses have been soluble in an acid medium [105]. As mentioned above, osteoclastic degradation of ceramic material is through acidification by proton pumping leading to a decrease in pH at the cell-material interface. In addition, other specialized cells (macrophages) remove from the body fluid through phagocytosis of foreign solid particulates with a size greater than 0.5 μm, including dead cells and invading microorganisms [86]. Phagocytosis environment is immediately initiated by creating signals following sensing of foreign particle, apoptotic cell, or microorganism [86]. Thus, all bioactive ceramic implants, including bioactive glasses, can suffer from macrophages attack resulting in a degraded material. Biomaterial resorption of up to 5 μm can result from phagocytosis of a single macrophage [106]. Bioactive glasses were phagocytized with the resorption rate increasing with decreased particle size [107,108]. Phagocytosis of silicate-based bioactive glass was reported after 24 hours of incubation with human macrophages [109].

6.5 BIOACTIVITY OF BIOCERAMICS

Bioactivity of a material is usually considered a materials' ability to bond with living tissues [110,111]. When in contact with the body fluid, bioactive material undergoes surface interactions through ionic exchange with the surrounding fluid. The ionic leaching (e.g., Ca^{2+}, Na^{+}, PO_4^{3-}), including ions from the surrounding fluid, supersaturate and precipitate, leads to the formation of a carbonated hydroxyapatite layer on the surface of the ceramic material. A ceramic material must have controlled ionic dissolution kinetics to produce a positive healing effect. Too slow ionic release kinetics results in inadequate ionic concentration to trigger cellular activities [5].

On the other hand, too high dissolution rates result in ineffective cellular responses [5]. The ability of a material to form a hydroxyapatite layer on its surface following incubation in simulated body fluid has been used as an *in vitro* measure of bioactivity. The formed layer of hydroxyapatite has similar chemical and structure composition to the inorganic phase of the natural bone condition that leads to the bond formation whose strength is greater or equal to the mechanical strength of the natural bone [112,113].

Some compositions of calcium-releasing ceramics such as calcium phosphate and glasses upon immersion in simulated body fluid or aqueous solutions are bioactive. The *in vitro* bioactivity can only serve as the initial screening stage toward *in vivo* bioactivity test, whose results may differ from *in vitro* assay due to the complex physiological environment. The *in vivo* bioactivity of material is evaluated through bioactivity index (I_B) defined via, $I_B = \frac{100}{t_{0.5bb}}$ where $t_{0.5bb}$ stands for the time required for more than half of the material interface to bond with the native tissue [15]. Under this consideration, materials with $I_B > 8$ bond with both soft and hard tissues while those with $0 < I_B < 8$ bonds with only hard tissues.

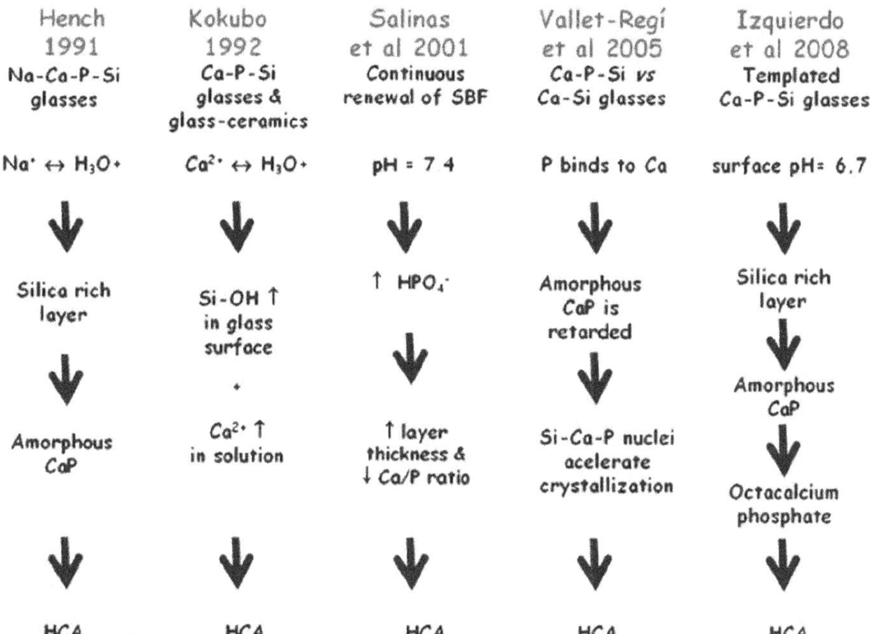

FIGURE 6.6 Various bioactivity mechanisms proposed by different authors. (Reproduced from Ref. [74] with permissions from the Royal Society of Chemistry.)

Some authors (Figure 6.6) have described the mechanism toward hydroxyapatite formation on the surface of some bioactive ceramics upon incubation in aqueous solutions. The formation of hydroxyapatite depends on the type and composition of the ceramic material, pH and ionic concentration of the incubation media, and the incubation duration.

6.6 BIOACTIVITY VERSUS BIODEGRADABILITY OF BIOCERAMICS

Bioactive and bioresorbable/biodegradable ceramics emerged as a second generation of biomaterials following the first generation of biologically and chemically inert ceramic materials. Therefore, a ceramic material combining bioactivity and biodegradability properties falls under the third generation of biomaterials. This includes bioceramic materials such as bioactive glasses and hydroxyapatite.

As mentioned above, the bioactivity response of a bioceramic material begins with an ionic exchange between the material and surrounding body fluid. Ionic leaching from the bioceramic material, including micro- and nanoparticulate wear responsible for specific biological activations, falls under physicochemical degradation. Thus, bioactive materials in particulate form become bioresorbable materials.

Bioactive ceramics *in vivo* follow a cascade of interactions toward material-tissue integration and finally are entirely replaced by the newly formed native tissues.

This crucial property of bioactive ceramics finds applications in regenerative medicine, particularly in bone healing related applications.

6.7 CONCLUSION

In the present chapter, biocompatibility and biodegradability of bioceramics have been briefly presented. Several key points can be drawn as follows:

1. The biocompatibility of a bioceramic material is evaluated in three consecutive test stages before clinical approval is granted. The stages in their orders are *in vitro* test, *in vivo* test (in animals), and clinical test (in patients). The first two stages are crucial and decisive to avoid possible side effects in a complex *in vivo* evaluation in humans (clinical test). Therefore, several *in vitro* tests are recommended on different *in vitro* assays before animal experiments since *in vitro* tests are cheap, fast, highly reliable, and relatively simple to handle compared to *in vivo* tests.
2. The overall mechanical degradation of the ceramic biomaterial is due to physicochemical and cellular degradations. Mechanical deterioration of bioceramic implants results in poor mechanical functional support of the damaged or diseased tissue. Therefore, biodegradation of a ceramic implant must match well with the rate of regeneration of the respective native tissue. Physicochemical degradation is crucial for specific cellular gene activation leading to both intracellular and extracellular activities. In addition, cell-mediated degradation creates the necessary space for simultaneous deposition of the native extracellular matrix.

REFERENCES

1. A.T. Khalaf, Y. Wei, J. Wan, J. Zhu, Y. Peng, S.Y. Abdul Kadir, J. Zainol, Z. Oglah, L. Cheng, Z. Shi, Bone tissue engineering through 3D bioprinting of bioceramic scaffolds: A review and update, *Life*. 12 (2022) 903. https://doi.org/10.3390/life12060903.
2. S. Pina, I.K. Kwon, R.L. Reis, J.M. Oliveira, Biocomposites and bioceramics in tissue engineering: Beyond the next decade, in: *Springer Ser. Biomater. Sci. Eng.*, Springer Science and Business Media Deutschland GmbH, 2022: pp. 319–350. https://doi.org/10.1007/978-981-16-7435-8_11.
3. F. Kermani, S. Kargozar, S. V. Dorozhkin, S. Mollazadeh, Calcium phosphate bioceramics for improved angiogenesis, *Biomater. Vasc. Angiogenes.* (2022) 185–203. https://doi.org/10.1016/B978-0-12-821867-9.00004-4.
4. S. Van Rijt, K. De Groot, S.C.G. Leeuwenburgh, Calcium phosphate and silicate-based nanoparticles: History and emerging trends, *Tissue Eng. – Part A*. 28 (2022) 461–477. https://doi.org/10.1089/ten.tea.2021.0218.
5. L.L. Hench, The story of bioglass(r), *J. Mater. Sci. Mater. Med.* (2006) 967–978. https://doi.org/10.1007/s10856-006-0432-z.
6. T. Yamamuro, Bioceramics, *Biomech. Biomater. Orthop.* (2004) 22–33. https://doi.org/10.1007/978-1-4471-3774-0_3.
7. A. Bhardwaj, L.M. Pandey, Biomaterials: Types and applications, in: *Nanoscale Eng. Biomater. Prop. Appl.*, Springer, Singapore, 2022: pp. 89–114. https://doi.org/10.1007/978-981-16-3667-7_4.

8. L.L. Hench, E.C. Ethridge, Biomaterials – The interfacial problem, *Adv Biomed Eng.* 5 (1975) 35–150. https://doi.org/10.1016/b978-0-12-004905-9.50007-4.
9. L.L. Hench, *An Introduction to Bioceramics*, 2nd edition, Imperial College Press, 2013. https://doi.org/10.1142/P884.
10. S.F. Hulbert, The use of alumina and zirconia in surgical implants, in: *An Introd. to Bioceram*, 2nd edition, Imperial College Press, 2013: pp. 27–47. https://doi.org/10.1142/9781908977168_0002.
11. C. Piconi, G. Maccauro, F. Muratori, E. Brach Del Prever, Alumina and zirconia ceramics in joint replacements, *J. Appl. Biomater. Biomech.* 1 (2003) 19–32. https://doi.org/10.1177/228080000300100103.
12. S.F. Hulbert, The use of alumina and zirconia in surgical implants, In: *An Introd. to Bioceram.*, World Scientific, 1993: pp. 25–40. https://doi.org/10.1142/9789814317351_0002.
13. T. Yamamuro, Bioceramics, in: *Biomech. Biomater. Orthop*, 2nd edition, Springer, London, 2016: pp. 21–33. https://doi.org/10.1007/978-1-84882-664-9_3.
14. M. Bohner, Bioresorbable ceramics, *Degrad. Rate Bioresorbable Mater. Predict. Eval.* (2008) 95–114. https://doi.org/10.1533/9781845695033.2.95.
15. L.L. Hench, Bioceramics: From concept to clinic, *J. Am. Ceram. Soc.* 74 (1991) 1487–1510. https://doi.org/10.1111/j.1151-2916.1991.tb07132.x.
16. G. Solarino, C. Piconi, V. De Santis, A. Piazzolla, B. Moretti, Ceramic total knee arthroplasty: Ready to go? *Joints.* 5 (2017) 224–228. https://doi.org/10.1055/s-0037-1607428.
17. P. Bergschmidt, R. Bader, D. Ganzer, C. Hauzeur, C. Lohmann, W. Rüther, D. Tigani, N. Rani, F.L. Prats, C. Zorzi, V. Madonna, S. Rigotti, F. Benazzo, S.M.P. Rossi, G. Kundt, H.R. Bloch, W. Mittelmeier, Ceramic femoral components in total knee arthroplasty – Two year follow-up results of an international prospective multi-centre study, *Open Orthop. J.* 6 (2012) 172–178. https://doi.org/10.2174/1874325001206010172.
18. C.Y. Hu, T.R. Yoon, Recent updates for biomaterials used in total hip arthroplasty, *Biomater. Res.* 22 (2018) 1–12. https://doi.org/10.1186/s40824-018-0144-8.
19. W. Christian, K. Max, The role of aluminium ceramics in total hip arthroplasty, in: *Adv. Ceram. – Electr. Magn. Ceram. Bioceram. Ceram. Environ.*, IntechOpen, 2011. https://doi.org/10.5772/18094.
20. H. Shao, M. Sun, F. Zhang, A. Liu, Y. He, J. Fu, X. Yang, H. Wang, Z. Gou, Custom repair of mandibular bone defects with 3D printed bioceramic scaffolds, *J. Dent. Res.* 97 (2018) 68–76. https://doi.org/10.1177/0022034517734846.
21. A. Constantinou, Thermo-responsive ABC triblock terpolymers for 3D printing, *J. Biotechnol. Biomater.* 06 (2016). https://doi.org/10.4172/2155-952x.c1.050.
22. L.A. van Dijk, R. Duan, X. Luo, D. Barbieri, M. Pelletier, C. Christou, A.J.W.P. Rosenberg, H. Yuan, F. Barrèrre-de Groot, W.R. Walsh, J.D. de Bruijn, Biphasic calcium phosphate with submicron surface topography in an Ovine model of instrumented posterolateral spinal fusion, *JOR Spine.* 1 (2018) e1039. https://doi.org/10.1002/jsp2.1039.
23. H. Li, X. Zou, Q. Xue, N. Egund, M. Lind, C. Bünger, Anterior lumbar interbody fusion with carbon fiber cage loaded with bioceramics and platelet-rich plasma. An experimental study on pigs, *Eur. Spine J.* 13 (2004) 354–358. https://doi.org/10.1007/s00586-003-0647-3.
24. L. Fu, H. Engqvist, W. Xia, Spark plasma sintering of biodegradable Si3N4 bioceramic with Sr, Mg and Si as sintering additives for spinal fusion, *J. Eur. Ceram. Soc.* 38 (2018) 2110–2119. https://doi.org/10.1016/j.jeurceramsoc.2017.10.003.
25. B. Ortega, C. Gardner, S. Roberts, A. Chung, J.C. Wang, Z. Buser, Ceramic biologics for bony fusion – A journey from first to third generations, *Curr. Rev. Musculoskelet. Med.* 13 (2020) 530–536. https://doi.org/10.1007/s12178-020-09651-x.

26. L. Fu, Y. Xiong, G. Carlsson, M. Palmer, S. Örn, W. Zhu, X. Weng, H. Engqvist, W. Xia, Biodegradable Si3N4 bioceramic sintered with Sr, Mg and Si for spinal fusion: Surface characterization and biological evaluation, *Appl. Mater. Today.* 12 (2018) 260–275. https://doi.org/10.1016/j.apmt.2018.06.002.
27. Ceramic Hip Joint Replacement Devices | Earl's View (n.d.). https://earlsview.com/2011/07/24/ceramic-hip-joint-replacement-devices/ (accessed September 2, 2022).
28. D.F. Williams, On the mechanisms of biocompatibility, *Biomaterials.* 29 (2008) 2941–2953. https://doi.org/10.1016/j.biomaterials.2008.04.023.
29. L.G. Donaruma, Definitions in biomaterials, D. F. Williams, Ed. Elsevier, Amsterdam, 1987, 72 pp, *J. Polym. Sci. Polym. Lett. Ed.* 26 (1988) 414–414. https://doi.org/10.1002/pol.1988.140260910.
30. D.F. Williams, *Definitions in Biomaterials: Progress in Biomedical Engineering*, Elsevier, 1987: pp. 216–238. https://doi.org/10.1016/j.biomaterials.2008.04.023.
31. W. Elshahawy, Biocompatibility, in: *Adv. Ceram. – Electr. Magn. Ceram. Bioceram. Ceram. Environ.*, InTech, 2011: pp. 359–378. https://doi.org/10.5772/18475.
32. FDA, Use of International Standard ISO 10993-1, "Biological evaluation of medical devices-Part 1: Evaluation and testing within a risk management process" Guidance for Industry and Food and Drug Administration Staff Preface Public Comment (2020). https://www.fda.gov/vaccines-blood-biologics/guidance-compliance-regulatory-information- (accessed December 22, 2021).
33. G. Thrivikraman, G. Madras, B. Basu, In vitro/in vivo assessment and mechanisms of toxicity of bioceramic materials and its wear particulates, *RSC Adv.* 4 (2014) 12763–12781. https://doi.org/10.1039/c3ra44483j.
34. W. Song, S. Li, Q. Tang, L. Chen, Z. Yuan, In vitro biocompatibility and bioactivity of calcium silicate-based bioceramics in endodontics (Review), *Int. J. Mol. Med.* 48 (2021) 1–22. https://doi.org/10.3892/ijmm.2021.4961.
35. M. Sari, P. Hening, Chotimah, I.D. Ana, Y. Yusuf, Bioceramic hydroxyapatite-based scaffold with a porous structure using honeycomb as a natural polymeric Porogen for bone tissue engineering, *Biomater. Res.* 25 (2021) 1–13. https://doi.org/10.1186/s40824-021-00203-z.
36. X. Zhang, Y. Liu, Y. Su, X. Fan, F. Hu, A study of the effects of hydroxyapatite bioceramic extract on Ang/Tie2 system of umbilical vein endothelial cells, In: *Technol. Heal. Care*, IOS Press, 2021: pp. S531–S538. https://doi.org/10.3233/THC-218050.
37. M. Sun, A. Liu, H. Shao, X. Yang, C. Ma, S. Yan, Y. Liu, Y. He, Z. Gou, Systematical evaluation of mechanically strong 3D printed diluted magnesium doping wollastonite scaffolds on osteogenic capacity in rabbit calvarial defects, *Sci. Rep.* 6 (2016). https://doi.org/10.1038/srep34029.
38. A. Intisar, N. Hussain, A. Ramzan, T. Sawaira, A. Roy, M. Bilal, Biomedical applications of inorganic biomaterials, in: *Funct. Biomater.*, Springer, Singapore, 2022: pp. 265–284. https://doi.org/10.1007/978-981-16-7152-4_10.
39. R. Depprich, M. Ommerborn, H. Zipprich, C. Naujoks, J. Handschel, H.P. Wiesmann, N.R. Kübler, U. Meyer, Behavior of osteoblastic cells cultured on titanium and structured zirconia surfaces, *Head Face Med.* 4 (2008) 1–9. https://doi.org/10.1186/1746-160X-4-29.
40. T. Munro, C.M. Miller, E. Antunes, D. Sharma, Interactions of osteoprogenitor cells with a novel zirconia implant surface, *J. Funct. Biomater.* 11 (2020). https://doi.org/10.3390/JFB11030050.
41. C. Hadjicharalambous, O. Prymak, K. Loza, A. Buyakov, S. Kulkov, M. Chatzinikolaidou, Effect of porosity of alumina and zirconia ceramics toward pre-osteoblast response, *Front. Bioeng. Biotechnol.* 3 (2015) 175. https://doi.org/10.3389/fbioe.2015.00175.
42. J. Marchi, C.S. Delfino, J.C. Bressiani, A.H.A. Bressiani, M.M. Marques, Cell proliferation of human fibroblasts on alumina and hydroxyapatite-based ceramics with different surface treatments, *Int. J. Appl. Ceram. Technol.* 7 (2010) 139–147. https://doi.org/10.1111/j.1744-7402.2009.02388.x.

43. J. Li, Y. Liu, L. Hermansson, R. Soremark, Evaluation o biocompatibility of various ceramic powders with human fibroblasts in vitro, *Clin. Mater.* 12 (1993) 197–201. https://doi.org/10.1016/0267-6605(93)90073-G.
44. O. Standard, Application of transformation-toughened zirconia ceramics as bioceramics (1997). https://elibrary.ru/item.asp?id=5400038 (accessed November 28, 2021).
45. Y.-Z. Huang, H.-Q. Xie, X. Li, Scaffolds in bone tissue engineering: Research progress and current applications, In: *Encycl. Bone Biol.*, Academic Press, 2020: pp. 204–215. https://doi.org/10.1016/b978-0-12-801238-3.11205-x.
46. M. Kamitakahara, C. Ohtsuki, T. Miyazaki, Review paper: Behavior of ceramic biomaterials derived from tricalcium phosphate in physiological condition, *J. Biomater. Appl.* 23 (2008) 197–212. https://doi.org/10.1177/0885328208096798.
47. H.J. Kang, P. Makkar, A.R. Padalhin, G.H. Lee, S. Bin Im, B.T. Lee, Comparative study on biodegradation and biocompatibility of multichannel calcium phosphate based bone substitutes, *Mater. Sci. Eng. C.* 110 (2020). https://doi.org/10.1016/j.msec.2020.110694.
48. H. Lu, Y. Zhou, Y. Ma, L. Xiao, W. Ji, Y. Zhang, X. Wang, Current application of beta-tricalcium phosphate in bone repair and its mechanism to regulate osteogenesis, *Front. Mater.* 8 (2021) 277. https://doi.org/10.3389/fmats.2021.698915.
49. D. dos S. Tavares, L.D.O. Castro, G.D. de A. Soares, G.G. Alves, J.M. Granjeiro, Synthesis and cytotoxicity evaluation of granular magnesium substituted β-tricalcium phosphate, *J. Appl. Oral Sci.* 21 (2013) 37–42. https://doi.org/10.1590/1678-7757201302138.
50. M. Tamai, R. Nakaoka, T. Tsuchiya, Cytotoxicity of various calcium phosphate ceramics, *Key Eng. Mater.* 309–311 (2006) 263–266. https://doi.org/10.4028/WWW.SCIENTIFIC.NET/KEM.309-311.263.
51. L.A. Dos Santos, R.G. Carrodéguas, S.O. Rogero, O.Z. Higa, A.O. Boschi, A.C.F. De Arruda, Alpha-tricalcium phosphate cement: "in vitro" cytotoxicity, *Biomaterials.* 23 (2002) 2035–2042. https://doi.org/10.1016/S0142-9612(01)00333-7.
52. H.A.I. Cardoso, M. Motisuke, A.C.D. Rodas, O.Z. Higa, C.A.C. Zavaglia, pH evolution and cytotoxicity of [Alpha]-tricalcium phosphate cement with three different additives, in: *Key Eng. Mater.*, 2012: pp. 403–408. https://doi.org/10.4028/www.scientific.net/KEM.493-494.403.
53. G. Kaur, G. Pickrell, G. Kimsawatde, D. Homa, H.A. Allbee, N. Sriranganathan, Synthesis, cytotoxicity, and hydroxyapatite formation in 27-Tris-SBF for sol-gel based CaO-P2O5-SiO2-B2O 3-ZnO bioactive glasses, *Sci. Rep.* 4 (2014) 1–14. https://doi.org/10.1038/srep04392.
54. H. Oudadesse, E. Dietrich, Y.L. Gal, P. Pellen, B. Bureau, A.A. Mostafa, G. Cathelineau, Apatite forming ability and cytocompatibility of pure and Zn-doped bioactive glasses, *Biomed. Mater.* 6 (2011) 035006. https://doi.org/10.1088/1748-6041/6/3/035006.
55. M. Rismanchian, N. Khodaeian, L. Bahramian, M. Fathi, H. Sadeghi-Aliabadi, In-vitro comparison of cytotoxicity of two bioactive glasses in micropowder and nanopowder forms, *Iran. J. Pharm. Res.* 12 (2013) 437–443. https://doi.org/10.22037/ijpr.2013.1348.
56. N.M. Possolli, D.F. da Silva, J. Vieira, N. Maurmann, P. Pranke, K.B. Demétrio, E. Angioletto, O.R.K. Montedo, S. Arcaro, Dissolution, bioactivity behavior, and cytotoxicity of 19.58Li2O·11.10ZrO2·69.32SiO2 glass-ceramic, *J. Biomed. Mater. Res. – Part B Appl. Biomater.* 110 (2021) 67–78. https://doi.org/10.1002/jbm.b.34889.
57. A.M.P. Magri, K.R. Fernandes, F.R. Ueno, H.W. Kido, A.C. da Silva, F.J.C. Braga, R.N. Granito, P.R. Gabbai-Armelin, A.C.M. Rennó, Osteoconductive properties of two different bioactive glass forms (powder and fiber) combined with collagen, *Appl. Surf. Sci.* 423 (2017) 557–565. https://doi.org/10.1016/j.apsusc.2017.06.152.
58. J. Jeong, J.H. Kim, J.H. Shim, N.S. Hwang, C.Y. Heo, Bioactive calcium phosphate materials and applications in bone regeneration, *Biomater. Res.* 23 (2019) 1–11. https://doi.org/10.1186/s40824-018-0149-3.

59. S. Agrawal, R. Srivastava, Osteoinductive and osteoconductive biomaterials, In: *Racing Surf. Antimicrob. Interface Tissue Eng.*, Springer, Cham, 2020: pp. 355–395. https://doi.org/10.1007/978-3-030-34471-9_15.
60. T. Tanaka, Y. Kumagae, M. Saito, M. Chazono, H. Komaki, T. Kikuchi, S. Kitasato, K. Marumo, Bone formation and resorption in patients after implantation of β-tricalcium phosphate blocks with 60% and 75% porosity in opening-wedge high tibial osteotomy, *J. Biomed. Mater. Res. – Part B Appl. Biomater.* 86 (2008) 453–459. https://doi.org/10.1002/jbm.b.31041.
61. R. Fujita, A. Yokoyama, T. Kawasaki, T. Kohgo, Bone augmentation osteogenesis using hydroxyapatite and β-tricalcium phosphate blocks, *J. Oral Maxillofac. Surg.* 61 (2003) 1045–1053. https://doi.org/10.1016/S0278-2391(03)00317-3.
62. M. Ebrahimi, M. Botelho, Biphasic calcium phosphates (BCP) of hydroxyapatite (HA) and tricalcium phosphate (TCP) as bone substitutes: Importance of physicochemical characterizations in biomaterials studies, *Data Br.* 10 (2017) 93–97. https://doi.org/10.1016/j.dib.2016.11.080.
63. X. Wang, M. Lin, Y. Kang, Engineering porous β-tricalcium phosphate (β-TCP) scaffolds with multiple channels to promote cell migration, proliferation, and angiogenesis, *ACS Appl. Mater. Interfaces.* 11 (2019) 9223–9232. https://doi.org/10.1021/acsami.8b22041.
64. M. Bohner, B.L.G. Santoni, N. Döbelin, β-tricalcium phosphate for bone substitution: Synthesis and properties, *Acta Biomater.* 113 (2020) 23–41. https://doi.org/10.1016/j.actbio.2020.06.022.
65. T. Tanaka, H. Komaki, M. Chazono, S. Kitasato, A. Kakuta, S. Akiyama, K. Marumo, Recherche fondamentale et application clinique du bêta-tricalcium phosphate (β-TCP), *Morphologie.* 101 (2017) 164–172. https://doi.org/10.1016/j.morpho.2017.03.002.
66. A. Szcześ, L. Hołysz, E. Chibowski, Synthesis of hydroxyapatite for biomedical applications, *Adv. Colloid Interface Sci.* 249 (2017) 321–330. https://doi.org/10.1016/j.cis.2017.04.007.
67. F. Ai, L. Chen, J. Yan, K. Yang, S. Li, H. Duan, C. Cao, W. Li, K. Zhou, Hydroxyapatite scaffolds containing copper for bone tissue engineering, *J. Sol-Gel Sci. Technol.* 95 (2020) 168–179. https://doi.org/10.1007/s10971-020-05285-0.
68. R.N. Granito, A.C.M. Renno, H. Yamamura, M.C. de Almeida, P.L.M. Ruiz, D.A. Ribeiro, Hydroxyapatite from fish for bone tissue engineering: A promising approach, *Int. J. Mol. Cell. Med.* 7 (2018) 80–90. https://doi.org/10.22088/IJMCM.BUMS.7.2.80.
69. H. Shi, Z. Zhou, W. Li, Y. Fan, Z. Li, J. Wei, Hydroxyapatite based materials for bone tissue engineering: A brief and comprehensive introduction, *Crystals.* 11 (2021) 1–18. https://doi.org/10.3390/cryst11020149.
70. L. Zhang, G. Jia, M. Tang, C. Chen, J. Niu, H. Huang, B. Kang, J. Pei, H. Zeng, G. Yuan, Simultaneous enhancement of anti-corrosion, biocompatibility, and antimicrobial activities by hierarchically-structured brushite/Ag3PO4-coated Mg-based scaffolds, *Mater. Sci. Eng. C.* 111 (2020) 110779. https://doi.org/10.1016/j.msec.2020.110779.
71. Y. Zhuang, Q. Liu, G. Jia, H. Li, G. Yuan, H. Yu, A biomimetic zinc alloy scaffold coated with brushite for enhanced cranial bone regeneration, *ACS Biomater. Sci. Eng.* 7 (2021) 893–903. https://doi.org/10.1021/acsbiomaterials.9b01895.
72. A. Laskus, A. Zgadzaj, J. Kolmas, Selenium-enriched brushite: A novel biomaterial for potential use in bone tissue engineering, *Int. J. Mol. Sci.* 19 (2018) 4042. https://doi.org/10.3390/ijms19124042.
73. A.J. Salinas, M. Vallet-Regí, Bioactive ceramics: From bone grafts to tissue engineering, *RSC Adv.* 3 (2013) 11116–11131. https://doi.org/10.1039/c3ra00166k.
74. C. Großardt, A. Ewald, L.M. Grover, J.E. Barralet, U. Gbureck, Passive and active in vitro resorption of calcium and magnesium phosphate cements by osteoclastic cells, *Tissue Eng. – Part A.* 16 (2010) 3687–3695. https://doi.org/10.1089/ten.tea.2010.0281.

75. Y. Li, W. Weng, K.C. Tam, Novel highly biodegradable biphasic tricalcium phosphates composed of α-tricalcium phosphate and β-tricalcium phosphate, *Acta Biomater.* 3 (2007) 251–254. https://doi.org/10.1016/j.actbio.2006.07.003.
76. H. Rojbani, M. Nyan, K. Ohya, S. Kasugai, Evaluation of the osteoconductivity of α-tricalcium phosphate, β-tricalcium phosphate, and hydroxyapatite combined with or without simvastatin in rat calvarial defect, *J. Biomed. Mater. Res. – Part A.* 98A (2011) 488–498. https://doi.org/10.1002/jbm.a.33117.
77. D.J. Indrani, B. Soegijono, W.A. Adi, N. Trout, Phase composition and crystallinity of hydroxyapatite with various heat treatment temperatures, *Int. J. Appl. Pharm.* 9 (2017) 87–91. https://doi.org/10.22159/ijap.2017.v9s2.21.
78. D. Heymann, Cellular mechanisms of calcium phosphate ceramic degradation, *Histol. Histopathol.* 14 (1999) 871–877. https://doi.org/10.14670/HH-14.871.
79. A.M. Gatti, T. Yamamuro, L.L. Hench, O.H. Andersson, In-vivo reactions in some bioactive glasses and glass-ceramics granules, *Cells Mater.* 3 (1993) 283–291. https://digitalcommons.usu.edu/cellsandmaterials (accessed December 16, 2021).
80. E.A.B. Effah Kaufmann, P. Ducheyne, S. Radin, D.A. Bonnell, R. Composto, Initial events at the bioactive glass surface in contact with protein-containing solutions, *J. Biomed. Mater. Res.* 52 (2000) 825–830. https://doi.org/10.1002/1097-4636(20001215)52:4<825::AID-JBM28>3.0.CO;2–M.
81. H. Saito, Y. Araki, H. Katsuno, T. Nakada, Phase transition of amorphous calcium phosphate to calcium hydrogen phosphate dihydrate in simulated body fluid, *J. Cryst. Growth.* 553 (2021) 125937. https://doi.org/10.1016/j.jcrysgro.2020.125937.
82. F. Barrère, M. Ni, P. Habibovic, P. Ducheyne, K. de Groot, Degradation of bioceramics, in: *Tissue Eng.*, Elsevier Inc., 2008: pp. 223–254. https://doi.org/10.1016/B978-0-12-370869-4.00008-2.
83. R.Z. Legeros, S. Lin, R. Rohanizadeh, D. Mijares, J.P. Legeros, Biphasic calcium phosphate bioceramics: Preparation, properties and applications, *J. Mater. Sci. Mater. Med.* 14 (2003) 201–209. https://doi.org/10.1023/A:1022872421333.
84. S. Yamada, D. Heymann, J.M. Bouler, G. Daculsi, Osteoclastic resorption of calcium phosphate ceramics with different hydroxyapatite/β-tricalcium phosphate ratios, *Biomaterials.* 18 (1997) 1037–1041. https://doi.org/10.1016/S0142-9612(97)00036-7.
85. S. Yamada, D. Heymann, J.M. Bouler, G. Daculsi, Osteoclastic resorption of biphasic calcium phosphate ceramic in vitro, *J. Biomed. Mater. Res.* 37 (1997) 346–352. https://doi.org/10.1002/(SICI)1097-4636(19971205)37:3<346::AID-JBM5>3.0.CO;2–L.
86. E. Uribe-Querol, C. Rosales, Phagocytosis: Our current understanding of a universal biological process, *Front. Immunol.* 11 (2020) 1066. https://doi.org/10.3389/fimmu.2020.01066.
87. A. Ibara, H. Miyaji, B. Fugetsu, E. Nishida, H. Takita, S. Tanaka, T. Sugaya, M. Kawanami, Osteoconductivity and biodegradability of collagen scaffold coated with nano-β-TCP and fibroblast growth factor 2, *J. Nanomater.* 2013 (2013). https://doi.org/10.1155/2013/639502.
88. P. Frayssinet, N.R. Bioland, T.F.T. Bioland, T.J. Fages Bioland, T.D. Hardy, Cells and materials cell-degradation of calcium phosphate ceramics (1993). https://digitalcommons.usu.edu/cellsandmaterials/vol3/iss4/6 (accessed December 18, 2021).
89. E.F. Eriksen, Cellular mechanisms of bone remodeling, *Rev. Endocr. Metab. Disord.* 11 (2010) 219–227. https://doi.org/10.1007/s11154-010-9153-1.
90. B. Boyce, Z. Yao, L. Xing, Osteoclasts have multiple roles in bone in addition to bone resorption, *Crit. Rev. Eukaryot. Gene Expr.* 19 (2009) 171–180. https://doi.org/10.1615/CritRevEukarGeneExpr.v19.i3.10.
91. F.D. Beaman, L.W. Bancroft, J.J. Peterson, M.J. Kransdorf, Bone graft materials and synthetic substitutes, *Radiol. Clin. North Am.* 44 (2006) 451–461. https://doi.org/10.1016/j.rcl.2006.01.001.

92. C.J. Damien, J.R. Parsons, Bone graft and bone graft substitutes: A review of current technology and applications, *J. Appl. Biomater.* 2 (1991) 187–208. https://doi.org/10.1002/jab.770020307.
93. G.J. Dias, P. Mahoney, N.A. Hung, L.A. Sharma, P. Kalita, R.A. Smith, R.J. Kelly, A. Ali, Osteoconduction in keratin-hydroxyapatite composite bone-graft substitutes, *J. Biomed. Mater. Res. – Part B Appl. Biomater.* 105 (2017) 2034–2044. https://doi.org/10.1002/jbm.b.33735.
94. M. Barbeck, M. Dard, M. Kokkinopoulou, J. Markl, P. Booms, R.A. Sader, C.J. Kirkpatrick, S. Ghanaati, Small-sized granules of biphasic bone substitutes support fast implant bed vascularization, *Biomatter.* 5 (2015) e1056943. https://doi.org/10.1080/21592535.2015.1056943.
95. Z. Gu, H. Wang, L. Li, Q. Wang, X. Yu, Cell-mediated degradation of strontium-doped calcium polyphosphate scaffold for bone tissue engineering, *Biomed. Mater.* 7 (2012) 065007. https://doi.org/10.1088/1748-6041/7/6/065007.
96. K. Devoe, S. Banerjee, M. Roy, A. Bandyopadhyay, S. Bose, Resorbable tricalcium phosphates for bone tissue engineering: Influence of Sro doping, *J. Am. Ceram. Soc.* 95 (2012) 3095–3102. https://doi.org/10.1111/j.1551-2916.2012.05356.x.
97. G.A. Fielding, A. Bandyopadhyay, S. Bose, Effects of silica and zinc oxide doping on mechanical and biological properties of 3D printed tricalcium phosphate tissue engineering scaffolds, *Dent. Mater.* 28 (2012) 113–122. https://doi.org/10.1016/j.dental.2011.09.010.
98. D. Ke, S. Tarafder, S. Vahabzadeh, S. Bose, Effects of MgO, ZnO, SrO, and SiO 2 in tricalcium phosphate scaffolds on in vitro gene expression and in vivo osteogenesis, *Mater. Sci. Eng. C.* 96 (2019) 10–19. https://doi.org/10.1016/j.msec.2018.10.073.
99. S. Vahabzadeh, S. Bose, Effects of iron on physical and mechanical properties, and osteoblast cell interaction in β-tricalcium phosphate, *Ann. Biomed. Eng.* 45 (2017) 819–828. https://doi.org/10.1007/s10439-016-1724-1.
100. S. Meininger, C. Moseke, K. Spatz, E. März, C. Blum, A. Ewald, E. Vorndran, Effect of strontium substitution on the material properties and osteogenic potential of 3D powder printed magnesium phosphate scaffolds, *Mater. Sci. Eng. C.* 98 (2019) 1145–1158. https://doi.org/10.1016/j.msec.2019.01.053.
101. J.R. Jones, L.L. Hench, Factors affecting the structure and properties of bioactive foam scaffolds for tissue engineering, *J. Biomed. Mater. Res. – Part B Appl. Biomater.* 68 (2004) 36–44. https://doi.org/10.1002/jbm.b.10071.
102. L. Wen, C. Rüssel, D.E. Day, G. Völksch, Bioactive comparison of a borate, phosphate and silicate glass, *J. Mater. Res.* 21 (2006) 125–131. https://doi.org/10.1557/jmr.2006.0025.
103. W.C. Lepry, S.N. Nazhat, A review of phosphate and borate sol-gel glasses for biomedical applications, *Adv. NanoBiomed Res.* 1 (2021) 2000055. https://doi.org/10.1002/anbr.202000055.
104. K. Schuhladen, X. Wang, L. Hupa, A.R. Boccaccini, Dissolution of borate and borosilicate bioactive glasses and the influence of ion (Zn, Cu) doping in different solutions, *J. Non. Cryst. Solids.* 502 (2018) 22–34. https://doi.org/10.1016/j.jnoncrysol.2018.08.037.
105. L. Björkvik, X. Wang, L. Hupa, Dissolution of bioactive glasses in acidic solutions with the focus on lactic acid, *Int. J. Appl. Glas. Sci.* 7 (2016) 154–163. https://doi.org/10.1111/ijag.12198.
106. K. Zheng, W. Niu, B. Lei, A.R. Boccaccini, Immunomodulatory bioactive glasses for tissue regeneration, *Acta Biomater.* 133 (2021) 168–186. https://doi.org/10.1016/j.actbio.2021.08.023.
107. M. Shi, Z. Chen, S. Farnaghi, T. Friis, X. Mao, Y. Xiao, C. Wu, Copper-doped mesoporous silica nanospheres, a promising immunomodulatory agent for inducing osteogenesis, *Acta Biomater.* 30 (2016) 334–344. https://doi.org/10.1016/j.actbio.2015.11.033.

108. H. Pohunková, M. Adam, Reactivity and the fate of some composite bioimplants based on collagen in connective tissue, *Biomaterials.* 16 (1995) 67–71. https://doi.org/10.1016/0142-9612(95)91098-J.
109. H. Tripathi, S. Kumar Hira, A. Sampath Kumar, U. Gupta, P. Pratim Manna, S.P. Singh, Structural characterization and in vitro bioactivity assessment of SiO2-CaO-P2O5-K2O-Al_2O_3 glass as bioactive ceramic material, *Ceram. Int.* 41 (2015) 11756–11769. https://doi.org/10.1016/j.ceramint.2015.05.143.
110. F. Baino, S. Yamaguchi, The use of simulated body fluid (SBF) for assessing materials bioactivity in the context of tissue engineering: Review and challenges, *Biomimetics.* 5 (2020) 1–19. https://doi.org/10.3390/biomimetics5040057.
111. J.P. Matinlinna, Handbook of oral biomaterials (2014). https://doi.org/10.4032/9789814463133.
112. L.L. Hench, H.A. Paschall, Direct chemical bond of bioactive glass-ceramic materials to bone and muscle, *J. Biomed. Mater. Res.* 7 (1973) 25–42. https://doi.org/10.1002/jbm.820070304.
113. L.L. Hench, C.G. Pantano, P.J. Buscemi, D.C. Greenspan, Analysis of bioglass fixation of hip prostheses, *J. Biomed. Mater. Res.* 11 (1977) 267–282. https://doi.org/10.1002/jbm.820110211.

7 Modeling and Simulations on Medical Implementations of Bioceramics

Ritambhara Dash, Poorti Yadav, Ram K. Singh, and A.S. Bhattacharyya
Central University of Jharkhand

7.1 INTRODUCTION

A biomaterial is a substance that interacts with biological systems mainly for medical purposes [1,2]. They can be natural or synthesized. Biocompatibility is a major factor for biomaterials and is application-specific [3]. Although biomaterials can be made from metallic components, polymers, ceramics or composite materials, bioceramics are the most accepted form [4–7]. There are four major types of bioceramics materials: bioinert ceramics, bioactive ceramics, bioresorbable ceramics, and piezoceramics. The use of a bioceramic can be passive or active in the form of a biomedical device or a replacement like HAp hip joints. Their focus can be partial or as a whole in a living structure performing augmentation or replacing a natural function. Bioceramics are used frequently in dental applications, bioinert implants surgery, and drug delivery [4–8].

Bioceramics and piezoceramics are mainly made of alumina or hydroxyapatite, calcium phosphate. Calcium phosphate or HA $\{Ca_{10}\,(PO_{46}\,(OH)_2)\}$ – the principal component of natural bone. Ceramics made from synthetic ceramic phosphate can be used in the body for bone replacement. They can bond to bone and promote bone growth at their surfaces. They can be also used as a coating on dental and orthopedic implants [9,10].

Calcium phosphate-based bioceramics are used in medicine and dentistry. Applications include dental implants, percutaneous devices, and use in periodontal treatment, alveolar ridge augmentation, orthopedics, maxillofacial surgery, otolaryngology, and spinal surgery. The different phases of calcium phosphate ceramics are used depending upon whether a resorbable or bioactive material is desired [11,12].

The computational modeling methods and software are on a rise in today's time. They are economical both in terms of time and cost and also provide an in-depth atomistic view giving a better understanding of the structure property correlation of the bioceramics make their more efficient production and use. The atomic modeling techniques for proteins and pharmaceuticals and organic molecules, inorganic

 DOI: 10.1201/9781003258353-9

materials, microporous catalysts (zeolites and metal-organic frameworks, MOF), high-temperature superconductors, ternary and quaternary oxides, and biomaterials are performed. There are three primary aspects on which the modeling is usually performed [13].

a. modeling crystal structures
b. modeling amorphous structures atomistic structures for noncrystalline solids
c. modeling inorganic surface chemistry

Many modeling and simulation techniques have been used in diverse areas, and it is not possible to cover every aspect. However, an attempt has been made in some of the important and recent computational techniques employed.

7.2 NANOBIOINTERFACE INTERACTIONS – MD SIMULATIONS

The interaction of nanoparticles with cells is the foundation of biocompatibility. Simulations in this regard were performed using all-atom molecular dynamics (AAMD), coarse-grained molecular dynamics (CGMD), and dissipative particle dynamics (DPD) methods [14]. Molecular dynamics (MD) simulation techniques are based on molecular structure and interactions. AAMD is commonly used in the case of biomembranes due to its high accuracy. However, due to the use of a lot of computing resources and time for simulations, they are not suitable for the mesoscopic phenomena of the lipid monolayers. CGMD is comparatively economical as it involves a cluster of atoms as one head interacting with each other. Martini force field, CGMD and DPD, are the two most popular MD simulations for biological systems. Martini force field simulations are suitable for complex lipid systems; however, it does not take into consideration the air–water surface tension. DPD and, more importantly, many-body DPD which is better in terms of the air–water interface due to the use of many-body potential forces are used for biomembrane simulations [15]. MD simulations have been used in other aspects of bioceramics including the use of shell model to study nanosegregation caused by fluorine ions in bioactive glasses [16–18].

Multiscale simulation methods where the output from one simulation is fed as the input to the other using boundary conditions provide a spatiotemporal coherence with the experiments. However, there is a loss in accuracy and a balance between the two depending upon the situation needs to be maintained where a POPC lipid molecule is simulated using the three techniques mentioned above [19].

7.3 DENTAL APPLICATIONS

Bioceramics can be also used as a coating on dental and orthopedic implants A tooth structure includes supporting bone, periodontal ligament (PDL), dentine, and pulp. A dental bridge with load distributed at the ceramic pontics was modeled using FEM. The crack initiation and propagation were studied for analyzing the fracture strength of bioceramic structures, thereby providing a means to design optimization.

The principal stress and strain energy density (SED) are determined. The crack was found to propagate toward the left pontic on the canine side [20].

In dentistry, increased use of calcium hydroxide weakens the dentin and increases the risk of root fracture. By apexification, a calcified barrier is created without using any apparent pathosis to induce root-end closure which decreases the use of calcium hydroxide and also reduces the duration of treatment. A simulated open apex was created on polyvinyl tubes filled with NeO MTA plus and MTA Angelus which are bioceramics used for dental applications [21]. The sealing ability of root canals using Gutta Percha bioceramic was evaluated based on a bacterial leakage test using microcomputed tomography. The GP/epoxy resin-based sealer showed good performance in terms of resistance to leakage and 3D compaction [22].

7.4 STATISTICAL OPTIMIZATION IN BIOCERAMICS

Bioceramic (mHAP) was capped with Mn-doped superparamagnetic iron oxide (SPIONS) nanoparticles for As(III) removal from water. The statistical optimization was done using *Response Surface Methodology* (RSM) which is a factual mathematical modelization technique used to evaluate the relationship between controlled experimental variables and outcomes [23]. Other statistical methods used in bioceramics are like *Taguchi optimization* methods where mechanical properties of bioceramics were improved buy the use of bioresorbable polymers and *Grey relational analysis (GRA)* in laser microgrooving on the surface of HAp by laser [24,25].

7.5 BONE GRAFTING

A comparatively new bioceramic α-calcium sulfate hemihydrate (α-CSH) and α-CSH/platelet-rich plasma (PRP) was used for bone grafting to accelerate bone healing and regenerative rabbit model. The images obtained from microcomputed tomography were reconstructed into a 2D image using NRecon software. Parameters like bone mineral density and new bone volume were evaluated using this software using the Hounsfield Unit. It was observed that the bone density increased and defect volume decreased after 2 weeks [26,27].

7.6 BIOACTIVE GLASS AND HAp

Bioceramics used to replace bones in the human body and are usually made of HAP and alumina. Especially, the use of HAp which is a calcium phosphate compound {Ca_{10} (PO_{46} $(OH)_2$) is on the rise as its chemical composition matches with the bones and even can also cause bone growth on their surfaces. HAP is used in both orthopedic and dental implants.

The first bioactive material, the melt-derived 45S5 Bioglass®, was discovered by Hench and coworkers in the 1970s [28]. Bioceramics in the form of bioactive glasses (45S5 Bioglass®) and hydroxyapatite undergo complex interactions with the biological environment, and their modeling is very crucial for processing and their applications. Modeling of bioceramics (both bioactive glasses as the 45S5 Bioglass® and hydroxyapatite) and their complex interactions with the biological environment are done.

The interactions take place at the surface, and therefore, modeling of the surface chemistry is a significant aspect. Water is a significant molecule in bioactive systems, and its interaction with HA in surface adsorption has been simulated by *the ab initio* method. The mechanical properties of the first-ever biomaterial 45S5 Bioglass® are improved by adding oxides as fillers. Molecular dynamics simulation is used for amorphous or glassy materials. Both 45S5 bioglass and HAP have been the subject of several experimental studies; there has been an attempt to understand the integration mechanism in the bones and teeth at the molecular level using classical dynamics and *ab initio* methods. The structural and dynamical properties of 45S5 bioglass as well as surface characteristics and reactivity of HAP have been looked into [28–32].

Simulations have been done on the carbonate defects found in the HAP lattice. The carbonate group in the HAP channel and at the position of the phosphate groups are the two types of defects observed. A combined defect can also exist. Simulation studies predicted that two OH groups replaced by a single carbonate group are the most energetically favorable. Charge compensation by substitution of sodium and potassium for a calcium ion makes the defect energetically favorable [33].

Simulation of HAP nanocrystals is done for HRTEM image calculations. High-resolution image simulations are done using MEGACELL software whose output files are mainly the raw data of atomic positions. The HRTEM experimental data images of HA particles oriented along different zone axes were interpreted applying this software to construct HA nanocrystals. This provides a better interpretation of phase-contrast images [34].

7.7 DENSITY FUNCTIONAL THEORY CALCULATIONS

DFT calculations were carried out considering (100) and (001) surfaces of HAP which were represented using $1\times2\times2$ and $3\times3\times1$ slab models. It was found that the adsorption of phosphate onto the two crystallographic surfaces is energetically favored. Calculations predicted that triphosphate only adsorbed on the (001) surface of HAP. A predominance of (001) surface was observed. The coexistence of the monoclinic hexagonal HAP phase at room temperature has been predicted. The DOS distribution for the initially optimized HAP lattice and all-atom positions and with changes of total charges for bulk unit cells were calculated. The forbidden energy gap was also obtained. Calculation of DOS led to the study of the influence of Nb lattice charges on the shift and changes of bandgap [35].

Beta-tricalcium phosphate (I^2 – TCP) are biomaterials with excellent biocompatibility and identical chemical composition to the natural teeth and bone. The electronic and optical properties of I^2 – TCP have been investigated using DFT using potential linear augmented plane wave method (PLAPW) with three types of approximations, viz., local density approx, general gradient approx., and modified Becke-Johnson (mBJ) approximations with bandgap values of 5.5, 5.9, and 6.8 eV, which are quite accurate [36].

Network connectivity (NC) which is the average of bridging oxygen (BO) atoms per glass-forming species is used to compare the bioactivity of different glass compositions. The value of NC is 4 for pure silica glass and 2 for a chain-like structure, It was found that bioactivity is shown in glasses with NC values lower than 3. A low value of NC indicates an open and fragmented glass structure. The value of NC is 1.9 for 45S5 bioglass [37,38].

For porous biomaterials, the fundamental mathematical operations, viz., dilation and erosion, were used which can be defined with the basic Minkowski set operation called addition and subtraction. Erosion of an image I by a structural element H is expressed as $I \ominus H$, which reduces the size of the pores by eroding its outline. The dilation on the other hand is expressed as $I \oplus H$, where pores smaller than H are filled [39].

7.8 BIOCERAMIC BLENDS

The mixture rule is applicable when we blend a bioceramic-like HAP with any other material like Ti. If P_{HAP} and P_{Ti} are any specific characteristics associated with pure HAP and Ti with the respective volume fraction V_{HAP} and V_{Ti}, then the combined characteristics of the Ti-HAP can be represented as

$$P = P_{HAP}V_{HAP} - P_{Ti}\left(1 - V_{HAP}\right)$$

Computer simulations have been performed on the mechanical behavior of biphasic calcium phosphate which is a blend of HAP and β-TCP and is used as a bone repair material. The simulated annealing molecular dynamics method was used to study the different mass ratio compositions. An increase in HAP concentration was found to increase the elastic modulus of the blend [40,41].

7.9 3D PRINTING IN BIOMATERIALS

3D printing is used nowadays for scaffold design and fabrication which requires an optimization among the mechanical properties, pore size, and interconnectivity. Finite element modeling is done to find the effective modulus to match the bone characteristics. Bioceramics filled with magnetite nanoparticles by 3D printing were made, and nonlinear bending and postbuckling characteristics have been studied. The maximum rupture strength in terms of tensile strength P, diameter D, and thickness t of the Brazilian disc model is given as [42]

$$\sigma_r = \frac{2P}{\pi Dt}$$

Poly-lactic acid (PLA) is a commonly used bioplastic in the packaging and fabrication of biomedical devices such as orthopedic implants, drug delivery systems, and scaffolds. A PLA filament was used to 3D print. Due to its poor mechanical property in the amorphous state as well as feeble potential for biomimic, it is either crystallized using heat or made into a composite using woof flour. 3D printing is made of composites of petroleum-based polymers and biofillers. Natural (collagen, gelatin, alginate) hyaluronic acid are used in bioprinting, For bioinks used in 3D printing, synthetic polymers such as PVA and PEG are used. Optical 3D printing of a single bio-based resin derived from soybeans can be processed even without the addition of a photoinitiator. The use of laser nanolithography and table top 3D printer was done in the fabrication process. Checkered patterns were made in an industry line production service [43–45].

7.10 MECHANICAL PROPERTIES OF BIOMATERIALS

Mechanical properties like tensile strength and Young's modulus of Ti-HAP were determined by the rule of mixture. Usually, ANSYS for CAD drawing is used to produce 3D modeling for a solid structure. In the case of bioimplants with very complex-layered composite structures, the shape is divided into small meshes connected at nodes. The higher is the number of mesh, the higher is the accuracy. The response of each mesh for each of the loading conditions is observed in FE analysis. The loading conditions include either external internal force function applied like displacement, pressure, and gravity in the structural discipline. System of solutions is solved using skyline; frontal and iteration schemes were employed.

FEM was used to simulate the femoral shaft with bone defects and treatment with the lateral plate. The peak von Mises stress and the plate displacement were determined aiming at bone reconstruction and stability of the medial femur where the damaged region is divided into several meshes. The FEM also tries to find out an alternative treatment method that prevents additional soft tissue and vascular damage.

Stress being the key indicator in fracture studies of bioceramics, continuum discrete element method (CDEM) is used which is based upon rotating crack tip model, extended finite element model (XFEM) has been developed which is more economical in terms of time. It consists of adding discontinuous basis functions to the standard polynomial shape function in elements intersected by a crack-for-crack opening displacements (COD) [46,47].

Artificial neural networks (ANN) of nanoindentation of Hap/PLLA composite were performed. The matrix phase properties of PLLA polymer were predicted. The Hap/PLLA composite has prosthetic applications and is used in replacing damaged hard tissues and filling bone defects. The advantage these composites have over other materials is the possibility of tailoring the mechanical and biological properties. ANN modeling involves the mapping of input and output values. Hardness is determined as the ratio of load and the projected area which is determined indirectly from the depth of penetration. The P-h curves were not coincident indicating a difference in the contact area. The P-h curve changes with the material characteristic changes were studied [48,49].

Bone tissue engineering (BTE) is done to treat osteoporosis and accelerate bone fracture healing. Scaffolds are used in BTE where permeability and mechanical properties are significant. Fluid flow is analyzed using FEM using computational fluid dynamics (CFD). Optimization of the parameters is done in the design of bone scaffolds to get properties similar to the region of implantation. Homogenization theory is used which deals with permeability and stiffness tensors [50].

7.11 SUMMARY

Although modeling and simulation in bioceramics is a vast area and its not possible to cover every aspect here, an attempt has been made to focus some of the significant methods. The molecular dynamics (MD) simulations which deal with the structure and molecular interactions to predict bulk properties and ab initio methods are highly used. The density functional theory calculating the lattice properties and density of states is beneficial in studying the effect of blending biocermaics on the properties like bandgap, etc. Simulation are performed in bone grafting and bone tissue engineering using finite element modeling and artificial neural network for optimization of mechanical properties in terms of biocompatibility. Statistical optimization and simulations in phase-contrast images obtained from microscopy are also done. A summary of the modeling and simulation techniques in bioceramics discussed in this chapter is given in the table below.

S. No.	Method	Applications	Reference
	MD simulations		
1	AA MD	Nanobiointerface interactions	[14]
2	CG MD		[14]
3	DPD	Biomembrane simulations	[14,15]
4	Shell model	Nanosegregation	[16–18,40,41]
	Statistical optimization		
5	Response surface methodology (RSM)	Statistical optimization, As (III) removal from water	[23]
6	Taguchi optimization	Mechanical properties	[24]
7	Grey relational analysis (GRA)	Laser grooving on HAp	[25]
8	Microcomputed tomographic (μ-CT)	Bone grafting	[26,27]
9	MEGACELL	Interpretation of HRTEM phase-contrast images	[34]
10	3D printing		[42–44]
	Finite element modeling (FEM)	**Dental applications**	[20]
11	Continuum discrete element method (CDEM)	Fracture studies	[45]
12	Extended finite element modeling (XFEM)		[46,47]
13	Artificial neural network (ANN)	Prosthetic applications	[48,49]
14	**Density functional theory (DFT)**		[35]
	Potential linear augmented plane wave method (PLAPW) • local density approximations, • general gradient approximations, • modified Becke-Johnson (m BJ) approximations.	Electronic and optical properties	[36]
15	Minkowski set operation	Porous biomaterials	[39]
16	*Ab initio*	Bioglass, surface interactions	[28–32]

REFERENCES

1. B. Okzale, M.S. Sakar, D.J. Mooney, *Biomaterials* 267, 2021, 120497.
2. M. Saini, Y. Singh, P. Arora, V. Arora, K. Jain, *World J Clin Cases* 3(1), 2015, 52–57
3. D.F. Williams, *Bioactive Mater* 10, 2022, 306–322
4. L.L. Hench, *J Am Ceram Soc* 81, 1998, 1705.
5. M. Vallet-Regi, *J Chem Soc Dalton Trans*, 2001, 97.
6. T. Kokubo et al., *Biomaterials* 24, 2003, 2161.
7. M. Cerruti, N. Sahai, *Rev Min Geochem* 64, 2006, 283.
8. L.L. Hench, *Biomaterials* 19, 1998, 1419.
9. J.C. Knowles, *J Mater Chem* 13, 2003, 2395.
10. N. Roveri, B. Palazzo, *Tissue, Cell and Organ Engineering*, Wiley-VCH, Weinheim, 2006.
11. B. Palazzo et al., *Adv Funct Mater* 17, 2007, 2180.
12. B. Palazzo et al., *Acta Biomaterialia* 5, 2009, 1241.
13. A. Pedone, M. Corno, Computer simulation techniques for modeling in bioceramics. *Critical reviews La Chimica*, L' Industria, 2010.
14. P.M. Pieczywek, W. Płaziński, A. Zdunek, *Sci Rep* 10, 2020, 14691.
15. Y. Zhu, X. Bai, G. Hu, arxiv. 2103.15315
16. A. Pedone, *J Phys Chem C* 113, 2009, 20773.
17. A. Tilocca et al., *Chem Mater* 19, 2007, 95.
18. G. Lusvardi et al., *J Phys Chem B* 106, 2002, 9753.
19. Zhang, X., Ma, G., Wei, W., *NPG Asia Mater* 13, 2021, 52.
20. C. Wu, J. Fang, S. Zhou, Z. Zhang, G. Sun, G.P. Steven, *Int J Num Meth Eng*, 2020, 6340
21. M.P. Paul, S. Amin, A. Mayya, R. Naik, S.S. Mayya, *The Open Dent J* 14(1), 2020, 698–703
22. K. Yanpiset, D. Banomyong, K. Chotvorrarak, R.L. Srisatjaluk, *Restor Dent Endod* 43(3), 2018, e30
23. N. Dhiman, Markandeya, F. Fatima, P.N. Saxsena, S. Roy, P.K. Routg, S. Patnaik, *RSC Adv* 7, 2017, 32866
24. Dorozhkin, V. Sergey, T. Ajaal, Trans Tech Publications, Ltd., December 2009.
25. N. Roy, A.S. Kuar, S. Mitra, *Mater Today: Proc* 18, 2019, 5540–5549
26. S. Syam et al., *Appl Sci* 11, 2021, 5271.
27. Lu et al. *BMC Musculoskelet Disord* 21, 2020, 421
28. L.L. Hench et al., *J Biomed Mater Res* 2, 1971, 117.
29. T. Kokubo et al., *J Biomed Mater Res Symp* 24, 1990, 721.
30. O.H. Andersson et al., *J Non Cryst Solids* 119, 1990, 290.
31. G. Lusvardi et al., *J Phys Chem B* 112, 2008, 12730.
32. G. Lusvardi et al., *J Biomater Appl.* 22, 2008, 505.
33. V. Bystrov, E. Paramonova, L. Avakyan, J. Coutinho, N. Bulina, *Nanomaterials (Basel)* 11(10), 2021, 2752.
34. C.A.O. Ramirez, J. Terra, A.J. Ramirez, D.E. Ellis, A.M. Rossi, *Key Eng Mater* 493–494, 2011, 763–767
35. M. Rivas, J. Casanovas, L.J. del Valle, O. Bertran, G. Revilla-López, P. Turon, J. Puiggalí, C. Alemán, *Dalton Trans* 44(21), 2015, 9980–9991.
36. A.M.A. Bakheeta, M.A. Saeeda, A.R.M. Isaa, R. Sahnoun, *Jurnal Teknologi (Sci & Eng)* 78(3–2), 2016, 159–164.
37. D.S. Brauer, N. Karpukhina, R.V. Law, R.G. Hilla, *J Mater Chem* 19, 2009, 5629–5636 | 5629.
38. H.R. Fernandes, A. Gaddam, A. Rebelo, D. Brazete, G.E. Stan, J.M.F. Ferreira, *Materials* 11, 2018, 2530
39. M. Ezzahmouly, A. Elmoutaouakkil, M. Ed-Dhahraouy, H. Khallok, A. Elouahli, A. Mazurier, A. El Albani, Z. Hatim, *Heliyon* 5(12), 2019, e02557.

40. N. Radhi, *Int J Civil Eng & Tech* 9(6), 2018, 28–39
41. X. Ma et al., *J Mol Model* 23, 2017, 156 (Ti- Hap)
42. A. Elghazel, R. Taktak, J. Bouaziz, Intech Open, 2016, DOI: 10.5772/63518.
43. D. Vaes, P. Van Puyvelde, *Prog Polym Sci* 118, 2021, 101411
44. V.I. Putlyaev et al., *Inorg Mater Appl Res* 10, 2019, 1101–1108
45. A. Jasemi, B.K. Moghadas, A. Khandan, S. Saber-Samandari, *Ceram Int* 48(1), 2022, 1314–1325
46. R. Barauskas, A. Sankauskaite, A. Abraitiene, *Text Res J* 88(3), 2018, 293–311.
47. R. Barauskas, A. Sankauskaite, A. Abraitiene, *Text Res J* 88(3), 2018, 293–311.
48. A.H. Montazeran, S.S. Saber, A. Khandan, *Nanomed J* 5(3), 2018, 163–171.
49. A. Mitra, S. Mehran, F. Arghavan, A. Noor, *Ceram Int* 40(8), 2014, 12439–12448
50. J.A. Sanz-Herrera, E. Reina-Romo, *Appl Sci* 9, 2019, 3674

Section C

Applications

8 Bioceramics
From Concept to Clinic

S. Chitra and S. Rajeshkumar
Saveetha Dental College & Hospitals, Saveetha Institute of Medical and Technical Sciences (SIMATS)

Nibin K. Mathew
University of Madras

8.1 INTRODUCTION

In this robotic world, extraordinary research in the field of medicine is building momentum to push human health forward. Generally, when a part of an organ is injured or damaged, tissues around it get disturbed, damaged, and collapsed; at that point, there is a need for engineered biomaterials to cure the problems. In this direction, invention of biomaterials has opened a plethora of possibilities to restore the task of natural living tissues and organs through 3R (repair or replace or regenerate) concept. For that, appropriate biomaterials with required structural characteristics need to be designed. Such biomaterials are generally defined as materials that are used to design devices to replace or regenerate the part or function of the organ in contact with living tissue [1]. The implementation of biomaterial or an implant mainly depends on biocompatibility, which means its favourable healthy interaction with cells, muscles/ligaments, fat, bones, and organs etc… Metals are prominent for their wear strength, high resistance, and ductility; however, their highly corrosive nature and low biocompatibility exterminate living tissues. In addition, due to the rapid metal-ion diffusion, metallic implants may cause allergic reaction to the host tissues [2]. Alternatively, soft polymers own distinct composition, are flexible, and are available in the form of gel and films, but they have restricted load bearing and orthopaedic applications owing to their poor mechanical strength. Similarly, hard polymers failed to induce osteoconductive as well as osteoinductive behaviours. Contradictorily, composites ensure high elastic strength and no corrosion issues but have short-term durability [3]. Further, bioceramics have revolutionized implant technology by their elevated compatible interaction with tissues and their unresponsiveness towards corrosion *in vivo*.

Alumina has been known as a medication since ancient times and is widely used to reduce water turbidity and in antibiotic/ haemostatic ointments. TCP, alumina, Hap, and ZrO_2 have been continuously roaming from the year 1910 until this era to treat bone substitution, hip and knee joints, dental coatings, and dental posted bridges [4]. In the ancient period, researchers used calcium phosphate-based ceramics for various biomedical applications; however, they were not aware of the crystalline phases of the derived

DOI: 10.1201/9781003258353-11

Ca-P [5]. In 1895, Rontgen discovered X-rays then Bragg introduced its application in the acquisition of crystal structure (X-ray diffraction [XRD]) in the year 1921. Emergence of XRD in the material world was important invention to understand the structural chemistry. Descriptive to predictive possibilities of calcium phosphates and their derivatives provided a new vision to researchers to design and fabricate better or required materials [6].

Plaster of Paris (calcium sulphate with phosphate compound) was the first bioceramic that was extensively analysed. Initially, research community hypothesized that calcium and phosphate mineral implantation may stimulate osseous repair towards the regeneration of bony defects. Several researchers investigated the impact of plaster of Paris on bone regeneration and reported that this material does not cause any adverse effects on osseous tissue and also initiates new bone generation. This material has bioactive features that trigger the regeneration potentiality and also compatible features were observed; however, this material was not approved for clinical applications due to poor mechanical stability [7]. In the same era, calcium hydroxide was exposed to the subcutaneous atmosphere and investigated the effectiveness towards bone formation and concluded that dystrophic bone formation with non-cellular reaction. Then, calcium carbonates and calcium oxide with alumina ($CaO\text{-}Al_2O_3$), titanium ($CaO\text{-}TiO_2$), and zirconium ($CaO\text{-}ZrO_2$) ceramics entered into the picture; these ceramics induce osseous ingrowths [8,9]. In these directions, ceramics are evolved followed by recent clinical bioceramics such as Hap, TCP, and bioactive glass [10].

8.2 BIOCERAMICS

Bioceramics are the main class of biomaterials especially designed for repair, regeneration, and reconstruction of damaged parts of the body [1,3] comprising of glasses, glass ceramics, and ceramics that are used as implant materials. Bioceramics have potential role in tissue replacement and have been used as an implant coating to reduce the rejection response and to enhance biocompatibility. They provide templates and a framework that dissolves and stimulates tissue regeneration, and they further function as resorbable lattices [11]. Bioceramics are one of the durable materials possessing appreciable chemical and thermal stability with wear resistance and high strength that paves its application in surgical implants. They act as a material that stimulates bone growth, and at the surface of joint prostheses, they have low coefficient of friction [12].

Implant materials should not create any toxic response with the surrounding tissues. It should be mechanically stable and not deform excessively or be rapid to fracture. It has to mimic natural host tissues and not to create any foreign body reaction. It should not stimulate unnecessary growth or chemical effect in body fluids as well as with neighbouring tissues. Implant material should not dissolve or corrode *in vivo* [13,14]. Bioceramics satisfy all the above-mentioned properties except for their mechanical stability. Rigid nature and brittleness of ceramics and glasses may cause fracture; also, reaction with water induces fatigue in some oxide ceramics. Therefore, ceramics with specific chemical properties are preferred for implant materials [15]. In general, tissue responses of bioceramics are categorized using three terms, namely, bioinert, bioresorbable, and bioactive [16]. The type of tissue response at implant interface and the mechanism of tissue attachment are directly interrelated. It is obvious that none of the material shows inert response to the living tissues; all those implants bring forth response to the living tissues either positively or negatively [1]. Four different types of interactions between implant and tissues are observed and are tabulated in Table 8.1 [1,17].

8.2.1 Bio-physico-chemical Properties of Bioceramics

Ceramics consists of enormous class of minerals, mostly non-metallic, whose types of chemical bonds, chemical composition, and behaviours are different from each other. Bioceramics are highly compatible to the host tissues than any other implant materials based on the bioactive, bioresorbable, and bioinert properties. In general, ceramics are a unique category of structure with fundamental physical factor that determines the characteristic feature of material's reaction with a bio-system; also, it can vary its properties under wide range of conditions [18].

8.2.2 Bioinert Ceramics

On insertion of bioinert material within the human body, it provides minimal interaction with the host tissue. In principle, they are anticipated to be non-allergic, non-carcinogenic, and non-toxic throughout the lifespan of the patients. Alumina, zirconia, and carbon are the classic examples for bioinert ceramics.

8.2.3 Bioresorbable Ceramics

Bioresorbable materials have the ability to replace natural tissues by degrading the ions gradually over time and lead to tissue regeneration instead of tissue replacement [19]. Degradation rate varies from material to material. Calcium phosphate derivatives are the unique examples for bioresorbable ceramic materials. Calcium-to-phosphate (Ca/P) ratio is used to quantify the stoichiometry of developed apatite. Solubility and pH are the fundamental criteria for apatite precipitation. The higher the ratio of Ca/P, the lower the solubility and acidity, and vice versa. Solubility of apatite is based on the amount of carbon present in the apatite that results in lattice distortion causing micro stresses and defects in the crystal structure [20]. Extensively used calcium phosphate-based ceramics are hydroxyapatite (HAP) and β- tricalcium

TABLE 8.1
Types of Tissue – Implant Interactions

Classification	Materials	Implant/ Tissue Bond	Tissue Response
Toxic	Lead oxide	None	Tissues die.
Bioinert	Alumina, zirconia	None	Formation of a non-adherent fibrous membrane around tissue.
Bioactive	Hydroxyapatite, bioglass, wollastonite embedded glass-ceramic (A-W)	Chemical	Formation of an interfacial bond with the implant.
Biodegradable	Tricalcium phosphate, calcium phosphate cements	Chemical	Replacement of the implant with tissue.

phosphate (β-TCP) with Ca/P ratio of 1.67 and 1.5, respectively. Clinical experimentation proved that, upon implantation of calcium phosphate and hydroxyapatite, it can withstand within the body for 7 years and is so called virtually inert [21].

8.2.4 Hydroxyapatite (HAP)

Hydroxyapatite with general formula $Ca_{10}(PO_4)_6(OH)_2$ possess close similarity to natural bone that constitutes inorganic components (collagen, fibrin) of very complex bone matrix tissue and so-called biological apatite [22]. Nevertheless, biological apatite at all times is nanocrystalline and nano-dimensional whose composition is non-stoichiometric. Also, it is a Ca^{2+} deficient variety of HAP holding trace elements, for instance Na^+, K^+, Mg^{2+}, F^-, CO_3^{2-}, Cl^-, and HPO_4^{2-}. Among all, the most general substituting ion is CO_3^{2-} that can replace PO_4^{3-} or OH^- ions [23,24]. Due to the close resemblance of HAP with natural bone, it exhibits its advantages in biomedical applications. Natural HAP possesses highly crystalline hexagonal structure with high mechanical stability. Synthetic HAP exhibits strong attraction and forms a powerful chemical bond with host hard tissues. Along with that, high biocompatibility, superior osteoconductivity, osteoinductivity, and gradual biodegradability make HAP an ideal material for implants in the field of orthopaedics and dentistry [25]. Alternatively, lower resorption rate of HAP is its drawback. Apparently, at the normal body temperature and physiological pH, HAP is thermodynamically stable so that, after implantation HAP graft serves as an implant for numerous years. Rarely, complications may result due to the mismatch between HAP and the bone tissue. To overcome this drawback, more soluble calcium phosphate-based material like β- TCP is combined with HAP (biphasic calcium phosphate) may be engineered with required phase compositions [26] (Figure 8.1).

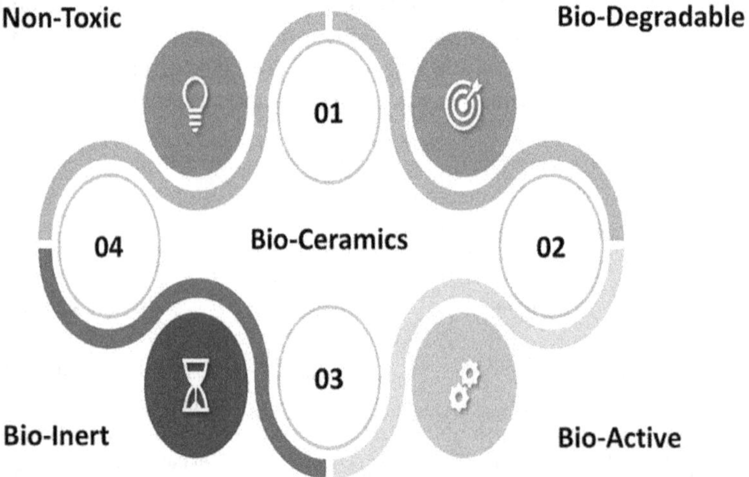

FIGURE 8.1 Shows the properties of bioceramics to categorize biological activities of the materials towards enhanced biomedical application.

8.3 METAL COMPOSITION IMPORTANCE IN PHYSIOLOGICAL ENVIRONMENT

8.3.1 Calcium

Human body comprises of calcium at the active region in natural bone and plays a significant role in blood vessels formation and neo-bone growth. It has been reported that low concentration of calcium (2–4 mmol) favoured osteoblast proliferation and differentiation. A medium concentration of Ca (6–8 mmol) stimulates mineralization, whereas a higher concentration of Ca (more than 10 mmol) is toxic to cells.

8.3.2 Silica

Silica plays a significant role in bone formation and regeneration; generally, Si is absorbed in the form of meta-silicates that are extensively distributed in connective tissue. Silica influences the metabolic activity of the host system in terms of calcification and stimulates the improved bone density and also inhibits osteoporosis. Silica components in bone matrix induce osteogenesis followed by the precipitation of hydroxyapatite as well as activate genes of osteoblasts and bone.

8.3.3 Magnesium

Next to potassium, sodium, and calcium, magnesium (Mg) ions are abundant cations in mammalian biological system and second predominant in intracellular fluid. In the early stages of osteogenesis, 6 mol% of magnesium possibly exist in cartilage and bone tissue; however, it tends to disappear in developed bone [27,28]. Magnesium plays a prime role in DNA stabilization and stimulates cellular growth [29,30]. Deficiency of magnesium components disturbs all important phases of skeletal metabolism in terms of osteoporosis and reduces bone growth and lower bone density [31,32]. Alternatively, supply of additional Mg^{2+} ions induces ionic leach in magnesium alloys to enhance the bone regeneration [33]. Hence, magnesium is vital component in bioceramics to enhance the physiological properties.

8.3.4 Zinc

Zinc (Zn) is one of the important trace elements in the living system. It acts as a cofactor in protein synthesis and also involved in RNA and DNA replication in cellular level. It provides favourable environment on the proliferation of osteoblast and improves the alkaline phosphatase activity towards bone tissue regeneration. It is well known that zinc oxide has the antimicrobial, and antifungal properties. Introducing zinc in the calcium silicate matrix initiates the new mineral phase formation of hardystonite ($Ca_2ZnSi_2O_7$), which is a potential crystalline phase of interest while substituting the Zn ion in the host calcium silicate material. Hardystonite showed considerably significant cell proliferation than wollastonite; on the other hand, prime drawback of this material is that the ionic dissolution rate is very minimal; therefore, no significant apatite formation could be identified [34]. Calcium silicate is a trending material for regeneration along with that infusion of Zn at appropriate concentration can influence the antibacterial and antioxidant properties that trigger faster healing.

8.3.5 Zirconium

Zirconium is traditional material for prosthetic devices; it has been used historically due to its acceptable biocompatibility and superior mechanical stability. Zirconium ions possibly introduced into several bioactive systems to upregulate biological properties and host material stability. Zirconia in the Ca-Si system initiates the formation of stable mineral phase of baghdadite ($Ca_3ZrSi_2O_9$) [35].

Research community has found that silicate-based bioceramics stimulate the vascularization and collagen deposition owing to the release of SiO_4^{4-}, which enhances the tissue regeneration. Cartilage defect is one of the serious causes of problem; therefore, therapeutic modalities are developed to cure this problem. Hence, it was found that hydroxyapatite/tri calcium phosphate supports cartilage layers with improved structural integration. Synergetic inter-reaction of SiO_4^{4-}, Ca^{2+}, and Li^+, Sr ions induce remarkable osteogenesis and activation of cartilage regeneration.

8.4 BIOCERAMIC MATERIALS ASPECT FOR BIOMEDICAL APPLICATIONS

Varity of metal oxide-based bioceramics being used for biomedical applications with respect to desire needs. Generally, Zirconia polycrystalline ceramics exhibits potential mechanical properties in terms of tensile strength and compression strength. Similarly, carbon-based bioceramics are also emerging nowadays; carbon allotropes such as carbon nanotubes, graphene, and diamond-like carbon materials have good chemical interaction and biocompatibility with better bio-physico-chemical properties. Among which graphene, an innovative as well as influential material, is being implemented in drug delivery, bio-sensors, and tissue engineering applications [36].

Calcium phosphate-based bioceramics are well-known materials; due to their similarity with bone and teeth structures, it has been used as a biocompatible efficient material for several biomedical applications. Bioactive, bioresorbable properties can be tailored based on the crystal structure with respect to the composition of the materials (Ca/P ratio) in terms of constructing calcium phosphate derivatives such as hydroxyapatite, tricalcium phosphate [α- $Ca_3(PO_4)_2$, β-$Ca_3(PO_4)_2$], tetracalcium phosphate [$Ca_4(PO_4)_2O$], and monocalcium phosphate ($Ca(H_2PO_4)_2$). In the early 1950s, properties of Hap have been analysed by researchers in various angle owing to its structural, morphological, mechanical, and compositional perspectives; initially, Hap was used as a grafting material that had no immunological reaction with neighbouring tissue. Later, the basic structure of HAp is changed based on recent desires, this material is used for bony ingrowths, and this kind of materials is considered a second-generation materials. Currently, emerging nanotechnology provided a plenty of dimensions for the developments of required structures. Nanocrystalline Hap is explicating an enhanced ionic dissolution, which influences

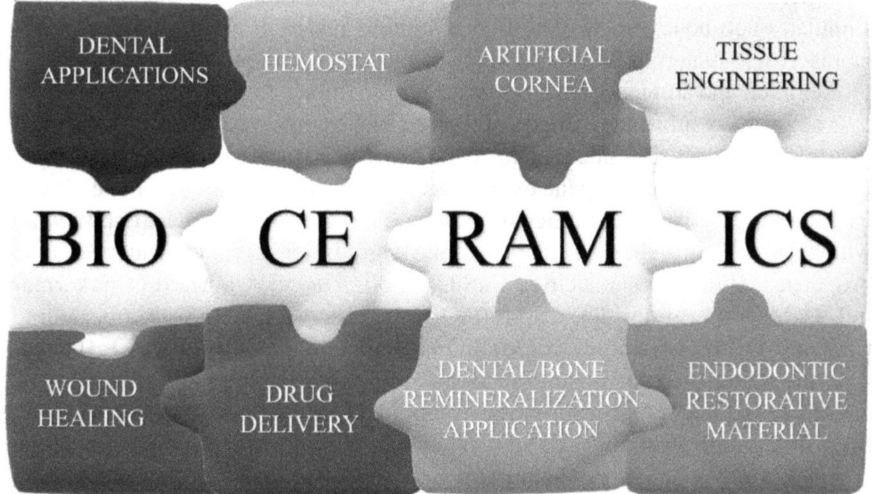

FIGURE 8.2 Represents the overview of currently demanding biomedical applications related with bioceramics.

the bioactivity. The materials property possibly altered in term of regulating crystallinity, morphology, particles distribution, particles size, and surface area, which has a strong impact on biological as well as mechanical performances. Material properties such as stability, porosity, bioresorbability, and bioactivity were mainly tailored based on basic characteristic features. Hydroxyapatite is also a suitable material as a therapeutic agent to carry drugs to targeted sites with the sustained release, which may stimulate osteoblastic cells [37].

Recently, natural calcium phosphates and calcium carbonates are in trend, which are derived from natural sources such as fish scales, fish bones, egg shell, snail shells, animal bones, sea weeds, teeth, and some of the leaves also containing calcium components. Natural calcium sources are very compatible to human system and economically cheap, but still there is no proper understanding about the reactivity of these kinds of materials. Several methodologies have been used to isolate the calcium phosphate ions/extract hydroxyapatite from biological sources. Burning of these natural materials at specific temperature is easy and cheap method to obtain natural hydroxyapatite [38]. On the other hand, reactivity, applicability, and repeatability are major drawbacks at the point of constricting/deriving these kinds of materials (Figure 8.2).

8.5 SILICATE-BASED BIOMATERIALS

Generally, biomaterials are divided into three categories (generations): First generation biomaterials consisting of alumina, zirconia, titanium, and polyethylene are bioinert in nature, which means these materials will not have much interaction with hosttissue. First generation materials will not generate any positive and negative effect due to the less interaction with host's immune response. Second-generation biomaterials include bioactive glasses and hydroxyapatite because of their bioactive biological responses that induce bonding between materials and living tissue, which stimulates neo-bone growth. Third-generation biomaterials are bioactive as well as bioresorbable materials and these kinds of materials promote cellular reaction and proliferation and stimulate the bone tissue regeneration.

The primary advantage of silicate-based bioceramics over calcium phosphate ceramics is mainly dependent on faster bone regeneration. Si plays a crucial role in mineral precipitation and gene expression towards the formation of bone. Several calcium silicate-based ceramics have been developed for biological application, which regulates the apatite mineralization, degradation, porosity, and mechanical properties. Bioceramics such as $CaSiO_3$ [wollastonite], $Ca_2MgSi_2O_7$ [akermanite], $CaMgSi_2O_6$ [diopside], $Ca_7Mg(SiO_4)_4$ [bredigite], $Ca_3MgSi_2O_8$ (merwinite), $Ca_2Zn(Si_2O_7)$ [hardystonite], $CaMgSiO_4$ [monticellite], and $Ca_3ZrSi_2O_9$ [baghdadite] are being used for bone tissue repair, regeneration, and replacement applications [39]. Calcium silicate-based compounds have been categorized with the ratio of Ca and Si, mainly $CaSiO_3$ (monocalcium silicate), Ca_2SiO_4 (dicalcium silicate), and Ca_3SiO_5 (tricalcium silicate). All these three-calcium silicate formulation may possess *in vitro* biocompatibility as well as bioactivity. Among which monocalcium silicate is extensively studied for bone repair and regeneration due to its productive bio-functionalities. This monocalcium silicate forms in two polymorphs with respect to thermal treatment β-$CaSiO_3$ (β-Wollastonite [formed at low temperature]) and α-$CaSiO_3$ (α-Wollastonite [formed at high temperature]). It is reported that [40] hydroxy carbonated apatite (HCA) formation is rapid in wollastonite rather than other bioactive glass ceramics. Similarly, *in vitro* cytocompatibility assessments explicate the attachment and proliferation of osteoblast-like cells and bone marrow mesenchymal stem cells. Wollastonite exhibited higher mechanical stability compared with calcium phosphate ceramics. Silicate-based ceramics trigger the osteogenic/cementogenic differentiation of periodontal ligament cells towards periodontal regeneration [41]. Release of Si ions from the matrix of the material possibly activates the signalling pathway of Wnt/β-catenin leading to the generation of alveolar bone and cementum by the differentiation of osteoblasts and cementoblast. Hence, glass ceramics greatly influence the periodontal regeneration with acceptable antibacterial efficacy.

8.5.1 Calcium Magnesium Silicates (Ca-Mg-Si) and Aluminosilicate

$Ca_2MgSi_2O_7$ [akermanite], $CaMgSi_2O_6$ [diopside], $Ca_7Mg(SiO_4)_4$ [bredigite], $Ca_3MgSi_2O_8$ (merwinite), and $CaMgSiO_4$ [monticellite] are pure phases of Ca-Mg-Si ceramic system. These calcium-magnesium-silicate ceramics express improved mechanical stability and slower degradation rate compared with $CaSiO_3$ (wollastonite) ceramics. Mg-O bonding energy is higher than Ca-O bonding energy; therefore, ionic dissolution rate of Ca-Mg-Si ceramics is minimal, which makes the crystal structure more stable and inhibit easy dissolution compared with Ca-SiO_3. Diopside can induce elevated apatite formation in simulated d body fluid than any other calcium-magnesium-silicate ceramics, which can closely bond with bone tissue [42].

$Al_2Si_2O_5(OH)_4.nH_2O$ (Halloysite) is naturally available aluminosilicate clay, and it has been considered a carrier to deliver anti-cancer drugs because of their tremendous biocompatibility. Laponite is also a kind of material like silicate clay, and this can be dissociated into non-toxic products such as Mg^{2+}, Li^+, Na^+, and $Si(OH)_4$. It was reported that up to the concentration of 1 mg/mL laponite is compatible with cells and induces osteogenic differentiation.

8.5.2 Bioactive Ceramics

The characteristic features exposed by bioactive material are intermediate between bioinert and bioresorbable materials. As per Larry Hench, who is known as the 'Father of Bioglass', "a bioactive material is one that elicits a specific biological response at the interface of the material, which results in the formation of a bond between the tissues and the material," that is, the biophysical and biochemical reactions at the interface generate strong chemical interfacial bonding to encourage bone growth and form a mineralized layer of hydroxyl apatite (HAP). HAP is a biologically active mineral that covers 70% of human natural bone and 95%–97% (by weight) in mature enamel [43,44]. The ions released from the bioactive materials accelerate apatite formation, further enhance vascularization, promote angiogenesis, osteoblastic differentiation, and osteoclastic response, and regulate bone resorption and bone tissue calcification, etc.

Thus, these bioactive materials create favourable and compatible environment for osteogenesis and also encourage bonding between living tissues and non-living materials resulting in mineralization interface [43]. Such bioactivity is found in many bioceramics including calcium phosphates, bioactive glasses, and their composites whose mechanism, density, and strength of the bond vary from one another. Bioglass is one of the revolutionary materials for biomedical applications; after implantation for 2 days silica-rich calcium phosphate layer grows on the peripheral layer of the materials. This is almost similar to natural apatite, with respect to incubation dense apatite layer possibly grow on the surface of material; after interaction with macrophages and cellular compounds, these bioactive glasses directly bond with natural bone [45]. Bioactive materials are a certain specific composition of glasses, ceramics, glass ceramics, and composites having the ability to form mechanically strong bond with the bone. Some have more specialized compositions like bioactive glass that has the ability to bond with soft tissues too [1,46].

Bioactive glass: Bioglass®, a boon material to the mankind that has revolutionized the medical society. After a conversation with Colonel Klinker in an Army Conference, Larry Hench learnt the composition of bone from the doctor. He rushed to the laboratory and started his innovative research towards the development of a biocompatible material for replacement and rebirth of organs for injured soldiers from Vietnam War [47,48]. He achieved the successful 45S5 Bioglass composition with 46 mol% SiO_2, 26.9 mol% CaO, 2.6 mol% P_2O_5, and 24.4 mol% Na_2O. 45S5 (wt.%) is equivalent to 46S5.2 (mol%) [49]. The word "bioactive glasses" refers to the glasses that own the ability to form hydroxyapatite layer on the materials' surface by a series of definite chemical reactions. On implanting the bioactive glass in the host tissue, the time-dependent kinetic modification results in the precipitation of biologically active hydroxycarbonate

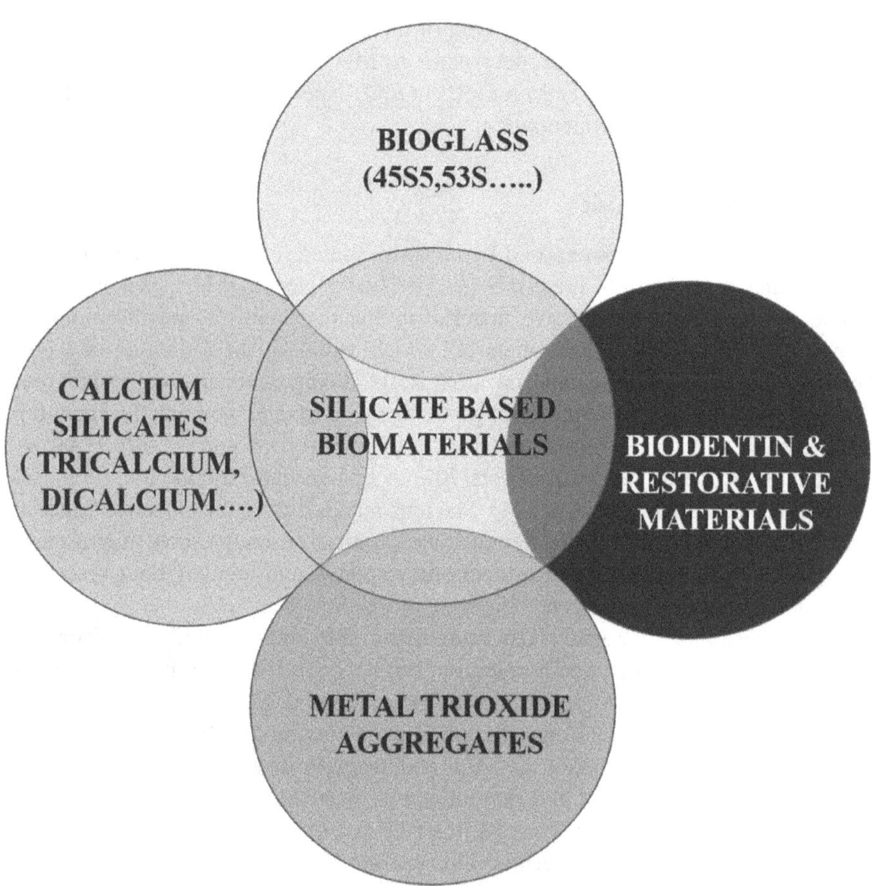

FIGURE 8.3 Predominant silicate and calcium silicate-based biomaterials for tissue engineering and dental applications.

apatite (HCA) that forms the bonding interface with the tissues [1,49]. The bioactive glass is put into practice to patients only when it is non-toxic, non-carcinogenic, non-antigenic, and non-mutagenic to avoid any unfavourable effect on the host cell (Figure 8.3).

8.6 RECENT ADVANCEMENT IN BIOCERAMICS

Generally, bioceramic powders are synthesized by various methodologies such as chemical-precipitation, solid state reaction, hydrothermal, sol-gel, microwave irradiation, microemulsion, and self-propagating combustion. Similarly, thin films are developed by spray pyrolysis and spin coating; and also, electro spun fibres are drawn from electro-spinning for the respective purposes [50]. Bioceramic powders are mainly made into different shapes in terms of the morphology of nanoparticulates, mainly spherical-shaped rods, ellipsoidal appearance, tetragons, and hexagon-like structures, depending upon the need. Recently, material scientists are focussing on the development of materials with exact requirement and are customized and fabricated; simultaneously, their behaviour was sequentially analysed in laboratory before entering into application. To elaborately understand the properties of the materials, various characterization tools have been implemented and with that it is possible to define the outcome or application of the respective bioceramics. Briefly, crystalline phase, cell volume, and grain size parameters were quantified by XRD, electron backscatter diffraction (EBSD), and selected area electron diffraction (SAED). Metal oxide bonding (functional group) attribution can be investigated through Fourier transform infrared spectroscopy (FT-IR) and Raman spectroscopy. Surface composition and binding properties were analysed via X-ray photoelectron spectroscopy (XPS); morphological as well as topographical aspects were analysed using scanning electron microscopy (SEM), transmission electron microscopy (TEM), and atomic force microscopy (AFM); in the same direction, porosity, pore diameter, and surface area were estimated through Brunauer-Emmett-Teller (BET) analysis; elemental compositions of the materials were analysed via X-ray fluorescence (XRF) and inductively coupled plasma optical emission spectrometry (ICP-OES & ICP-MS) [50]. Similarly, in biological part, detailed biocompatibility and immunohistochemistry, other biomolecule absorption and reaction mechanisms, and *in vivo* compatibility were also estimated before entering clinical trials. Recent technology helps to elaborately study the materials in the aspect of micro-structural properties along with atomic arrangements as well as biological properties that provide exact reaction mechanism of the materials towards the applications.

To regulate the properties of the ceramic materials, various components and polymers may be included in the host material. Hence, in this direction, hard polymers such as polyvinyl pyrrolidone (PVP), polyethylene glycol (PEG), polylactic acid (PLA), poly methyl methacrylate (PMMA), this initiates the mechanical stability of the host materials. Similarly, soft polymers for instance cellulose, alginate, gelatine, and chitosan also implemented with the bioceramics to modify the properties to gain the respective applications (soft tissue regeneration) [50]. Recently, bioceramics are developing and also established in plenty of formulations such as powders, thin films, nano-fibres, hard polymeric scaffolds, soft-hydrogel-based ceramics, combination of many biomaterials together can generate bioceramics mixtures, and growth factors assisted 3D-printed bioactive scaffolds for the more specific biomedical needs. In recent times, hydrogel-based bioactive ceramics and 3D-printed scaffolds are gaining enormous impact in wound healing and tissue engineering application. Thin films coatings are layered on the surface of sutures, screws, pins, fracture plates, and bone plates to avoid immune rejection and to induce osteoconduction. Fibre-like one-dimensional structures are mainly used to increase the surface properties that positively stimulate towards the applications. Hard and soft polymeric bio-active-scaffolds were designed and developed based on the requirement in terms of soft mimicking muscles and hard replacements. Nowadays, material engineering is one of the important areas in which different bioceramics such as hydroxyapatite, bioglass, tricalcium phosphate, and whitlockite are mixed in appropriate ratio's and developed as a biocomposites for relevant applications. Most important advanced 3D scaffolds possibly provide supporting properties to regenerate tissue and transplant cells with local delivery of therapeutic components. These kinds of engineered constructs can satisfy biochemical as well as functional requirements [51]. Bioceramics for sensing and 3D scaffolds are upcoming demanding application for the generation of healthier society.

8.7 BIOCERAMICS REACTION IN PHYSIOLOGICAL ENVIRONMENT

Bioglass is a semi-crystalline material, and Ca^{2+} and Na^{+} are crystallized on the silica network. While Bioglass is at *in vitro* or *in vivo* biological environment, initially sodium and calcium ions are leached from the glass network simultaneously; H^{+} and OH^{-} ions were replaced on that position. Released ions were re-precipitated and formed as amorphous calcium phosphate with the presents of calcium and phosphate and turned as hydroxyl carbonate apatite followed by hydroxyapatite on the superficial layer of bioactive glasses [52]. Amorphous silica network easily loses the calcium and sodium ions from the glass matrix; hence, this material exhibits rapid bioactivity rather than other bioceramics.

In case of hydroxyapatite, generally, it is strong hexagonal crystal structure; therefore, the release of ions from the crystalline material is not easy as compared to bioactive glasses. However, after particular interaction with physiological media, the ions tend to release and supersaturated minerals at biological environment also get precipitated on the surface of hydroxyapatite and turned as a mineral apatite. This precipitated mineral apatite is compatible with the cells and growth factors; grown apatite induces further reaction and stem cell proliferation [53]. Similarly, other bioceramics such as tricalcium phosphate, whitlockite, magnesium, zirconium, and alumina-based silicate materials stimulates ionic release and induce mineral deposition with respect to the resorption and degradation properties of respective materials [52].

Bioactive hydrogels are one of the important categories of polymeric compounds; recently, these materials are emerging as therapeutic vehicles as well as attractive biomaterials for wound healing applications with minimal immunological problems. However, mechanical stability of these materials was very poor; hence, this kind of materials is very much useful in drug delivery and external wound healing. Owing to the rapid degradability in bone tissue regeneration, this category of polymeric materials cannot sustain for longer duration; therefore, researchers have to develop the cross-linking methodologies to design and progress better material to promote bone/tissue regeneration [54].

Biocompatibility assessments of bioceramics were performed and reported on the *in vitro* as well as *in vivo* compatibility properties from the origin to the present. There is no end to conclude the investigations. Biocompatibility greatly varied based on the morphology, crystal structure, and ionic leach out; hence, research community was investigating each and every step of changes in cells, blood, protein, enzyme, and various types of biomolecules. Quan et al. [55] investigated the biocompatibility properties of hydroxyapatite-zirconia composite in rabbit model particularly in femur and hip muscles and reported that the bioceramics explicated superior osteoinductive properties. Similarly, Devi et al. [56] also elaborately analysed the bioceramics and discussed about the cytocompatibility as well as *in vivo* bone regeneration properties with rabbit model and concluded that the bioceramics exhibited synergetic proliferation and differentiation (MC3T3-E1 cells) with better bone regeneration. In case of dentistry also calcium silicate-based bioceramics exhibited improved bioactivity and biocompatibility with better efficiency at oral environment [57]. Reactive, kinetic, degradative, and dynamic properties of bioceramics influence the biological environment and impact the biocompatibility in terms of immunogenic, epigenetic, carcinogenic, and genotoxic effects. Even though, protection of fabricated material can be understood by the assessments of *in vitro* and *in vivo* results; however, exact material behaviour in human system can only be realized and concluded after clinical treatment in patients [58].

8.8 CLINICAL APPLICATIONS OF BIOCERAMICS

Bioceramics are generally used for wide variety of biomedical applications comprising replacement for hips, knees, tendons, ligaments, maxillofacial reconstruction, alveolar ridge augmentation, implantation, and bony dental arches. Hydroxyapatite is used for many biological applications, such as bone void fillers, tissue engineering for orthopaedic/dental implant coatings, periodontal, and endodontic defects. High strength silicon nitride ceramics is a potential implant material and has been used for arthrodesis (fused cage implantation). Bioceramic-based grafts stimulate osteoinductive properties that are promising spinal fusions. Hydroxyapatite, β-TCP, and calcium sulphates offer several advantages such as shape flexibility, biocompatible, and inert property, which drives the lumbar spine fusions [45].

Chemotherapy and radiotherapy can induce strong side effects; hence with the advent of nanotechnology, drug-loaded nanocarriers can act as therapeutic vehicle for the cancer treatment. Nanoparticulates can easily penetrate into the cells owing to their smaller size, consequently, release drug components to induce carcinogenic cell death [59,60]. Silica nanoparticles, iron oxide core shell nanostructures, and calcium phosphate-based ceramics are being used as a carrier to deliver certain antibiotic, anti-inflammatory drugs, and anti-tumour drugs [61]. Three-dimensional scaffolds act as a host tissue-like construction, in which cells and growth factors are adhered that allows easy tissue regeneration with minimal immunological reaction [62]. Bioactive glass ceramics are being used in orbital floor defects and ocular surgery owing to the transparency in visible light and potentiality to stimulate cells towards tissue regeneration. Calcium phosphate nanocarriers are providing effective delivery of gene (siRNA, pDNA), proteins, and growth factors [63].

8.8.1 Endodontic Sealing Materials

Endodontics is one of the import trust areas in dentistry; major problem in sealing materials is micro-leakage, which induces the bacterial invasion in between canal walls and root filling sealers. Hence, an ideal criterion for developing sealant materials includes biocompatibility, very minimal solubility, antimicrobial property, materials stability, and radiopacity. Bioceramic-based materials entered into endodontics in the year 1990s, in the order of retrograde filling materials, root canal sealers, root repair cements, and coated gutta-percha like cones. The primary advantages of bioceramics-based materials are physico-bio-chemical, biocompatible, and bioactive properties in terms of precipitation of hydroxyapatite and capability to generate a bond between the material and the dentin. Basic characteristic features required to be ideal sealants are short setting time (transition of powder form to paste like state), high mechanical stability, elevated alkaline pH for mineral precipitation, radiopacity to analyse the filling, biocompatibility, bioactivity for improved mineralization, and antimicrobial activity [64].

Bioceramics are potential candidate for endodontic sealing, such as mineral trioxide aggregate (MTA), Biodentine, ERRM (EndoSequence root repair material) Putty, ERRM Paste, and iRoot FS (preloaded paste in a syringe with material delivery tips for intracanal deliverance), BC Sealer and iRoot SP, MTA Fillapex, MTA Plus, gutta-percha, and bioactive glass. Mineral trioxide aggregate (MTA – tricalcium silicate, dicalcium silicate, tricalcium aluminate, tetracalcium aluminoferrite, gypsum, and bismuth oxide) is considered a gold standard for variety of clinical applications because of admirable short setting time (165 minutes), high mechanical stability, favourable alkaline pH, optimal porosity, antimicrobial, biological, and physico-chemical properties. Biodentine is kind of a ceramic sealant that contains tricalcium silicate, zirconium oxide (radiopaque agent), and calcium carbonate, calcium chloride liquid as a setting accelerator. Based on manufacturer's instruction, it is one of the early-setting (10–12 minutes) restorative materials, and its advantage over MTA is that this biodentine explicated clinically perceptible colour change [65,66]. ERRM Putty, ERRM Paste, and iRoot FS are composed of calcium silicates, calcium phosphates monobasic, tantalum oxide, and zirconium oxide. This ERRM has almost similar compressive strength compared to MTA; however, setting time is 4 hours [67]. BC Sealer and iRoot SP are pre-mixed ceramic-based endodontic sealer, which includes tricalcium silicate, dicalcium silicate, calcium silicate, colloidal silica, monocalcium phosphate, calcium hydroxide, and zirconium oxide. Hydroxyapatite crystals are precipitated in between the canal wall due to the compositions of calcium silicates that induces the bonding between dentin walls and sealer. MTA Fillapex and MTA plus are composed of mineral trioxide aggregates with natural and salicylate resin, bismuth and silica components, this material has high radiopacity and easy to handle. Highly soluble nature of this material may be beneficial to induce antibacterial activity. Gutta-percha is impregnated with bioceramics nanoparticles and can be used as sealing material owing to the unique stiffness and ease of handling inside the canal. Bioactive glasses are promising materials for restorative dentistry, due to their remarkable bioactivity and biocompatibility. These materials are used as a bonding material, bone regenerating material, and re-mineralizing agents in tooth pastes. By enhancing the materials stability as well as setting time, it could be an ideal material for restorative applications [66,68].

8.8.2 Dental Re-mineralization Application

Several biomaterials such as Novamin, nano-Hap, calcium sucrose phosphate [CaSP], Casein phosphor peptide amorphous calcium phosphate (CCP- ACP), tricalcium phosphate (TCP), and Pro-Argin are evaluated as re-mineralizing agents and reported the mineral deposition. Most articles [69] illustrated the survey of previously reported results and summarized stringent eligibility criteria as required to identify the appropriate materials for dental applications. In most of the research findings, Novamin is reported as an effective material for re-mineralization [70] and as is being used in some prevalent tooth pastes and dental products.

8.8.3 Tissue Regenerative Application

Bioceramics are well-known materials for tissue regeneration, bio-resorbable, and bioactive features, tunes this materials towards adoptable; incorporating new components such as different metal ions (Cu, Ag, Sr, Zn, and Mg etc.), flavonoids, and herbal content stimulate this as a fascinating material. In another dimension, new form of materials such as particles, fibres, scaffolds, and thin films was evolved to fulfil the current demands. Hydroxyapatite, tricalcium phosphate, and bioactive glasses being used in regenerative application. Along with host components, relevant metal ions like copper, strontium, zinc, aluminium, zirconium, fluoride, and boron, etc. were introduced in the latices of bioceramics to alter the crystal structure, morphology as a result of attain angiogenic, antimicrobial, regenerative, and mechanical properties. Natural healing takes time in case of diabetic and diseased people; therefore, these kinds of bioceramic materials may fasten the regenerative as well as wound healing behaviours.

8.8.4 Drug Delivery and Haemostat

Bioceramics are suitable materials for local drug delivery to treat osteoporotic fractures, bone infections, large bone defects, and even bone tumours. Mesoporous nano-ceramic-based materials can be act as a carrier to deliver drugs and opened a new prospective mainly for cancer therapy [71,72]. Haemorrhage is one of the serious problems in recent days due to accidents and various other traumatic situations. Blood loss is unavoidable while tooth extraction; in such a scenario, haemostat is unavoidable component to clot the blood and to preserve the socket. Rather than using other material, bioceramics in adaptable form may enrich the clotting efficiency along with regenerative properties (Figure 8.4) [73,74].

In this current chapter, we elaborate the properties of bioceramics towards the relevant biomedical application. All kinds of bioceramics were indicated followed by the respective materials property and their fabrication methodologies. Bioceramics demand in medical society in the clinical aspects was mentioned, and several silicate-based materials impact in dentistry, especially as restorative materials, was represented.

Various forms of bioceramics are being used for biomedical applications depending upon the need; generally, polymeric fibres, membranes, thin films, and nanoparticulates gain more attention for relevant applications [75–77]. Bioceramics stepped their foot prints in almost all the biomedical applications. Hap, TCP, and bioglass ceramics are available in the market for commercial usage. Hence, bioceramics are part of materials for therapeutic as well as diagnostic applications. However, tremendous researchers are going on in terms of fabrication and analysing the interaction between cells/biomolecules to exactly scrutinize the problem for the development of better remedy. In this chapter, it is elaborated on the importance and drawbacks of bioceramics towards clinical applications.

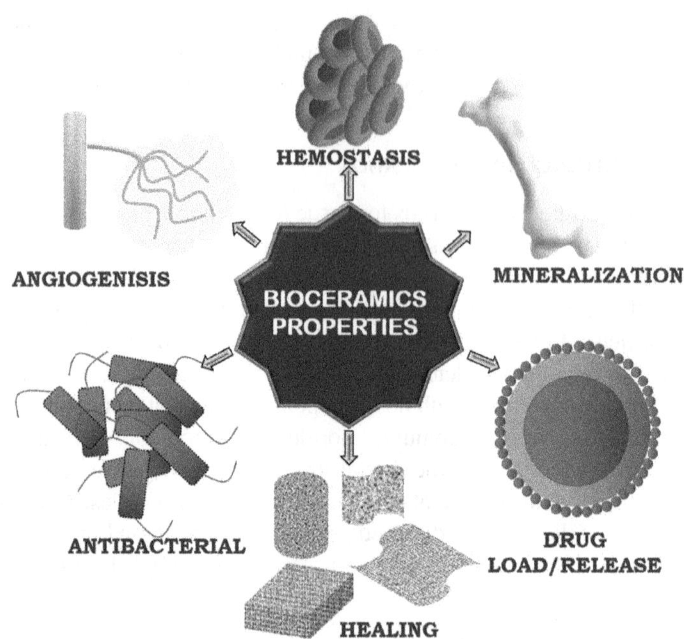

FIGURE 8.4 Pictorial representation of most attractive bioactive and biocompatible properties of bioceramics towards elevated biomedical applications.

8.9 CONCLUSION

Various types of bioceramics and their significance in clinical applications were systematically analysed. Recently, more changes have been implemented in the aspects of materials chemistry, and a detailed view of properties in atomic/molecular level was also analysed with the help of nano-characterization tools. Physico-chemical properties play a major role in biocompatibility to avoid immunological rejections. On the whole, bioceramics are modified towards the stabilization for clinical applications. However, poor mechanical stability and low durability are the major drawbacks of bioceramics. Research community is focussing on the material engineering with hard polymeric and/or metallic materials to develop a material with improved stability. Materials engineering may provide a direction to improve fracture toughness and tensile strength, which will be very much supportive for clinical point of view. However, currently, bioceramics are being used for dental/bone implant coating, sealants, and wound healing applications.

REFERENCES

1. LL Hench, Bioceramics, *J Am Ceram Soc*, 1998, 81(7), 1705–1728.
2. G Kaur, OP Pandey, S Kulvir, H Dan, S Brain, P Gary, A review of bioactive glasses; their structure, properties, fabrication and apatite formation, *J Biomed Mater Res A*, 2013, 102, 254–274.
3. G Kaur, OP Pandey, S Kulvir, H Dan, S Brain, P Gary, An introduction and history of the bioactive glasses, In: J. Marchi ed, *Biocompatible Glasses, Advanced Structured Materials*, Springer International Publishing: Switzerland, 2016, 53.

4. EAA Neel, A Aljabo, A Strange, S Ibrahim, M Coathup, AM Young, L Bozec, V Mudera, Demineralization-remineralization dynamics in teeth and bone, *Int J Nanomed*, 2016, 11, 4743–4763
5. S Nouri, MR Sharif, Y Panahi, M Ghanei, B Jamali, Efficacy and safety of aluminum chloride in controlling external hemorrhage: An animal model study, *Iran Red Crescent Med J*, 2015, 17(3), e19714.
6. RB Heimann, HD Lehmann. *Bioceramics – A Historical Perspective, Bioceramic Coatings for Medical Implants: Trends and Techniques*, First Edition. Wiley-VCH Verlag GmbH & Co. KGaA, 2015.
7. LL Hench, Bioceramics: From Concept to Clinic, *J Am Ceram Soc*, 1991, 74(7), 1487–1510.
8. A Marti, Inert bioceramics (Al_2O_3, ZrO_2) for medical application, *Injury Int J Care Injured*, 2000, 31, S-D33-36.
9. G Jose, Vargas-Hernandez, J Casarrubias-Vargas, H Gonzalez-Moran, CO Refugio-Garcia, EF Cuautle, A Jose de Jesus, Synthesis and effect of $CaTiO_3$ formation in $CaO{\cdot}Al_2O_3$ by solid-state reaction from $CaCO_3{\cdot}Al_2O_3$ and Ti, *Mater Chem Phys*, 2019, 232, 57–64.
10. SF Hulbert, LL Hench, D Forbers, LS Bowman, History of bioceramics, *Ceram Int*, 1982, 8(4).
11. SV Dorozhkin, Bioceramics of calcium orthophosphates, *Biomaterials*, 2010, 31, 1465–1485.
12. GP Jayaswal, Bioceramics in dental implants: A review, *J Ind Prosthodont Soc*, 2010, 10, 8–12.
13. KMR Nuss, B von Rechenberg, Biocompatibility issues with modern implants in bone – A review for clinical orthopedics, *The Open Orthopaed J*, 2008, 2, 66–78.
14. M Saini, Y Singh, P Arora, V Arora, K Jain, Implant biomaterials: A comprehensive review, *World J Clin Cases*, 2015, 3(1), 52–57
15. N Soumya, B Rajarshi, Fundamentals of medical implant materials, *Mater Med Dev*, ASM Handbook, 2012, 23.
16. G Heness, B Ben Nissa, Innovative bioceramics, *Mater Forum*, 2004, 27, 104–114.
17. J Huang, Design and development of ceramics and glasses, In: *Designing Smart Biomaterials to Mimic and Control Stem Cell Niche*, 2017, 320–327.
18. VA Dubok, Bioceramics: Yesterday, today, tomorrow, *Powder Metall Metal Ceram*, 2000, 39, 7–8.
19. M Jarcho, Calcium phosphate ceramics as hard tissue prosthetics, *Clin Orthop Relat Res*, 1981, 157, 259–278.
20. Palmer et al., Biomimetic systems for hydroxyapatite mineralization inspired by bone and enamel, *Chem Rev*, 2008, 108(11), 4754–4783.
21. Regi MV, Ceramics for medical applications, *J Chem Soc, Dalton Trans*, 2001, 2, 97–108.
22. DW Hutmacher, JT Schantz, CXF Lam, KC Tan, TC Lim, State of the art and future directions of scaffold based bone engineering from a biomaterials perspective, *J Tissue Eng Regenerat Med*, 2007, 1, 245–60.
23. N Eliaz, N Metoki, Calcium phosphate bioceramics: A review of their history, structure, properties, coating technologies and biomedical applications, *Materials*, 2017, 10, 334.
24. VM Regi, *Biomimetic Nanoceramics in Clinical Use: From Materials to Applications*. Royal Society of Chemistry: Cambridge, England, 2008.
25. Zhou et al., Nanoscale hydroxyapatite particles for bone tissue engineering, *Acta Biomater*, 2011, 7, 2769–2781.
26. Bellucci et al., Bioactive glasses and glass-ceramics versus hydroxyapatite: Comparison of angiogenic potential and biological responsiveness, *J Biomed Mater Res*, 2019, 107A, 2601–2609.

27. A Bigi, E Foresti, R Gregorini, A Ripamonti, N Roveri, JS Shah, The role of magnesium on the structure of biological apatites, *Calcif Tissue Int*, 1992, 50(5), 439–444.
28. FI Wolf, A Cittadini, Chemistry and biochemistry of magnesium, *Mol Aspects Med*, 2003, 24(1–3), 3–9.
29. ME Maguire, JA Cowan, Magnesium chemistry and biochemistry, *Biometals*, 2002, 15, 203–210.
30. FI Wolf, V Trapani, Cell (patho) physiology of magnesium. *Clin Sci*, 2008, 114(1), 27–35.
31. RK Rude, HE Gruber, LY Wei, A Frausto, BG Mills, Magnesium deficiency: Effect on bone and mineral metabolism in the mouse, *Calcif Tissue Int*, 2003, 72(1), 32–41.
32. RK Rude, FR Singer, HE Gruber, Skeletal and hormonal effects of magnesium deficiency, *J Am Coll Nutr*, 2009, 28(2), 131–141.
33. MP Staiger, AM Pietak, J Huadmai, G Dias, Magnesium and its alloys as orthopaedic biomaterials: A review, *Biomaterials*, 2006, 27(9), 1728–1734.
34. J Ma, BX Huang, XC Zhao, XH Hao, CZ Wang, Preparation and characterization of novel β-$CaSiO_3$.$Ca_2ZnSi_2O_7$ bioceramics with adjustable degradability and apatite-formation ability, *Mater Lett*, 2019, 236, 566–569.
35. TC Schumacher, E Volkmann, R Yilmaz, A Wolf, L Treccani, K Rezwan, Mechanical evaluation of calcium-zirconium-silicate (baghdadite) obtained by a direct solid-state synthesis route, *J Mech Behav Biomed Mater*, 2014, 34, 294–301.
36. M Vallet-Regi, Bioceramics: From bone substitutes to nanoparticles for drug delivery, *Pure Appl Chem*, 2019, 91(4), 687–706.
37. VS Kattimani, S Kondaka, KP Lingamaneni, Hydroxyapatite – Past, present, and future in bone regeneration, *Bone Tissue Regener Insights*, 2016, 7, BTRI.S36138.
38. AA Majhool, I Zainol, CNA Jaafar, HA Alsailawi, MZ Hassan, M Mudhafar, AA Majhool, A Asaad, A brief review on biomedical applications of hydroxyapatite use as fillers in polymer, *J Chem Chem Eng*, 2019, 13, 62–75.
39. P Srinath, P Abdul Azeem, K Venugopal Reddy, Review on calcium silicate-based bioceramics in bone tissue engineering.
40. P Siriphannon, Y Kameshima, A Yasumori, K Okada, S Hayashi, Influence of preparation conditions on the microstructure and bioactivity of α-$CaSiO_3$ ceramics: Formation of hydroxyapatite in simulated body fluid, *J Biomed Mater Res*, 2000, 52(1), 30–39.
41. Y Zhou, C Wu, Y Xiao, Silicate-based bioceramics for periodontal regeneration, *J Mater Chem B*, 2014, 2, 3907.
42. M Vallet-Regi, AJ Salinas, J Roman, M Gil, Effect of magnesium content on the in vitro bioactivity of CaO-MgO- SiO_2-P_2O_5 sol-gel glasses, *J Mater Chem*, 1999, 9(2), 515–518.
43. W Cao, L Hench, Bioactive materials, *Ceram Int*, 1996, 22, 493–507.
44. Palmer et al, Biomimetic systems for hydroxyapatite mineralization inspired by bone and enamel, *Chem Rev*, 2008, 108(11), 4754–4783.
45. H Oonishi, SC Kim, LL Hench, J Wilson, E Tsuji, H Fujita, H Oohashi, K Oomamiuda, Clinical application of hydroxyapatite. In: *Bioceramics and Their Clinical Applications*, Woodhead Publishing Series in Biomaterials, 2008, 606–687.
46. R Li, AE Clark, LL Hench, An investigation of bioactive glass powders by sol-gel processing, *J Appl Biomater*, 1991, 2, 231–239.
47. L Hench, The story of bioglass(r), *J Mater Sci Mater Med*, 2006, 17, 967–978.
48. DC Greenspan, Glass and medicine: The Larry Hench story, *Int J Appl Glass Sci*, 2016, 7(2), 134–138.
49. O Makata, C Fumio, Compositional dependence of the formation of calcium phosphate films on bioglass, *J Biomed Mater Res*, 1980, 14, 5544.
50. A Haider, S Haider, SS Hanb, I-K Kang, Recent advances in the synthesis, functionalization and biomedical applications of hydroxyapatite: A review, *RSC Adv*, 2017, 7, 7442.

51. MP Nikolovaa, MS Chavali, Recent advances in biomaterials for 3D scaffolds: A review, *Bioact Mater*, 2019, 4, 271–292.
52. L Hench, The story of bioglass(r), *J Mater Sci Mater Med*, 2006, 17, 967–978.
53. BR Rrioni, AAR de Oliveira, M de M Pereira, The evolution, control, and effects of the compositions of bioactive glasses on their properties and applications, In: *Biocompatible Glasses, Advanced Structured Materials.* Springer International Publishing: Switzerland, 2016, 53.
54. X Bai, M Gao, S Syed, J Zhuang, X Xu, X-Q Zhang, Bioactive hydrogels for bone regeneration, *Bioact Mater*, 2018, 3, 401–417.
55. R Quan, D Yang, X Wu, H Wang, X Miao, W Li, In vitro and in vivo biocompatibility of graded hydroxyapatite-zirconia composite bioceramics, *J Mater Sci Mater Med*, 2008, 19(1), 183–187.
56. KB Devi, B Tripathy, A Roy, B Lee, PN Kumta, SK Nandi, M Roy, In vitro biodegradation and in vivo biocompatibility of forsterite bio-ceramics: Effects of strontium substitution, *ACS Biomater Sci Eng*, 2018, acsbiomaterials.8b00788.
57. W Song, W Sun, L Chen, Z Yuan, In vivo biocompatibility and bioactivity of calcium silicate-based bioceramics in endodontics, *Front Bioeng Biotechnol*, 2020, 8, 580954.
58. G Thrivikraman, G Madrasb, B Basu, In vitro/in vivo assessment and mechanisms of toxicity of bioceramic materials and its wear particulates, *RSC Adv*, 2014, 4, 12763.
59. TM Oliveira, FCB Berti, SC Gasoto, B Schneider Jr, MA Stimamiglio, LF Berti, Calcium phosphate-based bioceramics in the treatment of osteosarcoma: Drug delivery composites and magnetic hyperthermia agents, *Front Med Technol*, 2021, 3, 700266.
60. C-Y Zhao, R Cheng, Z Yang, Z-M Tian, Nanotechnology for cancer therapy based on chemotherapy, *Molecules*, 2018, 23, 826.
61. Q Yu, J Chang, C Wu, Silicate bioceramics: From soft tissue regeneration to tumor therapy, *J Mater Chem B*, 2019, 7, 5449–5460.
62. V Selvarajan, S Obuobi, PLR Ee, Silica nanoparticles – A versatile tool for the treatment of bacterial infections, 2020, 8, 602.
63. TJ Levingstone, S Herbaj, J Redmond, HO McCarthy, NJ Dunne, Calcium phosphate nanoparticles-based systems for RNAi delivery: Applications in bone tissue regeneration, *Nanomaterials*, 2020, 10, 146.
64. Z Wang, Bioceramic materials in endodontics, *Endodont Top*, 2015, 32, 3–30.
65. KV Teja, S Ramesh, An update on bioceramic sealers, *Drug Invention Today*, 14(3), 2020.
66. A AL-Haddad, ZAC Ab Aziz, Bioceramic-based root canal sealers: A review, *Int J Biomater*, 2016, Article ID 9753210, 10 pages.
67. RM Walsh, KF Woodmansey, GN Glickman, J He, Evaluation of compressive strength of hydraulic silicate-based root-end filling materials, *J Endodont*, 2014, 40(7).
68. S Chitra, R Chandran, R Ramya, D Durgalakshmi, S Balakumar, Unravelling the effects of ibuprofen-acetaminophen infused copper-bioglass towards the creation of root canal sealant, *Biomed Mater*, 2022, 17(3), 035001.
69. S Khijmatgar, U Reddy, S John, AN Badavannavar, D Teena, Souza, Is there evidence for Novamin application in remineralization? A Systematic review. *J Oral Biol Craniofac Res*, 2020, 10, 87–92.
70. Z Abbasoglu, DA Bicak, DO Dergin, D Kural, I Tanboga, Is novamin toothpaste effective on enamel remineralization? An in-vitro study. *Cumhuriyet Dent J*, 2019, 22(1), 22–30.
71. S Chitra, P Bargavi, M Balasubramaniam, RR Chandran, S Balakumar, Impact of copper on in-vitro biomineralization, drug release efficacy and antimicrobial properties of bioactive glasses, *Mater Sci Eng C*, 2020, 109, 110598.
72. D Arcos, M Vallet-Regi, Bioceramics for drug delivery, *Acta Mater*, 2013, 61(3), 890–911.
73. S Chitra, S Balakumar, Insight into the impingement of different sodium precursors on structural, biocompatible, and hemostatic properties of bioactive materials, *Mater Sci Eng C*, 2021, 123, 111959.

74. S Chitra, P Bargavi, RR Chandran, S Balakumar, Thermal treatment stimulus on erythrocyte compatibility and hemostatic behaviour of one-dimensional bioactive nanostructures, *J Biomed Mater Res Part A*, 2020, 108(11), 2277–2290.
75. HA Heydarya, E Karamianb, J Heydaripourc, A Khandand, E Poorazizi, Electrospun Iranian Gum Tragacanth-polyvinyl alcohol/nanoclay as anew flexible nanocomposite for biomaterials applications, *Prog Nat Sci*, 2015, 3.
76. RN Oosterbeek, CK Seal, J-M Seitz, MM Hyland, Polymer-bioceramic composite coatings on magnesium for biomaterial applications, *Surf Coat Technol*, 2013, 236, 420–428.
77. S Chitra, NK Mathew, S Jayalakshmi, S Balakumar, S Rajeshkumar, Strategies of bioceramics, bioactive glasses in endodontics: Future perspectives of restorative dentistry, *BioMed Res Int*, 2022, 12, 2530156.

9 Bioceramics for Cosmetic Dentistry

Nikita Agrawal
Chirayu Medical College

Santosh Kumar Singh
Peoples College of Dental Sciences

Parimala Kulkarni
Chirayu Medical College

Divya Panday and Vatsal Chauhan
Peoples College of Dental Sciences

Brijesh Gangil
HNB Garhwal University

9.1 INTRODUCTION

The introduction of dental ceramics has revolutionized the concept of aesthetics in dentistry. Cosmetic dentistry aims at reconstructing the ideal form and function of the teeth and tissues without compromising oral health of the patient. Since its advent, ceramics have been popular because of their high biocompatibility and pleasing aesthetics [1].

The term ceramic came from the Greek word keramos which means 'potter's clay' [1]. Any product made from a non-metallic inorganic material when subjected to high temperature by firing can result in desirable ceramic properties [2]. These mostly comprise inorganic, non-metallic structures usually having compounds of oxygen and may comprise one or more metallic/semi-metallic elements [3]. Traditional ceramics that are used for pottery, porcelain, refractories (heat-resistant materials), and abrasives are made from clay, silica, and feldspar. These ceramics are composed of milled materials mostly inorganic pressed, fired, polished, refractory compounds of porcelains, glasses, ceramics, and glass ceramics (2013 version of the ADA Code on Dental Procedures and Nomenclature15).

The forming and burning of clay suspensions has been carried out since 5000B.C. In 1789, the first patent for porcelain tooth material was under the French dentist Dr. Chemant, followed later in 1903 by Dr. Charles Land as the first one to discover ceramic crowns in dental practice [4]. Weinstein et al. used feldspathic porcelain for the first time in 1962. But it was only in 1963 that the first commercially available porcelain VITA Zahnfabrik was introduced in the market. In 1965, McLean and Hughes introduced

DOI: 10.1201/9781003258353-12

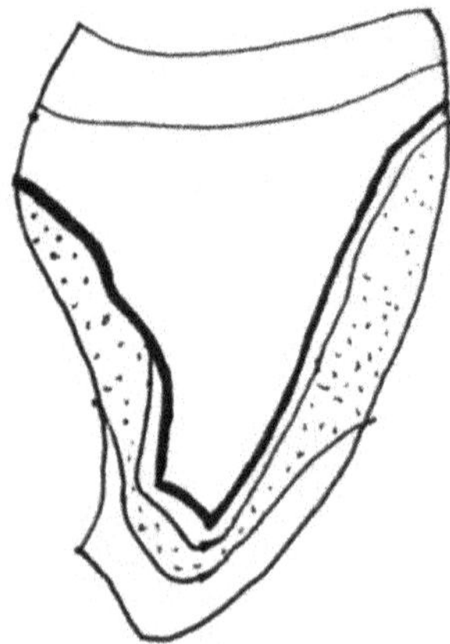
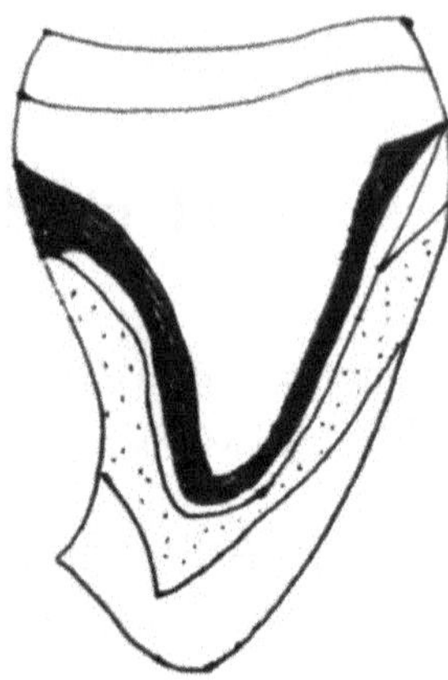

FIGURE 9.1 All-ceramic crown (a) from innermost surface of glass-infiltrate ceramic core; opaque dentine ceramic; dentine ceramic; enamel ceramic and porcelain fused to metal crown (b) from innermost surface of metal coping (veneered & un-veneered); opaque porcelain; body dentine porcelain, enamel porcelain.

aluminous core ceramics. A few years later in 1984, Adair and Grossman developed Dicor, by controlled crystallization method. In early 1990s when IPS Empress a pressable glass ceramic was introduced. In the late 1990s, IPS Empress 2- more fracture-resistant pressable glass ceramic was introduced (Figure 9.1) [5].

A compositional range of dental ceramics referred to as dental porcelains are mostly used for metal-ceramic restorations. Porcelain is referred to a mixture of the particular range of ceramic materials fired at high temperatures consisting of quartz (silica), kaolin (hydrated aluminosilicate), and feldspars (potassium and sodium aluminosilicates). Skilled technicians in commercial dental laboratory work with specialized equipment to shape and tint the ceramic restoration according to the dentist's specifications. However, CAD/CAM also provides chairside fabrication of a variety of machinable all-ceramic restorations [6].

9.2 CLASSIFICATION OF CERAMIC MATERIALS

9.2.1 According to Composition [2]

Ceramic consists of two phases, namely, glass and the crystalline phase. Glass phase forms the matrix, and the crystalline phase provides strength. The basic constituent is a naturally occurring mineral, feldspar, to which quartz is added as filler. Other substances are added to it, like glass modifier, pigmenting oxide, glaze, stain, opaquer, etc.

- Feldspathic porcelain
- Aluminous porcelain
- Leucite-reinforced ceramic
- Lithium disilicate–reinforced ceramic
- Glass-infiltrated ceramic
- Zirconia-reinforced ceramic.

9.2.2 According to Fabrication Process [2]

Ceramic could be fabricated in traditional ways consisting of burning out the wax patterns and pushing ceramic into the mould by a casting machine or under pressure. However, with the advent of newer technologies, computer assistance is put to use in the fabrication of ceramic.

- Sintering
- Casting
- Hot isostatic pressing
- Slip casting
- CAD/CAM milling and copy milling

9.2.3 Rosenblum and Sculman 1997: [7] Ceramic Could Be Used to Fabricate Metal-Ceramic or All-Ceramic Prosthesis

9.2.3.1 Metal-Ceramic Systems

- Cast Metal systems: e.g.Vita MetallKeramik
- Non-Cast Metal Systems: e.g. Renaissance

9.2.3.2 All-Ceramic Systems

- Conventional Powder: e.g. Alumina reinforced – Hi Ceram
- Castable Ceramics: e.g. Dicor, Cera Pearl
- Pressable Ceramics – IPS EMPRESS, CERESTORE
- Infiltrated Ceramics – In Ceram Alumina.

9.2.3.3 In Ceram Spinel: Machinable Ceramics – By Mechanical or Digital Methods

a. Analogous systems
    - Copy Milling – CELAY (mechanical)
    - Erosive techniques – Spark – Erosion – Procera
b. Digital systems – CAD – CAM systems
    - CEREC, CICERO

9.2.4 According to Gracis and Thompson [2]: Dental Ceramic and Ceramic-Like Materials

- Feldspathic ceramic
- Polycrystalline-based ceramic
- Fluorapetite-based ceramic
- Resin matrix ceramic

9.2.5 According to Hao Yu Shi et al. [8] Ceramic in Field of Operative Dentistry Can Be Classified as Follows

- Silicate ceramics: are non-metallic inorganic glass ceramics with superior optical properties.
  1. FELDSPAR PORCELAIN
  2. LEUCITE-REINFORCED GLASS CERAMICS
  3. LITHIUM DISILICATE CERAMICS (Figure 9.2)
- Polycrystalline ceramics: also known as sintering ceramics, mechanically stronger than above one usually used in chairside prosthesis done digitally.
  1. ALUMINA CERAMICS
  2. ZIRCONIA CERAMICS
- Resin matrix ceramics: is a newer type of ceramic material with a combination of ceramic and polymer containing a resin matrix on inorganic ceramic.
  1. VITA ENAMIC CERAMIC (PICN)
  2. RESIN NANOCERAMIC
  3. FLEXIBLE NANOCERAMIC

9.3 PROPERTIES OF DENTAL CERAMICS

A broad range of mechanical properties can be found in dental ceramics. The nature and amount of crystalline phase constituting the ceramic material impact the mechanical properties of the end product. Ceramics have properties like biocompatibility, aesthetics, chemical inertness; wear resistance, and thermal insulation that set them apart from the other restorative materials (Table 9.1).

9.4 FABRICATION TECHNIQUES

Various techniques have been employed to fabricate dental ceramics. Conventionally, the most popular technique used to fabricate aluminous core porcelain was the platinum foil technique. In this method, a fine layer of tin is put on the foil, which leads to the formation of tin oxide, which in turn enhances ceramic bonding. The fracture resistance of restoration is strengthened using aluminium foil [12].

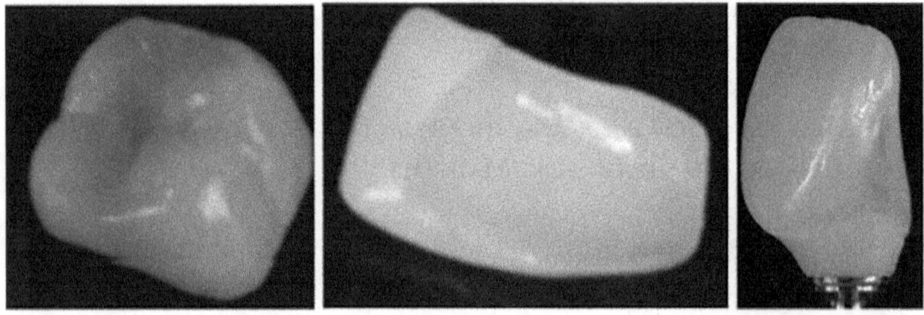

FIGURE 9.2 (From left to right) Feldspar porcelain, Leucite-reinforced glass ceramic veneers, lithium disilicate ceramic veneers.

TABLE 9.1
Properties of Dental Ceramics

SN	Properties	Description
1.	Flexural strength [6]	Feldspathic porcelains – 60 and 80 MPa. Leucite-reinforced ceramics – 120 MPa. Lithium disilicate–reinforced ceramics – 262–420 MPa Translucent cubic zirconia ceramics – 600–700 MPa Tetragonal zirconia (3Y-TZP) ceramics – 900–1,500 MPa
2.	Fracture toughness [7]	Feldspathic porcelains – 0.75 MPa/m2 Leucite-reinforced ceramics – 1.2 MPa·m0.5 Lithium disilicate–reinforced ceramics – 2.75 MPa·m0.5 3Y-TZP ceramics – greater than 6.0 MPa·m0.5
3.	Modulus of elasticity [9]	Feldspathic porcelain – 70 GPa lithium disilicate heat-pressed ceramics – 110 GPa 3Y-TZP ceramics – 210 GPa
4.	Thermal properties [10]	Feldspathic porcelain – 0.0030 cal/s/cm2 (°C/cm) is the value of conductivity, 0.64 mm2/s is diffusivity, and a linear CTE of about 12.0 × 10–6/°C in a range of 25° and 500°C. Aluminous ceramics, lithium disilicate ceramics, zirconia-based ceramics (3Y-TZP), and leucite-reinforced ceramics possess CTE – 10 × 10–6/°C, 10.5 × 10–6/°C and 14 to 18 × 10–6/°C, respectively.
5.	Optical properties [11]	Translucency- is one of the critical properties of ceramics. The metal substructure surface gets effectively masked due to the low translucency in the design makeup of opaque porcelain, which use Tin Oxide (SnO_2) and Titanium oxide (TiO_2) as their opacifying oxides. Enamel porcelain's translucency has a broad range between 45% and 50%, as compared to dentin porcelains whose translucency values are between 18% and 38% [14]. According to the nature and amount of the reinforcing crystalline phase, translucency of all –ceramic restoration also varies. Opaque ceramics constituted more of alumina and zirconia, whereas translucent ceramics were more Leucite reinforced and Lithium disilicate based. However, newly developed zirconia ceramics have also been made with increased translucency properties, using various methods. It is done by reducing the quantities of alumina additives in the 3Y-TZP composition, or by increasing the quantity of Yttrium oxide as a stabilizer. By doing so, the cubic phase amount gets higher which in turn results in significant translucency. Opalescence – a desirable optical property used to imitate the visual appearance of enamel. A light dispersion phenomenon occurs because the crystalline structure particle size is the same or less than the wavelength of light. A prismatic glass appearance is blue in either reflected or scattered light while transmitted light appears reddish-orange in colour. The light scattering effect as a result of zirconia oxide and Yttrium oxide enhances the opalescence in base dentin ceramics. Fluorescence – is acquired in dental ceramics using cerium oxide and other rare earth oxides (Figure 9.3).

Castable ceramic-like DICOR was fabricated using the lost wax technique, which is also known as the refractory die technique. Here, the ceramic is supplied as ceramic ingots (Figure 9.4) and cast via ceramming technique. Another class of ceramics is pressable ceramics like IPS EMPRESS, which are fabricated by melting

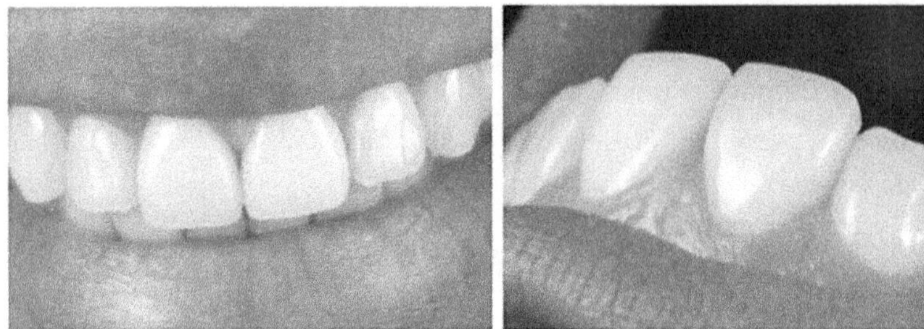

FIGURE 9.3 A resin matrix ceramics appearance.

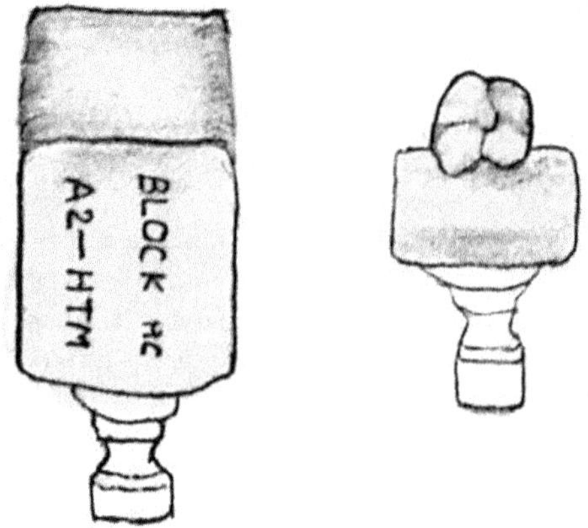

FIGURE 9.4 Ceramic blocks.

the ceramic ingots at a high temperature (1,150°C) and pressing them into a mould created by the lost wax technique [13].

The infiltrated ceramics, e.g. InCeram, InCeram spinel, and InCeram zirconia, are fabricated using the slip casting technique. The core material is alumina in case of InCeram, spinel in InCeram spinel, and tetragonal zirconia in case of InCeram zirconia. The water is mixed with the powder to form slurry which is painted on the refractory die and subjected to 1,120°C for 10 hours resulting in a porous and opaque core.

All these traditional methods were time-consuming and tedious. To overcome these disadvantages, machinable ceramics were introduced, which were supplied as ceramic blocks in various shades. Different methods namely (i) CAD/CAM or (ii) copy milling can be used for their fabrication. The first CAD/ CAM prototype was developed in 1975, by Dr. Francois Duret; later, in 1986 and 1987, CEREC and PROCERA systems were introduced, respectively [14].

CAD/CAM systems have two types for dental offices, which include acquisition type (digital impression) and scan and mill type [15].

CAD/CAM, i.e. computer aided design/computer aided machining systems exhibit three computer-linked functional components [16]:

1. A means of Data Acquisition
2. A means for Restoration Design
3. A means for Restoration Production

9.4.1 Copy Milling

Copy milling is a process where a structure is been made with the help of a device using a metal, ceramic, or polymer cast. This further reconstructs the spatial positions of the cast to guide the cutting and grounding like a key-cutting procedure as shown in Figure 9.5.

9.4.2 PROCERA and CEREC

In a digital scanner device, the die is kept on a high-speed rotating platform and a sapphire stylus scanning probe approaches the die at a **45° angle**. This milling process results in a duplicate preparation over which aluminium oxide is sparsely compacted. Moreover, to compensate for sintering shrinkage the duplicate preparation will be magnified by a factor of 0.2 (20%). This method is used especially for porcelains (AllCeram Porcelain, Ducera), which have the same coefficient of thermal expansion (7×10^{-6}°C) as that of the aluminium oxide [17].

9.5 TOOTH PREPARATION FOR CERAMIC RESTORATIONS

9.5.1 Veneer

Types of veneer preparation include partial, full veneer with window preparation, butt-joint incisal preparation, and incisal overlapping (Figure 9.6). In window preparation, stress concentration on incisal third leads to fracture of restoration. The incisal overlap distributes the occlusal forces over a larger area, hence providing enhanced support for restoration. There happens to be a controversy on stress distribution in the case of the palatal chamfer line [18,19].

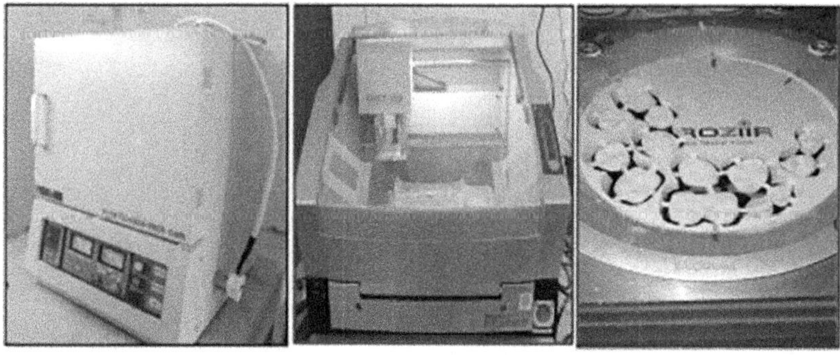

FIGURE 9.5 CAD/CAM unit. (Courtesy- Dr. Kartik Hegde, Zirconics Dental Lab.)

FIGURE 9.6 Types of veneer preparation (a) partial veneer, (b) window preparation full veneer, (c) butt-joint incisal preparation design full veneer, and (d) incisal overlapping full veneer.

9.5.2 INLAY

These are indicated in teeth where cast metal inlays can be done, with the additional requirement of tooth-coloured restoration [20]. When the isthmus is wide, direct composite restoration tends to fail. Ceramic inlays have superior mechanical properties to direct composite as the latter is subjected to polymerization shrinkage [21]. These types of inlays are contraindicated in teeth with higher occlusal loading like bruxers, clenchers, unfavourable occlusion, and group function occlusion. Isolation is of utmost importance for adequate bonding of ceramic inlays. If any clinical situation precludes proper isolation, then that constitutes a contraindication [22].

The tooth preparation (Figure 9.7) includes an occlusal reduction in a range of 1.5–2 mm with a minimum of 2 mm wide isthmus. The walls of the cavity are divergent to enable an easy path of insertion. The angle should be more than 2°–5° per wall. There should be no sharp angles in the preparation and the floor should be flat [7].

If the two-third distance from the central groove to the cusp tip is covered, then cusp capping is indicated. The occlusal reduction is 1–2 mm, and axial reduction is 1–1.5 mm required for cuspal capping. If there are any remaining caries, they should be excavated using a small round bur. It is then followed by liner and/or base placement, depending on the remaining dentin thickness. The cavosurface angle of all the margins should be 90° butt joint for marginal integrity. No occlusal, gingival, or proximal bevels are placed as ceramic tends to chip off in less thickness [7].

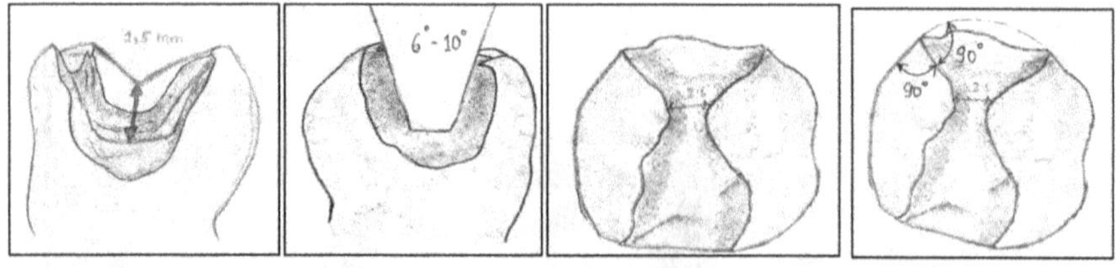

FIGURE 9.7 (a) Occlusal reduction, (b) bevel, (c) intercuspal distance, and (d) cavosurface angle.

9.6 SURFACE TREATMENT AND BONDING

9.6.1 Mechanical Methods

There are mainly two methods of surface treatment mechanically: (i) aluminium oxide air-abrasion and (ii) silica deposition method as shown in Figure 9.8 [23].

9.6.1.1 Aluminium Oxide Air-Abrasion

- Aluminium oxide when used at a high pressure to produce surface roughness and irregularities that enhances the bonding surface area [24]. In literature, many studies advocated lower bond strength using air abrasion unlike Kern et al. showed higher bond strength to zirconia crowns. Moreover, sandblasting increases the retention of the crown irrespective of any cement [25]. **Rocatec system** and **CoJet system** are two such commercially available products (Figure 9.9).

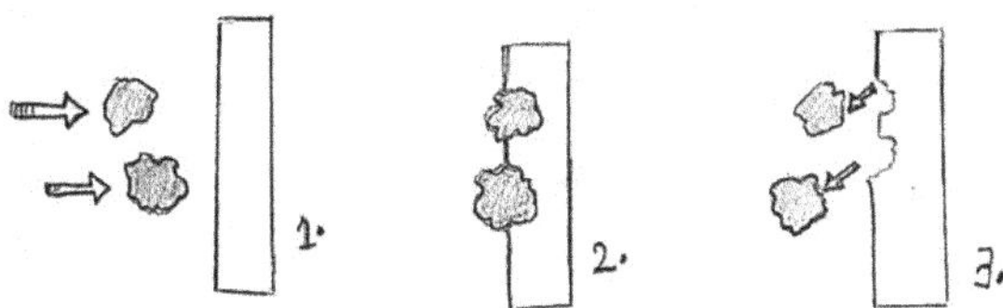

FIGURE 9.8 Air abrasion.

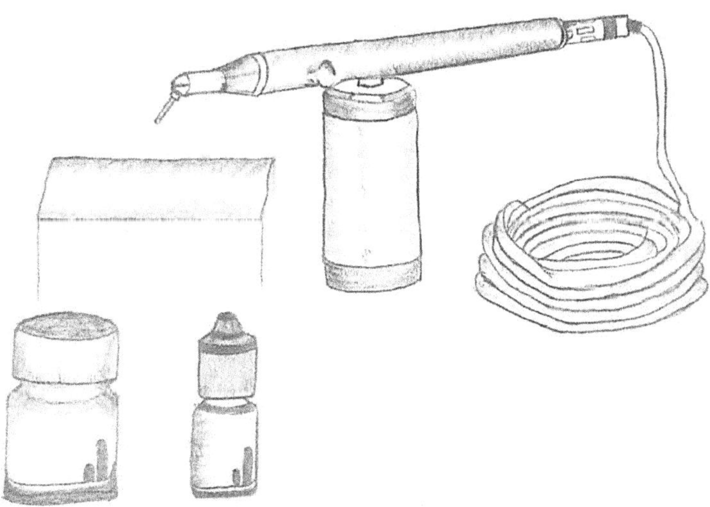

FIGURE 9.9 CoJet system.

9.6.1.2 Silica Deposition Method

The silica deposition method involves the blasting of silica-coated alumina particles over the surface of ceramics. This results in a reactive silica-rich outermost layer, which is induced to salinization (silane application) aiding in the cementation of ceramics [26].

9.6.2 Chemical Methods

Chemically, the surface can be treated by three products, namely, (i) hydrofluoric acid (ii) functional monomer, and (iii) silane coupling agent.

9.6.2.1 Hydrofluoric Acid Etching

The etching done with hydrofluoric acid (Figure 9.10) resulting in micro-porosities in glass ceramics and increased surface energy of the substrate. The micro-porosities result in penetration of resin, thus providing micromechanical interlocking [27].

9.6.2.2 Functional Monomer

Functional monomers such as MDP, i.e. **methacryloyloxydecyl dihydrogen phosphate** or phosphoric acid acrylate monomer, etc. when incorporated, enhance the bonding in zirconia ceramic. Moreover, some zirconia primers like a mixture of organophosphate and carboxylic acid monomers or phosphonic acid monomer (6-MHPA) have also reported better adhesion properties. However, the results of such studies were not statistically significant but the incorporation of primers and air-abrasion particles resulted in greater longevity and superior bond strength of the prosthesis [28].

9.6.2.3 Silane Coupling Agent

The coupling agent like silane will remove dirt from the surface which in turn increases the retention and bonding of cement to the ceramic surface. Silanes like trialkoxysilanes are the combination of inorganic-organic molecules that creates a siloxane network [29]. Contemporary many commercial products are available for silica coating.

9.6.3 Surface Treatment for Zirconia-Based Ceramics

Zirconia-based ceramics do not contain glassy matrix or silica; therefore, longevity and sustainability of bonding to the cement are still questionable.

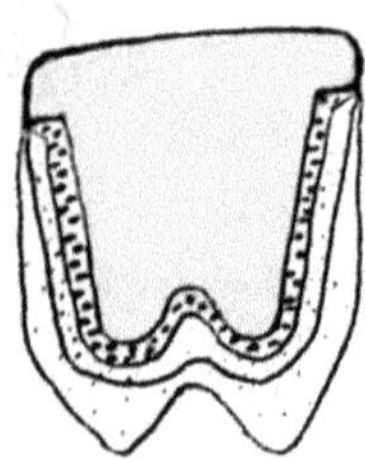
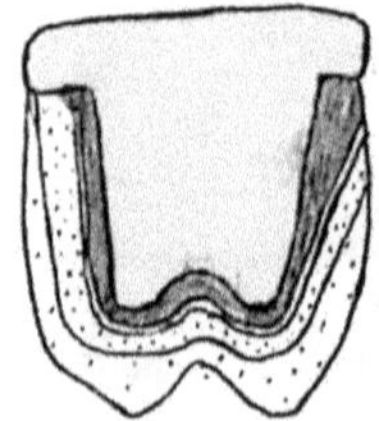

FIGURE 9.10 Types of ceramic restorations.

9.6.3.1 Selective Infiltration Etching

This is the recent technique by which the zirconia surface will be transformed into a retentive surface via inter-grain nanoporosity. These nanoporosities will aid in the mechanical interlocking of the resin cement. This technique is done using a specific glass infiltration agent, which will diffuse in the grains of zirconia resulting in nano-inter-grain porosities [26].

9.7 CLINICAL IMPLICATIONS

9.7.1 Selection of Ceramic

In a metal porcelain crown, the outermost surface consists of opaque ceramic (e.g. a titanium oxide glass), which aids in bonding and concealing the metal hue. Followed by an opaque layer is a body or dentin ceramic which mimics natural dentine. Then, an incisal ceramic is placed over the body ceramic to reciprocate the translucent effect like enamel. Then, the entire prosthesis is glazed using low-fusing or self-glazed ceramic followed by polishing [30,31].

The all-ceramic restorations primarily consist of crystalline structure ranging from 40% to 70% or even as high as 99.9% in purely polycrystalline ceramic form. All-ceramic prosthesis can be monolithic or bilayered depending on the fabrication as shown in Table 9.2 [31].

Apart from mechanical properties, one of the prime concerns is aesthetics. The crystal particle in ceramic is responsible for superior strength; however, it decreases light penetration leading to opacity [28]. Various types of commercially available ceramic based on of translucency are mentioned in the table (Table 9.3)

Zirconia is considered opaque, which poses a major disadvantage in the rehabilitation of anterior teeth. Various zirconia like zirconia-toughened alumina (ZTA), partially stabilized zirconia (PSZ), and tetragonal zirconia polycrystal (TZP) have superior mechanical properties but are not indicated for the aesthetic region. There is a newer types of zirconia like zirconia-toughened lithium silicate (ZTLS) and interpenetrating-phase composite zirconia (IPCZ), which are more translucent as compared to the earlier varieties, hence can be used in aesthetic areas [2].

TABLE 9.2
Types of All Ceramics on the Basis of Fabrication

Monolithic	Bilayered
• It is a single ceramic material. • Monolithic has the advantage of higher durability and simulation of actual tooth occlusal morphology. • They are unaesthetic in appearance. • These are indicated for posterior teeth.	• Ceramic core →strength, veneer→ final shape, shade, and aesthetic. • These ceramics are more aesthetic. • The weakest links at the interface of veneer and core lead to delamination and fracture. Also, it is difficult to achieve an ideal intercuspal relationship. • These are mostly indicated for anterior teeth.

TABLE 9.3
Types of All Ceramics on the Basis of Translucency

Translucent	Opaque
• Sintered feldspathic porcelain	• IPS e.max (lithium disilicate)
• IPS Empress Esthetic (pressable ceramics – leucite)	• Procera (alumina)
• Vitablocs Mark II (machinable feldspathic)	• Lava (zirconia)
	• Cercon (zirconia)

9.7.2 Clinical Scenarios

Amelogenesis imperfecta presents with discoloration of enamel accompanied by pitting and chipping of enamel in severe cases. Rehabilitation of aesthetics and function in such clinical scenarios can be achieved by composite or ceramic veneer. The advantage of using ceramic veneer is aesthetics, durability, biocompatibility, and less plaque accumulation. However, a certain amount of enamel removal (approx. 0.5 mm) is a must for the retention of the prosthesis. Whenever the dentin is exposed during tooth preparation, chances of microleakage and post-operative sensitivity should be taken care of during the phase between preparation and cementation. The ceramic veneer cannot be planned for patients with para-functional habits [32].

Aged composite restoration presents with marginal discoloration, which requires replacement. The longevity of composite restoration is less than that of ceramic veneer. The alteration in the shape of teeth can also be managed by ceramic veneer. During tooth preparation, the silicone putty index helps as a guide to calibrating the amount of reduction required for a particular tooth surface. This accounts for the prevention of over-reduction, which causes dentin exposure and under-reduction, which leads to an inadequate thickness of ceramic material [33].

Diastema between teeth can be treated by composite or ceramic veneer and orthodontic treatment. The latter is time-consuming and expensive whereas direct composite veneers cannot simulate proportion between various teeth. Ceramic veneer provides a chance to rehabilitate aesthetics keeping in mind the interrelationship between various teeth, with better longevity (Figure 9.11) [34].

Dental fluorosis presents with white to brown staining with pitting in severe cases. Bleaching, micro-abrasion, and macro-abrasion are some of the treatment modalities to tackle discoloration. However, sometimes more invasive techniques like veneer may be indicated to treat pitting (Figure 9.12).

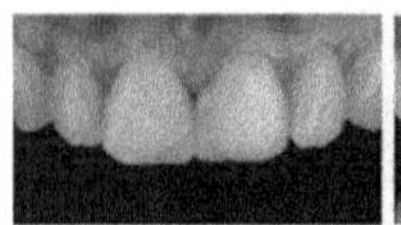
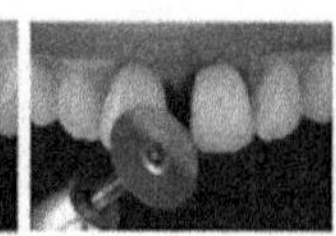
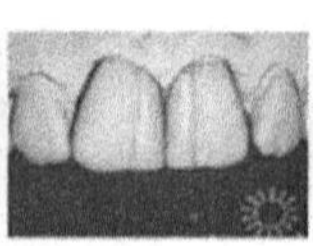
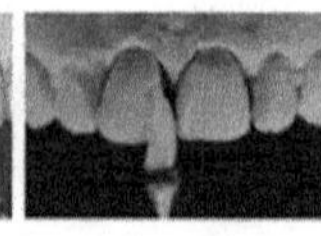
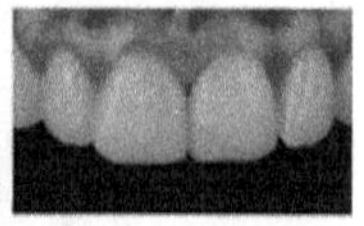

FIGURE 9.11 (From left upper to right lower) Aged composite restoration intra oral photograph, removal of old restoration, digital contouring of restoration, milled composite on model for evaluation, intraoral post restoration photograph [35].

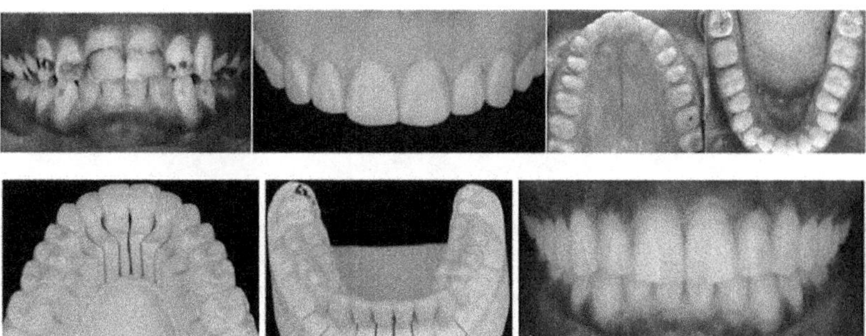

FIGURE 9.12 (From left upper to right lower) Intraoral photographs of severe dental fluorosis, diagnostic wax pattern, full mouth tooth preparation, maxillary arch pressed lithium disilicate crown, mandibular arch pressed lithium disilicate crowns, and intraoral photographs post cementation [36].

The establishment of stable occlusion and durable aesthetics requires ideal centric relation (CR) and vertical dimension of occlusion (VDO), which can be achieved by using CAD/CAM additives and restorations. This is done via "transitional bonding" where an interim bonded prosthesis like direct or indirect composite is used as a provisional restoration that is sequentially transitioned into definitive restorations. This can be exemplified with a case of generalized attrition (Figure 9.13) [37].

With the rapid advancement in CAD/CAM, additive manufacturing comes out to be a promising technique. One such technique is Stereolithography (SL) where the prosthesis is made by consecutive deposition of photosensitive material which auto polymerize immediately [38]. This technique helps treat cases of severe dental erosions. A case selection of exogenous erosion where the depth of dental erosion was scanned digitally followed by virtual wax-up and formation of the three-dimensional template to guide the tooth preparation proceeded by a porcelain veneer (Figure 9.14) [39].

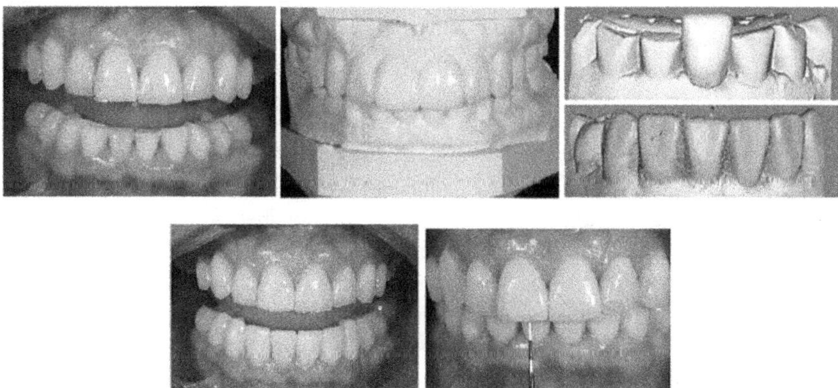

FIGURE 9.13 (From left upper side to right lower) attrition with the mandibular teeth; study model with wax-up showing new VDO, digital design using CEREC biogeneric copy and scanning the diagnostic wax-up as a provisional restoration; composite layered veneer in the mandibular arch, view of new VDO in CR [37].

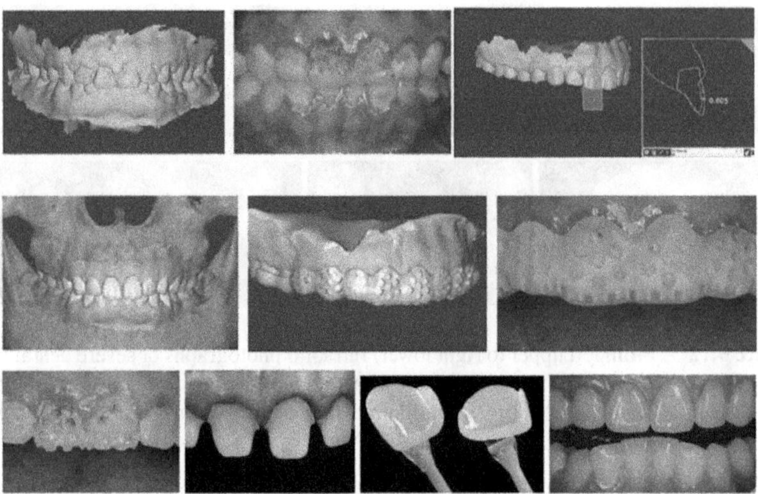

FIGURE 9.14 (From left upper to right lower) Pretreatment intraoral scan, pretreatment facial photograph, digital evaluation of the depth of erosion defects, matching of pretreatment scan, digital template to guide hole depth, after shade selection template seated, bottom of holes marked with pencil, holes removed and shoulder preparation done, evaluation and luting of restoration, and intraoral evaluation to check the marginal seal [39].

9.8 FINISHING AND POLISHING

According to recent literature, dental ceramics provide better aesthetics and superior wear resistance; however, colour stability and mechanical properties decide the longevity of the restorations. Moreover, the surface finishing also affects the colour stability and durability of the restoration, the finishing protocol of ceramic includes glazing and mechanical polishing [40].

If the surface of ceramic restoration is left rough, then there will be accentuated wear of opposite teeth and increased surface degradation. Surface smoothness can be increased by either polishing or glazing. Glazing may not adequately decrease the surface roughness since the glassy layer may be insufficient to fill the scratches and grooves; hence, polishing is preferred.

There are various commercially available systems for polishing ceramic-like EVE Diacera; JOTA; OptraFine-System; Sof-Lex; Brownie/Greenie/Occlubrush (BGO)

According to Flury et al., diamond-impregnated polishers are more effective than silica carbide polishers, and the performance of zirconia polishers is better [41].

Siddanna et al. evaluated three polishing systems of CAD/CAM ceramics that are spiral polisher, rubber cup polisher and brush paste polisher and evaluated through confocal microscope for surface roughness and in conclusion brush paste technique resulted in lowest surface roughness [40].

9.9 FAILURES IN CERAMIC RESTORATION

The commonest failure of veneers is a fracture [42]. Various studies indicate 5% failure; however, it may be 7%–14% in the case of parafunctional habit, unfavourable occlusion, extensive dentin bonding surface, and bonding to existing restorations.

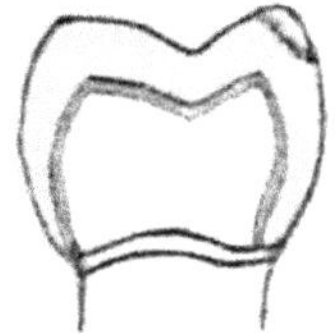
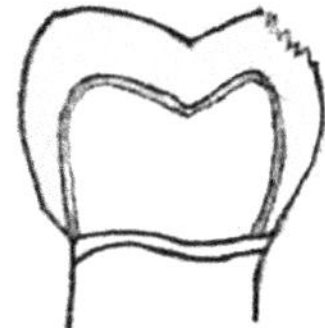
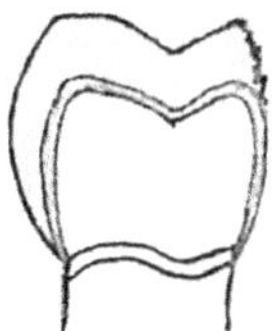

FIGURE 9.15 Failures in ceramic restoration.

Veneers that are predominantly luted on the dentinal surfaces tend to debond. After debonding, if the resin cement is present on the tooth structure, then the surface treatment of ceramic is considered improper. However, if remnants of cement are visible on the veneer, then the bonding substrate should be checked. Other causes include marginal staining and leakage through margin [43–46]. Aesthetic inlays and onlays fail because of bulk fracture of ceramic at the cuspal coverage or isthmus (Figure 9.15). At both these sites, lack of adequate tooth reduction causes inefficient stress distribution. Another reason for failure is the wear of resin cement leading to a marginal breakdown [47].

Fathy et al. reviewed recent research on the clinical performance of partial coverage ceramics categorized under **surface luster** and finish in which Lithium-discilicate ceramic restorations showed the superior results. While the CAD/CAM resin based showed loss of surface luster after a year or two follow-ups. The **colour stability** of Lithium-discilicate ceramic and resin base ceramic was good even after 3–5 years of follow-up; moreover, the marginal staining was intact even after 3 years of the interval. The **fracture and debonding** were commonly observed after 3–5 years of duration while **marginal adaptation** deterioration was minimal after 1-year duration. **Postoperative hypersensitivity** was reported in some cases with an improvement in **periodontal score** and no cases of **secondary caries** occurrence. The clinical success rate was around in the range of 100%–86% in overall studies [48].

The evidence-based dentistry revealed annual failure rate for ceramic veneers was 1.4% at 5 years interval and 1.2% after 10 years [49].

9.10 REPAIR OF CERAMIC RESTORATION

One of the major drawbacks of ceramic restoration is the tendency to fracture. Many times, repair is possible instead of replacing the restorations. Firstly, clean the fractured surface with pumice and bevel the surface with fine-grit diamond bur. Chairside air abrasion is done using 5% or 9.6% hydrofluoric acid for 20–90 seconds followed by rinsing and drying. This is followed by the application of a silane coupling agent and drying it gently. Then, the adhesive resin is applied, air-dried, and light-cured, followed by incremental placement of composite resin [50]. The procedure remains the same for the management of fracture of metal-ceramic restoration. However, the metal surface is air abraded and the ceramic surface can be air abraded or etched.

Jung and Ruttermann evaluated the types of pretreatments on the composite repair of hybrid ceramic under mechanical and chemical pretreatment. This includes diamond bur, airborne abrasion, and silica coating in mechanical pretreatment and self-cure bonding agent, silane agent, and adhesive primer (basically different types of adhesive systems) in chemical pretreatment. The suggestive outcome had no effect on mechanical pretreatment while chemical pretreatment showed variable results [51]. The bonding of polymer-based ceramics with repair composite was evaluated using different methods like the use of a self-etch bonding agent or hydrofluoric acid or sandblasting or ceramic primer or adhesive system. In case of repair of polymer-based ceramic restorations, pretreatment is required. Moreover, pretreatment application is done before the repair with composite despite the type of bonding applied. The properties of different types of ceramic and the composition of the adhesive system greatly affect the bonding strength [52].

9.11 CONCLUSIONS AND FUTURE DIRECTIONS

To commemorate the world of ceramics, Isamu Naguchi said, "The attractions of ceramics lie partly in its contraindications. It is both difficult and easy, with an element beyond our control. It is both extremely fragile and durable. Like 'Sumi' ink painting, it does not lend itself to erasures and indecision." This simplifies the widespread application of various ceramics in aesthetic dentistry. Likewise, the superior optical properties of silicate ceramics made it a better choice for veneers. However, the impeccable mechanical properties of polycrystalline and resin matrix ceramic favoured its application in crown preparations. Choosing the apt ceramic type in a clinical condition governs its longevity and durability. The overall success rate of ceramic prosthesis made it a superior choice of material when comes to aesthetic dentistry. The evolution of CAD-CAM processing techniques had added another feather to dental ceramics contributing to its biomechanical properties. Certain bonding protocols like preconditioning, and application of bonding are essential prerequisite during its fabrication. However, further researches fill paved the way for the development of ideal dental material.

Modern ceramics may seem to "cover" all clinical situations from replacing a single tooth to partial veneers although there are some drawbacks that should be taken into consideration. The complete ceramic replacement is not indicated in cases with short clinical crowns, subgingival preparations, inadequate oral hygiene, and patients with parafunctional habits.

Bioceramic materials have a very promising future. Contemporary newer composition, reduced grain size to nano size, and improved regulations for industrial production and laboratory processing aid in unique features that fulfil the aesthetic, mechanical, and biocompatible requirements.

This array of new generations of bioceramic materials offers interesting options in terms of material selection and production techniques. However, an appropriate understanding of the variety of a structure and their clinical application is needed for the longevity and durability of the restoration.

Moreover, a decision tree can help in the clinical application of aesthetic crowns, which can be as follows (Table 9.4).

TABLE 9.4
Decision Tree for Ceramic Crowns

Material	Anteriors	Premolars	Molars
Zirconia based	Layered zirconia	Layered/monolithic	Monolithic
Lithium disilicate based	• Bilaminate – full contour restoration/ crown • Hybrid with glass ceramic veneers	Bilaminate – full contour restoration/ crown	Not recommended

ACKNOWLEDGEMENT

Illustrations by Dr Aditi Mishra, BDS and Dr Anushree Maheshwari, BDS.

REFERENCES

1. V. Sukumaran, N. Bharadwaj, Ceramics in dental applications. *Trends in Biomaterials and Artificial Organs*. 20, no.1, (2006): 7–12.
2. S. Gracis, V.P. Thompson, J.L. Ferencz, N.R.F.A Silva, E.A. Bonfante, A new classification system for all-ceramic and ceramic-like restorative materials. *International Journal of Prosthodontics*. 28, no.3, (2015):
3. P.J. Babu, R.K. Alla, V.R. Alluri, S.R. Datla, A. Konakanchi, Dental ceramics: Part I – An overview of composition, structure and properties. *American Journal of Materials Engineering and Technology*. 3, no.1, (2015): 13–18.
4. S.V. Gopal, CAD-CAM and all ceramic restorations, current trends and emerging technologies: A review. *International Journal of Orofacial Research*. 2, no.2, (2017): 40–44.
5. J.V. Krishna, V.S. Kumar, R.C. Savadi, Evolution of metal-free ceramics. *The Journal of Indian Prosthodontic Society*. 9, no.2, (2009): 70–75.
6. F. Zarone, S. Russo, R. Sorrentino, From porcelain-fused-to-metal to zirconia: Clinical and experimental considerations. *Dental Materials*. 27, no.1, (2011): 83–96.
7. O.S. Abd El-Ghany, A.H. Sherief, Zirconia based ceramics, some clinical and biological aspects. *Future Dental Journal*. 2, no.2, (2016): 55–64.
8. H.Y. Shi, R. Pang, J. Yang, D. Fan, H.X. Cai, H.B. Jiang, J. Han, E-S. Lee, Y. Sun, Overview of several typical ceramic materials for restorative dentistry. *BioMed Research International*, (2022): 8451445.
9. A.D. Bona, P.H. Corazza, Y. Zhang, Characterization of a polymer-infiltrated ceramic-network material. *Dental Materials*. 30, no.5, (2014): 564–569.
10. E. Ruales-Carreraa, M.D. Bób, W.F. das Neves, M.C. Fredel, C.A.M. Volpato, D. Hotza, Chemical tempering of feldspathic porcelain for dentistry applications: A review. *Open Ceramics*. 9, (2022): 100201.
11. S. Kitouni, A. Harabi, Sintering and mechanical properties of porcelains prepared from Algerian raw materials. *Cerâmica*. 57, (2011): 453–460.
12. Y.K Lee, Translucency of human teeth and dental restorative materials and its clinical relevance. *Journal of Biomédical Optics*. 20, no.4, (2015): 045002.

13. V.A. Bilkhair, Fatigue behaviour and failure modes of monolithic CAD/CAM hybrid-ceramic and all-ceramic posterior crown restorations (Doctoral dissertation, Dissertation, Universität Freiburg, 2014).
14. B.E. Dahl. Mind the gap: Internal fit of fixed dental prostheses. (Doctoral thesis, University of Oslo Norway, 2020).
15. G. Davidowitz, P.G. Kotick, The use of CAD/CAM in dentistry. *Dental Clinics.* 55, no.3, (2011): 559–570.
16. A.O. Abdullah, F.K. Muhammed, B. Zheng, Y. Liu, An overview of computer aided design/computer aided manufacturing (CAD/CAM) in restorative dentistry. *Journal of Dental Materials and Techniques.* 7, no.1, (2018): 1–10.
17. E.A. McLaren, R.A. Giordano, Zirconia-based ceramics: Material properties, esthetics and layering techniques of a new veneering porcelain, VM9. *Quintessence of Dental Technology.* 28, (2005): 99–111.
18. S. Rinke, A-K. Pabel, M. Rödiger, D. Ziebolz. Chairside fabrication of an all-ceramic partial crown using a zirconia-reinforced lithium silicate ceramic. *Case Report in Dentistry*, (2016): 1354186
19. Y. Alothman, M.S. Bamasoud, The success of dental veneers according to preparation design and material type. *Open Access Macedonian Journal of Medical Sciences.* 6, no.12, (2018): 2402–2408.
20. C.D. Hopp, M.F. Land, Considerations for ceramic inlays in posterior teeth: A review. *Clinical, Cosmetic and Investigational Dentistry.* 18, no.2, (2013): 21–32.
21. N.B. Cramer, J.W. Stansbury, C.N. Bowman, Recent advances and developments in composite dental restorative materials. *Journal of Dental Research.* 90, no.4, (2011): 402–416.
22. S.F. Rosenstiel, M.F. Land, J. Fujimoto, *Contemporary Fixed Prosthodontics.* (St. Louis: Mosby, 2006), 223.
23. A.C. Cadore-Rodrigues, C. Prochnow, T.A.L. Burgo, J.S. Oliveira, S.L. Jahn, E.L. Foletto, M.P. Rippe, G.K.R. Pereira, L.F. Valandro, Stable resin bonding to Y-TZP ceramic with air abrasion by alumina particles containing 7% silica. *Journal of Adheives Dentistry.* 22, no.2, (2020): 149–159.
24. B. Ersu, B. Yuzugullu, A.R. Yazici, Senay CanaySurface roughness and bond strengths of glass-infiltrated alumina-ceramics prepared using various surface treatments. *Journal of Dentistry.* 37, no.11, (2009): 848–856.
25. R. Luthra, P. Kaur, An insight into current concepts and techniques in resin bonding to high strength ceramics. *Australian Dental Journal.* 61, no.2, (2016): 163–173.
26. J.Y. Thompson, B.R. Stoner, J.R. Piascik, R. Smith, Adhesion/cementation to zirconia and other non-silicate ceramics: Where are we now? *Dental Materials.* 27, no.1, (2011): 71–82.
27. R.D.L. Mattiello, T.M.K. Coelho, E. Insaurralde, A.A.K. Coelho, G.P. Terra, A.V.B. Kasuya, I.N. Favarão, L. de Souza Gonçalves, R.B. Fonseca, A review of surface treatment methods to improve the adhesive cementation of zirconia-based ceramics. *International Scholarly Research Notices.* Hindawi Publishing Corporation, ISRN Biomaterials, volume 2013, Article ID 185376, 10 pages. https://dx.doi.org/10.5402/2013/185376
28. Y. Oba, H. Koizumi, D. Nakayama, T. Ishii, N. Akazawa, Hideo Matsumura effect of silane and phosphate primers on the adhesive performance of atri-n-butylborane initiated luting agent bonded to zirconia. *Dental Materials Journal.* 33, no.2, (2014): 226–232.
29. J.P. Matinlinna, T. Heikkinen, M. Ozcan, L.V.J. Lassila, P.K. Vallittu, Evaluation of resin adhesion to zirconia ceramic using some organosilanes. *Dental Materials.* 22, no.9, (2006): 824–831

30. Ceramic veneer cementation. "Styleitaliano." Nov. 29, 2015. https://www.styleitaliano.org/ceramic-veneer-cementation/.
31. A. Warreth, Y. Elkareimi. All-ceramic restorations: A review of the literature. *The Saudi Dental Journal.* 32, no.8, (2020): 365–372.
32. S. Shibata, C.M.C. Taguchi, R. Gondo, S.C. Stolf, L.N. Baratieri, Ceramic veneers and direct-composite cases of amelogenesis imperfecta rehabilitation. *Operative Dentistry.* 41, no.3, (2016): 233–242.
33. B.T. Rotoli, D.A.N.L. Lima, N.P. Pini, F.H.B. Aguiar, G.D.S. Pereira, L.A.M.S. Paulillo, Porcelain veneers as an alternative for esthetic treatment: Clinical report. *Operative Dentistry.* 38, no.5, (2013): 459–466.
34. F. de Siqueira, A. Cardenas, Y.L. Gruber, C. Kose, Y.M. Pupo, G.M. Gomes, O. Gomes, J.C. Gomes, Using CAD/CAM-modified correlation mode to produce laminate veneers: A six-month case report. *Operative Dentistry.* 42, no.5, (2017): E139–E147.
35. W.F. Vasques, T.A. Sá, F.V. Martins, E.M. Fonseca, Composite resin CAD-CAM restorations for a midline diastema closure: A clinical report. *The Journal of Prosthetic Dentistry.* 127, no.2, (2022): 206–209.
36. J.D. Lee, N. Inoue, C. Lee, S. Park, S.J. Lee, Comprehensive management of severe dental fluorosis with adhesively bonded all-ceramic restorations. *Prosthesis.* 3, (2021): 194–208.
37. B.P. LeSage, CAD/CAM: Applications for transitional bonding to restoreocclusal vertical dimension. *Journal of Esthetic and Restorative Dentistry.* 32, no.2, (2020): 132–140.
38. A.D. Bona, V. Cantelli, V.T. Britto, K.F. Collares, J.W. Stansbury, 3D printing restorative materials using a stereolithographic technique: A systematic review. *Dental Material.* 37, no.2, (2021): 336–350.
39. T. Luo, J. Zhang, L. Fan, Y. Huang, J. Yu, H. Yu, A digital workflow with the virtual enamel evaluation and stereolithographic template for accurate tooth preparation to conservatively manage a case of complex exogenous dental erosion. *Journal of Esthetic and Restorative Dentistry.* 34, no.5, (2022): 733–740.
40. G.D. Siddanna, A.J. Valcanaia, P.H. Fierro, G.F. Neiva, D.J. Fasbinder, Surface evaluation of resilient CAD/CAM ceramics after contouring and polishing. *Journal of Esthetic Restorative Dentistry.* 33, no.5, (2021): 750–763.
41. S. Flury, A. Lussi, B. Zimmerli, Performance of different polishing techniques for direct CAD/CAM ceramic restorations. *Operative Dentistry.* 35, no.4, (2010): 470–481.
42. D. Layton, T. Walton, An up to 16-year prospective study of 304 porcelain veneers. *International Journal of Prosthodontics.* 20, no.4, (2007): 389–396.
43. M. Fradeani, M. Redemagni, Marcantonio Corrado Porcelain laminate veneers: 6-to 12-year clinical evaluation – A retrospective study. *International Journal of Periodontics & Restorative Dentistry.* 25, no.1, (2005): 9–17.
44. U.S.B.I. Kapferer, D. Burtscher, H. Dumfahrt, Clinical performance of porcelain laminate veneers for up to 20 years. *International Journal of Prosthodontics.* 25, no.1, (2012): 79–85.
45. C. D'Arcangelo, F. De Angelis, M. Vadini, M. D'Amario, Clinical evaluation on porcelain laminate veneers bonded with light-cured composite: Results up to 7 years. *Clinical Oral Investigations.* 16, no.4, (2012): 1071–1079.
46. C. D'Arcangelo, F. De Angelis, M. Vadini, M. D'Amario, Five-year clinical evaluation of 300 teeth restored with porcelain laminate veneers using total-etch and a modified self-etch adhesive system. *Operative Dentistry.* 34, no.5, (2009): 516–523.

47. A. Fabianelli, C. Goracci, E. Bertelli, C.L. Davidson, M. Ferrari, A clinical trial of Empress II porcelain inlays luted to vital teeth with a dual-curing adhesive system and a self-curing resin cement. *Journal of Adhesive Dentistry.* 8, no.6, (2006): 427–431.
48. H. Fathy, H.H. Hamama, N. El-Wassefy, S.H. Mahmoud, Clinical performance of resin-matrix ceramic partial coverage restorations: A systematic review. *Clinical Oral Investigations.* 26, no.5, (2022): 3807–3822.
49. T. Mazzetti, K. Collares, B. Rodolfo, P.A. da Rosa Rodolpho, F.H. van de Sande, M.S. Cenci. 10-year practice-based evaluation of ceramic and direct composite veneers. *Dental Materials.* 38, no.5, (2022): 898–906.
50. B. Loomans, M. Özcan, Intraoral repair of direct and indirect restorations: Procedures and guidelines. *Operative Dentistry.* 41, no.S7, (2016): S68–S78.
51. S.N. Jung, S. Rüttermann, Influence of mechanical and chemical pre-treatments on the repair of a hybrid ceramic. *Dental Materials.* 38, no.7, (2022): 1140–1148.
52. E.Ö. Bayazıt, Repair of aged polymer-based CAD/CAM ceramics treated with different bonding protocols. *The International Journal of Prosthodontics.* 34, no.3, (2021): 357–364.

10 Bioceramics for Hip and Knee Implants

Ikra Iftekhar Shuvo
Massachusetts Institute of Technology

Lovely Khandakar
University of Manitoba

Sara V. Fernandez and Canan Dagdeviren
Massachusetts Institute of Technology

10.1 INTRODUCTION

Reconstructive and regenerative orthopedic surgeries have generated considerable interest in fabricating artificial body parts for implants. Medical advancements and developments have heightened the use of biomaterials for reclamation of damaged body parts. Among the different categories of biomaterials, bioceramics have gained popularity in prosthetics (an artificial mechanical device designed to replace the biological part). Bioceramics are biocompatible to humans and other mammals and can therefore be used for repairing any unfixed parts. Since bioceramics closely resemble that of the host tissue, it can promote a regenerative response in the organism (Dorozhkin 2010). Notably, the bioceramics contribute to minimizing exposure to metallic surfaces, thereby augmenting the prosthetic experience of the user by reducing the source of potential sensitizing ions (Piconi and Maccauro 2015).

In orthopedic surgeries, total knee arthroplasty (TKA) and total hip arthroplasty (THA) outpace every other surgery and, therefore, incur high cost and outcome durability (Schwartz et al. 2020). Superior biocompatibility, endurance to a larger degree of torques, load-bearing capacity, low density, and high corrosion/wear resistance of the bioceramic implants have intensified their demand in THA/TKA surgical procedures. While THA requires replacement of the upper femur (thigh bone) and resurfacing/replacement of the mating pelvis (hip bone), TKA refers to the replacement of the diseased cartilage surface of the lower femur, tibia, and the patella (Joseph 2003). Due to the lower reactivity, early stabilization, and longer functional life, bioceramic implants demonstrate the potential to replicate the mechanical behavior of original bones (Shekhawat et al. 2021). Pragmatically, the finite lifespan of the ceramic implants could also necessitate revision surgeries for TKA/THA patients (rTKA/rTHA). In addition, any unexpected mechanical mismatch or infection from ceramic debris could cause premature failure of knee and hip joint implants (Shekhawat et al. 2021). A report by the Department of Orthopedic Surgery of the Emory University

DOI: 10.1201/9781003258353-13

School of Medicine forecasted an increase of 70% and 182% incidents in rTHA and rTKA, respectively, from 2014 to 2030 (Schwartz et al. 2020). Alarmingly, just from prosthetic joint infection, rTHA and rTKA are expected to rise by 176% (from 2,808 cases in 2002 to 16,169 cases in 2030) and 170% (from 9,089 cases in 2002 to 53,569 cases in 2030), respectively. Between 2002 and 2014, rTKA increased three times more than rTHA. Therefore, it is not at all a trivial matter to properly understand the selection criteria, properties, and evidence-driven cases of the bioceramics to help prepare for the growing trends of the knee and hip joint replacement surgeries.

10.2 MARKET SIZE

Dental industries comprise the lion's share of the use of bioceramics, which accounts for about 42% of the total application (Technavio Research 2017). After dental applications, the orthopedic surgeries account for the next largest segment of the global bioceramics market, followed by areas such as cardiovascular, drug delivery, and tissue engineering. The global market value of bioceramics was US$1 billion in 2001 (Vallet-Regí 2001), which reached US$14 billion in 2020 and has been projected to be US$23 billion by 2031 (Fact.MR 2021). The demand for bioceramics materials is propelled by the increase in life expectancy as well as advancements in biological implants. In the USA alone, ~0.5 million total hip arthroplasty (THA) surgeries were done in 2020, and the number is predicted to rise to 1.5 million surgeries/year by 2040 (Transparency Market Research 2022). Approximately 0.5 million total knee arthroplasty (TKA) surgeries/year were performed in the USA as of 2010 at the expense of $15,000 USD/patient, totaling an aggregate of US$9 billion/year (Cram et al. 2012). The number of surgeries surpassed 1 million/year in 2020, and a 401% increase is expected by 2040 to ~4 million replacements/year (Rheumatology Advisor 2019). A study conducted by Mayo clinic, presented at the American Academy of Orthopedic Surgeons (AAOS) annual meeting, revealed that both TKA and THA are comparatively more prevalent in women than in men (3 and 1.4 million women out of 4.7 and 2.5 million US people who underwent TKA and THA in 2014) (Mayo Clinic 2014).

10.3 BIOCERAMIC COMPONENTS FOR HIP/KNEE JOINTS

A recent report has predicted that 30% of hospital beds could soon become occupied by osteoporosis patients, i.e., patients with a porous bone disease that leads to weak and brittle bones. This "silent disease" slows down the body's natural new bone synthesis process (Habraken et al. 2016) to the point where the breakdown process of old bone tissue outpaces the new bone tissue formation process; consequently, bones become fragile and break easily. Unfortunately, 20% of the patients with an osteoporotic hip fracture die within the first year after surgery. Such an alarming number indicates the need for advanced materials for bone replacements.

There are both natural and artificial materials available for bone replacements (Vallet-Regí 2014). The natural option includes autologous bone (self-donor), homologous bone (tissue bank), and heterologous bone (animal sourced). The natural options became less appealing due to their risk of disease transmission or scarcity of materials. The artificial option includes bioceramics. Interestingly, bioceramics are considered materials that exhibit the best resemblance to the mineral components of the bone joints.

Knee and hip joints are among the largest joints in the human body, supporting body weight and locomotion. Unfortunately, hip and knee replacements are the most common arthroplasty surgery. Knee-joint pain can arise from wear and tear from daily activities like walking, jogging, or lifting. Joint fractures, torn ligaments, patellar instability, torn meniscus, or ligaments injuries are also a few of the common causes of knee surgeries. Recent years have seen the burgeoning applications of bioceramics and their composites in implants or orthopedic surgeries: bioactive glasses for cranial repair, zirconia in load-bearing components, alumina for keratoprostheses or orthopedic knee fixation devices, and so on. Bioceramics are also used in condyles and tibial plateau for knee replacement (Antoniac 2016).

Figure 10.1 illustrates the application of a non-metal implant for TKA using bioceramic composites based on alumina (Al_2O_3) and zirconia (ZrO_2). TKA implants typically consist of three main components: femoral, tibial, and patellar part (Piconi and Maccauro 2015). The tibial component is a flat platform with a cushion of wear-resistive solid plastic, polyethylene. Besides a metallic platform, a bioceramic platform could be used for the tibial component. By replacing metal condyles with bioceramic-made condyles, the wear-performance of the polyethylene insert could be augmented. This is due to the higher scratch hardness of bioceramics. Thus, it provides better resistance

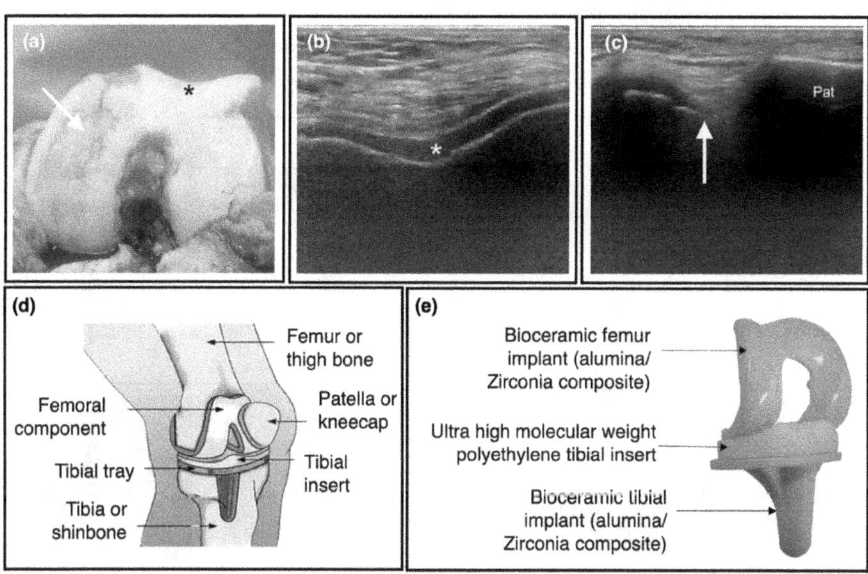

FIGURE 10.1 Cartilage degeneration in late-stage knee osteoarthritis observed during the total knee arthroplasty (TKA). (a) Intraoperative photography of the femoral cartilage exhibits intact (asterisk) and damaged cartilage (arrow) at the femoral sulcus and medial femoral condyle of the knee, respectively (Nevalainen et al. 2018). The corresponding ultrasonographic (b) and radiographic (c) images demonstrate the normal (asterisk) and degenerated (arrow) area of the femoral cartilage. The schematic (d) shows the implant components (femoral, tibial, and tibial insert components) during a TKA (Bahraminasab and Jahan 2011), and the digital photograph (e) displays the non-metal implantable total knee replacement device made of bioceramic composites (from alumina and zirconia) (Meier et al. 2016).

to damage and protects the polished surface of the articulating condyles. Knee arthroplasty is classified into two categories: total and partial knee arthroplasty (TKA/PkA). During TKA, all these three parts are replaced with prostheses. However, for partial knee arthroplasty (PKA), only the affected region of the knee is replaced.

Like knee arthroplasty, hip arthroplasty is also a surgical procedure performed to relieve pain and restore the functionality of the hip using an artificial implant. The need for a total hip arthroplasty may arise from several issues, like injuries/accidents, menopause in the case of women, age-related bone diseases, and bone degeneration among the older populations. Hip arthroplasty can be of two types, which are total and partial hip arthroplasty (THA/PHA). THA includes replacement of both ball (femur head) and socket, while PHA involves replacing only the ball.

Figure 10.2 exhibits a hip stem implant used in THA. The stems are typically made of different alloys of titanium (Ti) or cobalt-chromium (CoCr). The cups could also be made of Ti or polyethylene. For liner, tough plastic materials are generally used, UHMWPE, for instance, which has high wear and abrasion resistance. Alumina and its composites are mainly used for the femoral head; however, CoCr-based and metallic femoral heads are also available. Monoclinic zirconia is used as a coating on the surface of metallic ball heads to better the wear behavior of metal-on-polyethylene (MoPE) implant bearings.

While marketing any prostheses or bone implants, the morphological study of the target market is crucial for successful engineering design. For instance, a study

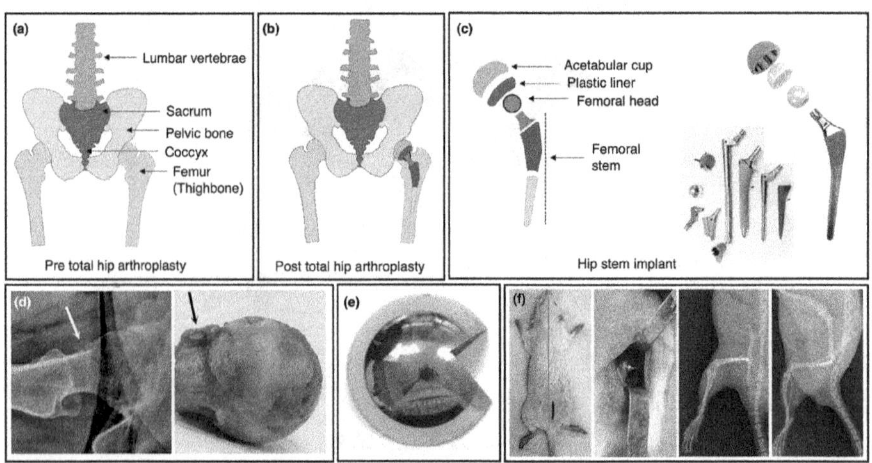

FIGURE 10.2 Hip stem prosthesis for total hip arthroplasty (THA) (or replacement) shown in the (a–c) schematics with photographic image of a commercial femoral hip implant (c) (Murr et al. 2012). Radiographic image demonstrates the presence of osteophytes (bone lumps causing painful joints) (white arrowhead) (d) in a patient with late-stage hip osteoarthritis, which is clearly distinguishable (black arrowhead) in the femoral head removed during THA (Nevalainen et al. 2020). As implants, metallic femoral heads (e) could be used in a metal-on-polyethylene (MoPE) implant framework (Cui et al. 2016). (f) Rat models are used in alumina or different material-based THA experiments to investigate osteolysis in aseptic loosening by implanting man-made prostheses (Li et al. 2018).

was conducted in 2007, which was aimed to compare the need for a revision TKA (rTKA) surgery between people of two ethnic background: 73 Japanese and 76 Americans with TKA (Iorio et al. 2007). This study showed that the longevity of the implants and the needs for revision surgeries for the two groups were different. The mean implant longevity for the Japanese patients was 6.6 years, with 4.1% of patients requiring revision surgery, while only 2.6% of the American patients needed revision surgery, demonstrating a mean 9 years of the longevity of their prostheses. The study hypothesized that the anomalies of the implant performance between two test subject groups could be attributed to the flawed marketing campaign of the prosthetic implants without considering their morphological differences.

10.4 CLASSIFICATION OF BIOCERAMICS

Biomaterials are any synthetic materials used for making devices to replace part of a living system or to function in direct contact with living tissue (Wong and Bronzino 2007; Agrawal 1998). In the field of regenerative medicine, biomaterials play a vital role in cell proliferation, adhesion, spreading, differentiation, and tissue formation in all three space dimensions (Antoniac 2016). Superior biocompatibility and relevant mechanical performance are the two critical reasons for which biomaterials are becoming popular in clinical applications (Kumar and Baino 2020). Biocompatibility of the biomaterials comes from their specific chemical compositions and topographical features which directs the cellular response toward tissue regeneration. Some other preferable qualities of biomaterials include osteo-inductivity (ability to induce osteo-genesis, i.e., bone formation), osteo-conductivity (ability to grow bone on a osteo-conductive surface and conform to it), and osteo-integration or osseo-integration (ability to fuse so strongly with the bone that it cannot be disintegrated without fracture) (Stevens 2008). In contrast to these highly biocompatible biomaterials, low-biocompatible prosthesis materials like Cu, Ag, or bone cements exhibit very low to zero osteo-conduction (Albrektsson and Johansson 2001). Generally, biomaterials are divided into four types: (i) biometals, (ii) biopolymers, (iii) bioceramics, and (iv) biocomposites (Dorozhkin 2011). The following discussion will focus mainly on bioceramics and their applications for knee and hip joint implants.

Typically, bioceramics can be categorized into three classes: (i) bioinert, (ii) bioactive, and (iii) bioresorbable materials. However, it must be noted that there are numerous studies where authors study bioactive and bioresorbable materials together – calcium phosphate (CaP) and hydroxyapatite (HAp) for instance – and categorize them as second-generation bioceramics, in contrast to first-generation bioinert and third-generation scaffolds for tissue engineering (Punj et al. 2021). Among the bioceramics, bioinert materials (e.g., alumina or zirconia) can co-exist with the tissues without causing much noticeable change; however, bioactive materials (e.g., glass ceramics) can form direct biochemical bonds with the tissue (Dubok 2000). Bioresorbable materials, on the other hand, undergo gradual dissolution in the biosystem of the organism and are replaced by bone tissues without toxicity or rejection. Table 10.1 gives a summary of these three categories of bioceramics, and Figure 10.3 illustrates a comparison of the different mechanical properties of ceramic and glass materials.

TABLE 10.1
Summary on the Three Categories of Bioceramics

Subjects	Bioinert Materials	Bioactive Materials	Bioresorbable Materials
Reactivity with the host	Physical and mechanical properties remain constant and do not exhibit any reactivity with the host tissues.	Undergoes osteo-conduction and able to form direct chemical bond with host tissue and, thus, enables fixation of the implant within host skeletal system	With time they get absorbed and replaced by bone in the bone tissue, i.e., the resorbed ceramics are replaced by endogenous tissue
Applications	Typically used as bearing surface for joint prostheses and in making bone plate, bone screw, femoral head, and parts of knee, hip, shoulder, wrist, elbow, tooth, etc.	Bone grafts and coating material for metallic prosthetics or implants	In bone defect or void fillers in the form of granules, bone grafts and replacement of the surrounding tissue
Examples[b]	Alumina, zirconia	Bioglass®, apatite-wollastonite (AW) containing glass ceramics	Calcium sulfates (CaSs), calcium phosphates (CaPs), hydroxyapatite (HA)
Incorporation into bone	Following the pattern of "contact osteogenesis"	Following the pattern of "bonding osteogenesis"	Similar to "contact osteogenesis"
Major advantage	High strength, non-toxicity, excellent corrosion resistance, superior stability, and in-vivo biocompatibility	In-vivo biocompatibility and rapid tissue bonding	Eliminates the need of surgical revisions or second surgery it
Major disadvantage	Material never transforms into bone, and sometimes may cause negligible foreign body reaction	Low fracture toughness and mechanical strength	Low interfacial stability between bone tissues and bioresorbable materials
Hardness (HV)[a]	High (e.g., 1,200 – 2,000)	Low (e.g., 350 – 600)	–
Tensile strength (GPa)[a]	High (250 – 400)	Low (e.g., 0.12 – 122)	Lower (0.03 – 0.2)
Compressive strength (MPa)[a]	1,600 – 4,000	600 – >2,000	20 – 900
Fracture toughness (MPa. $m^{1/2}$)[a]	5.0 – 12.0	0.6 – 1.0	< 1.0

[a] Shekhawat et al. (2021).
[b] Punj et al. (2021).

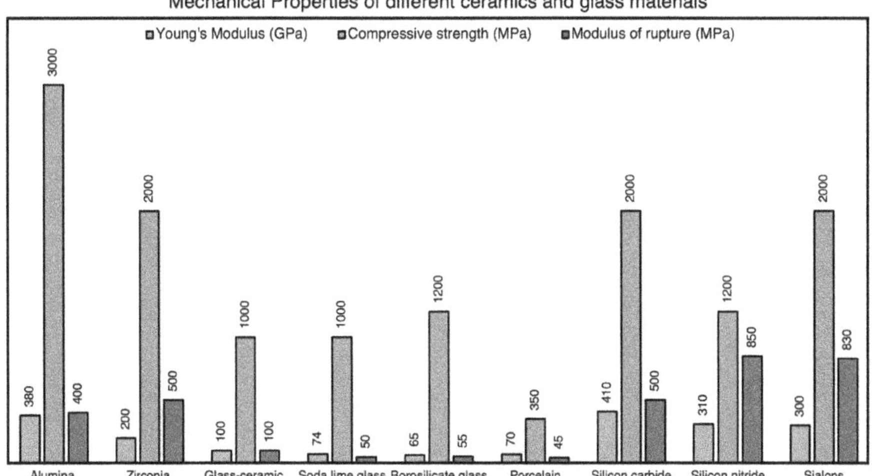

FIGURE 10.3 Mechanical properties of different ceramics and glasses. (Based on data from Park (2009).)

10.5 OVERVIEW OF DIFFERENT TYPES OF BIOCERAMICS

A critical review of different types of bioceramic materials used in hip and knee implants is discussed in the following sections.

10.5.1 Bioinert Ceramics

It is hard to say whether any material exists that is completely inert or 100% safe (to be used as body implant – can we use?) for body implants; however, bioinert ceramic materials do have comparatively stable physiochemical properties (Kumar et al. 2018). Oxide ceramic materials, for example, are stable, inorganic, bioinert materials: they do not undergo further oxidative processes and stay chemically inert the entire time they reside inside an organism (Piconi and Sprio 2021). Hence, the chemical stability of oxides makes them an ideal choice for bioceramics. Alumina (Al_2O_3), zircona (ZrO_2), and their composites are the major classes of bioinert materials widely used in orthopedics and have gained popularity for applications in arthroprosthetic surgeries for joint replacements. Table 10.2 synopsizes the results from different clinical trials and case studies that incorporated bioinert ceramics for hip and knee implants.

10.5.1.1 Alumina (Al_2O_3)

Al_2O_3 is the most widely used bioinert ceramic for THA (Vallet-Regí 2001). Alumina displays good performance under compression, although it is brittle under tension. The tensile strength of alumina is better at a higher density and smaller grain size. By incorporating low-melting magnesium oxide (MgO) into the ceramics, full density at a lower temperature can be reached, thus decreasing grain growth and increasing ceramic strength. Unfortunately, the addition of MgO reduces the hardness – a setback that could be solved by adding small amount of chromia (Cr_2O_3) (Piconi et al. 2003).

TABLE 10.2
Overview of Different Studies on Bioceramics Used in Clinical Trials and Experimental Research Works

Material	Application	Strength	Limitation	Additional Remarks	Ref.
Third-generation cementless alumina CoC (ceramic-on-ceramic) bearings	Total hip arthroplasty	Wear-resistance, excellent implant survival rate, and low osteolysis and ceramic fractures	Squeaking was identified due to edge loading and lubrication loss	94.2% survival rate at 20 years	Xu et al. (2022)
Alumina sandwich liner	Total hip arthroplasty	Stable formation of bone without any infections	High risk of liner fracture at a mean 7.3 years follow-up due to design defects	91.4% survival rate at 12 years	He et al. (2022)
Oxidized zirconium (Oxinium)	Femoral component for total knee arthroplasty	Applicable as an alternative to cobalt-chromium bearing surface that could undergo up to 1,000 lbf (68,400 psi) fatigue load reduces mechanical failures	Femoral component fracture and debonding due to poor osteotomy and cementing technique at the implant interface	First reported failure case of the Oxinium-based femoral implant	Ichimura et al. (2022)
Magnesia partially stabilized zirconia (MgPSZ)	Femoral component for revision total knee arthroplasty	Bearing surfaces made of MgPSZ are known to prevent the release of any metallic ions or debris	Less information is available to draw any conclusion because of the proprietary nature of the implant designed with the MgPSZ materials	A good candidate as implant components for patients displaying clinical signs of metal allergy or sensitivity	Whiteside (2022)
Alumina-toughened zirconia (ATZ)	Arthroplasty for hip resurfacing	Improved fixation stability due to increased contact area by 1.8 times with the bone material	Titanium-based inner layer is required to improve the stability	The fixation stability could be optimized by carefully choosing the bone material and specimen size	Vogel et al. (2022)

Alternatively, mechanical property could be improved through a hot isostatic pressing (HIP) process, which involves shaping at a high pressure and temperature and produces a high-density ceramic having limited grain growth (López 2014).

Al_2O_3 can also be added to different bioceramics to improve their performance; examples include β-tricalcium phosphate (β-$Ca_3(PO_4)_2$ or β-TCP), which displays excellent osteo-conductivity and biocompatibility with the physiological environment. Its bone-like chemical composition makes it a suitable alternative for bone graft. However, its application in the human body is limited by its reputation of having weak rupture resistance (Sprio et al. 2013). Barkallah et al. demonstrated the potential of Al_2O_3 to improve the overall mechanical properties of β-TCP-based composites (Barkallah et al. 2018). This collaboration between French and Tunisian researchers showed that the addition of Al_2O_3 with 10 wt% TCP and 5 wt% titania (TiO_2) powder improved the overall mechanical properties of the bioceramics, leading to a compressive strength of 352 MPa, flexural strength of 98 MPa, tensile strength 86.65 MPa, and fracture toughness of 3 MPa $m^{1/2}$. In 2021, these researchers studied the tribological (i.e., friction, wear, lubrication, and design) behaviors of the composites, using 2D profilometer and SEM analysis to measure wear volume and associated mechanism, respectively (Barkallah et al. 2021). The result showed that the combination of β-TCP, 10 wt% Al_2O_3, and 5 wt% TiO_2 produced the best composites: the best wear resistance and microhardness with the lowest friction coefficient (Figure 10.4).

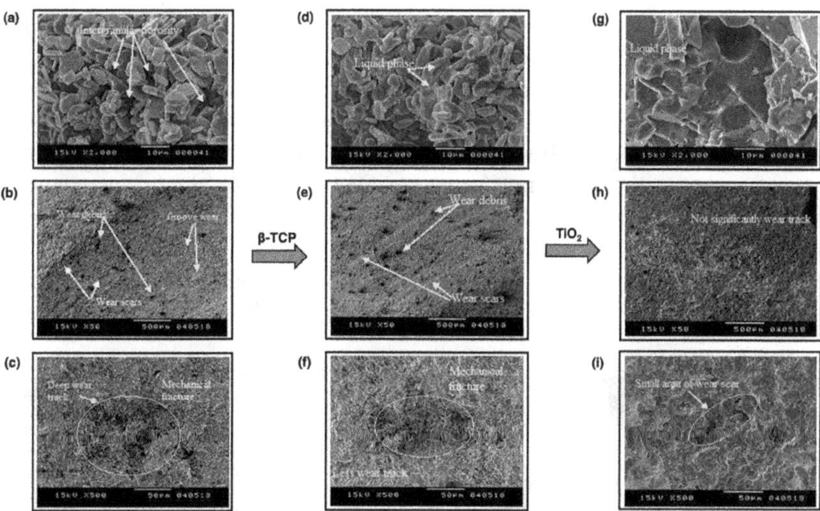

FIGURE 10.4 SEM micrographs of alumina Al_2O_3 and its composites with β-tricalcium phosphate (β-TCP) and titania (TiO_2) (Barkallah et al. 2021). (a) Unworn intergranular porous and (b, c) worn surface of 100% pure Al_2O_3. (d) Adding 10 wt% β-TCP with the Al_2O_3 produces composites with finer microstructures as a liquid phase emergence on the unworn surface, and consequently, a reduced widths of wear scars are seen on the (e, f) worn surface, improving its fracture toughness property. (g) Further addition of the 5 wt% TiO_2 enhances the liquid phase between TCP and TiO_2 as seen in the unworn surface of the composite. (h, i) As a result, the specimens become more dense, compact, and the debris are less deep, leading to an overall improvement of the tribological properties, i.e., lower wear volume and friction coefficient.

10.5.1.2 Zirconia (ZrO_2)

ZrO_2 exists in three different crystalline structures at ambient pressure: monoclinic, tetragonal, and cubic (Weng et al. 2021). At 1,000°C–1,200°C, zirconia undergoes an allotropic phase transition from monoclinic to tetragonal, and at 2,370°C, the phase changes from tetragonal to cubic (Park 2009a,b). During the manufacturing stages, ball milling hours can have significant effect on the crystallite size and lattice strain of the zirconia (Elsen et al. 2017). Prolonged ball milling results in reduced crystallite size. For example, ball milling of 4, 6, or 8 hours will produce a crystallite size of 34, 28, and 25 nm, respectively, with a corresponding lattice strain of 0.000236, 0.000157, and 0.000104 (unit paper also have no unit).

Pure ZrO_2 is not suitable for direct application due to the difficulty of transformation from one form to another (López 2014). Any change in shape and volume during the transformation process can easily lead to material degradation and cracking. Furthermore, ZrO_2 manifests comparatively higher level of wear during in-vivo studies compared to in-vitro studies (Dawson-Amoah et al. 2020). The underlying causes are attributed to the presence of proteins, pH of bodily fluids, and salts in contrast to the artificial ageing simulations using autoclaves. Figure 10.5 summarizes the THA case study of a 50-year-old female with a 5-year history of right-hip pain, a non-trivial family history of father with colon cancer and mother with breast cancer and Alzheimer's. The patient underwent a ceramic-on-polyethylene THA.

The mechanical performance of zirconia can be improved by adding stabilizing agents such as MgO, CaO (calcium oxide) or Y_2O_3 (yttrium oxide), during the fabrication process to limit the phase transformation. For instance, a high degree of flexural strength and fracture toughness of zirconia can be observed when it is partially stabilized with Y-TZP (yttrium tetragonal zirconium polycrystal) (Park 2009). The increase in the fracture toughness is a result of the cessation of crack propagation during the phase transformation. On the other hand, yttrium magnesium oxide-stabilized zirconia (Y-Mg-PSZ) could be added for higher Weibull modulus compared to Y-TZP.

The size of the pore is an important factor for bone growth: it should be sufficiently large to accommodate development of the organic and inorganic

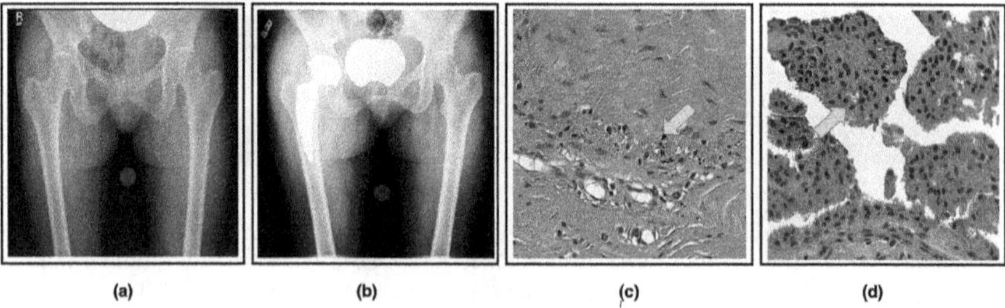

FIGURE 10.5 Radiographs of the THA implant on the right-hip of a female patient and pathologic evaluation during the revision period (Dawson-Amoah et al. 2020). (a, b) The radiographs exhibit the pre- and post-operative THA of the right-hip. (c, d) During the revision surgery, histological investigation of the synovium tissues confirms the macrophagic infiltration of ZrO_2 debris (shown in arrow) from the ceramic head.

components of the bone along with the bone cells (Klawitter and Hulbert 1971). Optimum pore size allows mineralization and provides space and a smooth path for growth of vascular tissue. Pore size of approximately 200 μ must be provided for proper development of osteons. Increasing the pore size and surface area of the bioceramics may increase bone-forming bioactivity by accelerating biological apatite deposition (Antoniac 2016). Research conducted in 2016 using 3Y-TZP and steric acid found that differing contents of stearic acid powder can be used to achieve the desired mechanical property and pore size in zirconia (Li et al. 2016). In addition to pore size, the inter-connectivity among the pores plays a major role in bone growth; therefore, both the pore size and the overall pore structure need to be taken into consideration.

Proper porosity plays a vital role in providing the template for cell attachment on the surface and allows formation of the three-dimensional spreading-out structure (Li et al. 2016). A study using 0, 5, and 10 wt% stearic acid with 3Y-TZP resulted in 1.1%, 5.8%, and 16% porosity, respectively. The 16% porosity proved to have superior biocompatibility as it allowed high cell proliferation. Both porosity and a lightweight structure of superior properties are essential. A study conducted in 2018 found that adding 15%–20% of silicon nitride to zirconia effectively reduced the density of the composite, lowering the weight of the finished product (Renoldelsen and Vivekananthan 2018). Interestingly, addition of silicon nitride to zirconia also improves the sintering property of zirconia.

10.5.1.3 Different Composites of ZrO_2 and Al_2O_3 Ceramics

Bioinert material has both benefits and drawbacks when considered for hip or knee arthroplasties. There are ways to enhance its beneficial characteristics, such as improving mechanical properties using bioceramic composites. For example, higher content of silicon oxide makes the composite bioinert in its behaviors: it induces the formation of a fibrous capsule at the interface of tissue and implant (Dubok 2000). Similarly, Homerin et al. studied two different fabrication methods that exhibited superior fracture toughness (FT) of biocomposites like zirconia toughened alumina (ZTA): (i) attrition milling and hot-pressing and (ii) electrochemical dispersion (Homerin et al. 1986). Such composite structures are used in joint replacement surgeries (Piconi et al. 2003). A study shows that an addition of ZrO_2 up to 25% (wt) into the alumina matrix results in increased fracture toughness as a result of the phase transformation of the zirconia particles. The performance of the ZTA composites could be further improved by introducing stabilizers as they prevent microcrack formations inside the composite structures (Trabelsi et al. 1989). Homerin et al. showed the impact of stabilizer concentrations (1 and 3 mol% Y_2O_3) on FT. The 3 mol% concentration showed a constant and steady increase in the FT; however, 1 mol% showed a dramatic effect in its fracture toughness properties, to a maximum at 10 vol% ZrO_2, and then decreasing (Homerin et al. 1986). While the improvement of the FT is due to the increase in phase transfer volume, the linkage between microcracks of neighboring ZrO_2 particles causes its drop after reaching the maximum. A separate study by Trabelsi et al. also revealed that when the amount of ZrO_2 exceeds 10 vol%, microcracks are formed in the sintered materials due to phase transformation (Trabelsi et al. 1989). Besides improved mechanical properties, combining ZrO_2 with Al_2O_3 also reduces the water corrosion in Al_2O_3; however, in this process, the wear resistance is reduced due to the reduced hardness, which arises from adding excessive ZrO_2.

Thermal fatigue resistance of bioceramics materials developed for joints or implants is another important attribute, especially when compliant with the human body temperature. ZTA composites, for instance, present better thermal fatigue resistance compared to the pure Al_2O_3 (Orange et al. 1992). Hence, the composition of the bioceramics, stabilizer concentration, and environmental parameters could be strategically selected to engineer the optimum composite performance for suitable orthopedic or biomedical end-applications.

10.5.2 Bioresorbable Ceramics

After implant, bioresorbable ceramics slowly disappear within a given period of time, while their physiochemical properties enable restoration of the target bone along with the growth of blood vessels and nerve fibers (Dubok 2000). Calcium sulfate (CaS) and calcium phosphate (CaP) are the major bioresorbable materials (Punj et al. 2021). While the CaS-based bone grafts degrade rapidly, CaPs degrade slowly (Ferguson et al. 2017). Among the widely used CaP ceramics (Figure 10.6a and b), tricalcium phosphate (TCP), hydroxyapatite (HA) (Figure 10.6c and d), and biphasic calcium phosphate (BCP) are the most common materials (Punj et al. 2021). α-TCP and β-TCP are two phases of TCP, but both of them dissolve faster than the HA. Nearly, all CaPs undergo biodegradation to varying degrees, but in an analogous form in the following order: α-TCP > β-TCP>>HA (Hench 1991).

The underlying mechanism of the resorption (biodegradation) process of this class of bioceramics is interesting. CaP ceramics, for instance, could be an ideal candidate to manifest such a phenomenon in three different stages: (i) physiochemical dissolution, (ii) physical disintegration into tiny particles, and (iii) biological factors (Hench 1991). At first, the pH of the surrounding environment and the solubility of the ceramic product propagate the physiological dissolution process which initiates a phase transition, e.g., amorphous CaP, dicalcium phosphate dihydrate, octa-CaP, and anionic hydroxyapatite (HA). Next, during the physical disintegration stage, the product breaks down into tiny particles as a result of the chemical attack of grain boundaries. Finally, biological factors such as phagocytosis cause a decrease in the surrounding pH.

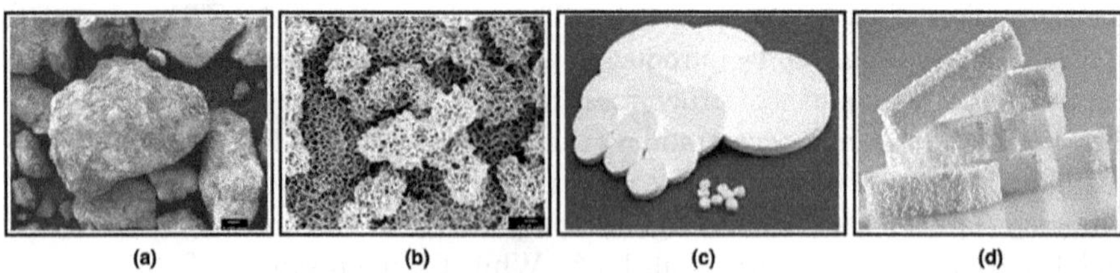

FIGURE 10.6 Calcium phosphate (CaP) ceramics. (a, b) SEM micrographs of β-tricalcium phosphate (β-TCP) (magnification level: 50×) and hydroxyapatite (HA) (magnification level: 5,000×) (Sheikh et al. 2015). (c, d) Dense (nonporous) and porous (by adding pore-generating additives) HAs produced by sintering the ceramic powders inside electric furnace at varying temperatures (Fiume et al. 2021).

CaP commonly refers to the calcium cations (Ca^{2+}) with negative anions of phosphates like orthophosphate (PO_4^{3-}), metaphosphate (PO^{-3}), or pyrophosphate ($P_2O_4^{-7}$). Bovine milk typically contains this principal form of calcium; 90% of tooth enamel is based on CaP. Hydroxyapatite (HA or HAP) (aka hydroxylapatite) is a type of CaP mineral that has 65% intrinsic compound resemblance to the mammalian bone structure (Fernando et al. 2016), with a Ca:P atomic ratio of 1.67. Bioresorbable materials form HA and promote bone tissue formation. The resorption rate could vary for different HA-based bioceramics. For instance, a 12-week slow resorption rate was reported after implanting femoral bone inside a rabbit (Tan et al. 2013); however, the resorption rate for bioresorbable materials could be accelerated by increasing their surface area and reducing the crystallinity or grain size (Hench 1991).

Since the bioresorbable materials take part in the formation and resorption process of the bone tissue, they are highly effective as scaffolds and filling spaces. For example, HA is used as bone filler for small defects that may arise from fractures in tibia (Quarto et al. 2001). Interestingly, bioresorbable bioceramics can be distinguished from bioactive bioceramics mainly by their structural factors (Antoniac 2016). A good example would be the nonporous HA, which is a bioactive material that is retained within the organism for at least 5–7 years without change. On the other hand, HA applied as a highly porous form-factor behaves as a bioresorbable ceramic that can be resorbed within a period of 1 year (Antoniac 2016). Several techniques are used for depositing bioresorbable coating into metal implants such as thermal spraying, sputter coating, pulse laser deposition, dynamic mixing method, dip coating, sol-gel technique, electrophoretic deposition, biomimetic deposition, hot isostatic pressing, and electrochemical deposition (Yang et al. 2005).

Highly porous composite ceramics are also employed for orthopedic applications by mixing bioresorbable HA with bioinert and bioactive ceramics: α-Al_2O_3-HA-bioactive glass, for instance. Wet chemical precipitation, sol-gel, and conventional melting-quenching processes could be employed to mix HA, α-Al_2O_3, and bioactive glass powders (Yelten and Yilmaz 2019). Such fabrication techniques often introduce unnecessary biproducts during the sintering process and lower the mechanical strength of the composites due to the highly porous structure of the sintered composites. However, researchers point out that this class of biocomposites performs better in terms of transmitting nutrient supply or body fluids due to the high (28%–30%) porosity. Further, the CaP molar ratio is around 1.65 for this new class of HA composite pellets, making them much more compatible with body fluids compared to traditional bioceramics (Ratner 1996).

10.5.3 Bioactive Ceramics

Bioactive materials, ones that react with bone tissue, could be considered "midway" between bioinert and bioresorbable materials. They have the capacity to react with the living cells and tissues inside the body and evoke a very specific biological response leading to the formation of a bond between the (introduced) material and the body tissue (Agrawal 1998). They are called osteo-conductive materials, as they stimulate the differentiation process of the stem cells to bone building osteoblast cells. Highly bioactive materials allow osteoprogenitor (i.e., the potential to form new bone) cells to colonize on its surface. Soluble ions released by the bioactive materials stimulate cell division and trigger growth factor and extracellular matrix protein production (Antoniac 2016).

The formula for bioactive glass ceramics was developed by Hench et al. in the 1970s and was named Bioglass® 45S5 (Hench et al. 1971). Hench et al. presented all the possible bonds formed between bone and biomaterial surfaces, including direct ionic covalent, electrostatic ionic, hydrogen, and van-der-Waals bond. The researchers divided the requirements of the biomaterial model into three criteria: chemical, crystallography, and microstructural. The study concluded that glass ceramics was the best-fit to meet all three requirements and had unique bone-forming properties, which is why it gained significant attention from researchers and scientists (Borden et al. 2021). Further, its chemical requirement can be achieved for end-use applications as it is possible to incorporate any element of the periodic table in any percentage into the glass. It displays a rapid rate of surface reactivity, giving it fast tissue-bonding properties (Ducheyne et al. 1993). Properties like in-vivo dissolution, ion release, and interparticle spacing can be used to determine the effectiveness of bioactive glass as bone graft (Borden et al. 2021). The study reported that a spherical shape is the optimum geometry for bioactive glass bone formation as the spherical particles displayed a more uniform shape and smooth surface compared to the irregularly shaped particles. Another study showed the evidence of new mineralized bone tissue formation surrounding the ceramic prosthesis after just 4 weeks of implanting (Barros et al. 2002). In the first week, there were appearances of bone mineralizing at the interface of the bioactive glass (Ducheyne et al. 1993). On the fourth week, the interface was completely bonded to the bone with no intervening fibrous tissue. Figure 10.7 displays the fundamental building blocks of new bone formation at the 45S5 bioactive glass material interface (Brézulier et al. 2021). The mineral component of bone includes $Ca_{8.3}(PO_4)_{4.3}(CO_3)_x(HPO_4)_y(OH)_{0.3}$ (Vallet-Regí 2001). These mineral component makes up two-third of the dry weight of the bone. Ionic substitution of the mineral component of bone includes CO_3^{2-}, Na^+, and Mg^{2+}. Collagen and water collectively make up the 43% of the remaining portion of bone.

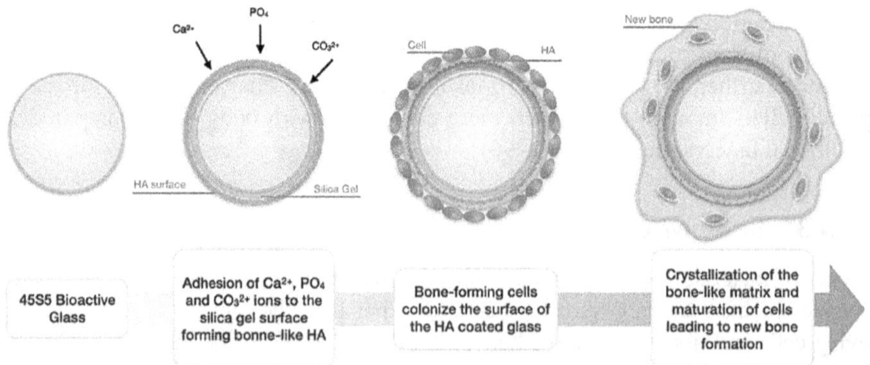

FIGURE 10.7 Bone formation mechanism after inserting bioglass 45S5 into a bone defect. The bone tissue minerals are formed on its surface, HCA (hydroxy-carbano-apatite) (HA, hydroxyapatite) (Brézulier et al. 2021).

Bioglass® 45S5 is considered the gold standard in bioactive materials for clinical applications, with the highest bioactivity index (I_B) of 12.5. On the other hand, 45S5 (NovaBone) or S53P4 (AbminDent1) have the highest level of bioactivity index (class A), indicating its ability to bond with bone and connecting soft tissues through osteo-conduction and osteo-stimulation; glass-ceramic materials (e.g., A/W glass ceramics) have a relatively lower level of bioactivity (class B) demonstrating the ability to bond only with the bone through osteo-conduction. Figure 10.8a displays the compositions of different bioactive glass and glass-ceramic materials for clinical applications. Gao et al. have outlined a comprehensive review of the applications of scaffolds made of different bioactive ceramics for bone repairs and regenerations (Gao et al. 2014). Additive manufacturing techniques like powder bed-selective laser processing (PBSLP), binder jetting, material extrusion, and sheet lamination are few of the fabrication techniques for bioactive ceramics (Kamboj et al. 2021).

A disadvantage of bioactive materials is mechanical weakness due to the low fracture toughness (FT) and crack growth from cyclic fatigue, resulting from the two-dimensional amorphous glass network. However, in terms of mechanical strength, it is weaker than bioinert ceramics (Poitout 2004). Therefore, they are not suitable for load-bearing applications (Ducheyne et al. 1993). Even though the mechanical weakness of bioactive glass does not allow use in repairing a large osseous defect, it is an excellent choice for filling small defects (Vallet-Regí 2014). Research conducted at the Universiti Sains, Malaysia, demonstrated that by varying the Al_2O_3 concentration at a high heat treatment of 950°C, mechanical compressive strength could be improved (from ~4 to 10.7 MPa) for SiO_2-CaO-Na_2O-P_2O_5 bioactive glass (Oh et al. 2020). X-ray diffraction (XRD) revealed that a new and larger crystalline phase was developed that was attributed to the formation of $Na_2CaSi_2O_6$ crystalline structures. Figure 10.8b demonstrates a comparison of mechanical properties among three different commercial bioactive glass ceramics: Cerabone® A/W, Ceravital®, and Bioverit®.

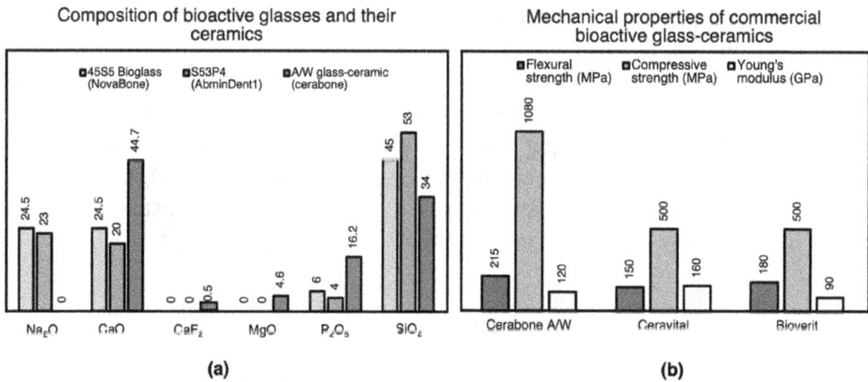

FIGURE 10.8 Bioactive glass materials and their properties. (a) Compositions of different bioactive glass and glass-ceramic materials for clinical applications. (Based on data from Hench (2016).) (b) Comparative performance among different bioactive glass ceramics. (Based on data from Siqueira and Zanotto (2011).)

10.6 IMPORTANCE OF BIOCOMPATIBILITY OF IMPLANTS

The reaction between the body and foreign materials is a critical challenge for the fixation of orthopedic devices (Hench et al. 1971). The chemical and physical nature of the bioceramics determine the kind and the extent of tissue response that will be triggered following implantation (Ravaglioli and Krajewski 1992). A study conducted in Tokyo, Japan, to test blood compatibility of sputter-deposited alumina films showed that incorporating alumina films is promising for developing blood-compatible and durable materials. The study found a 50% reduction in the platelet adhesion on the implant surface when the surface was coated with alumina film; there was a 50% reduction in the platelets adhesion on the implant surface and lowered intrinsic coagulation factor XI1 (Yuhta et al. 1994).

A serious challenge for joint arthroplasty is periprosthetic (body structure close to the implant) joint infection caused by microorganisms. A study found that Gram-positive *Cocci* are the most common infectious pathogen for periprosthetic infection, namely, *Staphylococcus aureus* and *Staphylococcus epidermis* (Pulido et al. 2008). The first step of infection is bacterial adhesion to the implant surface followed by the formation of biofilms leading to a complex interaction among the host-defense system, implant, microorganisms, and their by-products (Romanò et al. 2016). Therefore, it will be wise to find ways to eliminate infection from its root by finding ways to prevent microorganisms from adhering to implant surfaces. In such context, Pezzotti et al. developed a way to investigate the bacteriostatic response of ZTA and silicon nitride (Si_3N_4) using molecular biology characterization and advanced Raman Spectroscopy (Pezzotti et al. 2018). The research group concluded that non-oxide Si_3N_4 performs better at inhibiting bacterial infection due to its surface chemistry against bacterial loading.

There are several variables that determine bacterial adhesion and proliferation in the biomaterial implant surface, such as pathogen types, physiochemical properties, environmental factors, and surface morphology (Kumar et al. 2018). Bioceramics can prevent bacterial adhesion as they contain nanocrystals of a diameter between 1 and 3 nm. Incorporating fluoride ions into the bioceramics formulation can also improve the antibacterial property (Hermansson 2015). Ion doping mechanisms can further enhance certain properties of biomaterials like biodegradation abilities, biomechanical properties, and biocompatibilities (Xie et al. 2012). Potassium and strontium ions (K/Sr) doped into calcium polyphosphate (CPP) for bone tissue regeneration have better compatibility when compared to CPP and HA scaffolds. Another study found that incorporation of trace elements like Sr, zinc (Zn), magnesium (Mg), and silicon (Si) into bioactive materials will give improved ability to control the osteogenic property of bone-forming cells (Zhang et al. 2012).

10.7 BIOCOMPATIBILITY TESTS

Researchers at University of Leeds (UK) employed histological tests to analyze some retrieved tissues from an artificial ceramic-on-ceramic hip joint following a rTHA (Hatton et al. 2002). TEM (transmission electron microscopic) tests of the laser-captured micro-dissected tissues showed the presence of bioceramic particles in the size range of 5–90 nm, and SEM micrographs showed particles in the 0.05–3.2 μm size, presenting the possibility of two different size ranges of wear particles from the bioceramic prosthesis. This wear debris could cause health risks based on the level of reactivity or constituents and could limit their medical relevance. Hence, it is highly recommended to conduct biocompatibility tests for any medical devices that would come into contact with the patient (Ramakrishna et al. 2015). ISO 10993 is recognized by the FDA (Food and Drug Administration) for biocompatibility testing to ensure the safety of the medical devices. ISO 10993-1 lists the tests to be conducted for tissue and bone implants, considering the area of contact and duration of contact in the patient body. Initial evaluation tests for bone and tissue implant include cytotoxicity, sensitization, and irritation (intracutaneous reactivity). Additional tests such as systematic/acute toxicity, subacute and sub-chronic toxicity, and genotoxicity are required for prostheses with prolonged contact (1–30 days) or permanent implants.

Toxic materials are those that trigger a macro-scale rejection in the form of inflammatory or carcinogenic response or both (Hench et al. 1971). Hence, for safety of the patient and to avoid unnecessary revision surgeries following the TKA/THA, the toxicity level of the bioceramics should be tested at a cellular level prior to any clinical applications. Tests should be conducted on the implants for their biological, morphological, and phytochemical behaviors to avoid any traces of systematic or local toxicity to ensure safety of the patients (Kumar et al. 2018). Cytotoxicity test (i.e., tissue culture test) is test done in-vitro to determine if the medical device will cause any cell death from direct contact or as a result of leaching of a toxic substance (Ramakrishna et al. 2015). Some common cytotoxin assays include Trypan blue, MTT, MTS, XTT, WST-1, LDH, NRU, GSH, and AlamarBlue (Thrivikraman et al. 2014). Genotoxicity (i.e., toxic to DNA) tests are usually performed after the cytotoxicity tests (Thrivikraman et al. 2014). Genotoxins are chemical agents that have the potential to cause DNA or chromosomal damage (Phillips and Arlt 2009): damage of DNA can lead to malignant transformation (i.e., cancer). Nano particles resulting from implant wear get into the cytoplasmic space (Thrivikraman et al. 2014) and induce oxidative stress at the cellular level, which leads to the production of reactive oxygen species (ROS). These ROS disturb the intra- and inter-cellular signaling pathways (Zuberek and Grzelak 2018). As a result, cells start to behave abnormally and may cause cancer.

10.8 IMPLANT FAILURE PREVENTION

Success of bioceramic implants depends mainly on their biocompatibility, mechanical properties, and engineering design (Wong and Bronzino 2007). Implant failures, in general, depend on several factors as shown in Figure 10.9i and could include carcinogenicity or bacterial colonization. Hence, it is significant to ensure that the implant has sufficient load-bearing capacity for its purpose, and its design framework fits into the biological system properly. In-vivo degradation of prosthetic implants is considered one of the primary factors limiting the longevity of the total joint arthroplasty (TJA) (Jacobs et al. 1994). Such degradation could arise from wear and corrosion. Wear happens due to the loss of materials resulting from the relative motion between two surfaces (via adhesion, abrasion, or fatigue) (Jacobs et al. 1994). Wear generates debris that can trigger a local host response and may eventually cause osteolytic cavity due to osteolysis (i.e., periprosthetic bone loss or bone resorption

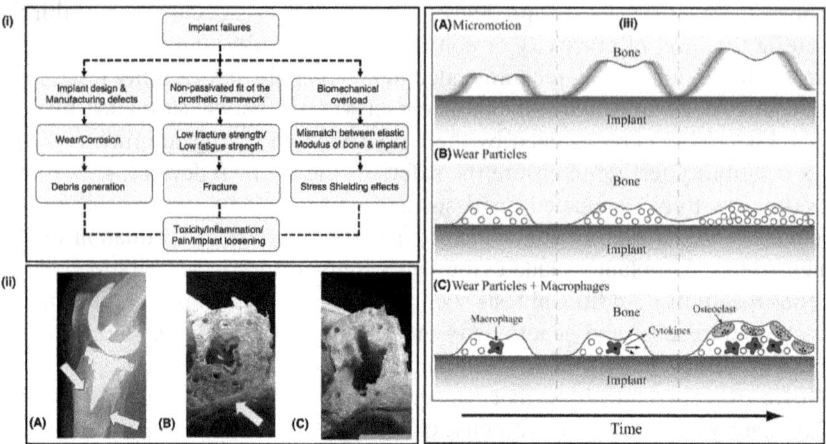

FIGURE 10.9 (i) Different factors leading to implant failures. (Modified from Priyadarshini et al. (2019).) (ii) (a–c) Screw track osteolytic cavity from osteolysis (marked with white arrows) after cementless total knee replacement (Klutzny et al. 2019). Radiograph exhibits tibial osteolysis in the screw fixation area (ii-a) and the intraoperative photographs shows the femoral stems before (ii-b) and after (ii-c) the revision surgeries demonstrate the extent of bone loss defect. (iii) (a–c) The aseptic loosening mechanism at the interfacial gaps between bone and implant (Raphel et al. 2016). As implant moves relative to the bone, the overall micromotion worsens gradually (iii-a) and becomes instable as wear debris consumes the interfacial space (iii-b), which subsequently activates the macrophages, cytokines, and consequently leads to osteoclast bone-resorbing cells (iii-c).

surrounding an implant) and could compromise the implant fixation (Purdue et al. 2006) (Figure 10.9ii), resulting in aseptic loosening (i.e., implant failures without any mechanical reason or evidence of infection, which typically arise from osteolysis) (Figure 10.9iii) or chronic inflammation (Abu-Amer et al. 2007). Wear can be reduced by improving the bearing characteristics of the femoral head, condyle counter face, and by improving the stability of the molecular connection.

Failures from fixture (or locking mechanism) fretting is another crucial aspect, which is often ignored while choosing the replacement devices. In 2021, a study reported the fretting of fixture pins due to mechanical mismatch with the tibial baseplate, leading to knee-joint instability and severe bone loss of a 46-year Caucasian woman and, thereby, necessitated a revision surgery (Figure 10.10) (Lamba et al. 2021). The rTKA revealed the total loss of lateral collateral ligament, femoral condyle, and popliteus tendon due to poor prosthetic design, material selection, and locking mechanism. Hence, sufficient micro-scale and ultra-scale bonding at the material–bone interface could solve the orthopedic fixation problem associated with the loosening of nails, screws, plates, and hip prostheses. Furthermore, the application of composite structures improves the tribological properties and mechanical strength of the implants (as discussed in the earlier segments). Other factors relating to success and failure of the implants are beyond the control of the engineering design, such as surgical technique during implanting, health condition of patient, mode of physical activities of the patient. Hence, a holistic knowledge framework

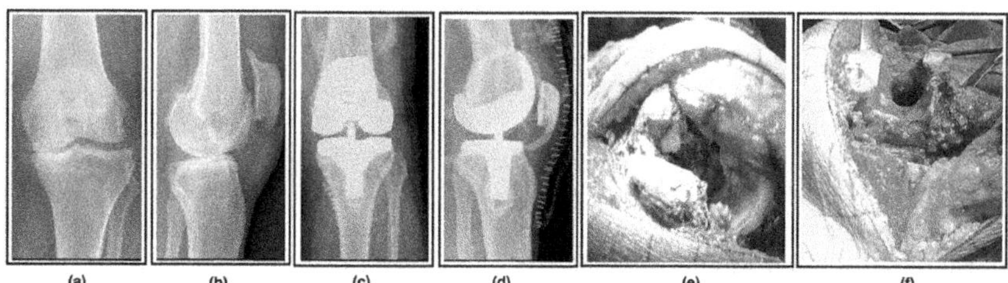

FIGURE 10.10 Failure in the bio-implantable TKA device (Lamba et al. 2021). TKA was conducted for a woman suffering from symptomatic osteoarthritis, which could be seen from the radiographs of pre-TKA anteroposterior and lateral views (a, b). Implanted TKA device is shown in the radiographs (anteroposterior and lateral views) of the post-operative TKA (c, d). While replacing the femoral implant, device residue (gray and white fibrous material) (e) was identified in the implanted region that caused the complete loss of lateral femoral condyle (f).

based on physiology of bone (composition and mechanics), material science, surgical expertise, immunology, design manufacturing, diet, and awareness should be implemented to improve the knee/hip joint arthroplasty procedure.

10.9 CONCLUSIONS AND FUTURE PROSPECTS

The application of bioceramics is expected to increase due to their unique functionalities. Bioceramic-based implants or prostheses have seen dramatic growth in knee/hip replacement surgeries in the last few decades due to their biocompatibility, low density, and ease of fabrication. However, it is also essential to understand the evolving nature of bioceramics and evaluate their medical performance through clinical trials. For this, scientists continue to investigate novel bioceramic composites. A conformable decoder based on piezoelectric bioceramic composites is an arena that engineers could explore to innovate new-generation-integrated medical devices for bone joints. As the world is moving toward the Internet of Health Things (IoHT), it could be naturally expected that the futuristic knee/hip bioceramic prostheses will soon have conformable decoding abilities that would connect with the patients for them to continuously monitor the health of their knee/hip prosthetic components. Such technology could ameliorate the implantable experience of the patients. The conformable decoding system could also host microfluidic actuators as drug carriers. By combining machine learning technologies with such smart knee/hip joint prostheses, doctors (or users) will soon predict and prevent periprosthetic knee/hip joint infections by triggering the actuators and administering on-demand drug delivery. Since the bone tissues intrinsically constitute piezoelectric components, the synergy between bone and piezoelectric bioceramics could also boost the antibacterial performance of the implant sites of bioceramic knee/hip joints and, ultimately, augment the implant lifespan.

REFERENCES

Abu-Amer, Y., I. Darwech, and J.C. Clohisy. 2007. Aseptic loosening of total joint replacements: Mechanisms underlying osteolysis and potential therapies. *Arthritis Research & Therapy* 9 (Suppl 1): S6. https://doi.org/10.1186/ar2170.

Agrawal, C.M. 1998. Reconstructing the human body using biomaterials. *JOM* 50 (1): 31–5. https://doi.org/10.1007/s11837-998-0064-5.

Albrektsson, T., and C. Johansson. 2001. Osteoinduction, osteoconduction and osseointegration. *European Spine Journal* 10 (October): S96–S101. https://doi.org/10.1007/s005860100282.

Antoniac, I.V. 2016. *Handbook of Bioceramics and Biocomposites*. edited by I.V. Antoniac. Cham: Springer International Publishing. https://doi.org/10.1007/978-3-319-12460-5.

Bahraminasab, M., and A. Jahan. 2011. Material selection for femoral component of total knee replacement using comprehensive VIKOR. *Materials & Design* 32 (8–9): 4471–77. https://doi.org/10.1016/j.matdes.2011.03.046.

Barkallah, R., R. Taktak, N. Guermazi, K. Elleuch, and J. Bouaziz. 2021. Mechanical properties and wear behaviour of alumina/tricalcium phosphate/titania ceramics as coating for orthopedic implant. Engineering Fracture Mechanics 241 (January): 107399. https://doi.org/10.1016/j.engfracmech.2020.107399.

Barkallah, R., R. Taktak, N. Guermazi, F. Zaïri, J. Bouaziz, and F. Zaïri. 2018. Manufacturing and mechanical characterization of Al_2O_3/β-TCP/TiO2 biocomposite as a potential bone substitute. *The International Journal of Advanced Manufacturing Technology* 95 (9–12): 3369–80. https://doi.org/10.1007/s00170-017-1434-3.

Borden, M., L.E. Westerlund, V. Lovric, and W. Walsh. 2021. Controlling the bone regeneration properties of bioactive glass: Effect of particle shape and size. *Journal of Biomedical Materials Research Part B: Applied Biomaterials*, December. https://doi.org/10.1002/jbm.b.34971.

Brézulier, D., L. Chaigneau, S. Jeanne, and R. Lebullenger. 2021. The challenge of 3D bioprinting of composite natural polymers PLA/bioglass: Trends and benefits in cleft palate surgery. *Biomedicines* 9 (11): 1553. https://doi.org/10.3390/biomedicines9111553.

Cram, P., X. Lu, S.L. Kates, J.A. Singh, Y. Li, and B.R. Wolf. 2012. Total knee arthroplasty volume, utilization, and outcomes among medicare beneficiaries, 1991–2010. *JAMA* 308 (12): 1227. https://doi.org/10.1001/2012.jama.11153.

Cui, Z., Y-X. Tian, W. Yue, L. Yang, and Q. Li. 2016. Tribo-biological deposits on the articulating surfaces of metal-on-polyethylene total hip implants retrieved from patients. *Scientific Reports* 6 (1): 28376. https://doi.org/10.1038/srep28376.

Dawson-Amoah, K.G., B.S. Waddell, R. Prakash, and M.M. Alexiades. 2020. Adverse reaction to zirconia in a modern total hip arthroplasty with ceramic head. *Arthroplasty Today* 6 (3): 612–6.e1. https://doi.org/10.1016/j.artd.2020.03.009.

Dorozhkin, S. 2011. Medical application of calcium orthophosphate bioceramics. *BIO* 1 (1): 1–51. https://doi.org/10.5618/bio.2011.v1.n1.1.

Dorozhkin, S.V. 2010. Bioceramics of calcium orthophosphates. *Biomaterials* 31 (7): 1465–85. https://doi.org/10.1016/j.biomaterials.2009.11.050.

Dubok, V.A. 2000. Bioceramics – Yesterday, today, tomorrow. *Powder Metallurgy and Metal Ceramics*. https://doi.org/10.1023/A:1026617607548.

Ducheyne, P., M. Marcolongo, and E. Schepers. 1993. Bioceramic composites. In *An Introduction to Bioceramics*, 281–97. World Scientific. https://doi.org/10.1142/9789814317351_0015.

Elsen, S.R., K. Jegadeesan, and J.R. Aseer. 2017. X-ray diffraction analysis of mechanically milled alumina and zirconia powders. *Nano Hybrids and Composites* 17 (August): 96–100. https://doi.org/10.4028/www.scientific.net/NHC.17.96.

Fact, M.R. 2021. Bioceramics market.

Ferguson, J., M. Diefenbeck, and M. McNally. 2017. Ceramic biocomposites as biodegradable antibiotic carriers in the treatment of bone infections. *Journal of Bone and Joint Infection* 2 (1): 38–51. https://doi.org/10.7150/jbji.17234.

Fernando, S., M. McEnery, and S.A. Guelcher. 2016. Polyurethanes for bone tissue engineering. In *Advances in Polyurethane Biomaterials*, 481–501. Elsevier. https://doi.org/10.1016/B978-0-08-100614-6.00016-0.

Fiume, E., G. Magnaterra, A. Rahdar, E. Verné, and F. Baino. 2021. Hydroxyapatite for biomedical applications: A short overview. *Ceramics* 4 (4): 542–63. https://doi.org/10.3390/ceramics4040039.

Gao, C., Y. Deng, P. Feng, Z. Mao, P. Li, B. Yang, J. Deng, Y. Cao, C. Shuai, and S. Peng. 2014. Current progress in bioactive ceramic scaffolds for bone repair and regeneration. *International Journal of Molecular Sciences* 15 (3): 4714–32. https://doi.org/10.3390/ijms15034714.

Habraken, W., P. Habibovic, M. Epple, and M. Bohner. 2016. Calcium phosphates in biomedical applications: Materials for the future? *Materials Today* 19 (2): 69–87. https://doi.org/10.1016/j.mattod.2015.10.008.

Hatton, A., J.E. Nevelos, A.A. Nevelos, R.E. Banks, J. Fisher, and E. Ingham. 2002. Alumina-alumina artificial hip joints. Part I: A histological analysis and characterisation of wear debris by laser capture microdissection of tissues retrieved at revision. *Biomaterials* 23 (16): 3429–40. https://doi.org/10.1016/S0142-9612(02)00047-9.

He, B., X. Li, R. Dong, P. Tong, and J. Sun. 2022. A multi-center retrospective comparative study of third generation ceramic-on- ceramic total hip arthroplasty in patients younger than 45 years with or without the sandwich liner: A ten-year minimum. *Journal of Orthopaedic Surgery* 30 (2): 102255362211099. https://doi.org/10.1177/10225536221109960.

Hench, L.L. 1991. Bioceramics: From concept to clinic. *Journal of the American Ceramic Society* 74 (7): 1487–510. https://doi.org/10.1111/j.1151-2916.1991.tb07132.x.

Hench, L.L. 2016. Bioactive glass bone grafts: History and clinical applications. In *Handbook of Bioceramics and Biocomposites*, 23–33. Cham: Springer International Publishing. https://doi.org/10.1007/978-3-319-12460-5_5.

Hench, L.L., R.J. Splinter, W.C. Allen, and T.K. Greenlee. 1971. Bonding mechanisms at the interface of ceramic prosthetic materials. *Journal of Biomedical Materials Research* 5 (6): 117–41. https://doi.org/10.1002/jbm.820050611.

Hermansson, L., ed. 2015. *Nanostructural Bioceramics Advances in Chemically Bonded Ceramics*. Jenny Stanford Publishing.

Homerin, P., F. Thevenot, G. Orange, G. Fantozzi, V. Vandeneede, A. Leriche, and F. Cambier. 1986. Mechanical properties of zirconia toughened alumina prepared by different methods. *Le Journal de Physique Colloques* 47 (C1): C1-717–C1-721. https://doi.org/10.1051/jphyscol:19861108.

Ichimura, R., T. Minamikawa, H. Nakagawa, A. Mori, K. Midorikawa, K. Sakuragi, and H. Minamikawa. 2022. Fracture of the oxidized zirconium femoral component after total knee arthroplasty. *The Knee* 36 (June): 27–32. https://doi.org/10.1016/j.knee.2022.03.014.

Iorio, R., S. Kobayashi, W.L. Healy, A.I. Cruz, and M.E. Ayers. 2007. Primary posterior cruciate-retaining total knee arthroplasty: A comparison of American and Japanese cohorts. *Journal of Surgical Orthopaedic Advances* 16 (4): 164–70.

Jacobs, J.J., A. Shanbhag, T.T. Glant, J. Black, and J.O. Galante. 1994. Wear debris in total joint replacements. *Journal of the American Academy of Orthopaedic Surgeons* 2 (4): 212–20.

Joseph, D., ed. 2003. *Handbook of Materials for Medical Devices*. ASM International.

Kamboj, N., A. Ressler, and I. Hussainova. 2021. Bioactive ceramic scaffolds for bone tissue engineering by powder bed selective laser processing: A review. *Materials* 14 (18): 5338. https://doi.org/10.3390/ma14185338.

Klawitter, J.J., and S.F. Hulbert. 1971. Application of porous ceramics for the attachment of load bearing internal orthopedic applications. *Journal of Biomedical Materials Research* 5 (6): 161–229. https://doi.org/10.1002/jbm.820050613.

Klutzny, M., G. Singh, R. Hameister, G. Goldau, F. Awiszus, B. Feuerstein, C. Stärke, and C.H. Lohmann. 2019. Screw track osteolysis in the cementless total knee replacement design. *The Journal of Arthroplasty* 34 (5): 965–73. https://doi.org/10.1016/j.arth.2018.12.040.

Kumar, A., and F. Baino. 2020. Editorial: Bioceramics and bioactive glasses for hard tissue regeneration. Frontiers in Materials 7 (September). https://doi.org/10.3389/fmats.2020.593624.

Kumar, P., B.S. Dehiya, and A. Sindhu. 2018. Bioceramics for hard tissue engineering applications: A review title. *International Journal of Applied Engineering Research* 13 (5): 2744–52.

Lamba, C., K. Denning, E. Ouellette, S. Kurtz, and M. Bullock. 2021. An interesting case of osteolysis with accompanying metallosis in a primary total knee arthroplasty. *Arthroplasty Today* 11 (October): 81–7. https://doi.org/10.1016/j.artd.2021.07.002.

Li, D., C. Wang, Z. Li, H. Wang, J. He, J. Zhu, Y. Zhang, et al. 2018. Nano-sized Al_2O_3 particle-induced autophagy reduces osteolysis in aseptic loosening of total hip arthroplasty by negative feedback regulation of RANKL expression in fibroblasts. *Cell Death & Disease* 9 (8): 840. https://doi.org/10.1038/s41419-018-0862-9.

Li, J., X. Wang, Y. Lin, X. Deng, M. Li, and C. Nan. 2016. In vitro cell proliferation and mechanical behaviors observed in porous zirconia ceramics. *Materials* 9 (4): 218. https://doi.org/10.3390/ma9040218.

López, J.P. 2014. Alumina, zirconia, and other non-oxide inert bioceramics. In *Bio-Ceramics with Clinical Applications*, 153–73. Chichester, UK: John Wiley & Sons, Ltd. https://doi.org/10.1002/9781118406748.ch6.

Mayo Clinic. 2014. First nationwide prevalence study of hip and knee arthroplasty shows 7.2 million Americans living with implants. https://www.mayoclinic.org/medical-professionals/orthopedic-surgery/news/first-nationwide-prevalence-study-of-hip-and-knee-arthroplasty-shows-7-2-million-americans-living-with-implants/mac-20431170.

Meier, E., K. Gelse, K. Trieb, M. Pachowsky, F.F. Hennig, and A. Mauerer. 2016. First clinical study of a novel complete metal-free ceramic total knee replacement system. *Journal of Orthopaedic Surgery and Research* 11 (1): 21. https://doi.org/10.1186/s13018-016-0352-7.

Murr, L.E., S.M. Gaytan, E. Martinez, F. Medina, and R.B. Wicker. 2012. Next generation orthopaedic implants by additive manufacturing using electron beam melting. *International Journal of Biomaterials* 2012: 1–14. https://doi.org/10.1155/2012/245727.

Nevalainen, M.T., K.V. Kauppinen, T. Niinimäki, and S.S. Saarakkala. 2020. Comparison of ultrasonographic, radiographic and intra-operative findings in severe hip osteoarthritis. *Scientific Reports* 10 (1): 21108. https://doi.org/10.1038/s41598-020-78235-z.

Nevalainen, M.T., K. Kauppinen, J. Pylväläinen, K. Pamilo, M. Pesola, M. Haapea, J. Koski, and S. Saarakkala. 2018. Ultrasonography of the late-stage knee osteoarthritis prior to total knee arthroplasty: Comparison of the ultrasonographic, radiographic and intra-operative findings. *Scientific Reports* 8 (1): 17742. https://doi.org/10.1038/s41598-018-35824-3.

Oh, H.J., D-K. Kim, Y.C. Choi, S-J. Lim, J.B. Jeong, J.H. Ko, et al. 2020. Fabrication of piezoelectric poly(l-lactic acid)/BaTiO3 fibre by the melt-spinning process. *Scientific Reports* 10 (1): 16339. https://doi.org/10.1038/s41598-020-73261-3.

Orange, G., G. Fantozzi, P. Homerin, F. Thevenot, A. Leriche, and F. Cambier. 1992. Preparation and characterization of a dispersion toughened ceramic for thermomechanical uses (ZTA). Part II: Thermomechanical characterization. effect of microstructure and temperature on toughening mechanisms. *Journal of the European Ceramic Society* 9 (3): 177–85. https://doi.org/10.1016/0955-2219(92)90003-V.

Park, J. 2009a. *Bioceramics: Properties, Characterizations, and Applications*. In *Lecture Notes in Economic and Mathematical Systems*. New York: Springer. https://books.google.com/books?id=bWMLpUyZCFgC.

Park, J. 2009b. *Bioceramics*. New York, NY: Springer. https://doi.org/10.1007/978-0-387-09545-5.

Pezzotti, G., R.M. Bock, B.J. McEntire, T. Adachi, E. Marin, F. Boschetto, W. Zhu, O. Mazda, and S.B. Bal. 2018. In vitro antibacterial activity of oxide and non-oxide bioceramics for arthroplastic devices: I. In situ time-lapse Raman spectroscopy. *The Analyst* 143 (15): 3708–21. https://doi.org/10.1039/C8AN00233A.

Phillips, D.H., and V.M. Arlt. 2009. Genotoxicity: Damage to DNA and its consequences. 87–110. https://doi.org/10.1007/978-3-7643-8336-7_4.

Piconi, C., and G. Maccauro. 2015. Perspective and trends on bioceramics in joint replacement. In *Handbook of Bioceramics and Biocomposites*, 1–37. Cham: Springer International Publishing. https://doi.org/10.1007/978-3-319-09230-0_41-1.

Piconi, C., G. Maccauro, F. Muratori, and E.B.D. Prever. 2003. Alumina and zirconia ceramics in joint replacements. *Journal of Applied Biomaterials & Functional Materials* 1 (1): 19–32. https://10.1177/228080000300100103.

Piconi, C., and S. Sprio. 2021. Oxide bioceramic composites in orthopedics and dentistry. *Journal of Composites Science* 5 (8): 206. https://doi.org/10.3390/jcs5080206.

Poitout, D.G., ed. 2004. *Biomechanics and Biomaterials in Orthopedics*. London: Springer. https://doi.org/10.1007/978-1-4471-3774-0.

Priyadarshini, B., M. Rama, Chetan, and U. Vijayalakshmi. 2019. Bioactive coating as a surface modification technique for biocompatible metallic implants: A review. *Journal of Asian Ceramic Societies* 7 (4): 397–406. https://doi.org/10.1080/21870764.2019.1669861.

Pulido, L., E. Ghanem, A. Joshi, J.J. Purtill, and J. Parvizi. 2008. Periprosthetic joint infection: The incidence, timing, and predisposing factors. *Clinical Orthopaedics & Related Research* 466 (7): 1710–15. https://doi.org/10.1007/s11999-008-0209-4.

Punj, S., J. Singh, and K. Singh. 2021. Ceramic biomaterials: Properties, state of the art and future prospectives. *Ceramics International* 47 (20): 28059–74. https://doi.org/10.1016/j.ceramint.2021.06.238.

Purdue, P.E., P. Koulouvaris, B.J. Nestor, and T.P. Sculco. 2006. The central role of wear debris in periprosthetic osteolysis. *HSS Journal(r): The Musculoskeletal Journal of Hospital for Special Surgery* 2 (2): 102–13. https://doi.org/10.1007/s11420-006-9003-6.

Quarto, R., M. Mastrogiacomo, R. Cancedda, S.M. Kutepov, V. Mukhachev, A. Lavroukov, E. Kon, and M. Marcacci. 2001. Repair of large bone defects with the use of autologous bone marrow stromal cells. *New England Journal of Medicine* 344 (5): 385–86. https://doi.org/10.1056/NEJM200102013440516.

Ramakrishna, S., L. Tian, C. Wang, S. Liao, and W.E. Teo. 2015. Safety testing of a new medical device. In *Medical Devices*, 137–53. Elsevier. https://doi.org/10.1016/B978-0-08-100289-6.00006-5.

Raphel, J., M. Holodniy, S.B. Goodman, and S.C. Heilshorn. 2016. Multifunctional coatings to simultaneously promote osseointegration and prevent infection of orthopaedic implants. *Biomaterials* 84 (April): 301–14. https://doi.org/10.1016/j.biomaterials.2016.01.016.

Ratner, B.D. 1996. Biomaterials science: An interdisciplinary endeavor. In *Biomaterials Science*, 1–8. Elsevier. https://doi.org/10.1016/B978-0-08-050014-0.50005-5.

Ravaglioli, A., and A. Krajewski. 1992. *Bioceramics*. Dordrecht: Springer Netherlands. https://doi.org/10.1007/978-94-011-2336-5.

Renoldelsen, S, and M Vivekananthan. 2018. A preliminary study on the physical and biocompatibility characteristics of zirconia-silicon nitride bio-ceramics. *IOP Conference Series: Materials Science and Engineering* 402 (September): 012031. https://doi.org/10.1088/1757-899X/402/1/012031.

Rheumatology Advisor. 2019. Increased rate of total joint replacements predicted from 2020 to 2040. *Rheumatology Advisor*. 2019. https://www.rheumatologyadvisor.com/home/topics/osteoarthritis/increased-rate-of-total-joint-replacements-predicted-from-2020-to-2040/.

Rocha Barros, V.M. da, L.A. Salata, C.E. Sverzut, S.P. Xavier, R. van Noort, A. Johnson, and P.V. Hatton. 2002. In vivo bone tissue response to a canasite glass-ceramic. *Biomaterials* 23 (14): 2895–900. https://doi.org/10.1016/S0142-9612(01)00417-3.

Romanò, C.L., D. Romanò, I. Morelli, and L. Drago. 2016. The concept of biofilm-related implant malfunction and 'low-grade infection'. 1–13. https://doi.org/10.1007/5584_2016_158.

Schwartz, A.M., K.X. Farley, G.N. Guild, and T.L. Bradbury. 2020. Projections and epidemiology of revision hip and knee arthroplasty in the United States to 2030. *The Journal of Arthroplasty* 35 (6): S79–85. https://doi.org/10.1016/j.arth.2020.02.030.

Sheikh, Z., S. Najeeb, Z. Khurshid, V. Verma, H. Rashid, and M. Glogauer. 2015. Biodegradable materials for bone repair and tissue engineering applications. *Materials* 8 (9): 5744–94. https://doi.org/10.3390/ma8095273.

Shekhawat, D., Amit Singh, M.K. Banerjee, T. Singh, and A. Patnaik. 2021. Bioceramic composites for orthopaedic applications: A comprehensive review of mechanical, biological, and microstructural properties. *Ceramics International* 47 (3): 3013–30. https://doi.org/10.1016/j.ceramint.2020.09.214.

Siqueira, R.L., and E.D. Zanotto. 2011. Facile route to obtain a highly bioactive SiO_2-CaO-Na_2O-P_2O_5 crystalline powder. *Materials Science and Engineering: C* 31 (8): 1791–99. https://doi.org/10.1016/j.msec.2011.08.013.

Sprio, S., S. Guicciardi, M. Dapporto, C. Melandri, and A. Tampieri. 2013. Synthesis and mechanical behavior of β-tricalcium phosphate/titania composites addressed to regeneration of long bone segments. *Journal of the Mechanical Behavior of Biomedical Materials* 17 (January): 1–10. https://doi.org/10.1016/j.jmbbm.2012.07.013.

Stevens, M.M. 2008. Biomaterials for bone tissue engineering. *Materials Today* 11 (5): 18–25. https://doi.org/10.1016/S1369-7021(08)70086-5.

Tan, L., X. Yu, P. Wan, and K. Yang. 2013. Biodegradable materials for bone repairs: A review. *Journal of Materials Science & Technology* 29 (6): 503–13. https://doi.org/10.1016/j.jmst.2013.03.002.

Technavio Research. 2017. High demand from emerging economies to boost the bioceramics market: Technavio. https://www.businesswire.com/news/home/20170918005941/en/High-Demand-from-Emerging-Economies-to-Boost-the-Bioceramics-Market-Technavio.

Thrivikraman, G., G. Madras, and B. Basu. 2014. In vitro/in vivo assessment and mechanisms of toxicity of bioceramic materials and its wear particulates. *RSC Advances* 4 (25): 12763. https://doi.org/10.1039/c3ra44483j.

Trabelsi, R., D. Treheux, G. Orange, G. Fantozzi, P. Homerin, and F. Thevenot. 1989. Relationship between mechanical properties and wear resistance of alumina-zirconia ceramic composites. *Tribology Transactions* 32 (1): 77–84. https://doi.org/10.1080/10402008908981865.

Transparency Market Research. 2022. Femoral head prostheses market – Global industry analysis, size, share, growth, trends, and forecast, 2019–2027. https://www.transparencymarketresearch.com/femoral-head-prostheses-market.html.

Vallet-Regí, M. 2001. Ceramics for medical applications. *Journal of the Chemical Society, Dalton Transactions* 2: 97–108. https://doi.org/10.1039/b007852m.

Vallet-Regí, M., ed. 2014. *Bio-Ceramics with Clinical Applications*. Chichester, UK: John Wiley & Sons, Ltd. https://doi.org/10.1002/9781118406748.

Vogel, D., P. Henke, A. Haenel, J. Mokros, M. Liebelt, and R. Bader. 2022. Experimental evaluation of the primary fixation stability of uncemented ceramic hip resurfacing implants. *Proceedings of the Institution of Mechanical Engineers, Part H: Journal of Engineering in Medicine* 236 (4): 496–503. https://doi.org/10.1177/09544119211070892.

Weng, W., W. Wu, M. Hou, T. Liu, T. Wang, and H. Yang. 2021. Review of zirconia-based biomimetic scaffolds for bone tissue engineering. *Journal of Materials Science* 56 (14): 8309–33. https://doi.org/10.1007/s10853-021-05824-2.

Whiteside, L.A. 2022. Clinical results of revision TKA in patients with presumed metal and cement allergy. *The Journal of Arthroplasty* 37 (6): S250–57. https://doi.org/10.1016/j.arth.2022.02.052.

Wong, J.Y., and J.D. Bronzino. 2007. *Biomaterials*. edited by J.Y. Wong and J.D. Bronzino. CRC Press. https://doi.org/10.1201/9780849378898.

Xie, H., Q. Wang, Q. Ye, C. Wan, and L. Li. 2012. Application of K/Sr Co-doped calcium polyphosphate bioceramic as scaffolds for bone substitutes. *Journal of Materials Science: Materials in Medicine* 23 (4): 1033–44. https://doi.org/10.1007/s10856-012-4556-z.

Xu, J., T. Oni, D. Shen, Y. Chai, W.K. Walter, and W.L. Walter. 2022. Long-term results of alumina ceramic-on-ceramic bearings in cementless total hip arthroplasty: A 20-year minimum follow-up. *The Journal of Arthroplasty* 37 (3): 549–53. https://doi.org/10.1016/j.arth.2021.11.028.

Yang, Y., K. Kim, and J. Ong. 2005. A review on calcium phosphate coatings produced using a sputtering process? An alternative to plasma spraying. *Biomaterials* 26 (3): 327–37. https://doi.org/10.1016/j.biomaterials.2004.02.029.

Yelten, A., and S. Yilmaz. 2019. A novel approach on the synthesis and characterization of bioceramic composites. *Ceramics International* 45 (12): 15375–84. https://doi.org/10.1016/j.ceramint.2019.05.031.

Yuhta, T., Y. Kikuta, Y. Mitamura, K. Nakagane, S. Murabayashi, and I. Nishimura. 1994. Blood compatibility of sputter-deposited alumina films. *Journal of Biomedical Materials Research* 28 (2): 217–24. https://doi.org/10.1002/jbm.820280212.

Zhang, M., C. Wu, K. Lin, W. Fan, L. Chen, Y. Xiao, and J. Chang. 2012. Biological responses of human bone marrow mesenchymal stem cells to Sr-M-Si (M = Zn, Mg) silicate bioceramics. *Journal of Biomedical Materials Research Part A* 100A (11): 2979–90. https://doi.org/10.1002/jbm.a.34246.

Zuberek, M., and A. Grzelak. 2018. Nanoparticles-caused oxidative imbalance. 85–98. https://doi.org/10.1007/978-3-319-72041-8_6.ssfffffffB

11 Bioceramics for Regenerative Medicine

Pugalanthi Pandian Sankaralingam and Poornimadevi Sakthivel
Bone Substitutes

Vijayakumar Chinnaswamy Thangavel
Kamaraj College of Engineering and Technology

11.1 BIOMATERIALS

11.1.1 Ideal Biomaterial Should Be

In the past few decades, many generations of biomaterials have come into use, and an ideal biomaterial has not yet been found in bone tissue engineering. The demands for an ideal biomaterial for bone tissue engineering are many. Based on the biological aspect, they should be biocompatible, osteoinductive, osteoconductive, bioconvertable, and nontoxic as a whole and also their dissolution products (Olszta et al. 2007, 89; Unal et al. 2021, 175–193; Hoppe et al. 2011, 2757–74). To avoid encapsulation, the biological reactions should not have an acidic front. The rate of dissolution of the products should also promote the rate of new bone synthesis during the biological reactions. On the physical aspect, the difference in the moduli between the implanted biomaterial and the normal bone will lead to stress shielding with the resultant consequences. Their density and corrosiveness on repeated loading are also other important factors (Zaman et al. 2015, 19–25; Amini et al. 2012).

11.1.2 Existing Biomaterials and Their Drawbacks

Various materials have been investigated for bone regeneration capability. In bone grafting and regenerative operations, autologous bone is still the preferred material. Autologous bone is frequently extracted and reimplanted from a secondary place inside the body (Ginebra et al. 2018, 173–183; Hayakawa et al. 2008, 53–77). Autologous bone possesses hydroxyapatite (HAp) (an inorganic mineral) and bone cell properties. Even though this bone is not a living material, it can remodel into new and functional bone. Since HAp can also be processed in the lab into nanoparticulates by blending HAp nanoparticles into biopolymeric composites, it can be used in structural support during bone remodelling (Xie et al. 2008, 1–6). Autografts can be used in conjunction with growth hormones or synthetic bone replacement materials. The limitation of using autologous bone is the secondary trauma site created that has

 DOI: 10.1201/9781003258353-14

to heal. Further, its usage is limited by availability. The supply of materials obtained from cadavers (allograft) is unrestricted. The aforementioned substance is frequently demineralized, leaving a collagenous scaffold for future bone formation. Further, it fails to function as an osteoinductive but only as an osteoconductive material. The risk of disease transmission is always present with this material (Lee et al. 2011, 153–170).

11.1.3 Bone Prosthetics

Materials such as polymers, metals, and ceramics are commonly used for fabricating biomaterials (Figure 11.1). In bone tissue engineering applications, though some polymeric materials are biodegradable and bioactive, they do not exactly match the strength of the bone (Alsharabasy 2018, 8–11). Metals can also be used in various forms, like bones, joints, teeth, screws, pins, plates, and stents. These metallic implants exhibit very high mechanical strength and also serve as temporary support. It will not be either converted into a bone or resorbed by the body. The possibility of having infections and secondary surgery is unavoidable in this case (Devi et al. 2018, 530–543).

11.1.4 Bioceramics

In orthopaedic applications, bioceramics are employed as bone, joints, and teeth, which are mainly used for the purpose of repair and replacement of damaged or diseased parts. Bioceramics possess properties such as low chemical reactivity, biocompatibility, osseointegration, regeneration, and mineral component deposition. Bioceramics are classified into bioconductive, bioactive, and bioresorbable ceramics by their physiological reaction.

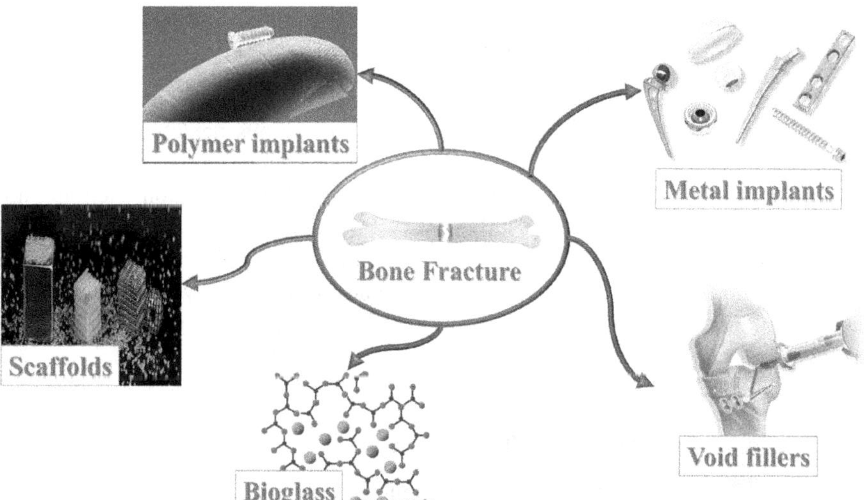

FIGURE 11.1 Schematic representation of different ways of healing bone fracture.

11.1.5 Generations of Bioceramics

The first generation of bioceramics is HAp, which is prepared by a high-temperature sintering process. The crystallization temperature could regulate the shape of HAp nanoparticle formation, and the different shapes of HAp could positively provide reinforcement to the biopolymeric composite (Xie et al. 2008, 1–6; De Silva et al. 2014, 807–818). The chemical composition of HAp resembles that of the bone mineral and is suitable for bone and dental implants (Farooq et al. 2012, 199–201). The sintered HAp was highly fragile and the materials were nonbiodegradable and only partly osteoconductive. The second generation of bioceramics are calcium phosphate, dicalcium phosphate, beta-tricalcium phosphate, and tetracalcium phosphate. When compared to HAp, the above-mentioned materials have a better calcium/phosphate (Ca/P) ratio, which enables the material to be both osteoconductive and osteoinductive (Salinas et al. 2013, 40–51). Bioglass (silica based) – the third generation of bioceramics is osteoconductive, osteoinductive, and bioconvertible – was developed by L.L. Hench in 1969. Bioglass generated a breakthrough in the biomaterial industry due to its bone-bonding ability between the tissues and implants in physiological fluids. Si, Ca, P, and Na ion exchange and dissolution products induce intracellular and extracellular responses in order to encourage the formation of a bone (osteogenesis). An extensive *in vitro* and *in vivo* study revealed that the 45S5 bioglass has angiogenic properties that regulate vascular endothelial growth factor expression and vascularization, among other things (Koons et al. 2020, 584–603). The clinical application of silica-based bioglass has been widely investigated in the areas of orthopaedic implants, middle ear surgery (small bone repair), and periodontology. To improve the antimicrobial and antibacterial properties, boron or silver is added to the silicate network. The deficit, the long duration taken for dissolution, in the third generation of bioglass was improved in the fourth generation of bioceramics or phosphate-based bioglass (Hench and Thompson 2010, S379–S391).

11.1.6 Methods of Bioceramic Preparation

Bioceramics are fabricated in different forms (bulk or porous materials), such as implants, prosthetics, and void fillers for repairing various defects in the human body. Bioceramics were produced by various methods (wet chemical method, hydrothermal process, solid-state reaction, hydrolysis, sol-gel, precipitation, etc.) and are available in the form of single crystal, polycrystalline, glass, glass bioceramics, and biocomposites (Salinas et al. 2013, 40–51). The drawbacks of using these methods are stoichiometric imbalance, agglomeration, and inhomogeneity. The microarchitecture and purity of the prepared bioceramics are assessed by morphology, granule size, shape, boundary, phase distribution, crystallinity, porosity, etc.

11.1.7 Metal Oxide–Doped Bioceramics

Bioceramics were brittle and, to improve their physical properties, they were doped with various metal oxides in different concentrations, leading to different compositions. The oxides are categorized into network formers (B_2O_3, SiO_2, and P_2O_5),

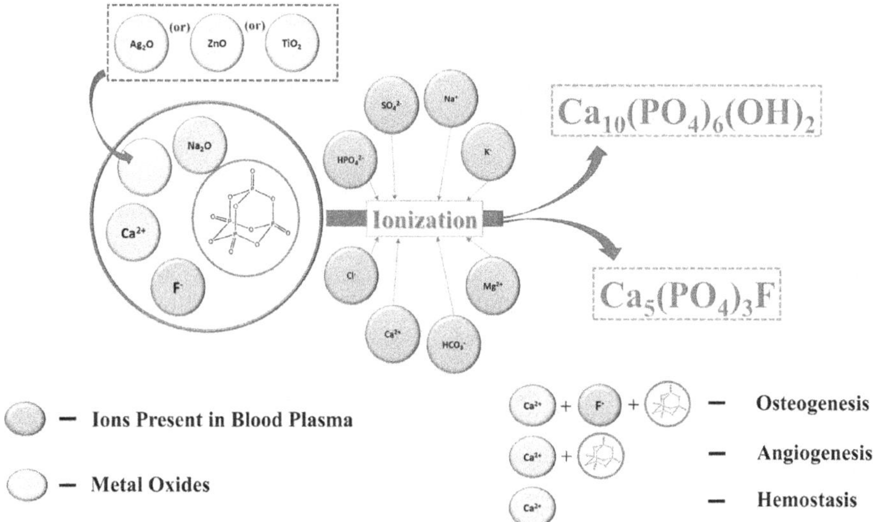

FIGURE 11.2 Schematic representation of components of metal oxide–doped fluorophosphate bioglass and its beneficial effects.

modifiers (Li_2O, Na_2O, K_2O, MgO, CaO, SrO, BaO, ZnO, and PbO), and intermediate oxides (TiO_2, VO, Ga_2O_3, SeO, MoO, TeO, WO_2, and Bi_2O_3) (Hoppe et al. 2011, 2757–74; Pantulap et al. 2022, 1–41). The biological characteristics were improved by developing silica-free phosphate glasses, and they were further improved by doping them with fluorides, which are well known to enhance bone formation (Figure 11.2).

11.1.8 Phosphate-Based Bioactive Glass Doped with Different Metal Oxides and Fluorides

The objective of this chapter is to prepare calcium phosphate glass doped with fluoride ions and metal ions in the glassy web and evaluate whether the addition of fluoride alters the physical properties of the glass system to a deleterious level, compromising the biological advantage achieved (Salinas et al. 2013, 40–51; Zaman et al. 2015, 19–25; Amini et al. 2012). The metal oxides were selected as per the FDI-approved concentration in composites such as titanium, zirconium, and silver. Both the physical and biological aspects of phosphate-based glass have been improved by incorporating metal oxides and bioconversion by incorporating fluoride ions.

11.1.9 Beneficial Effects of Metal Oxides (Ag_2O, TiO_2, and ZrO_2)

The infection resistance of the bioactive glass can be improved by adding silver molecules. The addition of Ag_2O to bioactive glasses helps minimize the risk of microbial infections through the potential antimicrobial property of silver ions, which inhibits replication of bacterial RNA, DNA, and biofilm formation. The presence of Ag^+ ions acts as a killing agent in bacterial strains such as *Pseudomonas aeruginosa*,

Staphylococcus aureus, and *Escherichia coli* (Kumar et al. 2018, 2744–52; Hayakawa et al. 2008, 53–77; Wang et al. 2022, 9291). The bioinert metal TiO_2 wouldn't elicit an inflammatory response. The incorporation of TiO_2 in bioglass composition enhances chemical durability because of Ti-O-P bonds' presence instead of P-O-P bonds (Rahaman 2017, 56–66; Lee et al. 2011, 153–170). Zirconium oxide (ZrO_2) interacts with the host tissue without gross reaction. A literature survey has demonstrated the durability of zirconium as the best implant material for articular surface replacements due to its non-corrosive nature, higher yield strength, and fracture toughness. The ZrO_2 glass ceramic development led to the creation of a high-strength material that can be used in a variety of biomedical applications such as abutments and dental posts (Akter 2016, 3–16; Grishchenko et al. 2022, 114–122).

11.1.10 Food and Drug Administration

By FDA (Food and Drug Administration) of the United States standards, TiO_2/ZrO_2/Ag_2O was added at an optimal mole percentage of 1%. To determine the appropriate fluoride dose for increasing bioconversion without destabilizing the physical properties of the glass system, calcium fluoride (CaF_2) was introduced in incremental doses up to 5 mol%. The elastic strength, thermal stability, morphology, and structural reorganization of the titanium, zirconium, and silver fluorophosphate glasses were compared and discussed.

11.2 EXPERIMENTAL STUDIES

11.2.1 Melt Quenching Technique

The required inorganic chemicals (Table 11.1) were procured from Sigma Aldrich, India. Silver oxide, titanium oxide, and zirconium oxide were measured in accordance with the FDA-approved 1 mol% of the composition. The chemical substances were ground well and it was heated in a heating furnace for 1 hour at 100°C–120°C. Once the materials were allowed to cool at room temperature, it was again ground in

TABLE 11.1
Chemical Composition of Metal Oxide–Doped Fluorophosphate Bioglasses and Fluoride-Doped Calcium Phosphate Glasses

	Sample Code and Composition (with Dopant Metal Oxide)				
Chemicals (mol%)	**M**Fp1**	**M**Fp2**	**M**Fp3**	**M**Fp4**	**M**Fp5**
P_2O_5	45.00	45.00	45.00	45.00	45.00
CaO	29.00	29.00	29.00	29.00	29.00
Metal oxide*	1	1	1	1	1
Na_2O	25.00	23.75	22.50	21.25	20.00
CaF_2	0.00	1.25	2.50	3.75	5.00

Metal oxide*-Ag_2O/TiO_2/ZrO_2; M**-Ag/Ti/Zr.

a ball mill and transferred to a rhodium-doped platinum crucible. In a preheated furnace, the crucible was placed at 1,050°C–1,400°C for an hour. The molten material was poured over the preheated graphite steel mould. The materials were annealed for 1 hour at 573°C and were then cooled to room temperature. Using a diamond cutter, the bioceramics were sliced to their desired sizes (Hench and Paschall 1973, 25–42) (Table 11.1).

11.2.2 Physicochemical and Biological Characteristics of Bioceramics

11.2.2.1 Density: Effects of Phosphate-Based Bioactive Glass Doped with Different Metal Oxides and Fluorides

The density of AgFp/TiFp/ZrFp was determined by Archimedes' principle using water as a buoyant liquid.

The density variation of three different fluorophosphate glasses is depicted in Figure 11.3, and the values are presented in Table 11.2. The density defines the effect of elemental packing in the prepared composition. Uniformly, a drop in density was identified by introducing fluoride into the glassy network (Maeda et al. 2017, 5433–38). The density of all the prepared glass samples lies between 2,707 and 2,600 kg/m^3. The density of bioglasses gets elevated, followed by the further addition of fluoride to 2.5 moles. For the three-metal oxide–doped fluorophosphate glass, the consistent addition of fluoride content till 5 mol% did not follow the linear density variation.

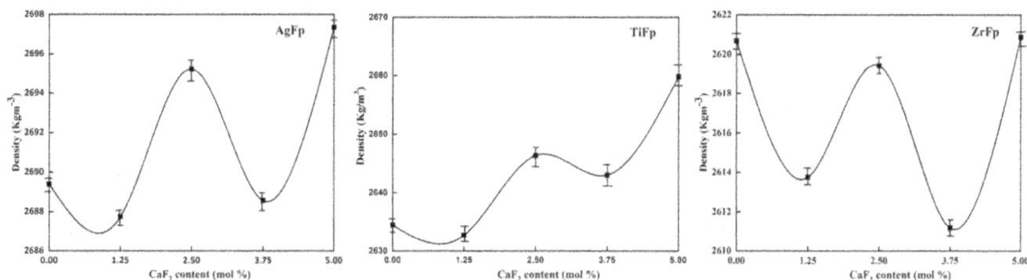

FIGURE 11.3 Densities measured for ZrFp, TiFp, and AgFp.

TABLE 11.2
Density Measurements of Metal-Doped Fluorophosphate Bioactive Glasses Having Varying Fluoride Content

	Density (ρ) (kg/m^3)				
Sample	**M**Fp1**	**M**Fp2**	**M**Fp3**	**M**Fp4**	**M**Fp5**
ZrFp	2,621	2,614	2,619	2,611	2,621
TiFp	2,634	2,632	2,646	2,643	2,660
AgFp	2,689	2,688	2,695	2,689	2,697

M**-Ag/Ti/Zr.

In a phosphate-based glassy network, the contribution of cations (Na^+, Ca^+) and fluoride breaks the P-O-P linkage to form the terminal oxygen.

With the consistent addition of fluoride in fluorophosphate glass, the non-bridging oxygen of the phosphate network and metal ions (ZrO_2, TiO_2, and Ag_2O) strengthens the ionic cross-linking and weakens the network-breaking cations. The results revealed that a high molar percentage of fluoride content exhibited high density compared to other proportions. The density of silver-doped fluorophosphate glass differed slightly from the TiFp and ZrFp groups (Salinas et al. 2013, 40–51; Maeda et al. 2017, 5433–38). Conventionally, Ag_2O has a higher molar mass (231.7 g/mol), which is capable of occupying more space in the network, whereas the molar masses of TiO_2 (79.9 g/mol) and ZrO_2 (123.2 g/mol) are considerably lower than that of Ag_2O (Devi et al. 2010, 2483–90).

11.2.2.2 Ultrasonic Measurements: Effects of Phosphate-Based Bioactive Glass Doped with Different Metal Oxides and Fluorides

Ultrasonic measurements of AgFp/TiFp/ZrFp were performed using the pulse echo method and cross-correlation approach.

Figure 11.4 represents the longitudinal, bulk, Young's, and shear moduli of ZrFp, TiFp, and AgFp with increasing fluoride content. The elastic property of the bioglass reveals the strength and interatomic potential of the biomaterial. Each fluorophosphate glass follows its own pattern due to the contribution of different ion and fluoride content (Kohles and Martinez 2000, 479–488). In ZrFp, a drastic drop in moduli was observed by the initial addition of fluoride content. The additional supplementation of fluoride content (2.5 mol%) elevated the moduli and the continuous increment of fluoride content could not alter the moduli much. The Young's and shear moduli of TiFp imitated the density pattern, whereas the longitudinal and bulk moduli slightly differed from the above pattern.

In AgFp, a drop in moduli was observed after the initial addition of calcium fluoride (1.25 mol%). The rising hump was noticed at 2.5 mol% and the further addition

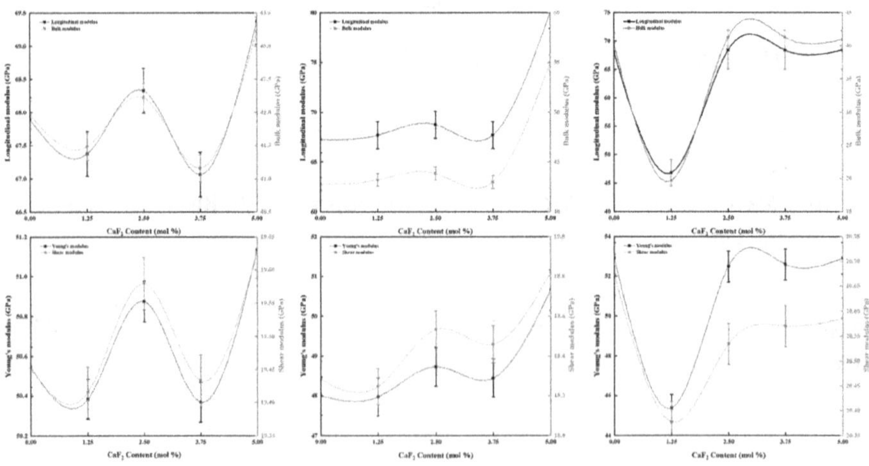

FIGURE 11.4 Longitudinal, bulk, Young's, and shear moduli of ZrFp, TiFp, and AgFp.

of fluoride content fluctuated like the density pattern. On the contrary, the results of ZrFp were different to those of AgFp and TiFp because the low atomic mass of ZrO_2 attains maximum moduli at 2.5 mol% itself. The moduli alteration defines the strength and weakness of the glassy network (Begum et al. 2006, 409–417). The maximum moduli were observed at the maximum calcium fluoride content of 5 mol% for AgFp and TiFp.

11.2.2.3 X-Ray Photoelectron Spectroscopy: Effects of Phosphate-Based Bioactive Glass Doped with Different Metal Oxides and Fluorides

The elemental composition of a glass sample with Al K as the source was analysed using an X-ray photoelectron spectroscope (XPS) (Model AXIS Ultra DLD, Kratos, Kyoto, Japan) at 210 W and X-ray as excitation radiation.

Samples of ZrFp3/TiFp5/AgFp5 were selected based on their density and moduli results. The elemental composition of the bioactive glass sample of ZrFp3/TiFp5/AgFp5 was investigated by XPS. The results were tabulated in Table 11.3 and presented in Figure 11.5. Calcium fluoride, sodium metaphosphate, phosphorus pentoxide, and sodium carbonate substances were observed at different spectral lines 3P, $2P^{3/2}$, 2S, and 1S with specific binding energies which all are essential for bone mineralization. The O1S spectrum is assigned as P-O-P sites that represent the bridging oxygen of the phosphate glasses and the nature of bonding between the oxygen and cations. The metal oxide–doped fluorophosphate glass was identified by the unique spectral lines

TABLE 11.3
Longitudinal Modulus (*L*), Bulk Modulus (*K*), Young's Modulus (*Y*), and Shear Modulus (*G*) of Metal-Doped Fluorophosphate Bioactive Glasses Having Varying Fluoride Content

	Elastic Moduli			
Sample Code	*L* (GPa)	*K* (GPa)	*Y* (GPa)	*G* (GPa)
ZrFp1	67.9	40.3	53.0	20.7
ZrFp2	46.7	19.6	45.4	20.4
ZrFp3	68.4	41.2	52.5	20.4
ZrFp4	68.4	41.1	52.6	20.4
ZrFp5	68.3	40.8	53.0	20.6
TiFp1	67.2	42.8	48.0	18.3
TiFp2	67.6	43.2	47.0	18.3
TiFp3	68.7	43.9	48.8	18.5
TiFp4	67.7	43.0	48.5	18.6
TiFp5	80.0	50.0	50.7	18.8
AgFp1	67.9	42.0	50.6	19.5
AgFp2	67.4	41.5	50.4	19.4
AgFp3	68.3	42.2	50.9	19.6
AgFp4	67.1	41.2	50.4	19.4
AgFp5	69.4	43.2	51.1	19.6

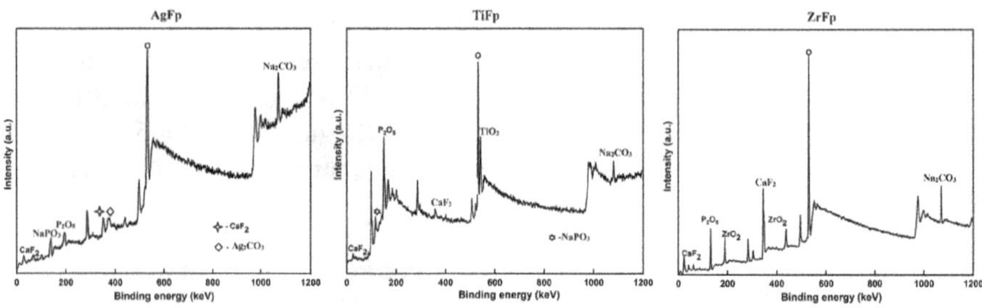

FIGURE 11.5 X-ray photon spectra of ZrFp3, TiFp5, and AgFp5.

noted at 184.4 eV (ZrO_2), 368.0 eV (Ag_2CO_3), and 539.2 eV (TiO_2). The Ti $2P^{3/2}$ peak is generally noted at 458.8 eV. This peak was found at 539.2 eV due to the entrapment of titanium ions in the fluorophosphate network (Moulder 1995, 230–232; Viornery et al. 2002, 2582–89; Siow et al. 2018, 1913–1922).

11.2.2.4 Thermal Properties: Effects of Phosphate-Based Bioactive Glass Doped with Different Metal Oxides and Fluorides

Differential scanning calorimetric (DSC) and thermogravimetric curves were recorded on an STA (Simultaneous Thermal Analysis) 449 *F3Nevio*. Under a nitrogen atmosphere, 3.5 mg of material was heated to 1,000°C at 50 K/min.

The DSC curve provides information regarding thermal properties like crystallization and melting of the material. The exothermic peak of ZrFp3, TiFp5, and AgFp5 was maximum at 540°C, 550°C, and 498°C, and the corresponding enthalpy of crystallization (ΔH_c) was identified as 36,151 and 210 J/g (Figure 11.6). The melting endotherm peak of ZrFp3, TiFp5, and AgFp5 was noted at 725°C, 734°C, and 730°C, respectively, and the associated enthalpies of fusion (ΔH_f) of ZrFp3, TiFp5, and AgFp5 were 171, 190, and 212 J/g accordingly. In the DSC curve, the crystallization exotherm is determined by the density of the material and the melting endotherm exhibits strength of the molecular interaction and purity of the material. Though different metal oxide–doped fluorophosphate glasses showed minimal variation in their results, the thermal results of AgFp5 had a higher enthalpy of crystallization (210 J/g) and enthalpy of fusion (212 J/g) than the rest. Upto 1,000°C, there is no discernible mass loss for all four samples which confirmed the thermal stability of the materials (Chatzistavrou et al. 2004, 944–951; Chatzistavrou et al. 2011, 118–129).

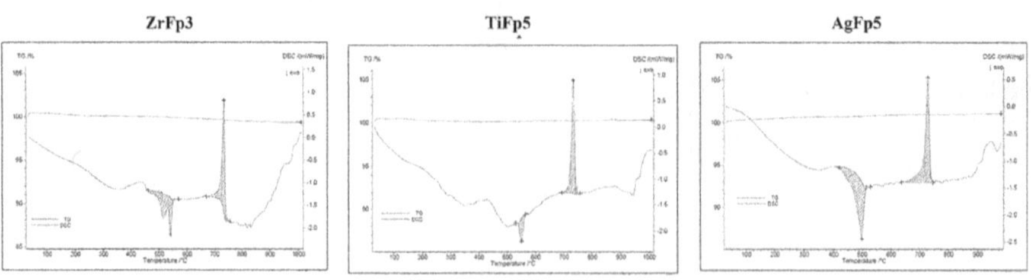

FIGURE 11.6 Thermal studies of ZrFp3, TiFp5, and AgFp5.

11.2.2.5 *In Vitro* Studies: Effects of Phosphate-Based Bioactive Glass Doped with Different Metal Oxides and Fluorides

The ionic concentration of the simulated body fluid (SBF) resembles human blood plasma and it was prepared by the standard kokubo protocol. The prepared glass samples (1 g) were soaked in 100 mL of sterile SBF which was placed in a CO_2 (5%) incubator at 37°C for 21 days. A Thermo ScientificTM OrionTM 3-Star Benchtop pH metre was used to record pH variation every day. After incubation for 21 days, the samples were removed from SBF, rinsed with double distilled water, and dried under laminar airflow.

During the stipulated period of 21-day immersion, the pH variation of ZrFp, TiFp, and AgFp wasn't linear (Figure 11.7). As a consequence of the phosphoric acid release, the pH value drops uniformly on the first day for all the metal oxide–doped fluorophosphate glass. Later, the alkaline earth metals like Na^+, Ca^{2+}, and ionic exchange with H^+ or H_3O^+ ions raise the pH. The pH fluctuation deviated on the series of days due to metallic ionic transition and ion leaching. Interestingly, the pH value of ZrFp, TiFp, and AgFp doesn't cross the critical level (both in acidic and basic) because of consistent apatite formation, and the respective pH value only lies in the range of 7.4–6.4. Substitutes of active cations Na^+ and/or Ca^{2+} with H^+ ions enhance the hydroxyl ion concentration in the glass network. After 18 days, the pH fluctuation is very minimal; it seems that the saturation point is much closer. Moreover, all three different fluorophosphate glasses follow different trends of pH (Macon et al. 2015, 115; Ren et al. 2012, 293–297; Wu and Xiao 2009, 25–29; Ranga et al. 2019, 75–81).

The following steps explain how fluoride salt reacts in SBF:

1. The P-O-P linkage split in physiological fluid; either cations (H^+) or anions (F^-) occupy the vacuum and form P-O-H^+ / P-O-F^-.
2. The countercations Na^+/ Ca^{2+} interact with water resulting in hydroxides NaOH/Ca $(OH)_2$.
3. The hydroxide groups of phosphoric acid along with calcium fluoride induce the deposition of fluorapatite.

$$9CaO(s) + 3P_2O_5(s) + CaF_2(s) \rightarrow Ca_{10}(PO_4)_6 F_2(s)$$

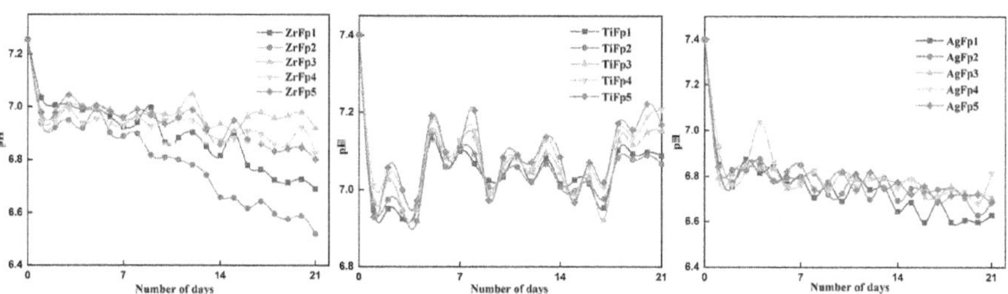

FIGURE 11.7 *In vitro* studies of ZrFp, TiFp, and AgFp.

11.2.2.6 Fourier Transform Infrared Spectral Study: ZrFp, TiFp, and AgFp

The FTIR spectra of samples were recorded using a Fourier transform infrared spectrophotometer (Model 8700; Shimadzu, Tokyo, Japan).

The structural characterization of ZrFp/TiFp/AgFp was analysed within the 4,000–400 cm^{-1} spectral range which is illustrated in Figure 11.8 and Table 11.4. The wide intense band is assigned as absorbed water molecules (O-H stretch) at the frequency range of 3,550–3,200 cm^{-1} in the material. The stretching vibration of the P-O-H bond was noticed around 2,900 cm^{-1} and their low intensity reveals the contribution of fluoride and carbonate salts (CO_3^{2-}) in the glass network. The absorption band at the frequency range of 1,650–1,642 cm^{-1} is assigned to P-O-H which bridges the hydrogen and non-bridging oxygen. The stretching mode of P=O at

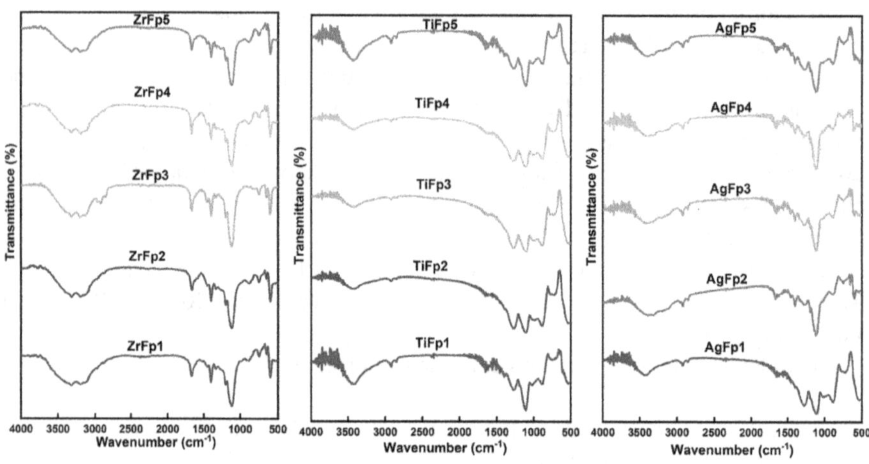

FIGURE 11.8 FTIR spectra of ZrFp, TiFp, and AgFp bioglasses (pre-immersion in SBF).

TABLE 11.4
XPS Analysis of Different Phosphate Glasses

Binding Energy (eV)				
ZrFp3	**TiFp5**	**AgFp5**	**Spectral Line**	**Compound**
28.0	28.0	31.8	3P	CaF_2
–	133.5	134.2	$2P^{3/2}$	$NaPO_3$
132.1	118.2,187.7	192.0	2S	P_2O_5
184.4	–	–	$3d^{5/2}$	ZrO_2
347.6	347.8	348.4	$2P^{3/2}$	CaF_2
–	–	368.0	$3d^{5/2}$	Ag_2CO_3
533.5	533.5	533.8	1S	O
–	539.2	–	$2P^{3/2}$	TiO_2
1,072.5	1,072.5	1,072.5	1S	Na_2CO_3

TABLE 11.5
FTIR Spectral Analysis of Different Bioglasses (Pre- and Post-immersion in SBF)

	Wave number (cm^{-1})			
Immersion	**ZrFp3**	**TiFp5**	**AgFp5**	**Assignments**
Pre-immersion	3,428	3,440	3,428	OH stretch
Post-immersion	3,257	3,444	3,428	
Pre-immersion	–	2,935	2,931	Stretching vibration of P-O-H
Post-immersion	–	2,908	–	
Pre-immersion	1,673	1,638	1,661	P-O-H bridge
Post-immersion	1,678	–	1,613	
Pre-immersion	1,394	–	1,394	P=O Stretching mode
Post-immersion	–	–	–	
Pre-immersion	–	1,281	1,278	Asymmetric stretching of P=O
Post-immersion	1,339	1,276	1,284	
Pre-immersion	1,119	1,116	1,118	Asymmetric stretching of PO_3
Post-immersion	1,117	1,116	1,119	
Pre-immersion	887	893	898	Asymmetric stretching mode of P-O-P/P-F stretch
Post-immersion	–	888	903	
Pre-immersion	748	729	746	Symmetric stretching mode of P-O-P
Post-immersion	751	739	733	
Pre-immersion	596	518	594	P-O bend
Post-immersion	590	518	526	

1,390–1,230 cm^{-1} and the asymmetric stretching mode of P=O at 1,300–1,200 cm^{-1} indicated the metaphosphate chains of the PO_2 group. The shoulder band at the 1,120–1,080 cm^{-1} range corresponded to the asymmetric stretching mode of PO_3. The band corresponding to the asymmetric and symmetric stretching modes of the P-O-P linkage lies in the 1,050–850 cm^{-1} and 780–740 cm^{-1} region, respectively. The P-F stretching band was observed in the 890–840 cm^{-1} range and may be attributed to the PF_3, POF_3, and PF_5. The band due to the P-O-P bending vibration was noted to be 580–460 cm^{-1} due to the deformation mode of the PO^- group (Gheisari et al. 2015, 5967–75; Rai et al. 2014; Nyquist and Kagel 2012). However, the presence of transition metal oxides (ZrO_2/ TiO_2/Ag_2O) entrapped in the phosphate matrix does not show any explicit variation in the absorption bands pertaining to the carbonate and phosphate groups of the bioglass (Table 11.5 and Figure 11.9) (Hench and Jones 2015, 194).

11.2.2.7 Role of Metal Oxides in Phosphate Glasses in *In Vitro* Study

The chosen Zr/Ti/Ag is considered the network modifier. During metal oxide–doped fluorophosphate glass involves ionic exchange; the elevated P/O ratio split the P-O-P linkage because of weaker interaction in the ZrFp/TiFp and AgFp network. Because of these circumstances, the decrease in the bridging oxygen level was attributed to

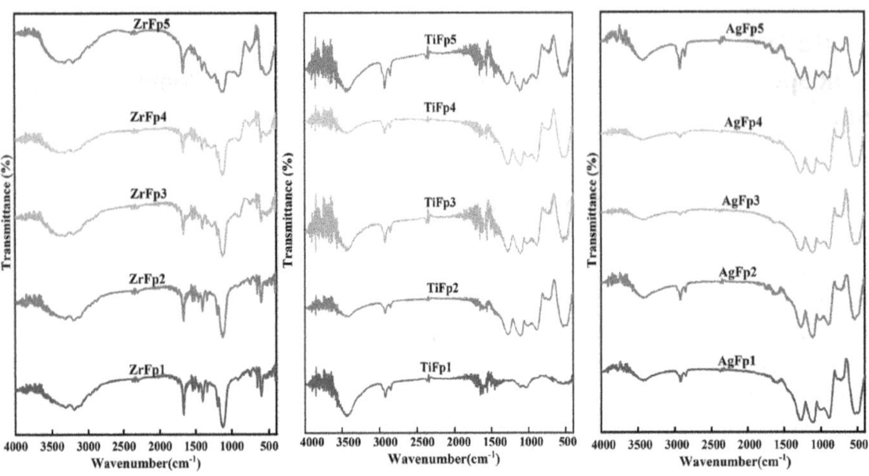

FIGURE 11.9 FTIR spectra of ZrFp, TiFp, and AgFp bioglasses (post-immersion in SBF).

an increase in non-bridging oxygen. The P-O-P splitting leads to P-O-Zr/P-O-Ti/P-O-Ag bond formation (Kaur et al. 2014, 254–274; Rabiee et al. 2015, 7241–51).

11.2.2.8 X-Ray Diffraction Studies: Effects of Phosphate-Based Bioactive Glass Doped with Different Metal Oxides and Fluorides

In the X-ray diffraction studies, ZrFp, TiFp, and AgFp have consistent shoulder peaks at 31.88°, 31.92°, and 32.17° and weaker peaks at 34.60°, 31.79°, and 31.79° respectively (Figure 11.10). More specifically, ZrFP3 with 2.5% of calcium fluoride content was intended for the slightly broadened peak. TiFp5 (5% CaF_2) was highlighted with the most diffraction. Precisely, AgFp is marked with a clear and sharp

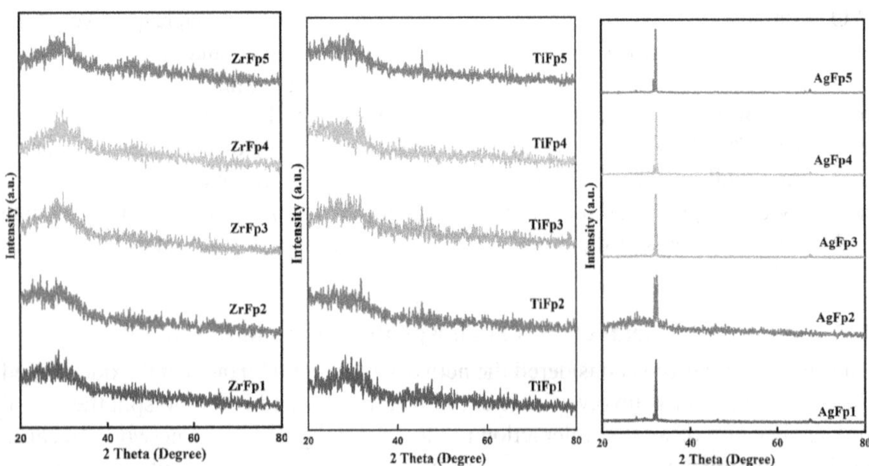

FIGURE 11.10 X-ray diffractograms of ZrFp, TiFp, and AgFp glasses.

peak throughout the five different CaF_2 contents. The observed diffraction peak is assumed to be the result of a HAp and fluoroapatite combination. (Yadav et al. 2017, 1321–1330; Xian 2009; Himanshu et al. 2016).

11.2.2.9 SEM Micrographs: ZrFp3, TiFp5, and AgFp5

The SEM (Scanning Electron Microscope) was used to investigate the morphology and the elemental composition of the surface using SEM (Model Ultra 55; Zeiss, Oberkochen, Germany).

The morphologies of ZrFp3, TiFp5, and AgFp5 were studied. All three-metal-doped fluorophosphate glasses exhibit a clear, smooth surface which expresses their amorphous nature. The SEM results indicated that the metal oxide doped in fluorophosphate glass did not alter the amorphous nature of the glass (Sankaralingam et al. 2021, 37–48). The morphology of the selected ZrFp3/TiFp5/AgFp5 is depicted in Figure 11.11. All three specimens were observed with a smooth surface and a

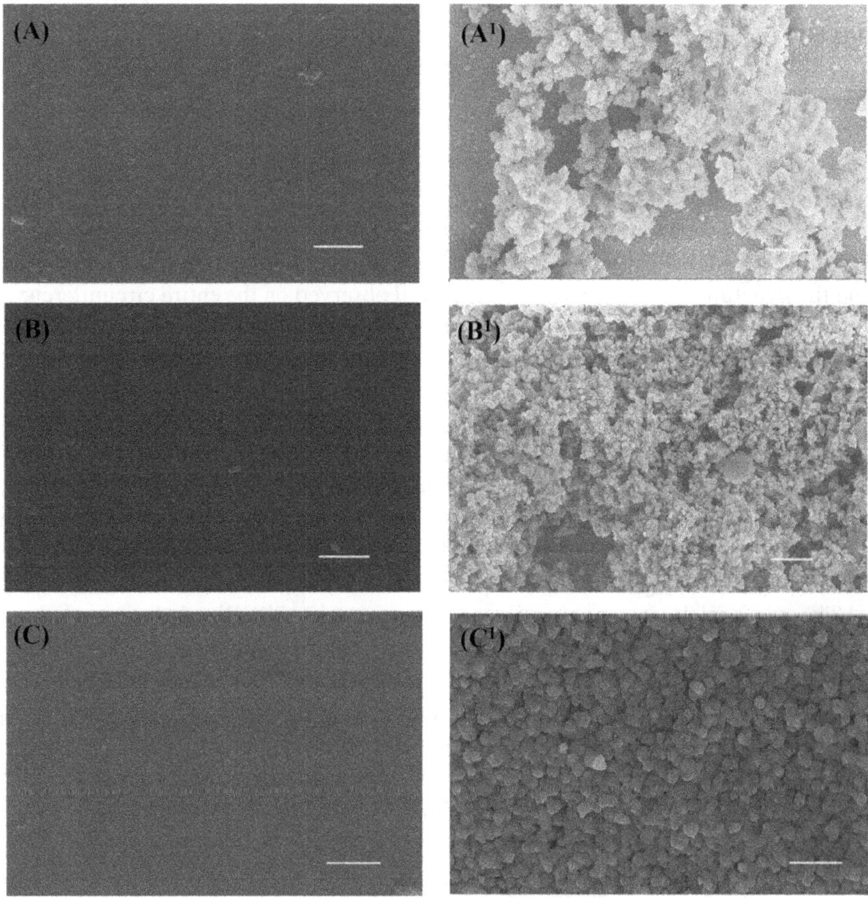

FIGURE 11.11 SEM images of ZrFp3 (A and A^1), TiFp5 (B and B^1), and AgFp5 (C and C^1) (at 1,000× magnification).

lack of precipitate formation. After *in vitro* degradation, the agglomeration of crystals exhibits the rough form and the size of the deposited particle in the range of 20–200 μm. Enormous deposition of crystal salts was observed in all the selected ZrFp3, TiFp5, and AgFp5 specimens evenly, whereas the aggregation pattern and the size of the crystals alone varied. ZrFp3 showed the flower-shaped apatite formation which perforated the specimen.

The morphology of TiFp5 showed the conglomeration of nanospheres in a closely packed structure. Micrographs of AgFp5 represented a hollow sphere with a rough surface arranged in an irregular pattern (Gouveia et al. 2021, 104–164; Yoder et al. 2016, 977–983). The deposited precipitates are assumed to be HAp and fluorapatite because of the elemental composition of OH, Ca, P, F, Zr, Ti, and Ag, and it is evident with the energy-dispersive X-ray analysis. Initially, the Ca/P ratio is calculated as 0.33 for all the metal oxide–doped fluorophosphate glasses, whereas the Ca/P ratio of the *in vitro* degraded specimen is 1.6 ± 0.7. If the material is subjected to dissolution, the Ca/P ratio would be less than 1.3. In contrast, the value of the Ca/P ratio beyond 1.3 reveals the saturation point of the material. The difference in the Ca/P ratio is a sign of ionic dissolution and apatite deposition (Prasad et al. 2017; Antoniac 2019). The SEM image of the section of the metal oxide–doped fluorophosphate glass is depicted in Figure 11.11.

11.2.2.10 *In vivo* Studies: ZrFp3, TiFp5, and AgFp5

After 10 weeks of implantation, the undecalcified section of the femoral condyle of the rabbit was examined. Micrographs of ZrFp3/TiFp5/AgFp5 are depicted in Figure 11.12. At 100× magnification, the glass size gets reduced from its initial size, and the structure of the implanted glass is well observed on the entire circumference. In the microenvironment, ionic dissolution of implanted glass (FP3-1.901 mm; ZrFp3-1.893 mm; TiFp5-2.000 mm; and AgFp5-2.106 mm) shrank to FP3-1.357 mm; ZrFp3-1.573 mm; TiFp5-1.630 mm; and AgFp5-1.845 mm, respectively. In Figure 11.12, at 1,000× magnification, the differential layering between the glass and bone in the form of an interface, binding the glass to the bone, is clearly shown. The energy-dispersive spectrometer spectrum of the glass–bone interface revealed the presence of minerals such as P, Na, Ca, Ti/Zr/Ag, O, and F at the bone–glass interface, which confirms the bioconversion of glass into bone (Mohamed 2008; Stanciu et al. 2006, 689–692; Sankaralingam et al. 2021, 1–9).

The confocal laser scanning image (Figure 11.13) visualized the morphology of the bone–glass interface. On Ar laser scanning, the brilliant wide fluorescence to a depth of 546 nm from the periphery indicates high adenosine triphosphatase activity, indicating bioconversion at the interface by the penetration of living cells within a period of 10 weeks. This confirms the transformation process of glass to bone (Thrivikraman et al. 2014, 12763–81; Gurusamy and Sankaralingam 2015 WO 2015/087344 Al, WO 2015/087347 Al, WO 2015/087346 Al; Jones 2013, 4457–4486). The bioconversion rate of ZrFp3/TiFp5/AgFp5 was calculated as 57%, 62%, and 66%, respectively.

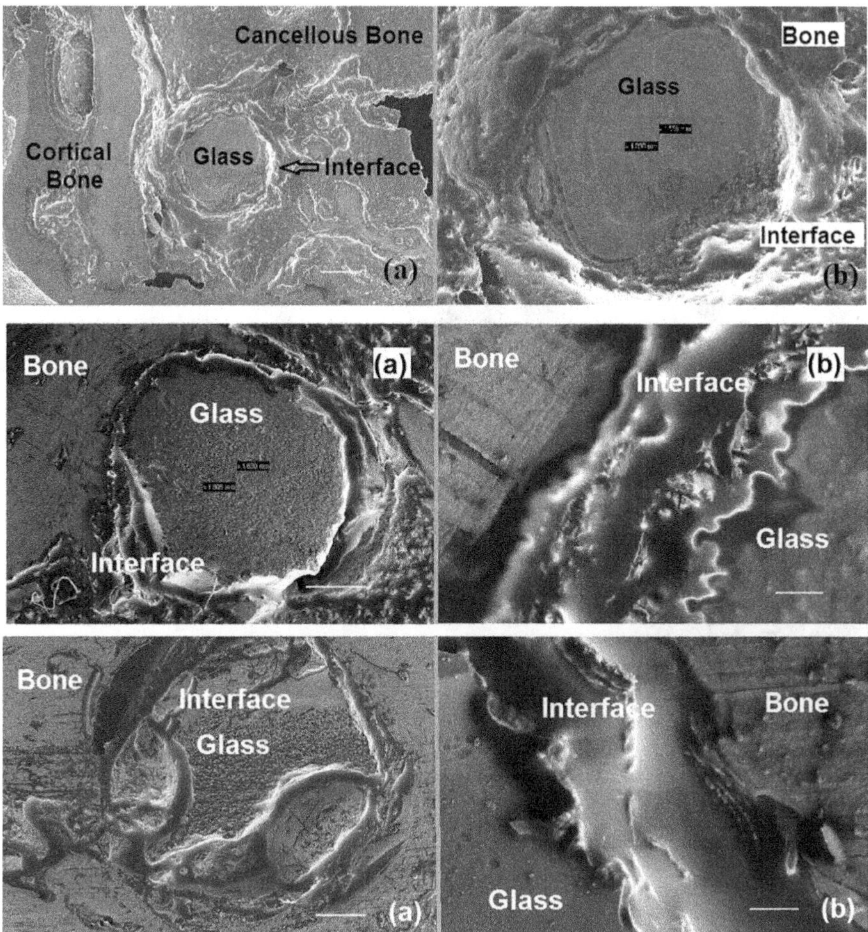

FIGURE 11.12 SEM images: in vivo studies of ZrFp3/TiFp5/AgFp4 (a) 100× and (b) 1,000× magnification.

11.3 CONCLUSIONS

The density, elastic strength, and structural properties of the fluorophosphate glass and zirconium/titanium/silver metal oxide–doped fluorophosphate glasses were studied and presented. The incremental fluoride content altered the density and modulus properties slightly. From the results, the optimum fluoride content for the metal oxide–doped fluorophosphates glasses in relation to their physical properties could not be confirmed, as every metal oxide had varying effects in their various moduli. Fluorophsphate glass is shown to be extremely useful in improving the biological potential of the calcium phosphate glass, but it has to be cautiously doped with metal oxides as they have varying effects on the physical properties of the composition. As all the glass samples have more than 40% of bone conversion within a short

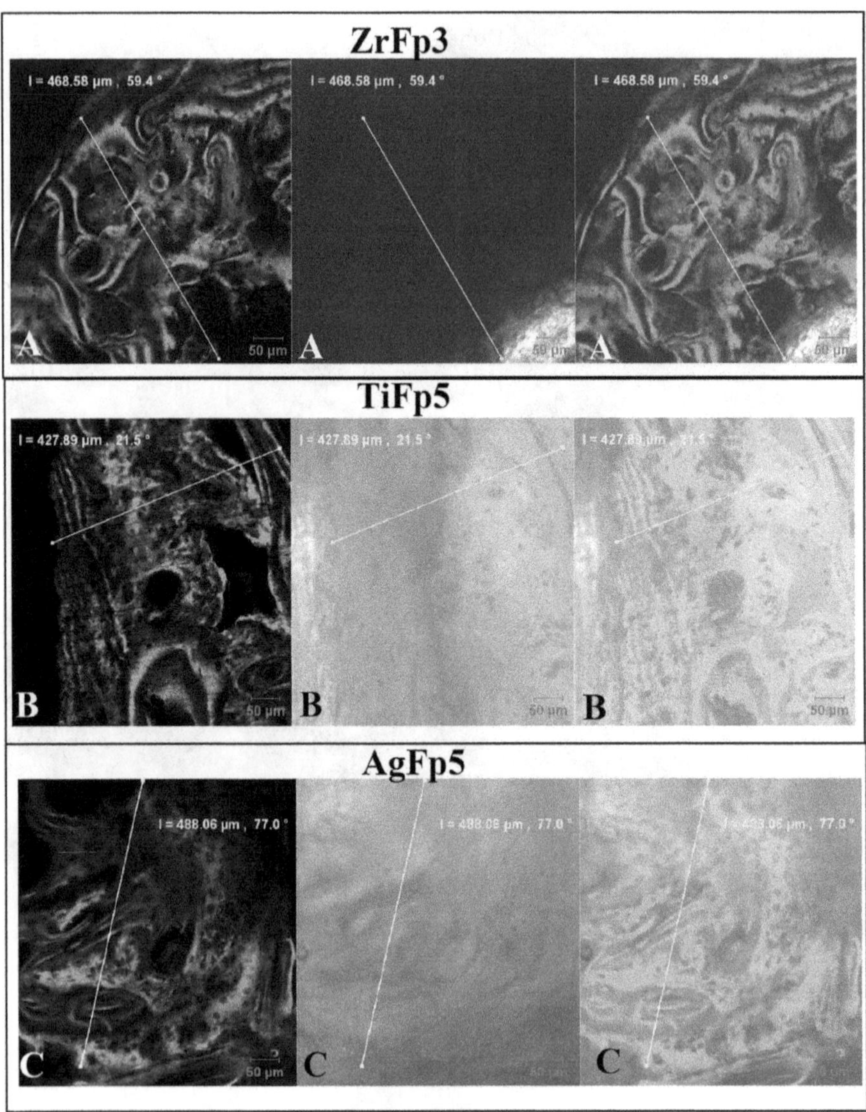

FIGURE 11.13 CLSM image of ZrFp3/TiFp5/AgFp5.

span of 10 weeks, it ensures that the prepared sample would be ideal for regenerative medicine.

ACKNOWLEDGEMENTS

The authors acknowledge M/S Pandian Advanced Medical Centre Private Limited, Madurai, India, for providing financial support for this study.

ABBREVIATIONS

β-TCP Beta-tricalcium phosphate
AgFp Silver fluorophosphate
Ag_2O Silver oxide
CaF_2 Calcium fluoride
Ca/P Calcium/phosphate
CO_3^{2-} Carbonate ion
DCP Dicalcium phosphate
DSC Differential scanning calorimetry
eV Electron volt
FDA Food and Drug Administration
FTIR Fourier transform infrared spectroscopy
HAp Hydroxyapatite
J/g Joules per gram
SBF Simulated body fluid
SEM Scanning electron microscopy
TG Thermogravimetry
TiFp Titanium fluorophosphate
TiO_2 Titanium oxide
TTCP Tetracalcium phosphate
XPS X-ray photoelectron spectroscopy
ZrO_2 Zirconium oxide
ZrFp Zirconium fluorophosphate

REFERENCES

Akter, F. Principles of tissue engineering. In *Tissue Engineering Made Easy*, pp. 3–16. Academic Press, 2016.

Alsharabasy, A.M. A mini-review on the bioactive glass-based composites in soft tissue repair. *Bioceramics Development and Applications* 8, no. 1 (2018): 8–11.

Amini, A.R., C.T. Laurencin, and S.P. Nukavarapu. Bone tissue engineering: Recent advances and challenges. *Critical Reviews(tm) in Biomedical Engineering* 40, no. 5 (2012).

Antoniac, I., ed. *Bioceramics and Biocomposites: From Research to Clinical Practice*. John Wiley & Sons, 2019.

Begum, A.N., V. Rajendran, and H. Ylänen. Effect of thermal treatment on physical properties of bioactive glass. *Materials Chemistry and Physics* 96, no. 2–3 (2006): 409–417.

Chatzistavrou, X., N. Kantiranis, E. Kontonasaki, K. Chrissafis, L. Papadopoulou, P. Koidis, A.R. Boccaccini, and K.M. Paraskevopoulos. Thermal analysis and in vitro bioactivity of bioactive glass-alumina composites. *Materials Characterization* 62, no. 1 (2011): 118–129.

Chatzistavrou, X., T. Zorba, E. Kontonasaki, K. Chrissafis, P. Koidis, and K. M. Paraskevopoulos. Following bioactive glass behavior beyond melting temperature by thermal and optical methods. *physica status solidi (a)* 201, no. 5 (2004): 944–951.

DeSilva, R., P. Pasbakhsh, A.J. Qureshi, A.G. Gibson, and K-L. Goh. Stress transfer and fracture in nanostructured particulate-reinforced chitosan biopolymer composites: Influence of interfacial shear stress and particle slenderness. *Composite Interfaces* 21, no. 9 (2014): 807–818.

Devi, A.V. Gayathri, V. Rajendran, and N. Rajendran. Ultrasonic characterisation of calcium phosphate glasses and glass-ceramics with addition of TiO2. *International Journal of Engineering Science and Technology* 2, no. 6 (2010): 2483–2490.

Devi, K.B., B. Tripathy, A. Roy, B. Lee, P.N. Kumta, S.K. Nandi, and M. Roy. In vitro biodegradation and in vivo biocompatibility of forsterite bio-ceramics: Effects of strontium substitution. *ACS Biomaterials Science & Engineering* 5, no. 2 (2018): 530–543.

Farooq, I., Z. Imran, U. Farooq, A. Leghari, and H. Ali. Bioactive glass: A material for the future. *World Journal of Dentistry* 3, no. 2 (2012): 199–201.

Gheisari, H., E. Karamian, and M. Abdellahi. A novel hydroxyapatite-Hardystonite nanocomposite ceramic. *Ceramics International* 41, no. 4 (2015): 5967–5975.

Ginebra, M-P., M. Espanol, Y. Maazouz, V. Bergez, and D. Pastorino. Bioceramics and bone healing. *EFORT Open Reviews* 3, no. 5 (2018): 173–183.

Gouveia, P.F., J. Mesquita-Guimarães, M.E. Galárraga-Vinueza, J.C.M. Souza, F.S. Silva, M.C. Fredel, A.R. Boccaccini, R. Detsch, and B. Henriques. In-vitro mechanical and biological evaluation of novel zirconia reinforced bioglass scaffolds for bone repair. *Journal of the Mechanical Behavior of Biomedical Materials* 114 (2021): 104–164.

Grishchenko, D.N., E.E. Dmitrieva, A.N. Fedorets, and M.A. Medkov. Bioglass 45S5 doped with zirconium dioxide: Preparation and properties. *Russian Journal of Inorganic Chemistry* 67, no. 1 (2022): 114–122.

Gurusamy, R., and Sankaralingam, P. *Bioactivity of Fluorophosphate Glasses and Method of Making Thereof.* Pandian Bio-Medical Research Centre, India, 2015. WO2015087345A1.

Gurusamy, R., and Sankaralingam, P. *Bioactivity of Silver added Fluorophosphate Glasses and Method of Making Thereof.* Pandian Bio-Medical Research Centre, India, 2015. WO 2015/087344 Al.

Gurusamy, R., and Sankaralingam, P. *Bioactivity of Titanium added Fluorophosphate Glasses and Method of Making Thereof.* Pandian Bio-Medical Research Centre, India, 2015.WO 2015/087347 Al.

Gurusamy, R., and Sankaralingam, P. *Bioactivity of Zirconium added Fluorophosphate Glasses and Method of Making Thereof.* Pandian Bio-Medical Research Centre, India, 2015.WO 2015/087346 Al.

Hayakawa, S., K. Tsuru, and A. Osaka. The microstructure of bioceramics and its analysis. In *Bioceramics and Their Clinical Applications*, pp. 53–77. Woodhead Publishing, 2008.

Hench, L.L., and H.A. Paschall. Direct chemical bond of bioactive glass-ceramic materials to bone and muscle. *Journal of Biomedical Materials Research* 7, no. 3 (1973): 25–42.

Hench, L.L., and I. Thompson. Twenty-first century challenges for biomaterials. *Journal of the Royal Society Interface* 7, no. suppl_4 (2010): S379–S391.

Hench, L.L., and J.R. Jones. Bioactive glasses: Frontiers and challenges. *Frontiers in Bioengineering and Biotechnology* 3 (2015): 194.

Himanshu, T., S.P. Singh, K.A. Sampath, M. Prerna, and J. Ashish. Studies on preparation and characterization of 45S5 bioactive glass doped with (TiO2+ ZrO2) as bioactive ceramic material. *Bioceramics Development and Applications* 6, no. 2 (2016).

Hoppe, A., N.S. Güldal, and A.R. Boccaccini. A review of the biological response to ionic dissolution products from bioactive glasses and glass-ceramics. *Biomaterials* 32, no. 11 (2011): 2757–2774.

Jones, J.R. Review of bioactive glass: From Hench to hybrids. *Acta Biomaterialia* 9, no. 1 (2013): 4457–4486.

Kaur, G., O.P. Pandey, K. Singh, D. Homa, B. Scott, and G. Pickrell. A review of bioactive glasses: Their structure, properties, fabrication and apatite formation. *Journal of Biomedical Materials Research Part A: An Official Journal of The Society for Biomaterials, The Japanese Society for Biomaterials, and The Australian Society for Biomaterials and the Korean Society for Biomaterials* 102, no. 1 (2014): 254–274.

Kohles, S.S., and D.A. Martinez. Elastic and physicochemical relationships within cortical bone. *Journal of Biomedical Materials Research: An Official Journal of The Society for Biomaterials and The Japanese Society for Biomaterials* 49, no. 4 (2000): 479–488.

Koons, G.L., M. Diba, and A.G. Mikos. Materials design for bone-tissue engineering. *Nature Reviews Materials* 5, no. 8 (2020): 584–603.

Kumar, P., B.S. Dehiya, and A. Sindhu. Bioceramics for hard tissue engineering applications: A review. *International Journal of Applied Engineering Research* 13, no. 5 (2018): 2744–2752.

Lee, K., E.A. Silva, and D.J. Mooney. Growth factor delivery-based tissue engineering: General approaches and a review of recent developments. *Journal of the Royal Society Interface* 8, no. 55 (2011): 153–170.

Macon, L.B. Anthony, T.B. Kim, E.M. Valliant, K. Goetschius, R.K. Brow, D.E. Day, A. Hoppe et al. A unified in vitro evaluation for apatite-forming ability of bioactive glasses and their variants. *Journal of Materials Science: Materials in Medicine* 26, no. 2 (2015): 115.

Maeda, H., T. Tamura, and T. Kasuga. Experimental and theoretical investigation of the structural role of titanium oxide in CaO-P2O5-TiO2 invert glass. *The Journal of Physical Chemistry B* 121, no. 21 (2017): 5433–5438.

Mohamed, A.M. An overview of bone cells and their regulating factors of differentiation. *The Malaysian Journal of Medical Sciences: MJMS* 15, no. 1 (2008): 4.

Moulder, J.F. Handbook of X-ray photoelectron spectroscopy. *Physical Electronics* (1995): 230–232.

Nyquist, R.A., and R.O. Kagel. *Handbook of Infrared and Raman Spectra of Inorganic Compounds and Organic Salts: Infrared Spectra of Inorganic Compounds*, vol. 4. Academic Press, 2012.

Olszta, M.J., X. Cheng, S.S. Jee, R. Kumar, Y-Y. Kim, M.J. Kaufman, E.P. Douglas, and L.B. Gower. Bone structure and formation: A new perspective. *Materials Science and Engineering: R: Reports* 58, no. 3–5 (2007): 77–116.

Pantulap, U., M. Arango-Ospina, and A.R. Boccaccini. Bioactive glasses incorporating less-common ions to improve biological and physical properties. *Journal of Materials Science: Materials in Medicine* 33, no. 1 (2022): 1–41.

Prasad, S., V.K. Vyas, K.D. Mani, Md. Ershad, and R. Pyare. Preparation, in-vitro bioactivity and mechanical properties of reinforced 45S5 bioglass composite with HA-ZrO2 powders. (2017).

Rabiee, S.M., N. Nazparvar, M. Azizian, D. Vashaee, and L. Tayebi. Effect of ion substitution on properties of bioactive glasses: A review. *Ceramics International* 41, no. 6 (2015): 7241–7251.

Rahaman, M.N., W. Xiao, and W. Huang. Review-bioactive glass implants for potential application in structural bone repair. *Biomed Glass* 3 (2017): 56–66.

Rai, V.N., B.N. Sekhar, D.M. Phase, and S.K. Deb. Effect of gamma irradiation on the structure and valence state of Nd in phosphate glass. *arXiv preprint arXiv:1406.4686* (2014).

Ranga, N., E. Poonia, S. Jakhar, A.K. Sharma, A. Kumar, S. Devi, and S. Duhan. Enhanced antimicrobial properties of bioactive glass using strontium and silver oxide nanocomposites. *Journal of Asian Ceramic Societies* 7, no. 1 (2019): 75–81.

Ren, F. Zeng, and Y. Leng. Carbonated apatite, type-A or type-B? In *Key Engineering Materials*, vol. 493, pp. 293–297. Trans Tech Publications Ltd, 2012.

Salinas, A.J., P. Esbrit, and M. Vallet-Regí. A tissue engineering approach based on the use of bioceramics for bone repair. *Biomaterials Science* 1, no. 1 (2013): 40–51.

Sankaralingam, P., P. Sakthivel, P.A. Subbiah, A. Periyasamy, J.B. Rahumathullah, and V.C. Thangavel. Fluorophosphate bio-glass for bone tissue engineering: In vitro and in vivo study. *Bioinspired, Biomimetic and Nanobiomaterials* 40, no. XXXX (2021): 1–9. https://doi.org/10.1680/jbibn.21.00025.

Sankaralingam, P.P., P. Sakthivel, P.A. Subbiah, J. Rahumathulla, and V.C. Thangavel. Preparation and physical-biological characterization on titanium doped fluorophosphate nanobioglass: Bone implants. *International Journal of Ceramic Engineering & Science* 3, no. 1 (2021): 37–48.

Siow, K.S., L. Britcher, S. Kumar, and H.J. Griesser. XPS study of sulfur and phosphorus compounds with different oxidation states. *SainsMalaysiana* 47, no. 8 (2018): 1913–1922.

Stanciu, G., S.G. Stanciu, C. Dan, K.M. Paraskevopoulos, X. Chatzistavrou, E. Kontonasaki, and P. Koidis. Surface topography characterization of apatite formation on bioactive glass modified dental ceramics using confocal laser scanning (CLSM) and environmental scanning electron microscopy (ESEM). In *Key Engineering Materials*, vol. 309, pp. 689–692. Trans Tech Publications Ltd, 2006.

Thrivikraman, G., G. Madras, and B. Basu. In vitro/in vivo assessment and mechanisms of toxicity of bioceramic materials and its wear particulates. *RSC Advances* 4, no. 25 (2014): 12763–12781.

Unal, S., F.N. Oktar, M. Mahirogullari, and O. Gunduz. Bone structure and formation: A new perspective. In *Bioceramics*, pp. 175–193. Elsevier, 2021.

Viornery, C., Y. Chevolot, D. Léonard, B-O. Aronsson, P. Péchy, H.J. Mathieu, P. Descouts, and M. Grätzel. Surface modification of titanium with phosphonic acid to improve bone bonding: Characterization by XPS and ToF-SIMS. *Langmuir* 18, no. 7 (2002): 2582–2589.

Wang, Y-C., S-H. Lin, C-S. Chien, J-C. Kung, and C-J. Shih. In vitro bioactivity and antibacterial effects of a silver-containing mesoporous bioactive glass film on the surface of titanium implants. *International Journal of Molecular Sciences* 23, no. 16 (2022): 9291.

Wu, C., and Y. Xiao. Evaluation of the in vitro bioactivity of bioceramics. *Bone and Tissue Regeneration Insights* 2, no. 1 (2009): 25–29.

Xian, W. Module V. Bioceramics: Hydroxyapatite. In *A Laboratory Course in Biomaterials*, pp. 181–206. CRC Press, 2009.

Xie, J.Z., S. Hein, K. Wang, K. Liao, and K.L. Goh. Influence of hydroxyapatite crystallization temperature and concentration on stress transfer in wet-spun nanohydroxyapatite-chitosan composite fibres. *Biomedical Materials* 3, no. 2 (2008): 025014.

Yadav, S.K., V.K. Vyas, S. Ray, Md. Ershad, A. Ali, S. Prasad, M.R. Majhi, and R. Pyare. In vitro bioactivity and mechanical properties of zirconium dioxide doped 1393 bioactive glass. *International Journal of Scientific and Engineering Research* 8, no. 3 (2017): 1321–1330.

Yoder, C.H., N.T. Landes, L.K. Tran, A.K. Smith, and J.D. Pasteris. The relative stabilities of A-and B-type carbonate substitution in apatites synthesized in aqueous solution. *Mineralogical Magazine* 80, no. 6 (2016): 977–983.

Zaman, H.A., S. Sharif, M.H. Idris, and A. Kamarudin. Metallic biomaterials for medical implant applications: A review. In *Applied Mechanics and Materials*, vol. 735, pp. 19–25. Trans Tech Publications Ltd, 2015.

12 Bioceramics for Drug Delivery

Tahrima Binte Rouf
University of Oklahoma

Md Enamul Hoque
Military Institute of Science and Technology (MIST)

12.1 INTRODUCTION

Biomaterials are of great significance in the world of biomedical engineering. The term biomaterials are commonly used to describe the biological materials involved in notable fields of engineering, health, and food. Their conception, research, and evaluation incorporate techniques and knowledge from the fields of biology, medicine, engineering, and science. Medicine has developed through time from being intuition reliant to being evidence based. In this regard, recent predictive medicine movement makes considerable use of data obtained from clinical studies and encourages the adoption of customized medicines. Furthermore, in the last 70 years, the field of biomaterials has seen a remarkable shift from application of inert substances as live tissue substitutes to the development of biodegradable bioactive materials for their replacements. Tissue and organ regeneration are the focus of the present third-generation biomaterials. Due to this quick evolution, several notions have evolved: the primary focus on replacement has switched to restoration, and now regeneration is preferred to restoration. While interacting with biological tissue was not the purpose of first-generation biomaterials, the main intent for fabrication of third-generation biomaterials is interacting with these tissues.

As a result, biomaterial surface parameters, e.g., surface topography, surface charge, and other aspects of surface chemistry, are critical for optimal results whenever these components are embedded amid living tissue (Zhang et al. 2021). Hence, effective surface functionalization is critical to facilitating cell differentiation, adhesion, and proliferation under ideal environment.

In the 1950s, inert ceramics were first utilized to replace broken portions of the human skeleton. Alumina and zirconium, two of the few ceramics utilized for this purpose, were not created exclusively for biological uses. Ceramics underwent a considerable transformation from the 1950s to the early 21st century.

Currently in clinical application, all bioceramics are developed to mend and regenerate human bone. Maxillofacial and orthopedic surgeons use a variety of commercial items to obtain various kinds of bioceramics. These commercial-grade bioceramics can be utilized in accordance with all applicable rules and homologations

DOI: 10.1201/9781003258353-15

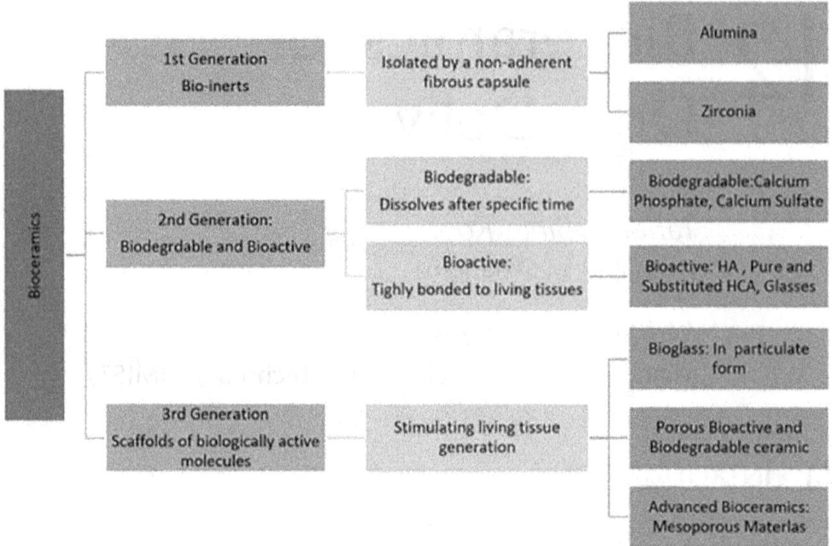

FIGURE 12.1 Bioactivity and biodegradability of three generations of bioceramics.

for this type of prosthesis, addressing actual and specific clinical demands. However, there are other intriguing novel forms of bioceramics that are at the forefront of innovation, specifically developed for a certain function (Vallet-Regi 2014; Arcos et al. 2014). The most common bioceramics utilized for bone healing are shown in Figure 12.1. Bioceramics of the third generation are used to construct scaffolds that support cells participating in the regeneration step. From the standpoint of tissue engineering, these scaffolds need to be biocompatible, mechanically supportive, avoid any unfavorable reactions, and have a deterioration rate comparable to the pace of tissue regeneration. Optimal pore size distribution with interlinked porosity is also necessary, as it promotes metabolite transport, cell and tissue colonization, as well as cell anchoring, while providing a large surface area. These prerequisites can be addressed via cutting-edge techniques like four-dimensional (4D) printing, a promising organ and tissue engineering technology based on multimaterial reprogramming that can change characteristics, function, and form to adapt to changing environmental circumstances. Biocompatible organ and tissue regeneration printing materials must be able to undertake dynamic operations in a physiological setting. In the future, 4D printing might be a useful tool in the biological research of functioning synthetic tissues and organs (Vallet-Regí and Salinas 2019; Vallet-Regí 2001; Vallet-Regí 2010a; Vallet-Regí and Ruiz-Hernández 2011).

12.2 NATURAL CERAMICS IN HUMAN BODY

Figure 12.2 depicts the structure of natural bone. As a multiscale structure system, natural bone can be classified into cancellous bone and cortical bone (Elias et al. 2013). Cortical bone, which is mainly the surface of the bone, contains 99% of calcium

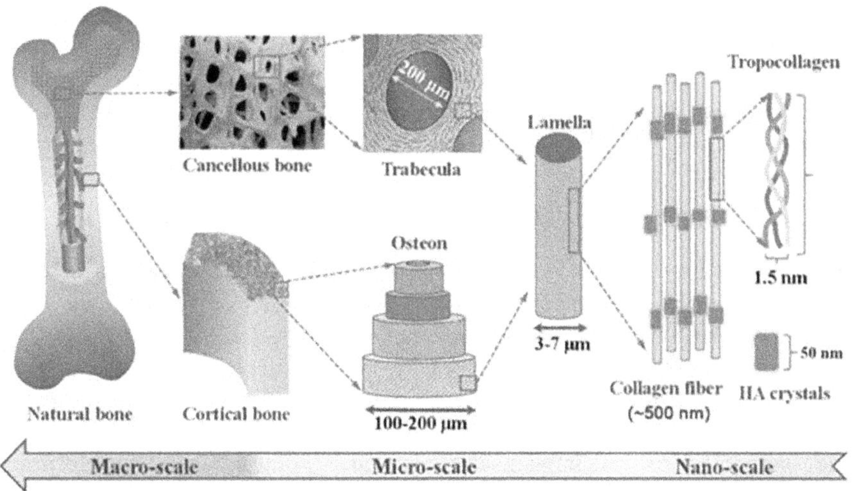

FIGURE 12.2 Multi-scale structural features and chemical composition of natural bone. (Reprinted with permission from Gao et al. (2017).)

and 90% of phosphate in the human body. Cancellous bone is a spongy substance that is dispersed throughout the bone. It is made up of adipose tissue, hematopoietic cells, and blood arteries that are intertwined lamellar trabeculae. All vertebrate species' bones and teeth are made of natural composite materials made up of organic and inorganic components (Andric et al. 2011). The organic elements are mostly tropocollagen-containing collagen fibers, which give the bone its toughness (Yunus Basha et al. 2015). Carbonate hydroxyapatite is the inorganic component of this composite, accounting for around 65% of total bone mass, while organic matter and water accounting for the remaining bulk (Nordin 2013; Glimcher and Lian 1989). The existence of perpetually functioning cells inside the bones qualifies these materials as "living biominerals." Osteoblasts, unique cells that generate and release osteoid, a protein combination primarily composed of type I collagen, initiate bone production processes. The osteoid is then mineralized in the second stage that involves the deposition of calcium phosphate in a controlled manner. The osteoblasts entrapped inside the mineral phase mature into osteocytes that oversee continuous bone formation. At the same time, bone catabolism and destruction are performed by a different type of cell called osteoclasts. Bone creation and disintegration is a continuous process that occurs during the body's development stages, allowing bone development while preserving their structure and homogeneity as well as assuring bone regeneration in the event of a fracture. This process also serves as a retention and transportation route for two critical components, calcium and phosphorous. An identical mechanism occurs in the teeth, except for the presence of an exterior surface layer, enamel, which distinguishes it from bone tissue. Dental enamel has a substantially higher inorganic content (up to 90%) than bone and is made up of massive, strongly aligned prismatic crystals. As a result, bone and dentine (which have similar properties) have considerably different crystallinity and carbonate concentration than enamel.

These traits are connected to enamel's mechanical properties, which make it the most durable and resistant biological component. Unlike bone tissue, adult dental enamel undergoes irreversible degradation and does not include cells. This trait emphasizes the importance of enamel-biocompatible materials in the treatment of tooth decay (Lee and Glimcher 1991; Vallet-Regí and González-Calbet 2004).

12.3 ARTIFICIAL CERAMICS FOR BONE REPLACEMENT

Glasses, glass ceramics, and calcium phosphates are the three types of ceramics employed as beginning components in bioactive composites (Manzano and Vallet-Regí 2020). The purpose of the resultant composites, which blend double or multiple components, is to increase the bioactive activity and/or mechanical responsiveness. For example, calcium phosphates are mixed with other inorganic salts to make bone cements. Another objective of this research is to establish shaping processes that will allow researchers to create implants in a specified size or form, with a specific porosity, based on the ceramic implant's intended usage. Implants intended for nanoapatite (a forerunner of bone formation) creation must be constructed as a porous component, with a specific level of macropores, to allow bone angiogenesis and oxygenation. Failure to adhere to this criterion during implant design phase results in no chemical reaction for inert material implants and limited external implant surface reaction for bioactive ceramic implants. In both types of implants, the solid inner core will help meet the bone replacement requirement, but it is incapable of bone-regenerative function. Inert ceramics' absence of regeneration properties supports their manufacturing in thick and solid forms, as shown in alumina and zirconia femoral head implants (Vallet-Regí et al. 2008). Figure 12.3 depicts the many bioceramics that can be utilized in the skeletal system, based on the role they must perform.

Template glasses (Salinas and Vallet-Regí 2016), star gels (Manzano et al. 2006), silica-based ordered mesoporous materials (Vallet-Regí 2010b), and organic-inorganic hybrids (Vallet-Regí et al. 2011) have all been proposed as bioceramics. Their potential for usage as bone graft substitutes materials is now being investigated.

Macroporous bioceramics with several microns pore sizes work well as tissue engineering scaffolds. Bioceramics surface topography is critical in the creation of artificial bone's complex structure (Díaz-Rodríguez et al. 2014). Figure 12.4 depicts a hierarchical macro-, micro-, and micro-/nanoporous structure of β-tricalcium phosphate (β-TCP) scaffolds with adjustable morphology created using a combination of in situ growing crystal method and digital light processing (DLP) printing. The SEM images show that the scaffolds have a three-dimensional pore structure with pore sizes ranging from 200 to 300 μm. This porous structure is better for bone tissue repair. The scaffold has a white surface and a well-constructed macroporous form.

Bioceramics with mesopores (2–50 nm pore diameters) are particularly suited to applications involving the loading and delivery of pharmaceuticals or biologically active compounds to assist bone regeneration. The regulated distribution capabilities of silica mesoporous bioceramics, together with their increased bioactive activity, make them promising candidates for fabricating 3D hierarchical structured bone tissue engineering scaffolds.

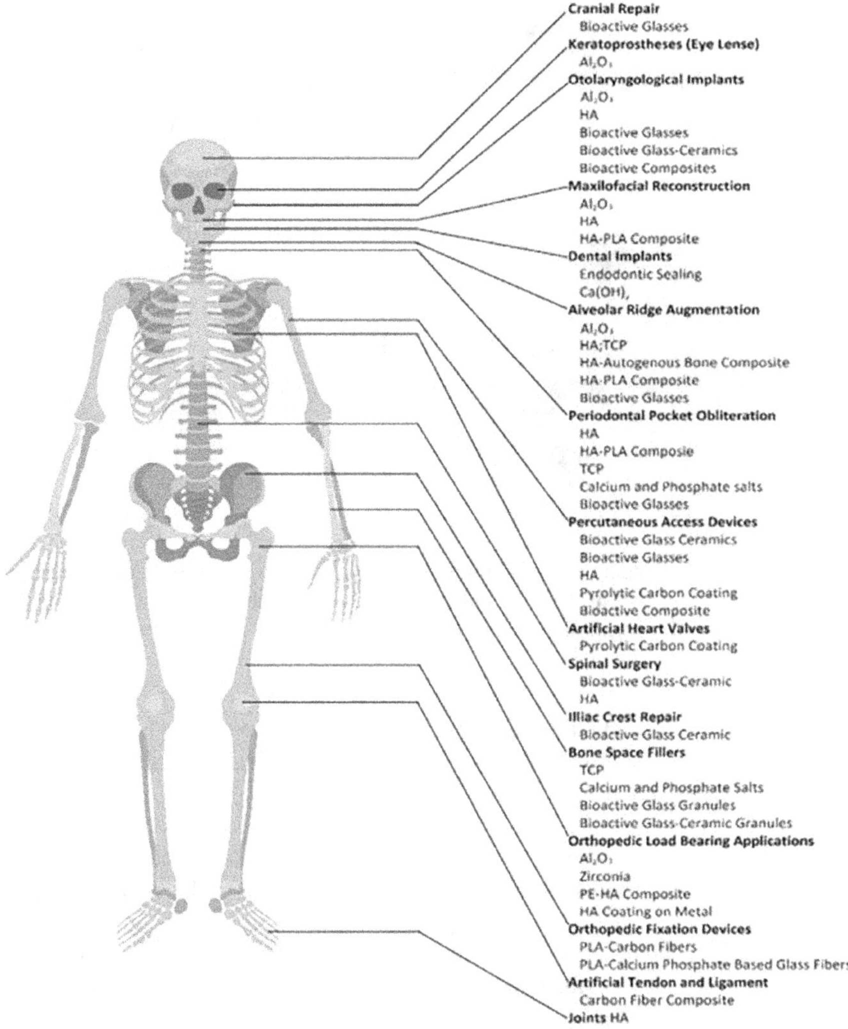

FIGURE 12.3 Clinical application of bioceramics depending on the role they serve in the skeletal system. (HA, Hydroxiapatite; PLA, polylactic Acid; TCP, Tricalcium Phosphate; PE, Polyethylene). (Dorozhkin 2018)

Ordered mesoporous ceramics were developed and explored for the catalysis sector, but due to their composition and porosity, they have found a wide range of applications in the biomedical field, including tissue engineering and drug delivery. Specifically, their importance in biomedicine arises from two distinguishing characteristics:

- Surface properties. The material constitutes of a silica network with silanol groups on the outside and is considered bioactive. As a result, mesoporous silica materials show similar behavior as bio glasses in that, when

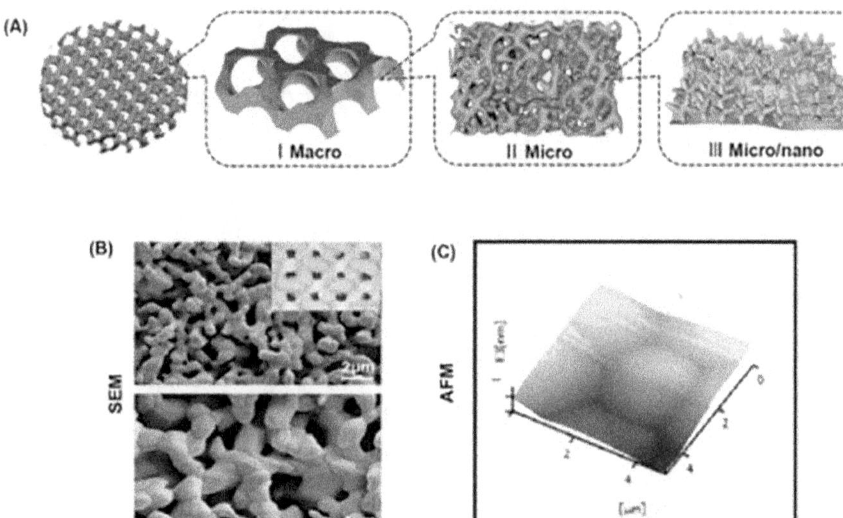

FIGURE 12.4 (A) Scheme of the hierarchical porous structure: macro, micro, and micro/nano. (B) SEM images of β-TCP scaffolds. (C) AFM images showing the surface topographies. (Reprinted with permission from Zhang et al. (2021).)

submerged in body fluid, an apatite-like layer resembling natural bone forms on their surface. For this reason, it is a material of interest for bone tissue regeneration.

- Textural characteristics. Different compounds, like medicine or biologically active species, can be placed into the pores that make up this organized mesoporosity, which can then be employed as local drug-delivery system.

The first reports of ordered mesoporous ceramics appeared in the 1990s, when researchers were looking for materials with wider pores than zeolites to increase their potential applications as catalyst supports, catalysts, and adsorbents. Surfactants were used as structure guiding agents by researchers (Beck et al. 1992) to create new types of materials, for example, the M41S and KSW-n families of mesostructured materials. Researchers (Horcajada et al. 2004; Vallet-Regi et al. 2000) found that these materials created new avenues of bone regeneration and drug-delivery application. Many studies have been conducted since then (Guo et al. 2021; Shukrun Farrell et al. 2020; Carmona et al. 2014). Bioceramics are available in nature in both porous and dense forms. On the other hand, inert ceramics are made using traditional processes that date back to the 1980s. Ceramics are traditionally made at high pressures and temperatures. These parameters are insufficient when trying to make nanometer scale biomimetic materials. Wet route synthesis techniques are now favored for creating silica mesoporous materials with concurrent existence of meso-, micro-, and nanoporosities as well as bioactive bioceramics designed as tissue engineering scaffolds (Philippart et al. 2017; García-Alvarez et al. 2017).

12.4 DRUG DELIVERY FROM SILICA MESOPOROUS NANOPARTICLES

Protein nanoparticles, lipid-based nanomedicines and liposomes, polymer-drug conjugates, polymeric nanoparticles and micelles are the most popular among the many distinct nanoparticles suggested as nanomedicines (Vallet-Regi et al. 2018; Sun et al. 2021). Mesoporous silica nanoparticles (MSNs) have been widely explored as drug-delivery nanocarriers due to their physicochemical properties (Vallet-Regi et al. 2000; Paris et al. 2017; Baeza et al. 2016). MSNs' thermal, chemical, and mechanical stability make them extremely durable. Their large surface area (ca. $1{,}000\,m^2/g$) narrow distribution of adjustable pore sizes (2–30 nm) and high pore volume (ca. $1\,cm^3/g$) contribute to their high loading capacity inside the porous system (Vallet-Regí et al. 2010a). Owing to these exceptional features, MSNs have been used in many biomedical fields such as nanosystems for gene transfection (González et al. 2011), nanosystems for diagnosis (Simmchen et al. 2012), and drug-delivery systems (Vallet-Regí et al. 2011).

As previously stated, mesoporous silica has several unique textural characteristics. It will now be discussed how they affect the loading and release kinetics of physiologically active drugs.

Pore diameter is a size-selective adsorption characteristic that also influences the release rate. Furthermore, molecule adsorption is mostly a surface phenomenon. As a result, the controlling characteristic in molecule adsorption is the effective surface area. Pore volume is the most important metric when it comes to confining extremely large molecules (e.g., proteins). When the drug chemistry allows, MSNs can be loaded with drug suspension using evaporation, diffusion, or suspension in a melting liquid drug (e.g., with ibuprofen) (Limnell et al. 2011). The self-assembly of an antibacterial drug (octenidine dihydrochloride, OCT) with silica to produce an ordered drug/MSN nanocomposite has also been investigated (Stewart et al. 2018). In Figure 12.5a and b, SEM and TEM micrographs of antimicrobial drug OCT-loaded mesoporous silica nanocomposite (OCT-MSN) sphere show clearly structured and segregated particles with a diameter of 424 nm. The rough, porous structure of post-drug-release OCT-MSNs (SEM in c, TEM in d) is seen with pores well resolved throughout the particle. In another study, MSN nanospheres with nonordered pore structure and fibrous morphology have shown great potential for anticancer drug delivery (AbouAitah and Lojkowski 2021).

The key determinant of release kinetics is mesoporous silica walls' organic functionalization with various organic groups. By enhancing host–guest interactions, this functionalization may also increase molecule adsorption. Silica mesoporous materials may also be generated in nanoparticle form using improved synthesis processes, bringing up new potential in medical applications. A revised method of the Stöber technique is followed for the fabrication of MSNs, which involves sol-gel mechanism under very diluted conditions to produce nanoparticles. The silica precursors condense over a surfactant template, which is then removed to produce a network of mesoporous cavities.

Most drug-delivery nanoparticles have been developed to treat complicated illnesses like cancer. When it comes to cancer therapy, the usage of nanoparticles

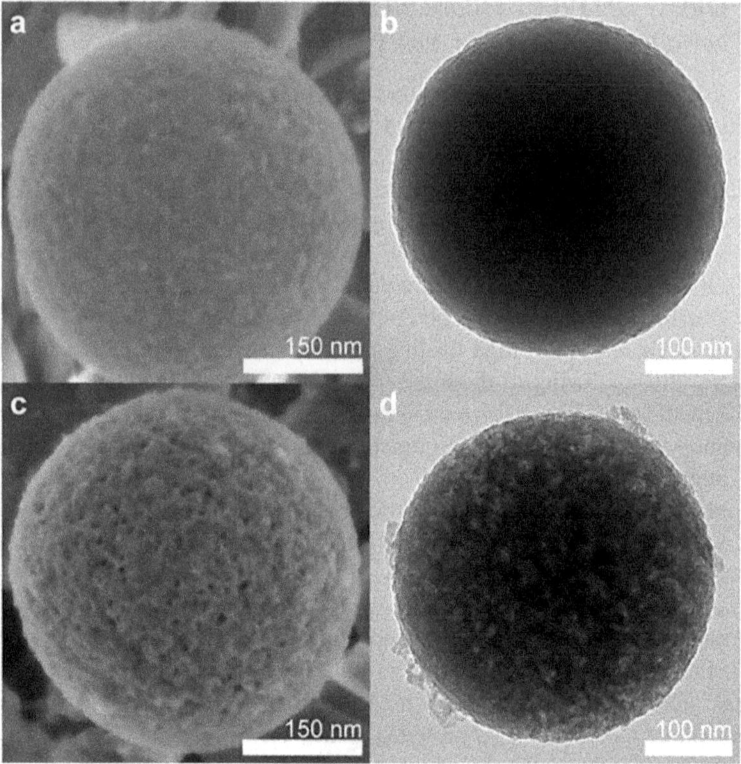

FIGURE 12.5 SEM and TEM micrographs of drug-templated MSNs before drug release (a and b, respectively) show well-defined and separated particles approximately 424 nm in diameter. Post-drug-release OCT-MSNs (SEM in c, TEM in d) show very clear surface pore openings and well-resolved pores throughout the particle. (Reprinted with permission from Stewart, Finer, and Hatton (2018).)

equipped with therapeutic medicines must guarantee that the nanoparticles reach the damaged region and discharge their medicines appropriately. The CAPIR cascade comprises five basic phases in the drug-delivery pathway to a solid tumor: "circulation in the blood, accumulation and penetration into tumors, cellular internalization, and intracellular release" (Pelaz et al. 2017). Several methods may be utilized to ensure a prolonged circulation period in the bloodstream; for example, extravasation can be prevented with near 100 nm nanoparticles size; moreover, blood components and reticuloendothelial system (RES) contact can be avoided with surface modification. Surface modification of nanoparticles allows them to be targeted at tumors in this approach. In comparison to other nanocarriers, MSNs are easy to functionalize because they can carry a variety of grafting reactions using various organic solvents, while also enduring elevated temperatures and numerous organosilane functionalizing agents. When their surface is positively charged, MSNs have been seen to permeate into several cell types (Paris et al. 2016).

The most difficult task for all forms of nanocarriers is to identify specific targets within the body effectively. Its significance stems from the fact that most traditional medications do not discriminate between healthy and malignant cells; therefore, by releasing and depositing these transported compounds in tumors rather than healthy cells via a selective nanocarrier might solve this problem. Following the groundbreaking work on MSNs (Manzano and Vallet-Regí 2018; Vallet-Regí et al. 2018), various research groups have looked into the topic of MSNs as drug-delivery carriers. Recently, the field of nanomedicine has seen a steady growth in the number of studies presenting MSNs as stimuli-responsive drug-delivery systems (Vallet-Regí and Ruiz-Hernández 2011). Despite the fact that various research on MSNs for drug delivery is available, staying current in such a dynamic and evolving area necessitates frequent updates.

Ceramic nanoparticles are critical in this context since each innovation in nanosystems with medicinal applications has posed new obstacles in the development of smart materials that can meet the ever-changing clinical demands (Chauhan et al. 2022). From a physiological standpoint, a common objective in medicine is to discover the optimal route to provide curative drugs. In many circumstances, the quantities suggested today are overly high, yet they are necessary to guarantee that the afflicted region receives the minimal dose required. Most of the dosage given to the patient is distributed throughout the body, impacting nontargeted areas. The cytotoxicity of the drugs to be used complicates the decision-making process in cancer treatments, where the benefit-risk balance of chemotherapy presents a difficulty. In order to solve this problem, researchers are now investigating localized and targeted medication release. In a recent study by Igaz et al., fluorescently labeled drug-delivery system based on MSNs was designed and tested where the monitoring of the intracellular presence and intracellular localization of the mesopore-released cargo ensured efficient delivery of mitomycin C (MMC)-loaded RhoB@MSNs (RhoB@MMC@MSNs) to multidrug-resistant cancer cells via MSNs (Igaz et al. 2022).

The greatest advantage of many nano- or microparticles, like MSNs, is their multifunctionality. Various functions, such as drug loading and subsequent release, attaching of biomolecules like nucleic acids or proteins to the outer membrane of the particle and facing curative targets, attaching of fluorophores or active complexes for magnetic resonance imaging (MRI) for optical monitoring, coating with polymers or gold, can all be accomplished at the same time. This adaptable, smart nanocarriers are intriguing candidates for usage in the clinic in the foreseeable future to resolve some of the difficulties of traditional therapy (González et al. 2018).

The silanol groups on the surface of MSNs make it relatively easy to chemically functionalize their surface, enabling the fabrication of multifunctional platforms with a wide range of properties. These characteristics of MSNs prompted numerous research groups to try encapsulating various therapeutic drugs in MSNs for delivery to the targeted site (Vallet-Regí and Ruiz-Hernández 2011; Li et al. 2012; Mamaeva et al. 2013). Many distinct mammalian cell lines have effectively endocytosed MSNs (Lu et al. 2009; Chen et al. 2013), and these nanoparticles are successfully absorbed in vitro at doses of less than 100 g/mL, according to toxicological testing (Hudson et al. 2008).

MSNs at doses below 200 mg/kg have shown satisfactory tolerance during in vivo biocompatibility tests on numerous animal models (Lu et al. 2010). As main administration route for MSN is through intravenous injection into the bloodstream, researchers have looked into their hemocompatibility, with promising findings (Zhao et al. 2011; Joglekar et al. 2013).

First-generation MSNs have proven to be highly beneficial for biomedical applications (Croissant et al. 2018). The goal for the second-generation MSNs, on the other hand, needs to be an efficient transition from experimental proof of concept to the clinic, based on favorable and consistent treatment outcomes. In this regard, before enrolling in clinical trials, pharmacodynamic and pharmacokinetic assessments, as well as biodistribution studies should clearly demonstrate, effectiveness and absence of toxicity (Prabhakar et al. 2013).

12.4.1 Stimuli-Responsive Drug Delivery

MSNs' open porosity allows therapeutic drugs to be loaded into their interconnected system of cavities, but it also allows those agents to be released when in solution, depending on the type of drug and solvent. As a result, effective pore closure is required to prevent premature discharge of the loaded drug while passing through the blood vessels, as an imprecise discharge of the medications might result in a variety of adverse effects (Mi 2020).

To block the pore openings, many options might be considered; arguably the most frequent is grafting stimuli sensitive valves to the pore openings (Castillo et al. 2018) or wrapping the entire nanoparticle with a divisible shell that allows the release to be triggered when it is removed. The closed pore entrance that may be opened based on a specific stimulus would be the end outcome in both situations. Internal triggering stimuli, which are often linked to a specific characteristics of the studied pathology, such as enzyme concentration, redox potential, and variations in pH, among others, or external triggering stimuli, which are applied remotely by the medical practitioner and consist of ultrasounds, light, electrical fields, and magnetic fields (Mura et al. 2013; Torchilin 2014; Baeza et al. 2015; Castillo et al. 2017) (Figure 12.6). Previous studies have concentrated on stimulus-responsive mesoporous silica in both bulk and nanoparticle form, including gate-like structures on pore entrances (Nadrah et al. 2014).

12.4.1.1 Magnetic Field Responsive Drug Delivery

Magnetic field is one of the most utilized stimuli in nanomedicine because of its dual effect. As shown in Figure 12.7, it may be applied to direct nanoparticles while adopting a constant magnetic field or to elevate the intrinsic temperature locally while utilizing an alternating magnetic field (Guisasola et al. 2018a, b). For magnetically controlled drug delivery, superparamagnetic microspheres having a mesoporous silica shell and Fe_3O_4@SiO_2 core were fabricated with high pore volume and magnetism (Deng et al. 2008). Similarly, for encapsulating ibuprofen, spheres with a magnetic Fe_3O_4/Fe core, homogeneous particle diameter of 270 nm, and a mesoporous silica shell were manufactured (Zhao et al. 2005). Integrating superparamagnetic iron oxide nanoparticles (ca. 5–10 nm) inside MSNs networks during their manufacturing

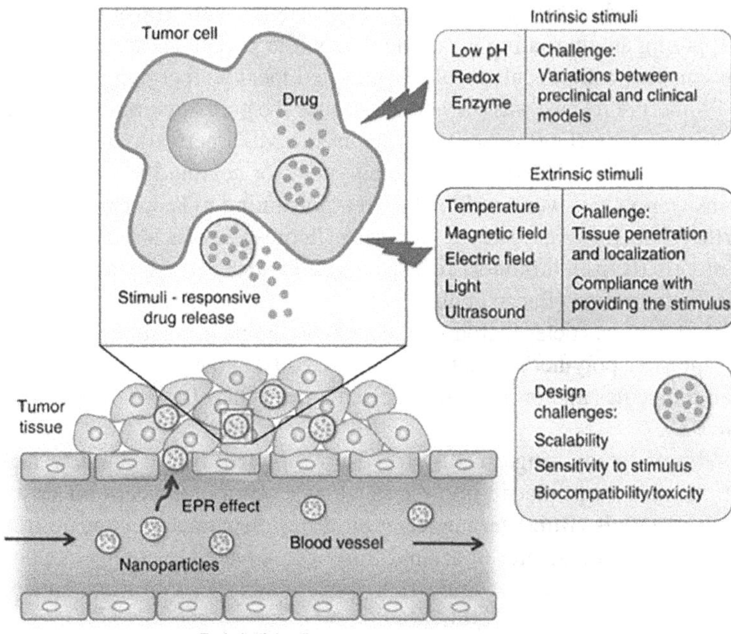

FIGURE 12.6 Schematic diagram summarizing various exogenous and endogenous stimuli for drug delivery. (Reprinted with permission from Rosenblum et al. (2018).)

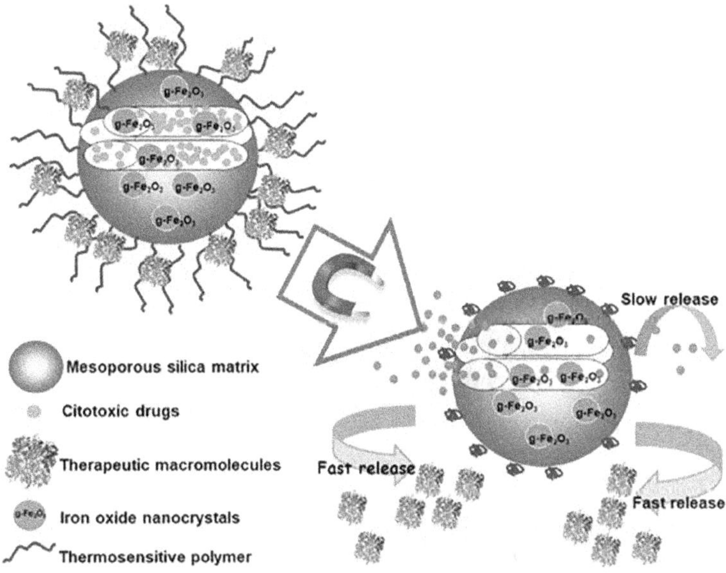

FIGURE 12.7 Mechanisms of drug release from magnetic field responsive DDS, which are guided by the magnetic field and magnetic hyperthermia. (Reprinted with permission from Vallet-Regí and Ruiz-Hernández (2011) and Baeza et al. (2012).)

is the most favored technique for using nanoparticles with MSNs (Guisasola et al. 2015). Applying an alternating magnetic field while pore holes are closed with temperature sensitive components results in elevated local temperature, prompting load release. Premature release was prevented because the pore openings were blocked by enclosing iron oxide (Fe_3O_4) nanoparticles in the MSN network and using poly(N-isopropylacrylamide), a thermosensitive polymer, for coating the surface. The thermoresponsive polymer was combined with a polyamine to keep proteins within the nanoparticles' shells (Figure 12.7). The drug (fluorescein) enclosed inside the pores and the protein (trypsin inhibitor) kept in the shell may both be released concurrently with this configuration (Baeza et al. 2012).

The Fe_3O_4 nanoparticles increased the ambient temperature to the point where the thermoresponsive polymer's structure changed, causing the pore openings to expand and small molecules and proteins to be discharged with different kinetics, as shown in Figure 12.7.

In a different study, single DNA strand was applied to functionalize the MSNs and then the cargo was placed into the pores as a proof of concept for this class of responsive materials (Ruiz-Hernández et al. 2011). Separately, magnetic iron oxide nanoparticles with a diameter of around 5 nm were bonded to complementary DNA sequences. Then, as shown in Figure 12.8, MSNs and DNA-iron oxide nanoparticles were combined to facilitate DNA hybridization.

Because the DNA sequence used had a melting temperature of 47°C, the Fe_3O_4 nanoparticles enclosed in the MSNs network raised the ambient temperature once alternating magnetic field was applied. As a result, the double-stranded DNA melted,

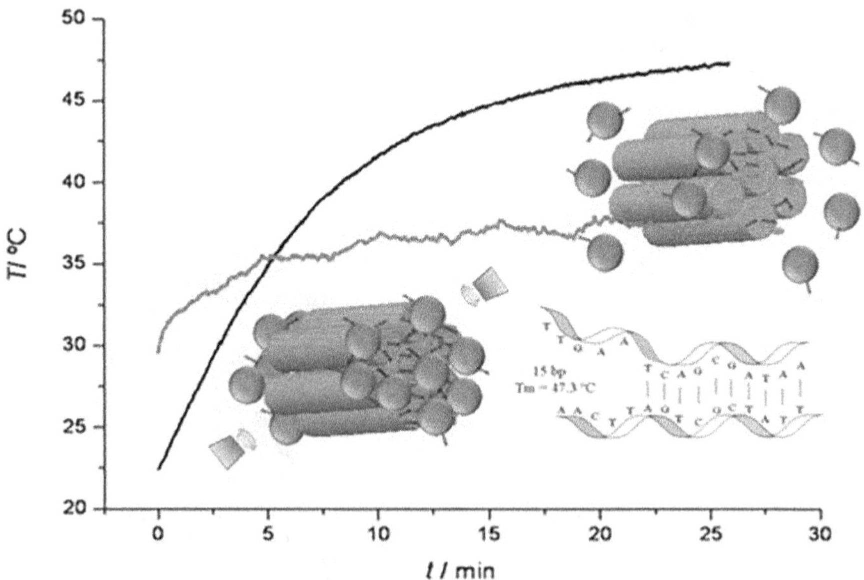

FIGURE 12.8 Pulsatile release from MSNs in response to magnetic fields. (Reprinted with permission from Ruiz-Hernández, Baeza, and Vallet-Regí (2011).)

causing the pore opening and cargo release. The reversibility of DNA linking is the most surprising feature of this proof of concept: When the magnetic field is removed and the temperature of the system drops, the DNA strands hybridize again, closing the pores. Upon reapplication of magnetic field, the pores reopened, resulting in an on-and-off release mechanism.

Superparamagnetic Fe_3O_4 nanocaps have been applied in closing of mesoporous silica nanorods: the load release is triggered by the antioxidants generated by the cells when they are exposed to an external magnetic field.

12.4.1.2 Light-Responsive Drug Delivery

In recent years, the application of various wavelengths of light (visible, near infrared, or ultraviolet) to trigger load release from MSNs has grown increasingly common. The application of light to promote MSN discharge is simple and may be directed on the specific area by the medical practitioner. Due to its bond-breaking strength, UV is the most common wavelength for stimulating load release from MSNs (Mal et al. 2003). MSNs were encapsulated with a protein shell that had a photosensitive surface that could be disrupted by 366 nm light. As a proof of concept, transferrin was used to functionalize the shell around the nanoparticles (Martínez-Carmona et al. 2015). As a result, UV light will enable the drug to be released after the MSNs have been absorbed into the tumor cells; this is a good therapy option for light-sensitive cancers like melanomas.

The high intensity of UV light, in addition to its poor penetration, is a possible disadvantage. Using visible light as an alternative is a viable option as it is safer and has a better tissue penetrability.

12.4.1.3 Ultrasound-Sensitive Drug Delivery

Because of its pain-free and deep penetration into living tissues, ultrasound is a particularly appealing stimulus for application in nanomedicine (Mo et al. 2012; Paris et al. 2018a). A drug-delivery system for ultrasound-sensitive MSNs has been devised, in which the drug-delivery system (DDS) is triggered using commercial-grade ultrasound equipment commonly found in recovery centers (Paris et al. 2018b, c). To seal the pores and prevent premature load release, this approach uses a copolymer comprised of ultrasound-sensitive and thermosensitive components to adorn the MSNs surface (Paris et al. 2015). When ultrasound triggers MSNs, it cleaves a section of the copolymer, causing hydrophobicity changes which results in copolymer conformation change at physiological temperatures, allowing the MSNs pores to open and the cargo to be released.

MSNs have also been used to construct transdermal medication delivery devices that are temperature and ultrasound stimuli responsive at the same time (Anirudhan and Nair 2018). Following the technique outlined above, these MSNs were adorned with an ultrasound-sensitive polymer.

Ultrasound was used as an imaging and guiding vector for Au nanoparticle-coated PEGylated MSNs encapsulated in perfluoro hexane (Wang et al. 2013). In this setup, ultrasound radiation efficiently initiated drug discharge while stimulating contrast-intensified ultrasound imaging as well as increasing ablation effectiveness owing to ultrasound guiding.

12.4.1.4 pH-Sensitive Drug Delivery

Internal stimuli make use of specific characteristics of the condition being treated. Because the pH of most tumors is less than that of healthy tissues, pH is one of the most used triggers for drug release from nanomedicines. This discrepancy is related to cancer cells' high glycolysis rate, which results in a large generation of lactic acid and a drop in pH. Various methods have been explored to take advantage of these pH changes (Martínez-Carmona et al. 2018); most of them are focused on restricting the pore openings of MSNs with various components attached to the nanoparticles' surface employing responsive linkers, such as imine bonds (Gao et al. 2012), boronate ester (Gan et al. 2011), acetal linkers (Liu et al. 2010), aromatic amines (Meng et al. 2010), ferrocenyl linkers (Xu et al. 2013), or acidic pH soluble calcium phosphates (Rim et al. 2011). Researchers created a pH-sensitive MSN-based approach for transportation and distribution of topotecan, a strong cytotoxic drug that frequently degrades at physiological pH (Martínez-Carmona et al. 2016). To target the nanocarriers onto tumor cells that overexpress folic acid receptors, the exterior surface of the topotecan-loaded MSNs was studded with a gelatin sensitive to acidic pH, followed by the addition of folic acid components on the external surface. In keeping with the pH-sensitive MSNs trend, another proof of concept was produced employing self-immolative polymers (SIPs) to adorn the nanoparticles' exterior surface (Gisbert-Garzaran et al. 2017). SIPs are made up of a polyurethane chain with an acidic pH-sensitive molecule at one end and a linear polymer at the other. This molecule is cleaved, causing the polymer to disintegrate from head to tail, releasing the initial monomers in a process known as self-immolation. MSNs containing a model molecule were covered with a SIP that was pH stable, limiting premature release on healthy tissues. The polymer was successfully decomposed when the system was subjected to acidic pH, opening the pores and activating load discharge in the acidic environment.

12.4.2 MSNs Loaded with Prodrugs

The potential therapy of cancer is undoubtedly the most significant function of nanoparticles for drug delivery. However, because the delivered load in such a medical therapy is usually a highly cytotoxic chemical, it is vital to prevent untimely discharge before accessing the tumor tissue. This is the primary goal for stimuli-responsive systems. However, designing entirely safe "zero release" systems, in which no release happens prior to approaching the target location, is exceedingly challenging. Another option is to inject dormant cytotoxic chemicals into the nanocarriers and then activate them once they reach the tumor tissue, resulting in extremely toxic compounds being formed solely in the targeted tissue. This method was created using nanoparticles that can carry a prodrug capable of being triggered by enzymes overexpressed in tumor tissue (Bildstein et al. 2011). The technique's main drawbacks include the lack of preexisting enzymes necessary to trigger particular prodrugs as well as the low quantity of activation enzymes in the tumor.

MSNs have been utilized in this paradigm to create a system that can produce lethal pharmaceuticals in situ within tumor cells, by conveying both the nontoxic prodrug, which is inserted in the pore structure, and the enzyme required for drug

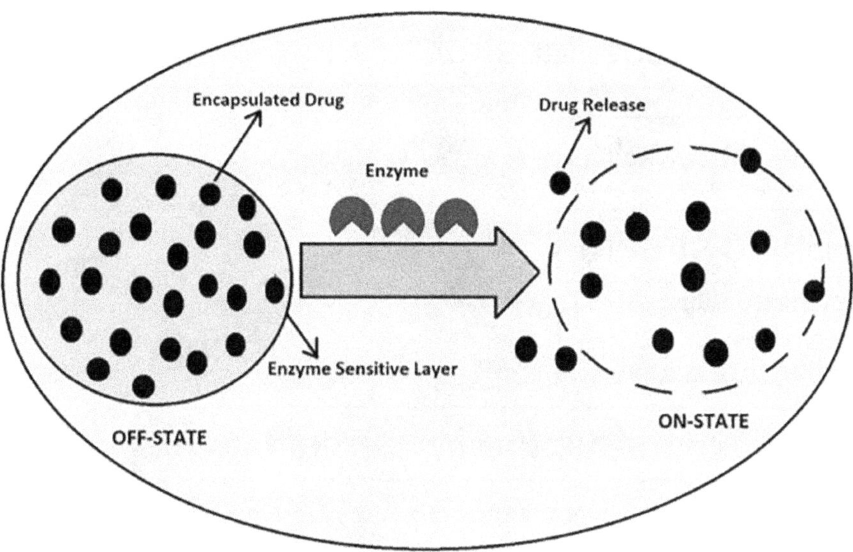

FIGURE 12.9 Enzyme responsive drug delivery. (Raza et al. 2019)

stimulation, which is implanted to the surface of the MSNs (Baeza et al. 2014). Researchers investigated a new nanocarrier made of functionalized mesoporous silica nanoparticles that can carry the nontoxic prodrug indol-3-acetic acid as well as the enzyme horseradish peroxidase that activates it. Horseradish peroxidase enzyme grafted on the MSNs' surface may convert the encapsulating prodrug indol-3-acetic acid to indole-3-carbinol, creating harmful chemicals including hydroxyl and reactive oxygen species. Polymeric capsule coating of enzyme helps maintains their stability. Figure 12.9 illustrates a simplified work mechanism of enzyme-responsive drug-delivery system.

12.4.3 Selective Targeting for Drug Delivery

When MSNs are considered for nanomedicine, the localization of nanocarriers into the target tissues is of utmost importance. In cancer treatments, this becomes even more important, as to avert side effects and harm to healthy cells. Owing to increased penetration and passive targeting or retention (EPR) effect, MSNs, or nanoparticles in general, tend to collect in solid tumors after being injected into the bloodstream (Matsumura and Maeda 1986; Nakamura et al. 2015). The process is based on anomalies found in tumor blood arteries, such as large interendothelial junctions, transendothelial holes with dimeters of several hundred nanometers, and a high number of fenestrations (Figure 12.10). Nanoparticles in the circulation will extravasate through the formerly indicated fenestrations in tumor arteries and concentrate in the tumor interstitium in this scenario. Moreover, due to the inadequate lymphatic drainage caused by the tumor's rapid development, those nanoparticles currently within the tumor will likely stay there.

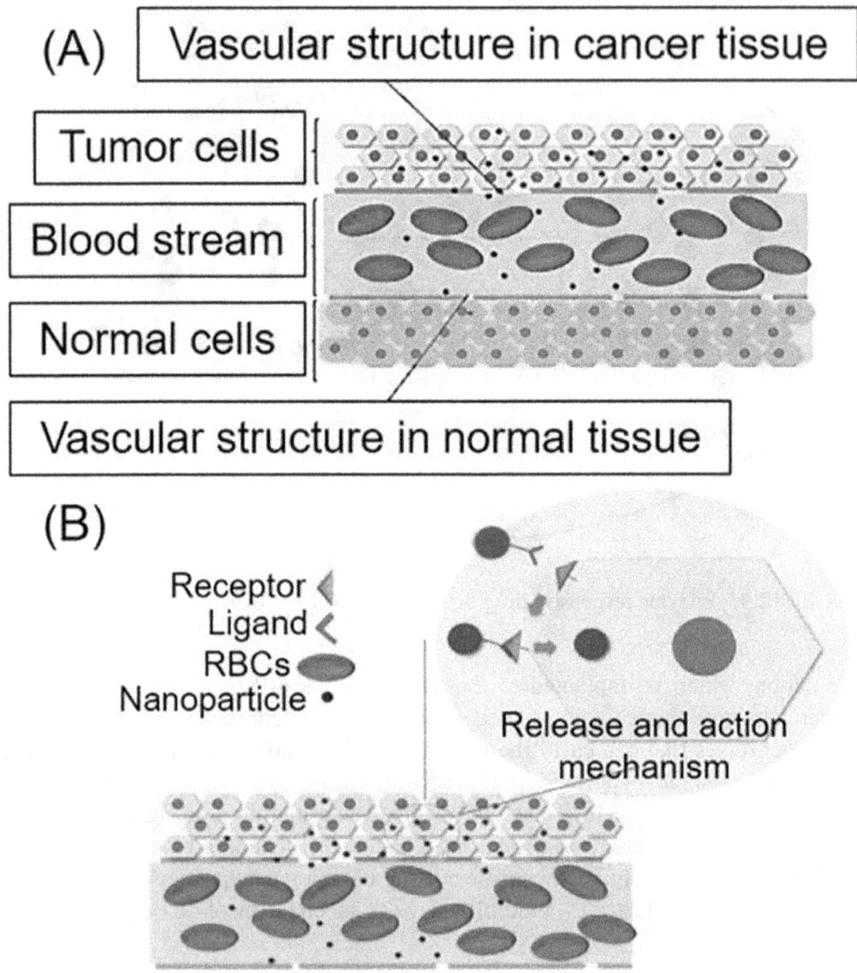

FIGURE 12.10 Scheme illustrating the passive targeting (EPR) and the active targeting into a tumor. (Reprinted with permission from Attia et al. (2019).)

Meanwhile, active targeting makes use of the fact that certain tumor cells have an overabundance of specific receptors on its exterior. The selective retaining and absorption of nanoparticles by cancer cells can be improved if they are functionalized with ligands that have a high affinity for certain receptors. This method is particularly intriguing since tumor aggregates are made up of very diverse tissues including a wide range of cell types. As a result, nanoparticles need to discriminate between tumor cells and nontumor cells present in the neoplastic tissue. Various methods have been studied to functionalize the surface of MSNs with particular ligands that are capable of selectively interacting with specific cellular receptors overexpressed in tumor cells. Transferrin (Martínez-Carmona et al. 2017; Fang et al. 2014), transactivator of transcription peptides (Li et al. 2014), methotrexate (Rosenholm et al. 2010), epidermal growth factor (Mickler et al. 2012), folic acid (Martínez-Carmona et al. 2016; Porta et al. 2013; López et al. 2017), antiepidermal growth factor receptor (Tsai et al. 2009), antiherceptin (Milgroom et al. 2014), and interleukin-13 peptide (Wang et al. 2013) are some examples of targeted ligands attached on MSN.

When it comes to cancer treatment using nanoparticles, there is one more issue to consider: the nanoparticles' low penetration throughout the tumor mass. This is caused by the collagen in the tumor mass's extracellular matrix, which enhances its density as compared to healthy tissue matrices. Nanoparticle penetration into the tumor mass is severely hampered by this high density. A proteolytic enzyme collagenase, which can dissolve the collagen-rich extracellular matrix, was decorated on the surface of MSNs as a viable remedy for this high-density matrix (Villegas et al. 2015). This enzyme was encapsulated in pH-sensitive polymeric capsules to prevent early biodegradation before reaching the tumor tissue. As a result, the nanocapsules would break down in the acidic environments of tumors and release collagen that dissolve the tumor's extracellular matrix, enhancing the MSNs' penetrability deep into the tumor mass.

12.5 CONCLUSIONS AND FUTURE DIRECTIONS

In the last 70 years, biomaterials have advanced at a rapid pace that shows no signs of slowing down anytime soon. Intensive efforts in regenerative medicine have been made over the last three decades to induce autonomous regeneration of an injured limb in the body, which has already been proven in other species – like the salamander – though not in humans. Future breakthroughs in biomaterials will necessitate concurrent research of all four size scales: MACRO, MICRO, NANO, and PICO, while cell and molecular biology will provide therapeutic answers. In this regard, the permeability of biomaterials needs to be investigated at all scales to completely comprehend their performance and to provide novel answers to specific problems. There are a few major obstacles that must be overcome before clinical translation can occur. To achieve reproducibility in the synthesis of MSNs, it is crucial to standardize the production protocols; it is also crucial that the produced nanoparticles exhibit the appropriate dispersibility and stability, and standardization of all surface functionalization method needs to be ensured prior to entering the clinic. More significantly, more biodistribution research of MSNs on various animal models must be executed in order to determine MSNs' ultimate fate. Innovative and upcoming technology will give new solutions, and the usage of cell-free organ scaffolds might be the solution to numerous issues in the future. From a broad view, there has been a significant advancement in the design and development of bioceramics, especially MSNs for biological applications. Thanks to discoveries in molecular and cell biology, we are, without a doubt, on the right track.

REFERENCES

AbouAitah, K., and W. Lojkowski. 2021. Delivery of natural agents by means of mesoporous silica nanospheres as a promising anticancer strategy. *Pharmaceutics* 13 (2). Multidisciplinary Digital Publishing Institute: 143. doi:10.3390/pharmaceutics13020143.

Andric, T., A.C. Sampson, and J.W. Freeman. 2011. Fabrication and characterization of electrospun osteon mimicking scaffolds for bone tissue engineering. *Materials Science and Engineering: C, Novel Structures for Tissue Engineering* 31 (1): 2–8. doi:10.1016/j.msec.2010.10.001.

Anirudhan, T.S., and A.S. Nair. 2018. Temperature and ultrasound sensitive gatekeepers for the controlled release of chemotherapeutic drugs from mesoporous silica nanoparticles. *Journal of Materials Chemistry B* 6 (3). The Royal Society of Chemistry: 428–39. doi:10.1039/C7TB02292A.

Arcos, D., A.R. Boccaccini, M. Bohner, A. Díez-Pérez, M. Epple, E. Gómez-Barrena, A. Herrera, J.A. Planell, L. Rodríguez-Mañas, and M. Vallet-Regí. 2014. The relevance of biomaterials to the prevention and treatment of osteoporosis. *Acta Biomaterialia* 10 (5): 1793–805. doi:10.1016/j.actbio.2014.01.004.

Baeza, A., M. Colilla, and M. Vallet-Regí. 2015. Advances in mesoporous silica nanoparticles for targeted stimuli-responsive drug delivery. *Expert Opinion on Drug Delivery* 12 (2): 319–37. doi:10.1517/17425247.2014.953051.

Baeza, A., E. Guisasola, E. Ruiz-Hernández, and M. Vallet-Regí. 2012. Magnetically triggered multidrug release by hybrid mesoporous silica nanoparticles. *Chemistry of Materials* 24 (3). American Chemical Society: 517–24. doi:10.1021/cm203000u.

Baeza, A., E. Guisasola, A. Torres-Pardo, J.M. González-Calbet, G.J. Melen, M. Ramirez, and M. Vallet-Regí. 2014. Hybrid enzyme-polymeric capsules/mesoporous silica nanodevice for in situ cytotoxic agent generation. *Advanced Functional Materials* 24 (29): 4625–33. doi:10.1002/adfm.201400729.

Baeza, A., M. Manzano, M. Colilla, and M. Vallet-Regí. 2016. Recent advances in mesoporous silica nanoparticles for antitumor therapy: Our contribution. *Biomaterials Science* 4 (5): 803–13. doi:10.1039/c6bm00039h.

Beck, J.S., J.C. Vartuli, W.J. Roth, M.E. Leonowicz, C.T. Kresge, K.D. Schmitt, C.T.W. Chu, et al. 1992, December 1. A new family of mesoporous molecular sieves prepared with liquid crystal templates. Research-article. *ACS Publications*. American Chemical Society. World. doi:10.1021/ja00053a020.

Bildstein, L., C. Dubernet, and P. Couvreur. 2011. Prodrug-based intracellular delivery of anticancer agents. *Advanced Drug Delivery Reviews* 63 (1–2): 3–23. doi:10.1016/j.addr.2010.12.005.

Carmona, D., F. Balas, and J. Santamaria. 2014. Pore ordering and surface properties of FDU-12 and SBA-15 mesoporous materials and their relation to drug loading and release in aqueous environments. *Materials Research Bulletin* 59 (November): 311–22. doi:10.1016/j.materresbull.2014.07.039.

Castillo, R.R., M. Colilla, and M. Vallet-Regí. 2017. Advances in mesoporous silica-based nanocarriers for co-delivery and combination therapy against cancer. *Expert Opinion on Drug Delivery* 14 (2): 229–43. doi:10.1080/17425247.2016.1211637.

Castillo, R.R., D. Hernández-Escobar, S. Gómez-Graña, and M. Vallet-Regí. 2018. Reversible nanogate system for mesoporous silica nanoparticles based on Diels-Alder adducts. *Chemistry (Weinheim an Der Bergstrasse, Germany)* 24 (27): 6992–7001. doi:10.1002/chem.201706100.

Chauhan, N., K. Saxena, and U. Jain. 2022, August. Smart nanomaterials employed recently for drug delivery in cancer therapy: An intelligent approach. *BioNanoScience*. doi:10.1007/s12668-022-01022-9.

Chen, Y., H. Chen, and J. Shi. 2013. In vivo bio-safety evaluations and diagnostic/therapeutic applications of chemically designed mesoporous silica nanoparticles. *Advanced Materials (Deerfield Beach, Fla.)* 25 (23): 3144–76. doi:10.1002/adma.201205292.

Croissant, J.G., Y. Fatieiev, A. Almalik, and N.M. Khashab. 2018. Mesoporous silica and organosilica nanoparticles: Physical chemistry, biosafety, delivery strategies, and biomedical applications. *Advanced Healthcare Materials* 7 (4): 1700831. doi:10.1002/adhm.201700831.

Deng, Y., D. Qi, C. Deng, X. Zhang, and D. Zhao. 2008. Superparamagnetic high-magnetization microspheres with an Fe_3O_4@SiO_2 core and perpendicularly aligned mesoporous SiO_2 shell for removal of microcystins. *Journal of the American Chemical Society* 130 (1): 28–9. doi:10.1021/ja0777584.

Díaz-Rodríguez, P., P. González, J. Serra, and M. Landin. 2014. Key parameters in blood-surface interactions of 3D bioinspired ceramic materials. *Materials Science and Engineering: C* 41 (August): 232–9. doi:10.1016/j.msec.2014.04.058.

Elias, C.N., M.A. Meyers, R.Z. Valiev, and S.N. Monteiro. 2013. Ultrafine grained titanium for biomedical applications: An overview of performance. *Journal of Materials Research and Technology* 2 (4): 340–50. doi:10.1016/j.jmrt.2013.07.003.

Fang, W., Z. Wang, S. Zong, H. Chen, D. Zhu, Y. Zhong, and Y. Cui. 2014. pH-controllable drug carrier with SERS activity for targeting cancer cells. *Biosensors & Bioelectronics* 57 (July): 10–5. doi:10.1016/j.bios.2014.01.042.

Gan, Q., X. Lu, Y. Yuan, J. Qian, H. Zhou, X. Lu, J. Shi, and C. Liu. 2011. A magnetic, reversible pH-responsive nanogated ensemble based on Fe_3O_4 nanoparticles-capped mesoporous silica. *Biomaterials* 32 (7): 1932–42. doi:10.1016/j.biomaterials.2010.11.020.

Gao, Y., C. Yang, X. Liu, R. Ma, D. Kong, and L. Shi. 2012. A multifunctional nanocarrier based on nanogated mesoporous silica for enhanced tumor-specific uptake and intracellular delivery. *Macromolecular Bioscience* 12 (2): 251–9. doi:10.1002/mabi.201100208.

García-Alvarez, R., I. Izquierdo-Barba, and M. Vallet-Regí. 2017. 3D scaffold with effective multidrug sequential release against bacteria biofilm. *Acta Biomaterialia* 49 (February): 113–26. doi:10.1016/j.actbio.2016.11.028.

Gisbert-Garzaran, M., D. Lozano, M. Vallet-Regí, and M. Manzano. 2017. Self-immolative polymers as novel pH-responsive gate keepers for drug delivery. *RSC Advances* 7 (1). Royal Society of Chemistry: 132–6. doi:10.1039/C6RA26771H.

Glimcher, M.J., and J.B. Lian. 1989. *The Chemistry and Biology of Mineralized Tissues: Proceedings of the Third International Conference on the Chemistry and Biology of Mineralized Tissues*, Held in Chatham, Massachusetts on October 16–21, 1988. CRC Press.

González, B., M. Colilla, J. Díez, D. Pedraza, M. Guembe, I. Izquierdo-Barba, and M. Vallet-Regí. 2018. Mesoporous silica nanoparticles decorated with polycationic dendrimers for infection treatment. *Acta Biomaterialia* 68 (March): 261–71. doi:10.1016/j.actbio.2017.12.041.

González, B., E. Ruiz-Hernández, M.J. Feito, C.L. de Laorden, D. Arcos, C. Ramírez-Santillán, C. Matesanz, M.T. Portolés, and M. Vallet-Regí. 2011. Covalently bonded dendrimer-maghemite nanosystems: Nonviral vectors for in vitrogene magnetofection. *Journal of Materials Chemistry* 21 (12). The Royal Society of Chemistry: 4598–604. doi:10.1039/C0JM03526B.

Guisasola, E., L. Asín, L. Beola, J.M. de la Fuente, A. Baeza, and M. Vallet-Regí. 2018a. Beyond traditional hyperthermia: In vivo cancer treatment with magnetic-responsive mesoporous silica nanocarriers. *ACS Applied Materials & Interfaces* 10 (15): 12518–25. doi:10.1021/acsami.8b02398.

Guisasola, E., A. Baeza, L. Asín, J.M. de la Fuente, and M. Vallet-Regí. 2018b. Heating at the nanoscale through drug-delivery devices: Fabrication and synergic effects in cancer treatment with nanoparticles. *Small Methods* 2 (9): 1800007. doi:10.1002/smtd.201800007.

Guisasola, E., A. Baeza, M. Talelli, D. Arcos, M. Moros, J.M. de la Fuente, and M. Vallet-Regí. 2015. Magnetic-responsive release controlled by hot spot effect. *Langmuir: The ACS Journal of Surfaces and Colloids* 31 (46): 12777–82. doi:10.1021/acs.langmuir.5b03470.

Guo, Q-Y., X-Y. Yan, W. Zhang, X-H. Li, Y. Xu, S. Dai, Y. Liu, B-X. Zhang, X. Feng, and J. Yin. 2021. Ordered mesoporous silica pyrolyzed from single-source self-assembled organic-inorganic giant surfactants. *Journal of the American Chemical Society* 143 (33). ACS Publications: 12935–42.

Horcajada, P., A. Rámila, K. Boulahya, J. González-Calbet, and M. Vallet-Regí. 2004. Bioactivity in ordered mesoporous materials. *Solid State Sciences* 6 (11): 1295–300. doi:10.1016/j.solidstatesciences.2004.07.026.

Hudson, S., R.F. Padera, R. Langer, and D.S. Kohane. 2008. The biocompatibility of mesoporous silicates. *Biomaterials* 29 (30): 4045–55. doi:10.1016/j.biomaterials.2008.07.007.

Igaz, N., P. Bélteky, D. Kovács, C. Papp, A. Rónavári, D. Szabó, A. Gácser, Z. Kónya, and M. Kiricsi. 2022. Functionalized Mesoporous Silica Nanoparticles for Drug-Delivery to Multidrug-Resistant Cancer Cells. *International Journal of Nanomedicine* 17 (July): 3079–96. doi:10.2147/IJN.S363952.

Joglekar, M., R.A. Roggers, Y. Zhao, and B.G. Trewyn. 2013. Interaction effects of mesoporous silica nanoparticles with different morphologies on human red blood cells. *RSC Advances* 3 (7). The Royal Society of Chemistry: 2454–61. doi:10.1039/C2RA22264G.

Lee, D.D., and M.J. Glimcher. 1991. Three-dimensional spatial relationship between the collagen fibrils and the inorganic calcium phosphate crystals of pickerel (Americanus Americanus) and herring (Clupea Harengus) bone. *Journal of Molecular Biology* 217 (3): 487–501. doi:10.1016/0022-2836(91)90752-r.

Li, Z., J.C. Barnes, A. Bosoy, J.F. Stoddart, and J.I. Zink. 2012. Mesoporous silica nanoparticles in biomedical applications. *Chemical Society Reviews* 41 (7): 2590–605. doi:10.1039/c1cs15246g.

Li, Z., K. Dong, S. Huang, E. Ju, Z. Liu, M. Yin, J. Ren, and X. Qu. 2014. A smart nanoassembly for multistage targeted drug delivery and magnetic resonance imaging. *Advanced Functional Materials* 24 (23): 3612–20. doi:10.1002/adfm.201303662.

Limnell, T., H.A. Santos, E. Mäkilä, T. Heikkilä, J. Salonen, D.Y. Murzin, N. Kumar, T. Laaksonen, L. Peltonen, and J. Hirvonen. 2011. Drug delivery formulations of ordered and nonordered mesoporous silica: Comparison of three drug loading methods. *Journal of Pharmaceutical Sciences* 100 (8): 3294–306. doi:10.1002/jps.22577.

Liu, R., Y. Zhang, X. Zhao, A. Agarwal, L.J. Mueller, and P. Feng. 2010. pH-responsive nanogated ensemble based on gold-capped mesoporous silica through an acid-labile acetal linker. *Journal of the American Chemical Society* 132 (5): 1500–1. doi:10.1021/ja907838s.

López, V., M.R. Villegas, V. Rodríguez, G. Villaverde, D. Lozano, A. Baeza, and M. Vallet-Regí. 2017. Janus mesoporous silica nanoparticles for dual targeting of tumor cells and mitochondria. *ACS Applied Materials & Interfaces* 9 (32): 26697–706. doi:10.1021/acsami.7b06906.

Lu, J., M. Liong, Z. Li, J.I. Zink, and F. Tamanoi. 2010. Biocompatibility, biodistribution, and drug-delivery efficiency of mesoporous silica nanoparticles for cancer therapy in animals. *Small (Weinheim an Der Bergstrasse, Germany)* 6 (16): 1794–805. doi:10.1002/smll.201000538.

Lu, F., S-H. Wu, Y. Hung, and C-Y. Mou. 2009. Size effect on cell uptake in well-suspended, uniform mesoporous silica nanoparticles. *Small (Weinheim an Der Bergstrasse, Germany)* 5 (12): 1408–13. doi:10.1002/smll.200900005.

Mal, N.K., M. Fujiwara, and Y. Tanaka. 2003. Photocontrolled reversible release of guest molecules from coumarin-modified mesoporous silica. *Nature* 421 (6921): 350–3. doi:10.1038/nature01362.

Mamaeva, V., C. Sahlgren, and M. Lindén. 2013. Mesoporous silica nanoparticles in medicine – Recent advances. *Advanced Drug Delivery Reviews* 65 (5): 689–702. doi:10.1016/j.addr.2012.07.018.

Manzano, M., D. Arcos, M.R. Delgado, E. Ruiz, F.J. Gil, and M. Vallet-Regí. 2006. Bioactive star gels. *Chemistry of Materials* 18 (24). American Chemical Society: 5696–703. doi:10.1021/cm0615370.

Manzano, M., and M. Vallet-Regí. 2018. Mesoporous silica nanoparticles in nanomedicine applications. *Journal of Materials Science. Materials in Medicine* 29 (5): 65. doi:10.1007/s10856-018-6069-x.

Manzano, M., and M. Vallet-Regí. 2020. Mesoporous silica nanoparticles for drug delivery. *Advanced Functional Materials* 30 (2). Wiley Online Library: 1902634.

Martínez-Carmona, M., A. Baeza, M.A. Rodriguez-Milla, J. García-Castro, and M. Vallet-Regí. 2015. Mesoporous silica nanoparticles grafted with a light-responsive protein shell for highly cytotoxic antitumoral therapy. *Journal of Materials Chemistry B* 3 (28). The Royal Society of Chemistry: 5746–52. doi:10.1039/C5TB00304K.

Martínez-Carmona, M., D. Lozano, A. Baeza, M. Colilla, and M. Vallet-Regí. 2017. A novel visible light responsive nanosystem for cancer treatment. *Nanoscale* 9 (41): 15967–73. doi:10.1039/c7nr05050j.

Martínez-Carmona, M., D. Lozano, M. Colilla, and M. Vallet-Regí. 2016. Selective topotecan delivery to cancer cells by targeted pH-sensitive mesoporous silica nanoparticles. *RSC Advances* 6 (56). Royal Society of Chemistry: 50923–32. doi:10.1039/C6RA07763C.

Martínez-Carmona, M., D. Lozano, M. Colilla, and M. Vallet-Regí. 2018. Lectin-conjugated pH-responsive mesoporous silica nanoparticles for targeted bone cancer treatment. *Acta Biomaterialia* 65 (January): 393–404. doi:10.1016/j.actbio.2017.11.007.

Matsumura, Y., and H. Maeda. 1986. A new concept for macromolecular therapeutics in cancer chemotherapy: Mechanism of tumoritropic accumulation of proteins and the antitumor agent smancs. *Cancer Research* 46 (12 Pt 1): 6387–92.

Meng, H., M. Xue, T. Xia, Y-L. Zhao, F. Tamanoi, J.F. Stoddart, J.I. Zink, and A.E. Nel. 2010. Autonomous in vitro anticancer drug release from mesoporous silica nanoparticles by pH-sensitive nanovalves. *Journal of the American Chemical Society* 132 (36): 12690–7. doi:10.1021/ja104501a.

Mi, P. 2020. Stimuli-responsive nanocarriers for drug delivery, tumor imaging, therapy and theranostics. *Theranostics* 10 (10): 4557–88. doi:10.7150/thno.38069.

Mickler, F.M., L. Möckl, N. Ruthardt, M. Ogris, E. Wagner, and C. Bräuchle. 2012. Tuning nanoparticle uptake: Live-cell imaging reveals two distinct endocytosis mechanisms mediated by natural and artificial EGFR targeting ligand. *Nano Letters* 12 (7): 3417–23. doi:10.1021/nl300395q.

Milgroom, A., M. Intrator, K. Madhavan, L. Mazzaro, R. Shandas, B. Liu, and D. Park. 2014. Mesoporous silica nanoparticles as a breast-cancer targeting ultrasound contrast agent. *Colloids and Surfaces. B, Biointerfaces* 116 (April): 652–7. doi:10.1016/j.colsurfb.2013.10.038.

Mo, S., C-C. Coussios, L. Seymour, and R. Carlisle. 2012. Ultrasound-enhanced drug delivery for cancer. *Expert Opinion on Drug Delivery* 9 (12): 1525–38. doi:10.1517/17425247.2012.739603.

Mura, S., J. Nicolas, and P. Couvreur. 2013. Stimuli-responsive nanocarriers for drug delivery. *Nature Materials* 12 (11): 991–1003. doi:10.1038/nmat3776.

Nadrah, P., O. Planinšek, and M. Gaberšček. 2014. Stimulus-responsive mesoporous silica particles. *Journal of Materials Science* 49 (2): 481–95. doi:10.1007/s10853-013-7726-6.

Nakamura, H., J. Fang, F. Jun, and H. Maeda. 2015. Development of next-generation macromolecular drugs based on the EPR effect: Challenges and pitfalls. *Expert Opinion on Drug Delivery* 12 (1): 53–64. doi:10.1517/17425247.2014.955011.

Nordin, B.E.C. 2013. *Calcium in Human Biology*. Springer Science & Business Media.

Paris, J.L., M.V. Cabañas, M. Manzano, and M. Vallet-Regí. 2015. Polymer-grafted mesoporous silica nanoparticles as ultrasound-responsive drug carriers. *ACS Nano* 9 (11): 11023–33. doi:10.1021/acsnano.5b04378.

Paris, J.L., M. Colilla, I. Izquierdo-Barba, M. Manzano, and M. Vallet-Regí. 2017. Tuning mesoporous silica dissolution in physiological environments: A review. *Journal of Materials Science* 52 (15): 8761–71. doi:10.1007/s10853-017-0787-1.

Paris, J.L., P. de la Torre, M. Manzano, M.V. Cabañas, A.I. Flores, and M. Vallet-Regí. 2016. Decidua-derived mesenchymal stem cells as carriers of mesoporous silica nanoparticles. In vitro and in vivo evaluation on mammary tumors. *Acta Biomaterialia* 33 (March): 275–82. doi:10.1016/j.actbio.2016.01.017.

Paris, J.L., C. Mannaris, M.V. Cabañas, R. Carlisle, M. Manzano, M. Vallet-Regí, and C.C. Coussios. 2018a. Ultrasound-mediated cavitation-enhanced extravasation of mesoporous silica nanoparticles for controlled-release drug delivery. *Chemical Engineering Journal, Smart Nanomaterials and Nanostructures for Diagnostic and Therapy* 340 (May): 2–8. doi:10.1016/j.cej.2017.12.051.

Paris, J.L., M. Manzano, M.V. Cabañas, and M. Vallet-Regí. 2018b. Mesoporous silica nanoparticles engineered for ultrasound-induced uptake by cancer cells. *Nanoscale* 10 (14): 6402–8. doi:10.1039/C8NR00693H.

Paris, J.L., G. Villaverde, M.V. Cabañas, M. Manzano, and M. Vallet-Regí. 2018c. From proof-of-concept material to PEGylated and modularly targeted ultrasound-responsive mesoporous silica nanoparticles. *Journal of Materials Chemistry B* 6 (18). The Royal Society of Chemistry: 2785–94. doi:10.1039/C8TB00444G.

Pelaz, B., C. Alexiou, R.A. Alvarez-Puebla, F. Alves, A.M. Andrews, S. Ashraf, L.P. Balogh, et al. 2017. Diverse applications of nanomedicine. *ACS Nano* 11 (3): 2313–81. doi:10.1021/acsnano.6b06040.

Philippart, A., N. Gómez-Cerezo, D. Arcos, A.J. Salinas, E. Boccardi, M. Vallet-Regi, and A.R. Boccaccini. 2017. Novel ion-doped mesoporous glasses for bone tissue engineering: Study of their structural characteristics influenced by the presence of phosphorous oxide. *Journal of Non-crystalline Solids* 455 (January): 90–7. doi:10.1016/j.jnoncrysol.2016.10.031.

Porta, F., G.E.M. Lamers, J. Morrhayim, A. Chatzopoulou, M. Schaaf, H. den Dulk, C. Backendorf, J.I. Zink, and A. Kros. 2013. Folic acid-modified mesoporous silica nanoparticles for cellular and nuclear targeted drug delivery. *Advanced Healthcare Materials* 2 (2): 281–6. doi:10.1002/adhm.201200176.

Prabhakar, U., H. Maeda, R.K. Jain, E.M. Sevick-Muraca, W. Zamboni, O.C. Farokhzad, S.T. Barry, A. Gabizon, P. Grodzinski, and D.C. Blakey. 2013. Challenges and key considerations of the enhanced permeability and retention (EPR) effect for nanomedicine drug delivery in oncology. *Cancer Research* 73 (8): 2412–7. doi:10.1158/0008-5472.CAN-12-4561.

Rim, H.P., K.H. Min, H.J. Lee, S.Y. Jeong, and S.C. Lee. 2011. pH-tunable calcium phosphate covered mesoporous silica nanocontainers for intracellular controlled release of guest drugs. *Angewandte Chemie (International Ed. in English)* 50 (38): 8853–7. doi:10.1002/anie.201101536.

Rosenholm, J.M., E. Peuhu, L.T. Bate-Eya, J.E. Eriksson, C. Sahlgren, and M. Lindén. 2010. Cancer-cell-specific induction of apoptosis using mesoporous silica nanoparticles as drug-delivery vectors. *Small (Weinheim an Der Bergstrasse, Germany)* 6 (11): 1234–41. doi:10.1002/smll.200902355.

Ruiz-Hernández, E., A. Baeza, and M. Vallet-Regí. 2011. Smart drug delivery through DNA/magnetic nanoparticle gates. *ACS Nano* 5 (2): 1259–66. doi:10.1021/nn1029229.

Salinas, A.J., and M. Vallet-Regí. 2016. Glasses in bone regeneration: A multiscale issue. *Journal of Non-crystalline Solids, Glasses in Healthcare*, 432 (January): 9–14. doi:10.1016/j.jnoncrysol.2015.03.025.

Shukrun Farrell, E., Y. Schilt, M.Y. Moshkovitz, Y. Levi-Kalisman, U. Raviv, and S. Magdassi. 2020. 3D printing of ordered mesoporous silica complex structures. *Nano Letters* 20 (9). ACS Publications: 6598–605.

Simmchen, J., A. Baeza, D. Ruiz, M.J. Esplandiu, and M. Vallet-Regí. 2012. Asymmetric hybrid silica nanomotors for capture and cargo transport: Towards a novel motion-based DNA sensor. *Small (Weinheim an Der Bergstrasse, Germany)* 8 (13): 2053–9. doi:10.1002/smll.201101593.

Stewart, C.A., Y. Finer, and B.D. Hatton. 2018. Drug self-assembly for synthesis of highly-loaded antimicrobial drug-silica particles. *Scientific Reports* 8 (1). Nature Publishing Group: 895. doi:10.1038/s41598-018-19166-8.

Sun, T., Y.U. Shrike Zhang, B.O. Pang, D.C. Hyun, M. Yang, and Y. Xia. 2021. Engineered nanoparticles for drug delivery in cancer therapy*. In *Nanomaterials and Neoplasms*. Jenny Stanford Publishing.

Torchilin, V.P. 2014. Multifunctional, stimuli-sensitive nanoparticulate systems for drug delivery. *Nature Reviews. Drug Discovery* 13 (11): 813–27. doi:10.1038/nrd4333.

Tsai, C-P., C-Y. Chen, Y. Hung, F-H. Chang, and C-Y. Mou. 2009. Monoclonal antibody-functionalized mesoporous silica nanoparticles (MSN) for selective targeting breast cancer cells. *Journal of Materials Chemistry* 19 (32). The Royal Society of Chemistry: 5737–43. doi:10.1039/B905158A.

Vallet-Regí, M. 2001. *J. Chem. Soc., Dalton Trans.* The Royal Society of Chemistry.

Vallet-Regí, M. 2010a. Nanostructured mesoporous silica matrices in nanomedicine. *Journal of Internal Medicine* 267 (1): 22–43. doi:10.1111/j.1365-2796.2009.02190.x.

Vallet-Regí, M. 2010b. Evolution of bioceramics within the field of biomaterials. *Comptes Rendus Chimie* 13 (1): 174–85. doi:10.1016/j.crci.2009.03.004.

Vallet-Regi, M. 2014. *Bio-ceramics with Clinical Applications*. John Wiley & Sons.

Vallet-Regí, M., M. Colilla, and B. González. 2011. Medical applications of organic-inorganic hybrid materials within the field of silica-based bioceramics. *Chemical Society Reviews* 40 (2): 596–607. doi:10.1039/c0cs00025f.

Vallet-Regí, M., M. Colilla, I. Izquierdo-Barba, and M. Manzano. 2018. Mesoporous silica nanoparticles for drug delivery: Current insights. *Molecules* 23 (1). Multidisciplinary Digital Publishing Institute: 47. doi:10.3390/molecules23010047.

Vallet-Regí, M., and J.M. González-Calbet. 2004. Calcium phosphates as substitution of bone tissues. *Progress in Solid State Chemistry* 32 (1): 1–31. doi:10.1016/j.progsolidstchem.2004.07.001.

Vallet-Regí, M., M. Manzano, J.M. González-Calbet, and E. Okunishi. 2010. Evidence of drug confinement into silica mesoporous matrices by STEM spherical aberration corrected microscopy. *Chemical Communications (Cambridge, England)* 46 (17): 2956–58. doi:10.1039/c000806k.

Vallet-Regí, M., D.A. Navarrete, and D. Arcos. 2008. *Biomimetic Nanoceramics in Clinical Use: From Materials to Applications*. Royal Society of Chemistry.

Vallet-Regi, M., A. Rámila, and R.P. del Real, and J. Pérez-Pariente. 2000, December 29. A new property of MCM-41: Drug delivery system. Research-article. *ACS Publications*. American Chemical Society. World. doi:10.1021/cm0011559.

Vallet-Regí, M., and E. Ruiz-Hernández. 2011. Bioceramics: From bone regeneration to cancer nanomedicine. *Advanced Materials* 23 (44): 5177–218. doi:10.1002/adma.201101586.

Vallet-Regí, M., and A.J. Salinas. 2019. 6: Ceramics as bone repair materials. In *Bone Repair Biomaterials* (Second Edition), edited by Kendell M. Pawelec and Josep A. Planell, 141–78. Woodhead Publishing Series in Biomaterials. Woodhead Publishing. doi:10.1016/B978-0-08-102451-5.00006-8.

Vallet-Regi, M., A. Salinas, A. Baeza, and M. Manzano. 2018. Smart nanomaterials and nanostructures for diagnostic and therapy. *Chemical Engineering Journal* 340. Elsevier Science SA, Lausanne, Switzerland: 1–1.

Villegas, M.R., A. Baeza, and M. Vallet-Regí. 2015. Hybrid collagenase nanocapsules for enhanced nanocarrier penetration in tumoral tissues. *ACS Applied Materials & Interfaces* 7 (43): 24075–81. doi:10.1021/acsami.5b07116.

Wang, X., H. Chen, Y. Zheng, M. Ma, Y. Chen, K. Zhang, D. Zeng, and J. Shi. 2013. Au-nanoparticle coated mesoporous silica nanocapsule-based multifunctional platform for ultrasound mediated imaging, cytoclasis and tumor ablation. *Biomaterials* 34 (8): 2057–68. doi:10.1016/j.biomaterials.2012.11.044.

Wang, Y., K. Wang, J. Zhao, X. Liu, J. Bu, X. Yan, and R. Huang. 2013. Multifunctional mesoporous silica-coated graphene nanosheet used for chemo-photothermal synergistic targeted therapy of glioma. *Journal of the American Chemical Society* 135 (12): 4799–804. doi:10.1021/ja312221g.

Xu, C., Y. Lin, J. Wang, L. Wu, W. Wei, J. Ren, and X. Qu. 2013. Nanoceria-triggered synergetic drug release based on CeO(2) – Capped mesoporous silica host-guest interactions and switchable enzymatic activity and cellular effects of CeO(2). *Advanced Healthcare Materials* 2 (12): 1591–99. doi:10.1002/adhm.201200464.

Yunus Basha, R., T.S. Sampath Kumar, and M. Doble. 2015. Design of biocomposite materials for bone tissue regeneration. *Materials Science and Engineering: C* 57 (December): 452–63. doi:10.1016/j.msec.2015.07.016.

Zhang, H., H. Zhang, Y. Xiong, L. Dong, and X. Li. 2021. Development of hierarchical porous bioceramic scaffolds with controlled micro/nano surface topography for accelerating bone regeneration. *Materials Science and Engineering: C* 130 (November): 112437. doi:10.1016/j.msec.2021.112437.

Zhao, W., J. Gu, L. Zhang, H. Chen, and J. Shi. 2005. Fabrication of uniform magnetic nanocomposite spheres with a magnetic core/mesoporous silica shell structure. *Journal of the American Chemical Society* 127 (25): 8916–17. doi:10.1021/ja051113r.

Zhao, Y., X. Sun, G. Zhang, B.G. Trewyn, I.I. Slowing, and V.S-Y. Lin. 2011. Interaction of mesoporous silica nanoparticles with human red blood cell membranes: Size and surface effects. *ACS Nano* 5 (2): 1366–75. doi:10.1021/nn103077k.

Section D

ISO/ASTM Specifications

13 Standard Terminologies and Definitions in Bioceramics

Mahbub Ahmed
Southern Arkansas University

13.1 INTRODUCTION

Bioceramics are the special types of ceramics used in medical and dental fields for the potential treatment of the human body. Bioceramics fall under the special subcategory of biomaterials in general. Bioceramic materials have great potential to repair the bone tissues in the human body. It was first discovered in the early nineties (Hench, 1991; Yamamuro et al., 1990) that certain kinds of ceramics can bond with living bones. This discovery unleashed the potential use of such ceramics in the medical field as possible implants. However, to be able to use as bone replacements, they need to have similar mechanical properties as the original bone component in the body. Furthermore, it is noted that most of the bones in the human body have a certain degree of load capacity, especially in terms of compression loads. Besides the mechanical properties, these implants need to meet the requirement of minimum pore size. A study (Prakasam et al., 2017) shows that the minimum pore size of 100 μm is required for the nutrition to diffuse through; similarly, for the successful growth of the bone tissues within the implant, there needs to be a minimum size of 200–350 μm of pores.

Bioceramics have practical uses in the dental field as well. Bioceramic-based silicates have been successfully used as root canal sealers. A study (Endodontic Practice: Case Studies, 2022) shows how bioceramics can repair the surrounding tissues around the root of a tooth. Today, bioceramics are also used in the area of cochlear, maxillofacial, spinal discs, and otolaryngology treatments (LeGeros et al., 2003).

The advent of the additive manufacturing process opened up tremendous opportunities in bioceramic research. Complex-shaped medical scaffolds can easily be manufactured through a high-precision 3D printing process where the pore density of the scaffold can be easily controlled through the slicing process.

13.2 COMMON TERMINOLOGIES AND DEFINITIONS

Bioceramics is a vast scientific area with many chemical and medical terms and definitions, procedures, relevant unique properties, and manufacturing processes. In this section, some common terminologies frequently used in the medical and research fields will be highlighted.

DOI: 10.1201/9781003258353-17

13.2.1 Medical Implants

According to the US Food and Drug Administration (Implants and Prosthetics, 2019), medical implants are devices or tissues placed inside or on the body's surface. Some implants are used as prosthetics in the body, while others are used for drug delivery or to assist in the functioning of different body parts. The use of implants in the medical field has a demand in general. According to a study, by 2030 (Kurtz et al., 2007) the number of implants will grow by 174% in the USA alone. Metal-based implants are commonly used today, including for hip replacements, as shown in Figure 13.1. However, these implants can corrode inside the body, which can release metal particles and potentially cause adverse effects on patients' health. As the implants will be directly in contact with the tissues, blood, and human cells, their wear particles must be biologically compatible and non-toxic. Several studies (Han et al., 2021; Hua et al., 2022) show that bioceramic-based dental implants have high biocompatibility, corrosion resistance, and low thermal conductivity compared with their metallic and polymer counterparts. The bioceramic implants are predominantly made through sintering processes, using bioceramic powders as raw materials. Usually, the powdered ceramics and chemical binders are hydraulically pressed in a die of the desired shape to form the shape of the implant, commonly known as the greenbody. This greenbody is sintered at a high temperature, allowing the particles to bond chemically to form a dense bioceramic implant. Hot isostatic processes can further improve the grain structure and some common mechanical properties in manufacturing the implants. The hot isostatic process involves applying uniform pressure from all directions through an inert gas. Manufacturing bone implants through 3D printing is also gaining popularity among many researchers today. With precise control and the repeatability of the 3D printing process, it has great potential to satisfy the future worldwide manufacturing needs of medical implants.

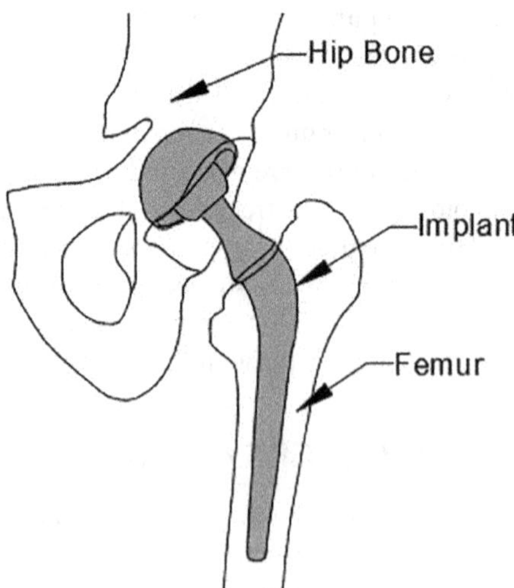

FIGURE 13.1 Metal implant schematic.

13.2.2 Alumina (Al_2O_3)

Alumina is popular implant material for having excellent wear resistance and corrosive properties. They are considered very stable oxidizers for possessing high ionic interaction between the aluminum and oxygen atoms. The use of alumina being dental implants has been started in the 1960s. The orthopedic use of alumina was first reported (Smith, 1963) in the 1960s when a porous alumina-based ceramic compound was infused with epoxy resin to form the bone substitute. A successful dental fixture was reported (British Patent No. 1083.769, 1965) to be developed by a scientist in Switzerland during the 1960s. Later in 1976, a group of scientists developed a dental implant made of alumina ceramic material (Schulte & Heimke, 1976) to restore an extracted or lost tooth. Alumina exhibits a high degree of chemical inertness in the human body. Having higher hardness than most metal alloys, alumina has great potential to be used in making implants in the area of joint replacements. However, there are some challenges to using alumina as a successful medical implant. As per the ASTM F603 standard, an orthopedic implant should have a bone density greater than 3.94 g/cc and uniform grain size of less than 4.5 μm (Cai & Yan, 2010). A strong chemical bond yields a relatively high melting point in alumina, which makes it difficult to manufacture through the traditional casting process. Moreover, the high hardness can create complexity in machining alumina-based materials. Some mechanical properties such as density, toughness, brittleness, and bending strength can be improved through sintering processes where the grain sizes are reduced.

13.2.3 Zirconia (ZrO_2)

Zirconia is a crystalline substance whose chemical name is Zirconium dioxide or ZrO_2. At room temperature, it stays as a monoclinic crystal structure. In contrast, at a higher temperature, it transforms into a tetragonal or cubic crystal structure. Cubic zirconia, a synthetic form of zirconia, is widely used as a gemstone comparable to diamonds. As pure zirconia is susceptible to change in crystalline structure upon temperature changes, it is stabilized with yttria or magnesia to improve the mechanical properties. The use of zirconia as a bioceramic has been reported as early as 1969 when scientists suggested (Helmer & Driskell, 1969) using zirconia as an alternative to titanium or aluminum prosthetics. Zirconia-based orthopedic hip implants have shown superior wear resistance over other systems (Chen et al., 2016). Zirconia has been a preferred material to be used in fixed prosthodontics (Kvam & Karlsson, 2013). A study (Ichikawa et al., 1992) has shown that zirconia is biocompatible with bones and tissues in a body and possesses high mechanical stability and a high opacity to the radio wave. The mechanical properties of zirconia can be enhanced through further processing. The fracture resistance, flexural strength, and resistance to crack inhibition of zirconia (Lughi & Sergo, 2010; Juntavee & Attashu, 2018) can be improved by sintering at specific conditions.

13.2.3.1 ZTA

ZTA is an acronym for zirconia toughened alumina. It is a composite of fine-grained alumina matrix with zirconia particles of tetragonal phase between

10% and 20% in volume. The mechanical properties of ZTA, such as hardness, compression strength, and wear resistance, are much superior to that of pure alumina. The toughness and fracture strength of ZTA are also superior to that of pure alumina (Wang & Stevens, 1989).

13.2.3.2 TZP

Tetragonal Zirconia Polycrystal, or TZP, is a 3% Yttria blended ceramic material with a fine grain structure. It generally exhibits excellent mechanical characteristics at room temperature. The flexural strength of the TZP ceramic can be as high as 1 GPA (Nettleship & Stevens, 1987). TZP at room temperature also exhibits excellent hardness and fracture toughness. These properties make it a good choice to be used as a medical implant material. TZP-made femoral heads are widely used for total hip replacements. However, excessive tensile loads can be detrimental to TZP-based products. An excessive tensile load may initiate a crack and cause an eventual failure (Stawarczyk et al., 2013) to a TZP-made implant. Also, degradation of mechanical properties for the TZP material can occur due to aging within the human body. Such degradation usually happens due to the transformation (Christel et al., 1989) of crystal structure from tetragonal to monoclinic in the presence of water.

13.2.4 HA or HAp

Hydroxyapatite, also known as HA or HAp, is a type of calcium phosphate that has a similar chemical crystallographic structure to the human bone tissue (Kokubo et al., 2004). The chemical formula for hydroxyapatite is $Ca_{10}(PO_4)_6(OH)_2$. The stoichiometric ratio of calcium to phosphorus is 1.67, the same as the calcium to phosphorus ratio in human bone apatite (Wu et al., 2014). Thermodynamically, hydroxyapatite is more stable than other calcium phosphate compounds under varying physiological conditions such as temperature, pH, and composition of the body fluids (Kalita et al., 2007). However, it decomposes at a temperature higher than 800°C. It has been known to scientists that hydroxyapatite is a non-toxic substance that can take part in the bone healing process and can form a strong chemical bond with the human bone tissue (Hench & Best, 2013; Jurczyk et al., 2011). It has a wide range of applications in the medical field, such as orthopedic and dental implant coating, restoration of periodontal defects, drug delivery, and bone fillers to fill the void because of bone defects (Kantharia et al., 2014). However, studies suggest (Samavedi et al., 2013; Mishra et al., 2016) that hydroxyapatite material has insufficient mechanical strength compared with human bone. Thus, using hydroxyapatite as a load-bearing implant in the human body remains a critical challenge.

13.2.5 TCP, α-TCP, and β-TCP

TCP stands for tricalcium phosphate. It is part of the calcium phosphate family. Calcium phosphate has been extensively explored as a bone repair and substitution materials since the past century (Dorozhkin, 2008). Among the TCPs, α-tricalcium phosphate or α-TCP is a type of calcium phosphate that is highly soluble in aqueous solutions. It is often used as bone cement to fill bone defects. The α-TCP cement is eventually transformed into new bone tissues. Another type of TCP is β-TCP or β-tricalcium phosphate. It is one of the widely used bone-repairing materials. It has good biocompatibility and can eventually be replaced by new bone cells (Yuan et al., 2010). β-TCP transforms into α-TCP above 1,120°C. It is noted that even though the TCPs have excellent biocompatibility in general, the mechanical load-bearing capacity of the TCPs are limited.

13.2.6 Biphasic Calcium Phosphate (BCP)

Biphasic calcium phosphate is a mixture of hydroxyapatite and β-tricalcium phosphate with a fixed proportion of each component. The compound is frequently used as a bone substitute in the medical field because of having similar properties to bones. BCP is obtained when a synthetic or biological calcium-deficient apatite (phosphate minerals) is sintered at temperatures above 700°C (LeGeros et al., 2003). The biological response to the BCP ceramics depends on the chemical composition and the physical properties. Though the BCP material has an excellent bone regeneration capability, the low load-bearing capacity and the poor sintering properties limit their uses in the biomedical field (Ma et al., 2018; Lu et al., 2004).

13.2.7 Mineral Trioxide Aggregate (MTA)

MTA stands for mineral trioxide aggregate. It was developed at Loma Linda University in the 1990s. It is a calcium silicate-based material composed of tricalcium silicate, dicalcium silicate, tricalcium aluminate, tetracalcium aluminoferrite, gypsum, and bismuth oxide (Tu et al., 2018). MTA is used worldwide for root-end filling material, especially in apical plugs used in apexification, repair of root perforations, and pulp capping materials (Lin et al., 2016). MTA is available in either gray or white form; the difference is that white MTA lacks iron (Camilleri et al., 2005).

13.2.8 BG

Bioactive glasses, or BGs, are amorphous materials. They were introduced in orthopedic applications in the late 1960s (Hench et al., 1971). BGs are considered biocompatible and non-toxic to living cells. By controlling the chemical compositions of the BGs, the degradation rate can be regulated (Q. Fu). BGs are generally brittle materials. Having low mechanical strength and low fracture toughness, they are not suitable for load-bearing bone applications. However, they can be combined with polymers to form composites with better mechanical properties. Bioglasses also have the potential use of being coatings on metallic and dental implants. Several studies (Schrooten & Helsen, 2000; Xiao et al., 2011; Zhang et al., 2019) show that bioglass coatings on the implants exhibit high osseointegration ability, excellent bioactive behavior, and good bonding with the bone tissue.

13.3 BIOCERAMIC STANDARDS

As the intended use of bioceramic materials in most cases is to develop medical implants, the standard testing procedure must be carried out to obtain the relevant properties. ASTM and ISO standards are available to obtain the physical properties, specifically the mechanical properties of bioceramics. Some ASTM and ISO standards commonly used for advanced ceramics, including bioceramics, are briefly mentioned below. To learn more about these standards, the reader is referred to the corresponding sources (ASTM Standards & Publications, 2022; ISO Standards, 2022).

13.3.1 ASTM C1161-02

This test method is used to obtain the flexural strength (i.e., the ultimate strength in bending) of advanced ceramics materials at ambient temperature. This test method is applicable for ceramics with 50 MPa or greater. It is assumed that the ceramic specimen is isotropic in nature. Both four-point and three-point loading configurations are included in the specification. The specimens with three geometric configurations can be used, with the width, depth, and length being $2 \times 1.5 \times 25$ mm, $4 \times 3 \times 45$ mm, and $8 \times 6 \times 90$ mm for configurations A, B, and C, respectively. The loading schematics for both four-point and three-point beams are shown in Figure 13.2. In the four-point arrangement, the specimen is symmetrically loaded at two locations that are situated one-quarter of the overall span away from the outer two support bearings. There should be an equal amount of overhangs beyond the outer bearings for both configurations. The specimen should be directly centered below the axis of the applied load. Specimens may be loaded in any suitable testing machine as long as a uniform rate of direct loading is maintained. The accuracy of the testing machine shall be within 0.5%. The maximum permissible stress in the specimen due to the initial load should be within 25% of the mean strength. The specimens should be loaded in a manner such that they are aligned with fixtures.

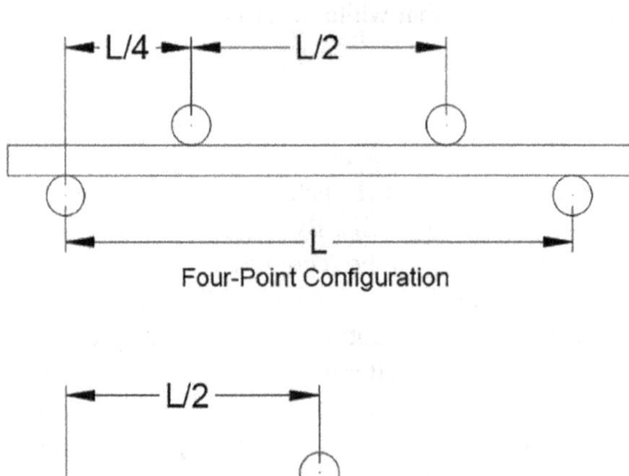

FIGURE 13.2 Four-point and three-point loading schematics.

13.3.2 ASTM D0149-20

This test method covers the measurement procedure to obtain the dielectric properties of solid insulating materials. It is used to determine the dielectric breakdown voltage across the thickness of the test specimen. The test can be run between 25 and 800 Hz range, with the most common at 60 Hz. The method is suitable for a wide range of temperatures, including in both liquid and gaseous mediums. The procedure does not cover the safety concerns in detail; however, the user must establish standard safety, health, and environmental practices while implementing the method.

13.3.3 ASTM C0372-94R20

This is a standard test method to determine the linear thermal expansion of some ceramic materials using the dilatometer technique. This standard is pertinent to the accuracy and repeatability of the test procedure while using the relevant instrument. The test method is relevant to the equipment with an accuracy of 3% or better for the percent linear expansion.

13.3.4 ASTM F2009-20

This is a standard test method to determine the axial force required to disassemble the taper connections of the implants under laboratory conditions. This is a common test procedure to test the femoral head for a total or partial hip replacement, where a self-locking taper secures the head into the base component. A schematic of the test procedure is given in Figure 13.3. The method applies to both ceramic

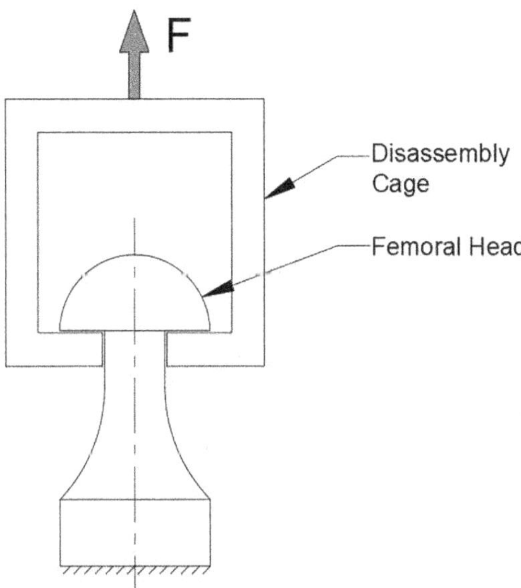

FIGURE 13.3 Femoral head test schematic.

and metal heads. The values stated in the standard are in SI units. The users must practice the standard safety and regulatory procedures while conducting tests by following this ASTM standard, as the testing method involves applying a significant amount of force.

13.3.5 ASTM F2393-12R20

This standard comprises the material requirements for surgical implants made of highly pure and dense zirconia that is partially stabilized with magnesium oxide. The standard used the SI unit system. The user is required to follow the standard safety practices while using this standard.

13.3.6 ASTM F1609-08R14

Because of their excellent biocompatibility, calcium phosphate-based biomaterials are often coated on load-bearing medical implants. This standard is relevant to such medical implants that use calcium phosphate coating. It specifies the material requirements of the coatings that include both hydroxyapatite coatings and tricalcium phosphate coatings and their combination. However, it does not consider the organic coatings containing the ionic species of calcium or calcium phosphates. The standard includes various coating processes, including mechanical, plasma spray, sintering, sputtering, and electrophoretic deposition. The coated surface can be smooth, textured, porous, or other topographical forms. The standard details the fabrication of the test specimens. The material grades can be established by the standard material testing methods, including physical and crystallographic characterizations as well as chemical analysis.

13.3.7 ASTM F1538-03R17

Glass-based ceramics have great potentials to be used as biomaterials. However, before being used in the human body, glass-based ceramic materials must be evaluated carefully for their biocompatibility. This standard details the material requirements and characterization methods for such glass-based bioceramics used as surgical implants or as the coating on surgical tools. However, the drug delivery systems are not included in the specification. The specification also excludes certain types of biomaterials, such as synthetic hydroxyapatite, aluminum oxide, alpha- and beta-tricalcium phosphate, and whitlockite. Material properties such as density, composition, particle size, hardness, flexural strength, Young's modulus, and thermal expansion can be determined as per the specification.

13.3.8 ASTM F1088-18

Beta-tricalcium phosphate is often used for surgical implant applications. This standard covers the chemical and crystallographic requirements for such biocompatible materials. The bioceramics must conform to this specification before being considered a medical grade β-tricalcium phosphate. X-ray fluorescence method can be used to determine the calcium and phosphorus content. While performing the elemental analysis, the concentration of calcium and phosphate should be in the expected stoichiometric proportion as in the β-tricalcium phosphate. Other metals or oxides present in the specimen with a concentration greater than 0.1% should be noted in the material descriptions.

13.3.9 ASTM C1424-15R19

This specification covers the determination of the compressive strength of advanced ceramic materials at ambient temperature under uniaxial monotonic loading. The compression test includes the stress-strain behavior of the material. A schematic diagram for the test specimen setup, along with the fixture blocks, is given in Figure 13.4. As well the geometric dimensions are specified in the specification as shown in Figure 13.5. Besides the geometries, the specimen fabrication process, force and displacement modes, allowable bending, data collection, and recording are addressed in the standard. The testing rates, including the rate of applied force, displacement, and strain rate, are also covered. The ceramic materials macroscopically isotropic, homogeneous, and continuous in nature are usually ideal for compression testing following this standard. Other advanced ceramic materials, including the whisker- or particle-reinforced or short fiber-reinforced composite ceramics, can also be tested using this standard as long as they exhibit similar macroscopic behaviors of being isotropic, homogeneous, and continuous. However, the composite ceramics reinforced with continuous fibers do not qualify to fall under this standard since they do not exhibit the macroscopic behaviors stated earlier. The SI unit system has been used in the specification. Safety is not covered in detail in this standard. The user is required to establish standard safety, health, and environmental practices while implementing this method.

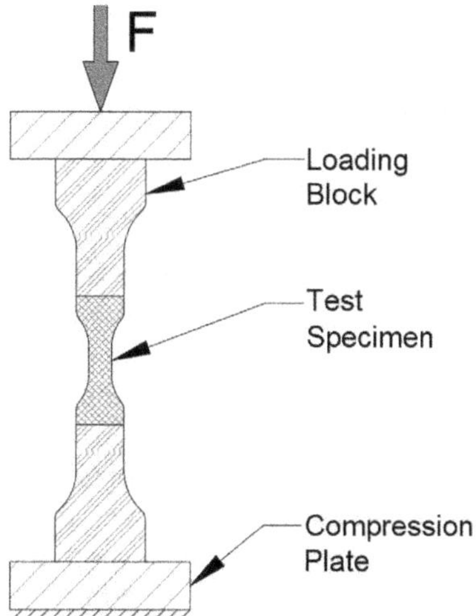

FIGURE 13.4 Test specimen setup for compressive strength measurement.

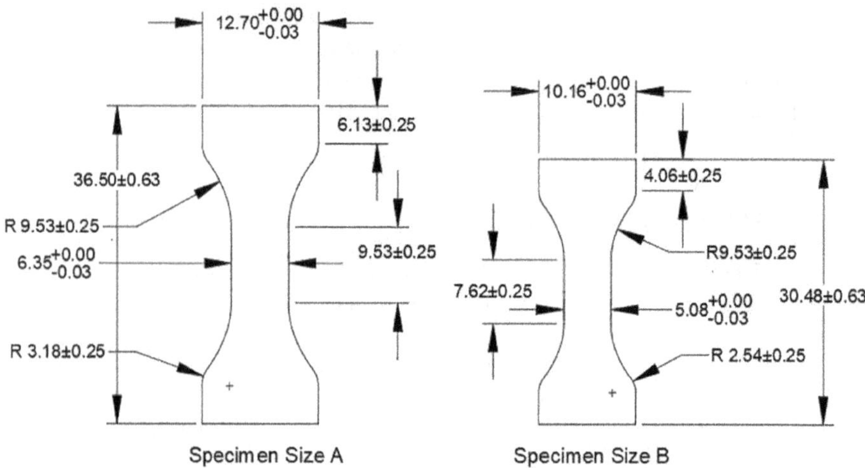

FIGURE 13.5 Test specimen configurations for compressive strength measurement.

13.3.10 ASTM C1684-18

This standard is intended for testing cylindrical specimens to determine the flexural strength at ambient temperature for advanced ceramics. In many cases, it is suitable to use rod-shaped specimens since they are already fabricated in that shape. The recommended geometries of the specimens are between 1.5 and 8 mm for the diameters and between 25 and 85 mm for the lengths. The standard four-point testing configuration is recommended. However, the three-point testing setup is also allowed. The standard details the apparatus, test fixtures, specimen geometries and preparation, relevant formulas, and reporting of results. The method is appropriate for monolithic and composite ceramics reinforced with particulate, whiskers, or short fibers. Glass materials can also be tested using this standard. However, it is not suitable for composite ceramics reinforced with continuous fibers. The SI unit system has been used in the specification. Some relevant safety concerns are addressed in the standard.

13.3.11 ASTM C1273-18

This standard details the procedure for obtaining tensile strength for monolithic advanced ceramics at ambient temperature under uniaxial loading conditions. A wide variety of geometric configurations for the test specimens are allowed. It also covers the fabrication of the test specimen, the modes of testing (including the force, displacement, and strain), allowable bending, and the data collection and reporting methods. This standard primarily applies to advanced materials that are isotropic, homogeneous, and continuous macroscopically. Certain whisker or particle-reinforced ceramic composites and some short fiber-reinforced composites can also be tested under this standard, considering that they exhibit similar behavior macroscopically. However, the continuous fiber-reinforced ceramic composites are not isotropic, homogeneous, and continuous macroscopically and are not recommended to use this standard. The SI unit system has been used in the standard. Only relevant safety concerns have been addressed in the standard. The user must comply with the test-specific safety rules.

13.3.12 ASTM C1366-19

This standard details the procedure to obtain tensile strength for monolithic advanced ceramics at an elevated temperature under uniaxial loading conditions. This standard is similar to ASTM C1273-18, which has been discussed earlier. However, the main difference between the two standards is the temperature at which the tests are carried out. While the standard C1273-18 uses the ambient temperature, C1366-19 uses an elevated temperature. The specimen geometries and fabrication methods as well as the testing modes, and data collection methods are similarly addressed in both standards.

13.3.13 ASTM C1291-18

This standard details the determination of creep strain, the rate of creep strain, and the time to failure under a tensile load at an elevated temperature for the advanced ceramic materials. The temperature is varied in the range between 1,073 and 2,073 K. It covers a wide range of geometric configurations for the test specimen. At a given temperature, the creep strain is evaluated based on the measurements of the extensions of the gauge length during the testing period. The minimum creep strain rate is evaluated as a function of temperature and the applied stress. This standard also includes the creep time to failure at a given temperature. The test method is applicable for advanced ceramic materials that are isotropic, continuous, and homogenous macroscopically. Some composite ceramics, including the short fiber-reinforced and the whisker- or particulate-reinforced, are also included in this standard if they exhibit similar homogeneous nature macroscopically. The ceramic composites reinforced with continuous fibers are not included in the standard due to their non-homogeneous nature. The units are given primarily in the SI system. The user is recommended to practice the appropriate safety measure while performing tests using this standard.

13.3.14 ASTM C1499-19

Determination of equibiaxial strength of advanced ceramic materials under monotonic uniaxial loading and at ambient temperature is covered in this standard. The specimens used in the method are concentric rings in shape. The standard also details the fabrication methods for the specimens, the modes of testing, testing rates, maximum deflection, and data collection and reporting techniques. Two types of specimens, machined and as-fired test specimens, are considered. The standard includes advanced ceramic materials with macroscopically homogeneous, continuous, and isometric characteristics. Some ceramic composites reinforced with short fibers and whiskers or particulates are also included in the standard. The ceramic composites with continuous fibers are not included in the standard because of being non-homogenous and non-isotropic in nature. The user is required to establish the appropriate safety, health, and environmental practices.

13.3.15 ASTM C1674-16

This standard is used to determine the flexural strength of advanced ceramics with two-dimensional honeycomb channel configurations. The tests are performed at ambient temperature. The test method focuses on ceramic components with hollow honeycomb channels. Ceramics with honeycomb structures have a wide variety of applications, and being used in the medical field is one of them. A study shows (Wang et al., 2021) that using honeycomb structure in biomimetic bone scaffold design is feasible and promising. The test method is applicable for ceramics that are linear-elastic in nature under a tensile load. The honeycomb-structured specimens may have a porosity of 30% or more, with the honeycomb channel's cross-sectional dimension being 1 mm. Both four-point and three-point bending test methods are described for a wide range of geometric configurations. The breaking load, the specimen and the cell dimensions, and the loading geometries are used to determine the nominal strength of the beam, the wall fracture strength, and the strength of the honeycomb structure. The SI system has been used in the standard as the primary unit system.

13.3.16 ASTM C1862-17

This standard includes the specifications for testing end plug specimens with a wide range of geometries and configurations. The specifications cover the push-out force, nominal joint strength, and nominal burst pressure of bonded ceramic end plugs at ambient as well as elevated temperatures. The test specimen should have similar geometric dimensions and shapes to the intended application and product design. The most common types of joints tested are adhesively bonded end plugs. The test method covers the capabilities, apparatus, specimen geometries and preparation methods, test specimen mounting, modes of testing, and the calculation and reporting of the result. The test includes the calculation and reporting of the longitudinal failure stress. The values stated in this standard are in accordance with the SI unit system. The user is required to establish a standard and relative safety practices while using this standard.

13.3.17 ASTM C1368-18

This standard covers measuring slow crack growth parameters of advanced ceramic materials at a constant stress rate. The strength is determined as a function of the applied stress rate at the ambient temperature and in a controlled test environment. The strength degradation shown with decreasing applied stress rate in a given environment is the foundation of this test procedure which enables the estimation of slow crack growth parameters of the material. The SI system has been used in the standard.

13.3.18 ASTM C1239-13R18

This standard covers the evaluation of uniaxial strength data and the estimation of the Weibull probability distribution parameters for the ceramic materials that are brittle in nature. These estimated Weibull distribution parameters are used to compare the quality among two or more data sets and to predict the probability of failure of a ceramic structure statistically.

13.3.19 ASTM C1683-10R19

This standard provides a way to convert the fracture strength parameters, including the mean strength and the Weibull characteristics strength of advanced ceramics obtained by one test geometry, to strength parameters for other geometries. It addresses uniaxial as well as biaxial strength data. This standard primarily focuses on evaluating Weibull probability distribution for the ceramic materials that are brittle in nature. The SI system has been used in the standard.

13.3.20 ASTM F603-12R20

This standard details the material requirements for high-purity and dense aluminum oxide in load-bearing surgical implant applications. This standard does not cover finished parts, such as femoral heads, acetabular inserts, and dental implants. This standard can be used as a qualification of the materials used in implants that are intended to be manufactured. The values stated in the standard are from SI unit system only.

13.3.21 ISO 6474-1:2019

This standard covers the test method and the characteristics of high-purity alumina-based bio-stable ceramic bone substitute materials. Such biomaterials are generally used for making bone spacers, orthopedic joint prostheses, or other bone replacement components. The standard does not specify the biocompatibility of the materials.

13.3.22 ISO 7206-10:2018

This standard covers the determination of a compressive or a tensile load required to cause the failure of the femoral head at specific laboratory conditions. As the document includes both metallic and non-metallic materials, the bioceramic-based femoral heads of modular construction for partial or total hip-joint replacements can also potentially be tested using this standard. This standard does not include the inspection and reporting of test specimens.

13.3.23 ISO 13779-2:2018(EN)

This standard covers the requirements for single-layer thermally sprayed coatings of hydroxyapatite ceramics applied on surgical implants. However, it does not cover the coating of other ceramics such as glass ceramics, α- and β-tricalcium phosphate, or other types of calcium phosphates. The requirements mainly include the properties of the materials. Though the document was primarily developed for plasma-sprayed hydroxyapatite coating, it can also be applicable to other types of thermally sprayed coatings of hydroxyapatite.

13.3.24 ISO/DIS 18531(EN)

This standard covers the determination of the physicochemical characteristics of calcium phosphate bioceramics used as bone cement materials. The specimen is prepared by mixing the calcium phosphate powder and liquid agents that are used in making artificial bones or similar applications.

REFERENCES

ASTM Standards & Publications. (2022, September). Retrieved from ASTM International: https://www.astm.org/products-services/standards-and-publications.html

Cai, Y.-z., & Yan, S.-g. (2010). Development of ceramic-on-ceramic implants for total hip arthroplasty. *Orthopaedic Surgery*, 175–181.

Camilleri, J., Montesin, F. E., Silvio, L. D., & Ford, T. R. (2005). The chemical constitution and biocompatibility of accelerated Portland cement for endodontic use. *International Endodontic Journal*, *38*(11), 834–842.

Chen, Y.-W., Moussi, J., Drury, J. L., & Wataha, J. C. (2016). Zirconia in biomedical applications. *Expert Review of Medical Devices*, *13*(10).

Christel, P., Meunier, A., Heller, M., Torre, J. P., & Peille, C. N. (1989). Mechanical properties and short-term in-vivo evaluation of yttrium-oxide-partially-stabilized zirconia. *Journal of Biomedical Materials Research*, *23*(1), 45–61.

Dorozhkin, S. V. (2008). Calcium orthophosphate cements for biomedical application. *Journal of Materials Science*, *43*, 3028–3057.

Endodontic Practice: Case Studies. (2022, September). Retrieved from Endodontic Practice: https://endopracticeus.com/the-use-of-bioceramic-sealer-in-endodontic-retreatment/

Han, J., Zhang, F., Meerbeek, B. V., Vleugels, J., Braem, A., & Castagne, S. (2021). Laser surface texturing of zirconia-based ceramics for dental applications: A review. *Materials Science and Engineering: Biomaterials Advances*, *123*.

Helmer, J. D., & Driskell, T. D. (1969). Research on bioceramics. *Symposium on Use of Ceramics as Surgical Implants*. Clemson University, Clemson, SC.

Hench, L. L. (1991). Bioceramics: From concept to clinic. *Journal of the American Ceramic Society*, 1487–1510.

Hench, L. L., & Best, S. M. (2013). Ceramics, glasses, and glass-ceramics. *An Introduction to Materials in Medicine*, 153–170.

Hench, L. L., Splinter, R. J., Allen, W. C., & Greenlee, T. K. (1971). Bonding mechanisms at the interface of ceramic prosthetic materials. *Journal of Biomedical Materials Research*, *5*(6), 117–141.

Hua, S.-B., Yuan, X., Wu, J.-M., Su, J., Cheng, L.-J., Zheng, W., et al. (2022). Digital light processing porous TPMS structural HA & akermanite bioceramics with optimized performance for cancellous bone repair. *Ceramics International*, 3020–3029.

Ichikawa, Y., Akagawa, Y., Nikai, H., & Tsuru, H. (1992). Tissue compatibility and stability of a new zirconia in vivo. *Journal Prosthetic Dentistry*, *68*(2), 322–326.

Implants and *Prosthetics*. (2019, September 30). Retrieved from US Food and Drug: https://www.fda.gov/medical-devices/products-and-medical-procedures/implants-and-prosthetics

ISO Standards. (2022, September). Retrieved from International Organization for Standardization (ISO): https://www.iso.org/standards.html

Juntavee, N., & Attashu, S. (2018). Effect of different sintering process on flexural strength of translucency monolithic zirconia. *Journal of Clinical and Experimental Dentistry*, *10*(8), e821–e830.

Jurczyk, K., Niespodziana, K., Jurczyk, M. U., & Jurczyk, M. (2011). Synthesis and characterization of titanium-45S5 Bioglass nanocomposites. *Materials & Design*, *32*(5), 2554–2560.

Kalita, S. J., Bhardwaj, A., & Bhatt, H. A. (2007). Nanocrystalline calcium phosphate ceramics in biomedical engineering. *Materials Science and Engineering C*, *27*(3), 441–449.

Kantharia, N., Naik, S. D., Apte, S., Kheur, M., Kheur, S., & Kale, B. (2014). Nano-hydroxyapatite and its contemporary applications. *Journal of Dental Research and Scientific Development*, *1*(1), 15–19.

Kokubo, T., Kim, H. M., Kawashita, M., & Nakamura, T. (2004). Bioactive metals: Preparation and properties. *Journal of Materials Science – Materials in Medicine*, *15*(2), 99–107.

Kurtz, S., Ong, K., Lau, E., Mowat, F., & Halpern, M. (2007). Projections of primary and revision hip and knee arthroplasty in the United States from 2005 to 2030. *The Journal of Bone & Joint Surgery*, 780–785.

Kvam, K., & Karlsson, S. (2013). Solubility and strength of zirconia-based dental materials after artificial aging. *Prosthetic Dentistry*, *110*(4), 281–287.

LeGeros, R. Z., Lin, S., Rohanizadeh, R., Mijares, D., & LeGeros, J. P. (2003). Biphasic calcium phosphate bioceramics: Preparation, properties and applications. *Journal of Materials Science: Materials in Medicine*, 201–209.

Lin, J. C., Lu, J. X., Zeng, Q., Zhao, W., Li, W. Q., & Ling, J. Q. (2016). Comparison of mineral trioxide aggregate and calcium hydroxide for apexification of immature permanent teeth: A systematic review and meta-analysis. *Journal of Formosan Medical Association*, *115*(7), 523–530.

Lu, J., Blary, M. C., Vavasseur, S., Descamps, M., Anselme, K., & Hardouin, P. (2004). Relationship between bioceramics sintering and micro-particles-induced cellular damages. *Journal of Materials Science – Materials in Medicine*, *15*(4), 361–365.

Lughi, V., & Sergo, V. (2010). Low temperature degradation -aging- of zirconia: A critical review of the relevant aspects in dentistry. *Dental Materials*, *26*(8), 807–820.

Ma, H., Feng, C., Chang, J., & Wu, C. (2018). 3D-printed bioceramic scaffolds: From bone tissue engineering to tumor therapy. *Acta Biomaterialia*, *79*, 37–59.

Mishra, V. K., Bhattacharjee, B. N., Kumar, D., Rai, S. B., & Parkash, O. (2016). Effect of a chelating agent at different pH on the spectroscopic and structural properties of microwave derived hydroxyapatite nanoparticles: A bone mimetic material. *New Journal of Chemistry*, *40*(6), 5432–5441.

Nettleship, I., & Stevens, R. (1987). Tetragonal zirconia polycrystal (TZP) – A review. *International Journal of High Technology Ceramics*, *3*(1), 1–32.

Prakasam, M., Locs, J., Salma-Ancane, K., Loca, D., Largeteau, A., & Berzina-Cimdina, L. (2017). Biodegradable materials and metallic implants – A review. *Journal of Functional Biomaterials*.

Samavedi, S., Whittington, A. R., & Goldstein, A. S. (2013). Calcium phosphate ceramics in bone tissue engineering: A review of properties and their influence on cell behavior. *Acta Biomaterialia*, *9*(9), 8037–8045.

Sandhaus, S. (1965). British patent no. 1083.769.

Schrooten, J., & Helsen, J. A. (2000). Adhesion of bioactive glass coating to Ti6A14V oral implant. *Biomaterials*, *21*(14), 1461–1469.

Schulte, W., & Heimke, G. (1976). The Tübinger immediate implant. *Die Quintessenz*, 17–23.

Smith, L. W. (1963). Ceramic-plastic material as a bone substitute. *Archives of Surgery*, *87*(4), 653–661.

Stawarczyk, B., Ozcan, M., Hallmann, L., Ender, A., Mehl, A., & Hämmerlet, C. H. (2013). The effect of zirconia sintering temperature on flexural strength, grain size, and contrast ratio. *Clinical Oral Investigations*, *17*(1), 269–274.

Tu, M.-G., Ho, C.-C., Hsu, T.-T., Huang, T.-H., Lin, M.-J., & Shie, M.-Y. (2018). Mineral trioxide aggregate with mussel-inspired surface nanolayers for stimulating odontogenic differentiation of dental pulp cells. *Journal of Endodontics*, *44*(6), 963–970.

Wang, J., & Stevens, R. (1989). Zirconia-toughened alumina (ZTA) ceramics. *Journal of Materials Science*, *24*, 3421–3440.

Wang, S., Shi, Z., Liu, L., Huang, Z., Li, Z., Liu, J., & Hao, Y. (2021). Honeycomb structure is promising for the repair of human bone defects. *Materials & Design*, *207*.

Wu, S., Liu, X., Yeung, K. W., Liu, C., & Yang, X. (2014). Biomimetic porous scaffolds for bone tissue engineering. *Materials Science and Engineering: R: Reports*, *80*, 1–36.

Xiao, Y., Song, L., Liu, X., Huang, Y., Huang, T., Chen, J., et al. (2011). Bioactive glass-ceramic coatings synthesized by the liquid precursor plasma spraying process. *Journal of Thermal Spray Technology*, *20*(3), 560–568.

Yamamuro, T., Hence, L., & Wilson, J. (1990). *CRC Handbook of Bioactive Ceramics*. CRC Press.

Yuan, H. P., Fernandes, H., Habibovic, P., Boer, J. d., Barradas, A. M., Ruiter, A. d., et al. (2010). Osteoinductive ceramics as a synthetic alternative to autologous bone grafting. *Proceedings of the National Academy of Sciences of the USA*, *107*, 13614–13619.

Zhang, M., Pu, X., Chen, X., & Yin, G. (2019). In-vivo performance of plasma-sprayed CaO-MgO-SiO2-based bioactive glass-ceramic coating on Ti-6Al-4V alloy for bone regeneration. Heliyon, 5(11).

14 Standards of Mechanical, Physical, Chemical, and Biological Properties of Bioceramics

A. Sen, N. Banerjee, A.R. Biswas, T.K. Ghosh, A. Samanta, M. Kumar, S. Das, U. Srivastava, and D. Sengupta
Calcutta Institute of Technology

S.R. Maity
NIT Silchar

14.1 INTRODUCTION

Ceramics is a term used for hard, brittle, and heat-resistant inorganic materials, mostly made from clay hardened by heat, like earthenware, porcelain, figurines, etc. Ceramics are inorganic materials, most of which are manufactured from clay hardened by heat, such as earthenware, porcelain, and figurines [1]. Through the decades, the usage and development of ceramics have evolved in various fields.

Clay, silica (quartz), and feldspar are the three main ingredients in traditional ceramics. However, with recent advancements, the addition of different compounds/elements results in new properties (while maintaining the basic composition) and contradicts traditional aspects. Ceramics are usually solid substances having high hardness and are brittle in nature. Once synthesized, they have a high melting point, medium tensile strength, low ductility, and negligible electrical and thermal conductivity. On the contrary, they can withstand high temperatures and dimensional stability along with optical transparency depending on composition. Materials that are implanted into the human body are exposed to a hostile environment while also being exceedingly sensitive. The body's tissues are particularly sensitive to foreign material. It can be easily stimulated into presenting signs of poisoning and rejection in bioceramics. As a result, the effectiveness of an implant in the body is determined by a variety of factors, including material quality, design and biocompatibility. Bioceramics have gained acceptance as a group of materials that are used in medical applications, primarily for orthopedic, maxillofacial, and dentin fields.

DOI: 10.1201/9781003258353-18

TABLE 14.1
Types and Classification of Bioceramics

Bioceramics		
Bio-inert	**Bioactive**	**Biodegradable (Bioresorbable)**
Al_2O_3 ZrO_2 SiC Pyrolytic carbon	Bioglass Glass-ceramic Hydroxyapatite HCA	$CaSO_4$ $CaCo_3$ Calcium phosphates (Di/Tri/Tet/Octa) Tet-CP

In this chapter, the details of mechanical, physical, chemical, and biological properties are discussed based on the classification of bioceramics. Table 14.1 represents types and classifications of bioceramics.

14.2 MECHANICAL PROPERTY

The typical characteristics of any material under numerous mechanical loading circumstances are referred to as a material's mechanical properties. While discussing characteristic values, refer to the numerical measurements of mechanical parameters at which particular material variations might take place. For instance, strain and stress are mechanical factors that are frequently used to describe a material but are not actual characteristics of the material. In this subsections, mechanical properties of various bioceramics have been studied and discussed.

14.2.1 Bio-inert

Bio-inert ceramics serve as a long-term replacement between dental and bone implants by remaining chemically inert, that is, not interacting with the environment of the human body. Mechanical characteristics of several bioceramics have been covered in the subsections that follow.

14.2.1.1 Aluminum and Zirconia

Alumina (as Al_2O_3) and zirconia (as ZrO_2) are the most extensively used dental bioceramic materials. Because of their better mechanical qualities, alumina (Al_2O_3) and zirconia (ZrO_2) were developed as ceramic materials for implant abutments. In addition, ZrO_2 is inert in the physiological environment and has higher flexural resistance and toughness along with a lower Young's modulus [2].

14.2.1.2 Silicon Carbide

SiC fabrics are often used as a possible supplement in matrix composites due to their advantageous mechanical and thermal qualities. A research study demonstrated the electro spinning of submicrometer SiC fibers using poly-carbosilane as a precursor, which were then air-cured and pyrolyzed at high temperatures. SiC fiber mats were produced at 1,300°C [3]. Enhancing the attitude of precursor fibers reduces defects in fibers, and the arrangement of polymeric structure in precursor fibers are the variables that contribute to increased mechanical strength following hot-drawing.

14.2.1.3 Pyrolytic Carbon

Pyrolytic carbon is part of turbostratic carbon, which has a graphite-like structure. The layers of pyrolytic carbon and other turbostratic carbons are disorganized, causing wrinkles and distortions. In comparison with graphite, pyrolytic carbon has higher durability. By altering the pyrolytic carbon processing parameters, it is possible to change the mechanical properties of pyrolytic carbon, including strength, elastic modulus, toughness, hardness, wear resistance, and fatigue resistance [4,5].

14.2.2 Bioactive

The ceramic materials are utilized as phase two in composites or as a protective layer on substrates, fusing their respective qualities to create a new material that has greater mechanical, biochemical, physical, and biological capabilities.

14.2.2.1 Bioactive Ceramics

For decades, ceramics (bioactive) have been used for repairing broken bones and teeth. However, as a result of their intrinsic bitterness and low fracture toughness, inorganic ceramic materials are still not widely used. Glass-ceramics and bioceramic/polymer composites are promising for boosting both longevity and strength. Specifically, interfacial interaction links and mechanical compatibility between bioactive implantation and the host tissue are essential [6–13]. There has been a lot of interest in using bioactive glasses in contact with bone and dental tissue because of their both inorganic origin and mechanical abilities, which are similar to "hard" bone tissue.

14.2.2.2 Bioglass Ceramics

Biologically active compounds glasses partially crystallize after thermal expansion heating, resulting in glass-ceramic materials with better mechanical properties than the original glass. Crystalline phases are present in glass-ceramics, which have an amorphous substrate. When contrasted to the source glass, the crystalline structures provide glass-ceramics greater strength and enhanced fatigue strength. The most effective toughening processes in this regard were discovered to be crack spanning and crack redirection [13–18]. Orthopedics and dentistry are the primary fields in which bioactive glass-ceramics are used.

14.2.3 Bioresorbable (Biodegradable)

Bio-absorbable polymers are utilized in implanted medical devices to provide a physical structure that aids healing and/or acts as a delivery system for medications. These polymers have the benefit of being able to dissolve once their therapeutic goal has been met [19–21].

14.2.3.1 Calcium Sulfate

Calcium sulfate hemihydrate bioceramic has a long history of use in medicine due to its exceptional properties. These applications include skeletal transplants, core rupture repair, ingredients for three-dimensional manufacturing, medication vesicles, and root-end filler. They also hold great promise for the production of scaffolds in bone tissue engineering. Other therapeutic advantages of calcium sulfate include its potential to produce local hemostasis following tooth extractions in anticoagulant-treated patients without requiring them to stop taking their medicine before surgery, as well as for local hemostasis in endodontic microsurgery. Compressive strength of wet and dry $CaSO_4$ is found to be 10–15 and 20–30 MPa, whereas the tensile strength ranged between 2–4 and 4–6 MPa, respectively [22].

14.2.3.2 Calcium Carbonate

The coral has been clinically used as bone substitute. Admittedly, the calcium carbonate composite ceramics are forged in sintering process at a low temperature for a brief period of time, while, at low temperature the biocompatible phosphate-based glass (PG) emanated liquid, as the cause of sintering in the sintering process. According to phase analyses, the calcium carbonate composite ceramics produced novel compounds rather than calcium oxide. Thus, with the increase of the additions of PG, the compressive strength also rose accordingly [23].

14.2.3.3 Calcium Phosphate

Phosphate cements originated more than a century ago for specific applications to the field of dentistry. These cements have other applications like as bioceramics with bone regeneration applications, stabilization of hazardous waste, and structural cements. Natural calcium phosphates derived from fish wastes are a promising material for biomedical application. However, their sintered ceramics are not fully characterized in terms of mechanical and biological properties [24]. CaP biomaterials are commonly used nowadays to enhance bone tissue regeneration [25].

14.2.3.4 Tricalcium Phosphate

Tricalcium phosphate (TCP; $Ca_3(PO_4)_2$) is a calcium phosphate with the Ca/P ratio of 1.5 and is splitted into α-phase and β-phase; α -TCP has the crystalline structure of a monoclinic spatial group, and β-TCP has the crystalline structure of a rhombohedral space group [26,27]. Temperatures between 900°C and 1,100°C are required for the formation of β-TCP, and α-TCP can occur at 1,125°C or greater [28]. In contrast to α-TCP, β-TCP possesses a more stable form as well as a faster rate of biodegradation. As a result, β-TCP is frequently employed in bone regeneration. Although less stable than HAP, β-TCP degrades more quickly and is more soluble, used to promote biocompatibility and has a high rate of resorption [29,30]. Osteoblasts as well as bone marrow stromal cells, two types of osteoprecursor cells, are encouraged to proliferate by β-TCP [31,32]. In bone cements including bone replacement, β-TCP is frequently employed [33,34]. Since each component is thoroughly and intimately combined at the submicron level, the biphasic and multi-phasic calcium phosphates do not exist in a form that is separable [35,36]. The biphasic calcium phosphates have mostly been assessed in terms of their bioactive components, bioresorbability, and tissue

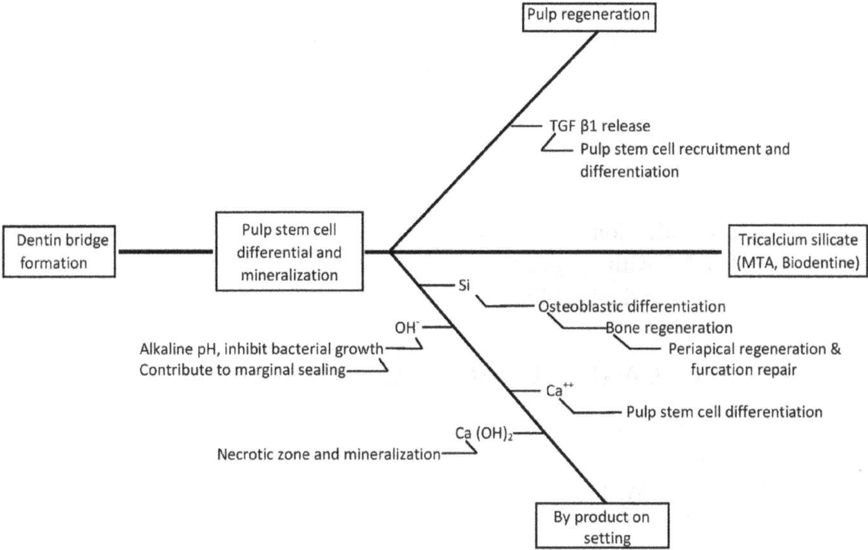

FIGURE 14.1 Schematic representation of silicium ions in tricalcium silicates to directly cause cell differentiation tricalcium silicates.

regeneration [37,38]. Biphasic calcium phosphates are used as dental materials, bone substitutes, and bone transplants. The combination of HAP and β-TCP (gel-polymer techniques have been used to create) has been actively used to promote mesenchymal stem cells' osteogenic differentiation, boost cell adhesion, connect growth factors, and improve mechanical properties [39–43]. It was covered that the bi-layer calcium phosphate scaffolds had microporous structures that affected vascularization and cell growth. The most reliable and almost twice as high composite scaffold yield strengths were found for Mg-Zn/5TCP and Mg-Zn/10TCP [44]. Figure 14.1 shows how silicium ions in tricalcium silicates directly cause cell differentiation. During hydration (OH), Ca^{++} ions, $Ca(OH)_2$, and hydroxyl are formed. These ions have a big impact because they help with mineralization and the eradication of microorganisms. When MTA and biodentine come into touch with the pulp, TGF-1, which controls the motility and division of pulp stems, is created.

14.3 PHYSICAL PROPERTIES

Bioceramics are highly resistant to corrosion and chemical attack and possess both ionic and covalent bonds, and their density lies between polymers and metals. Because of their capacity to work in close proximity to living tissue, bioceramics have revolutionized medical applications and research. As MTA (calcium silicate-based bioceramics) is asserted to be prominent in endodontic restorative operations and biodentine owing to its physio-biochemical properties [45]. When compared to pure Al_2O_3, the ZTA (zirconia-toughened alumina) composite has a 22% high density and microstructure that are extremely homogeneous and fine along with reduced porosity [2]. W/TCP composite ceramics has inherent qualities that are essential to the mechanism of bone neoformation [46].

The crystal phase of all the CaP bio ceramics obtained was biphasic and composed of hydroxyapatite (HA) and tricalcium phosphate (TCP). The density and microhardness of the CaP bioceramics increased in the temperature, while at temperatures higher than 1,100°C, these properties were not significantly altered, while the highest compressive strength is 116 MPa [47]. The characteristics results from the superior biomineralization and lattice provided by the β-TCP's nanoporous structure [48]. TCP-TiNP with the greatest TiNP content increases surface roughness and porosity by more than double [49].

14.4 CHEMICAL AND BIOCHEMICAL PROPERTIES

Ceramic is an inorganic, nitrate, carbide, or nitrite from a biological perspective. A metal cation and an oxide anion are both present in crystalline solids known as metal oxides. SiO_2 may make up around 70% of a typical composition, along with Al_2O_3, Fe_2O_3, CaO, K_2O, and Na_2O [50]. The number of applications increases as new ceramic varieties are created by researchers for several uses. The following are a few examples from the viewpoint of chemical characteristics.

14.4.1 Silicon Base Ceramics

Sialon (silicon aluminum oxynitride) is a high-strength, thermally shock-resistant, chemically and wear-resistant material with a low density. Bulletproof vests made from ceramics such as alumina, boron carbide, and silicon carbide are used to protect against small arms rifle fire. Silicon carbide (SiC), a refractory material, is employed as a susceptor in microwave furnaces. The abrasive powder silicon nitride (Si_3N_4) is also used to make ceramic ball bearings.

14.4.2 Zinc-Copper-Barium Base Ceramics

Yttrium barium copper oxide ($YBa_2Cu_3O_7$) is a high-temperature superconductor, while zinc oxide (ZnO) is a semiconductor used to make varistors. Steatite (magnesium silicates) acts as an electrical insulator. The ceramic uranium oxide (UO_2) is utilized as nuclear reactor fuel. Ceramic knife blades, diamonds, fuel cells, and oxygen sensors are all produced of zirconia (zirconium dioxide, ZrO_2). Superconductors include bismuth strontium copper oxide (Cu_2O) and magnesium diboride (MgB_2). As an electrical component, steatite (magnesium silicate, $MgO.XSiO_2.H_2O$) is utilized. Heating elements, capacitors, transducers, and data storage devices are all made of barium titanate ($BaTiO_3$), which has ferroelectric properties. There are a lot of such ceramics which are synthesized by different processes, and one major section is the bioceramics [51].

14.4.3 Area of Application

Bioceramics are an important subset of biomaterials [52]. In recent advancements, a new bredigite-magnetite nanocomposite scaffold for biomedical applications has been discovered, with good in vitro bioactivity and great biocompatibility. It can be useful for the bone restoration and cancer therapy. Bredigite $Ca_7MgSi_4O_{16}$, an essential silicate bio ceramic, was introduced to control apatite production and the biodegradation rate of physiological saline [53]. Figures 14.2 and 14.3 represent various field of applications of bioceramics in various medical implants.

The fact that implants come into direct contact with living tissue is one of the key applications for bioceramics. The deterioration of dental health with age has grown to be a serious issue. Particularly in light of the ageing population around the world,

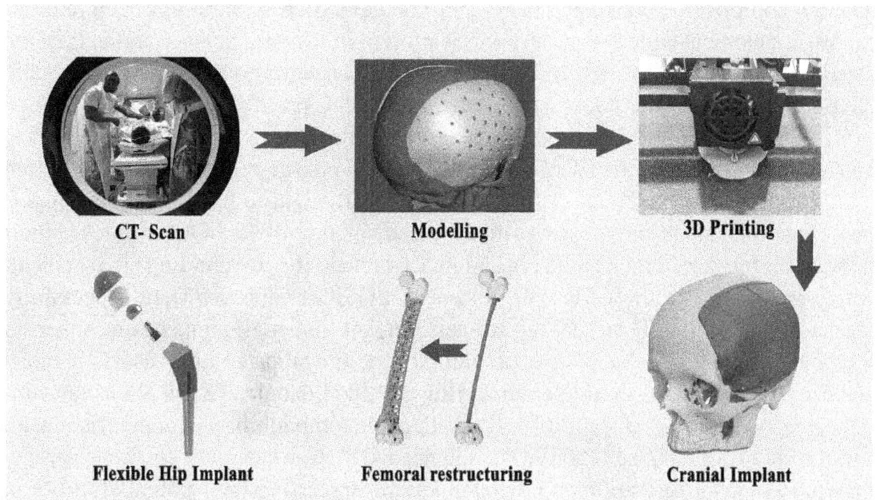

FIGURE 14.2 Various applications of bioceramics.

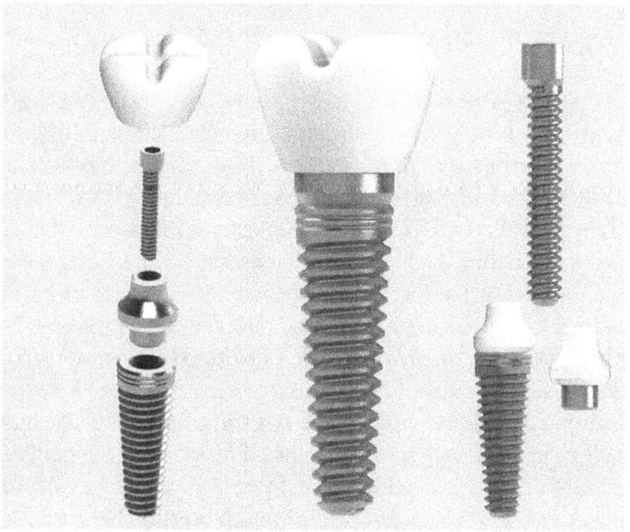

FIGURE 14.3 Dental applications of bioceramics.

oral disease can have a substantial impact on a person's quality of life. Particularly in light of the ageing population around the world, oral disease can have a substantial impact on a person's quality of life. Zirconia (ZrO_2)-based ceramics are extensively used in dental implants and total joint replacements due to their superior mechanical properties and biocompatibility [54]. Compared to titanium implants, ZrO_2-based ceramics stimulate soft tissue more, leading to healthier surrounding tissue and less bacterial development [55]. ZrO_2-based ceramics have a rather low ageing resistance [56] despite these benefits. $CaSiO_3$ ceramics have been identified as a bioactive substance for bone regrowth. As a bioactive bone regeneration material, wollastonite ($CaSiO_3$ composition consisting of 48.28% CaO and 51.72% SiO_2) has been referred to. A trace element called strontium (Sr) is present in the human body and aids in the development of strong bones. Even at low extract concentrations (ranging from 12.5, 25 to 50 mg/mL), Sr-$CaSiO_3$ is seen to promote HBDC proliferation. Sr can be added to $CaSiO_3$ to form this compound. In the last three decades, bioglass, glass-ceramics, and bioceramics containing CaO and SiO_2 for bone tissue regeneration have gotten a lot of interest [57]. Calcium silicate ($CaSiO_3$) is thought to have bioactive properties. However, one of the most significant disadvantages of $CaSiO_3$ ceramics is their fast-dissolving rate, which results in a high pH value in the surrounding environment, which is harmful to the cells [58]. Sphene ($CaTiSiO_5$) ceramics were successfully produced by adding Ti in $CaSiO_3$ sol-gelsolin–gel and high-temperature sintering methods. It was found that adding titanium to calcium silicate improves its chemical stability and biological characteristics. However, its biological applications are limited by its poor chemical stability and excellent biocompatibility. In comparison with $CaSiO_3$ ceramics, sphene ($CaTiSiO_5$) enhanced HBDC attachment and considerably increased their proliferation and differentiation, suggesting their potential utility in skeletal tissue regeneration as well as a coating onto currently available orthopedic/dental implants. In recent years, many researchers have been working on enhancing the efficiency of implant materials through surface modification utilizing ceramic coatings [59].

Metallic implants are protected by a variety of coatings, including hydroxyapatite (HA), biopolymers, and transition metal nitrides (such as TiN, TiON, and TiAlN) [12]. Transition metal carbides (TMC) have recently demonstrated appealing features such as high bodily fluid inertness, great biocompatibility, and outstanding tribological properties [60]. TiC and ZrC have greater biocompatibility when compared to other carbide systems [61]. Hydroxyapatite (HA), fluorapatite (FA), and β-tricalcium phosphate (β-TCP) are the most common bioactive ceramic coatings. The coating's calcium and phosphate ions are vital in osteogenesis, as they can trigger and encourage the formation of new bone tissue on its surface [62]. The piezoelectric effect of biological piezoelectric materials also promotes bone development. Slip casting was used to create hydroxyapatite (HA)/barium titanate ($BaTiO_3$) composite materials, and polarization was used to determine their piezoelectric characteristics. Additionally, human bone tissue is a sort of piezoelectric material, which means that biological electricity can be produced by the displacement of electrons in a local electric field caused by a deformation force [63]. Utilizing calcium phosphate-based ceramics, tissue engineering scaffolds and bioresorbable materials have been produced. Hydroxyapatite (HAP, $Ca_{10}(PO_4)_6$-$(OH)_2$) and b-tricalcium phosphate (b-TCP, b-$Ca_3(PO_4)_2$) are the two most common calcium phosphate compositions [64]. Because of its biological activity and likeness to the inorganic components of human bone, hydroxyapatite (HA) is commonly used in the creation of bone repair materials [65]. The mineral hydroxyapatite (HA; $Ca_5(PO_4)_3OH$), which makes up the hard tissues and bones of animals, has garnered a lot of attention lately, particularly in the study of biomaterials. This is as a result of its significance in clinical applications, such as in the development of medical implants and devices, as well as more recently in the more general field of tissue engineering [66]. The piezoelectric ceramics with a 10% HA/90% $BaTiO_3$ composition exhibited the best biocompatibility and bone-inducing activity.

Akermanite, diopside, and baghdadite, among other silicate bioceramics, have recently been used in the biomaterials sector to treat bone cancer disorders [67–70]. In comparison with calcium phosphate ceramics, several unique published researches indicated that various glass-ceramics containing MgO, CaO, and SiO_2 had suitable biocompatibility and biological behavior [71]. Silicate bioceramics have the potential for apatite production on their surface due to chemical reactivity in the simulated bodily fluid [72]. By forming robust oxide passive layers on substrates coated with ceramic carbide, which may improve corrosion resistance properties over uncoated stainless steel substrates, surface coatings with superior corrosion resistance properties are essential for the biomedical implant industry to reduce cytotoxicity issues. Bacterial growth was drastically decreased by carbide coatings (*P. aeruginosa*). The microplasma method, also known as plasma electrolyte oxidation or micro-arc oxidation in aqueous electrolyte solutions, is a quick and effective way to coat metals like titanium with bioactive compounds. Coating metallic biomaterials, mainly titanium alloys, with calcium phosphate is of interest. Furthermore, zirconium alloyed with niobium is a potential implant material since it is both tougher and more biocompatible than titanium [73]. Metallic implant materials include austenitic 316L stainless steel (SS), titanium (Ti), and titanium alloys ($Ti_6A_{14}V$), which are commonly utilized to make prosthetic hips, knees, and dental replacements [74]. Because of its ease of production, cost-effectiveness, and biocompatibility, the use of stainless steel is still essential [75]. Stainless steel materials are frequently utilized, although they are weak to pitting corrosion, limiting their variety of uses [76].

Ceramics reinforced with carbon nanotubes (CNTs) are biocompatible and have an elastic modulus similar to bone tissue [30]. Graphene, for example, has a fracture resistance of 125 GPa, making it a better prospect for usage as a reinforcing phase in composite ceramics [77,78,79].

14.5 BIOLOGICAL PROPERTIES OF BIOCERAMICS

Biological implants may still have issues with a paucity of bone supplies, transporting infections, and/or causing immunological rejection, despite the fact that autologous, allogeneic, and xenogeneic bone grafts have usually been allowed to be used for bone repair [80,81]. The initial reaction of either the surface or the entire resource to the biomedium, molecular convergence of a biocompatible implant with bone tissues, or total replacement by bio-absorbable material of healthy bone tissues are the most significant differences between bioactive bioceramics and all other implanted composites. Using bone replacement from corpses or animals, for example, leads the organism to have highly negative immune reactions, significantly restricting their use. In this case, biologically bioceramic implants, particularly calcium orthophosphates, can be entirely resorbed over time [82]. A foreign substance's complete compatibility with a living body has yet to be demonstrated. Only autogenous (tissue-produced chemicals) and exogenous (host–tissue reaction) compounds are completely compliant. In a time-dependent manner, the mechanisms that occur at perishable materials contacts modify the surface options of the deep-rooted material and therefore the fibroblasts that live at the interface [83]. Understanding in vivo host reactions is critical for creating unique things. Biomaterials and bioceramics, like all living things, catalyze the reaction to their environment, and they should not cause any changes or trigger any unwanted responses in nearby or distant tissues, in principle.

A bioactive substance will partially dissolve but will also cause the development of a physiological apatite outer surface while interacting with tissue at the nanoscale, resulting in direct chemical interactions with bone [84]. A bioresorbable substance degrades, allowing newly generated tissue to grow into any surface abnormalities, although it may or may not interact with the material directly. As a result, bioabsorbable materials serve as clear media or fillers that allow tissues to penetrate and migrate [85]. A biofluid-filled gap emerges close to the implant surface so a bioceramic implantation is anchored to the body. Ideally, the implantation and the neighboring cells develop a solid connection [86].

Many of the properties of CaP bioceramic materials may be found in bone cements, composites, scaffolds, and coatings, to name a few. Cement-free hip implants and dental implants are the most essential CAP criteria for medical applications. Oral surgery and craniofacial applications for bone replacement have shown promise using CPCs. CPC is also used to encourage bone growth in osteonecrotic areas of the body [87].

Bioceramics products of biphasic calcium phosphates (BCP) are the favored fabric for quite a few orthopedic and maxillofacial bone restoration operations. Their biodegradation rate, which can be altered through changing the proportionate ratio of the thing phases, is certainly considered one of their key advantages. The bioactivity of BCP needs to be promoted through optimizing their physicochemical characteristics for advanced bone tissue regeneration [88]. Bioceramic materials with nanoparticles may be more suitable for future applications [89].

Because of their excessive bioactivity, silicate bioceramics have obtained a variety of interests for several packages in gentle tissue regeneration. Silicate bioceramics are without difficulty included into silicate bioceramics to endow them with greater organic properties, consisting of improved angiogenesis, antibiosis, osteogenesis, and antitumor effects, which notably ensures the utility of silicate bioceramics primarily based on their inherent capacity to restore tissue. Furthermore, tumor therapy has made extensive use of nanosilicate bioceramics with distinctive architectures. The number of unique uses of silicate bio ceramics for tissue regeneration and tumor therapy has increased dramatically in recent years [90].

Titanium is the fabric of desire for dental implants and abutment additives because of its biocompatibility, strength, and corrosion resistance. Dental cement is used in cemented implant restorations to keep the crown attached to the abutment and to provide access to the implant. Due to its persistent presence in subgingival tissues, dental cement has been linked to inflammation and, in some circumstances, peri-implant [91].

Eggshell trash is one of nature's most considerable herbal wastes, and it is produced in large portions through the daily ingesting of eggs. Despite the reality that it has interdisciplinary packages, the bulk of this cloth is discarded. The extra specialized packages along with the use of ball-milled eggshell include drug transport agent or the advent of antibacterial energetic species [92].

Biowaste from fish scales contains tons of aluminous and metallic element phosphates; thus, it is a lot of promise as a staple for added biomaterials like hydroxyapatite (HAP). Bioceramic materials are created from HAP that is that the principal constituent part of laborious tissue within the human body, like bone and teeth [93].

Tricalcium salt compounds are proverbial to supply a waterproofing barrier that protects the underlying pulp from the restorative material's potential toxicity like residual microorganism and their toxins. The alkaline pH produced locally by the pulp capping substance may be responsible for this impact [94].

These novel calcium silicate-based restorative materials cements have shown that they have the ability to sustain tissue viability throughout the development of dental organs and the apical root's closing. Calcium hydroxide has been used for a long time [95]. HiFlow/CW is more effective at sealing than the iRoot SP/SC approach in the apical third due to its favorable biological characteristics and enhanced production of osteo/cementogenic genes [96].

14.6 CONCLUSION

Since the biodegradable nature permits avoiding the second operation and lessening in the suffering and cost for patients, biodegradable materials for bone regeneration and repair are actively sought after and create a lot of interest in the area of bioceramics research. Biodegradable materials, such as natural and artificial polymers, bioceramics, and metals based on magnesium, are currently being developed as well as being used in therapeutic settings. Different bioceramics have varying mechanical characteristics, biological behaviors, and biodegradation mechanisms. As the stress elongation and tensile strength of magnesium alloys are higher than those of polymers and bioceramics, the ceramic materials are the ones that have the highest amount of brittleness. It has been demonstrated that bioceramics and mg alloys promote the formation of new bone more than polymers do from a biological standpoint. This is due to both the bioactive nature of magnesium alloys and the osteoconductive and occasionally osteoinductive features of ceramics. An inflammatory reaction is usually induced by the acidic breakdown products of different polymeric materials; however, this is not the case when using bioceramics. Among the most crucial properties for biomaterials that degrade is the pace and amount of degradation. Bioceramics deteriorate and exhibit in vivo resorption through cell-mediated and solution-driven mechanisms, and they gradually give way to lamellar real bone. Enzymolysis and hydrolysis are the main methods used by biodegradable polymers to break down from macromolecules to molecules, and ultimately to water and carbon dioxide.

At the beginning of polymeric deterioration, the mechanical strength declines gradually; however, during bulk degradation, it drops quickly. Magnesium and magnesium alloys corrode in bodily fluid, and while this process starts off with a very fast deterioration rate, it gets increasingly slower with time. Due to the fact that their internal structures do not alter during deterioration, magnesium alloys retain their mechanical strength. Due to their inherent drawbacks, traditional metallic prostheses made from non-biodegradable materials are quickly losing popularity. Highly efficient implant biomaterials in osteologic repair applications are being developed using biodegradable and bioactive composites. Based on the knowledge obtained from previous research, it is anticipated that the next generations of biodegradable materials would display substantial advances in implant and biological implant site. Future biomaterials for bone regeneration and repair must be developed via extensive study.

REFERENCES

1. "Ceramic history". depts.washington.edu. Archived from the original on 2020-11-06. Retrieved 2020-03-02.
2. Parvin, T. Structural and mechanical properties of Al_2O_3-ZrO_2 dental bioceramics (2012).
3. Yu, P., Z. Lin, Y. Mu, and J. Yu. Highly flexible and strong SiC fibre mats prepared by electrospinning and hot-drawing. *Adv Appl Ceram* 120, no. 3 (2021): 144–155.
4. Ross, M., C. James, G. Couzens, and J. Klawitter. Pyrocarbon small joint arthroplasty of the extremities. In *Joint Replacement Technology*, pp. 628–673. Woodhead Publishing, 2014.
5. Boehm, R. D., C. Jin, and R. J. Narayan. Carbon and diamond. *Compr Biomater* 1 (2011): 109–126.
6. Molla, A. R., and B. Basu. Microstructure, mechanical, and in vitro properties of mica glass-ceramics with varying fluorine content. *J Mater Sci: Mater Med* 20, no. 4 (2009): 869–882.
7. Filho, O. P., G. P. La Torre, and L. L. Hench. Effect of crystallization on apatite-layer formation of bioactive glass 45S5. *J Biomed Mater Res* 30, no. 4 (1996): 509–514.
8. Kothiyal, G. P., A. Ananthanarayanan, and G. K. Dey. Glass and glass-ceramics. In *Functional Materials Preparation, Processing and Applications*, pp. 323–386. Elsevier Inc., London, 2012.
9. Turck, C., G. Brandes, I. Krueger, P. Behrens, H. Mojallal, T. Lenarz, and M. Stieve. Histological evaluation of novel ossicular chain replacement prostheses: An animal study in rabbits. *Acta oto-laryngologica* 127, no. 8 (2007): 801–808.
10. Vogt, J. C., G. Brandes, N. Ehlert, P. Behrens, I. Nolte, P. P. Mueller, T. Lenarz, and M. Stieve. Free Bioverit(r) II implants coated with a nanoporous silica layer in a mouse ear model – A histological study. *J Biomater Appl* 24, no. 2 (2009): 175–191.
11. Crovace, M. C., M. T. Souza, C. R. Chinaglia, O. Peitl, and E. D. Zanotto. Biosilicate(r) – A multipurpose, highly bioactive glass-ceramic. In vitro, in vivo and clinical trials. *J Non-Crystal Solids* 432 (2016): 90–110.
12. Hill, R. G. Bioactive glass-ceramics. In: Ducheyne P., ed. *Comprehensive Biomaterials, Volume 1: Metallic, Ceramic and Polymeric Biomaterials*, pp. 181–186. Elsevier, Netherlands, 2011.
13. Hill, R., and D. Wood. Apatite-mullite glass-ceramics. *J Mater Sci: Mater Med* 6, no. 6 (1995): 311–318.
14. El-Meliegy, E., and R. Van Noort. *Glasses and Glass Ceramics for Medical Applications*. Springer Science & Business Media, 2011.

15. Moisescu, C., C. Jana, S. Habelitz, G. Carl, and C. Rüssel. Oriented fluoroapatite glass-ceramics. *J Non-Crystal Solids* 248, no. 2–3 (1999): 176–182.
16. Habelitz, S., G. Carl, C. Rüssel, S. Thiel, U. Gerth, J. D. Schnapp, et al. Mechanical properties of oriented mica glass ceramic. *J Non-Crystal Solids* 220, no. 2–3 (1997): 291–298.
17. Apel, E., W. Höland, and V. Rheinberger. U.S. patent No. 7,074,730. Washington, DC: U.S. Patent and Trademark Office, 2006.
18. Höland, M., A. Dommann, W. Holand, E. Apel, and V. Rheinberger. Microstructure formation and surface properties of a rhenanite-type glass-ceramic containing 6.0 wt% P2O5. *Glass Sci Technol (Frankfurt)* 78, no. 4 (2005): 153–158.
19. Kasuga, T. Bioactive calcium pyrophosphate glasses and glass-ceramics. *Acta Biomater* 1, no. 1 (2005): 55–64.
20. Onwubu, S. C., P. S. Mdluli, S. Singh, and S. Thakur. Biodegradable natural materials in dentistry: Fiction or real? In *Applications of Advanced Green Materials*, pp. 77–88. Woodhead Publishing, 2021.
21. Sheikh, Z., S. Najeeb, Z. Khurshid, V. Verma, H. Rashid, and M. Glogauer. Biodegradable materials for bone repair and tissue engineering applications. *Materials* 8, no. 9 (2015): 5744–5794.
22. Syam, S., Y. C. Cho, C. M. Liu, M. S. Huang, W. C. Lan, B. H. Huang, et al. (2020). An innovative bioceramic bone graft substitute for bone defect treatment: In vivo evaluation of bone healing. *Appl Sci* 10, no. 22, 8303.
23. He, F., F. Yang, J. Zhu, Y. Peng, X. Tian, and X. Chen. Fabrication of a novel calcium carbonate composite ceramic as bone substitute. *J Am Ceram Soc* 98, no. 1 (2015), 223–228.
24. Bas, M., S. Daglilar, N. Kuskonmaz, C. Kalkandelen, G. Erdemir, S. E. Kuruca, et al. Mechanical and biocompatibility properties of calcium phosphate bioceramics derived from salmon fish bone wastes. *Int J Mol Sci* 21, no. 21 (2020): 8082.
25. Piconi, C., and S. Sprio. Editorial for the special issue on bioceramic composites. *J Compos Sci* 6, no. 3 (2022): 65.
26. Dickens, B., L. W. Schroeder, and W. E. Brown. Crystallographic studies of the role of Mg as a stabilizing impurity in β-Ca3 (PO4) 2. The crystal structure of pure β-Ca_3 (PO_4) 2. *J Solid State Chem* 10, no. 3 (1974): 232–248.
27. Mathew, M., L. W. Schroeder, B. Dickens, and W. E. Brown. The crystal structure of α-Ca_3 (PO_4) 2. *Acta Crystal Sect B: Struct Crystal Cryst Chem* 33, no. 5 (1977): 1325–1333.
28. Yubao, L., Z. Xingdong, and K. De Groot. Hydrolysis and phase transition of alpha-tricalcium phosphate. *Biomaterials* 18, no. 10 (1997): 737–741.
29. Horch, H-H., R. Sader, C. Pautke, A. Neff, H. Deppe, and A. Kolk. Synthetic, pure-phase beta-tricalcium phosphate ceramic granules (Cerasorb(r)) for bone regeneration in the reconstructive surgery of the jaws. *Int J Oral Maxillofac Surg* 35, no. 8 (2006): 708–713.
30. Yamada, S., D. Heymann, J-M. Bouler, and G. Daculsi. Osteoclastic resorption of calcium phosphate ceramics with different hydroxyapatite/β-tricalcium phosphate ratios. *Biomaterials* 18, no. 15 (1997): 1037–1041.
31. Yao, C-H., B-S. Liu, S-H. Hsu, Y-S. Chen, and C-C. Tsai. Biocompatibility and biodegradation of a bone composite containing tricalcium phosphate and genipin crosslinked gelatin. *J Biomed Mater Res Part A* 69, no. 4 (2004): 709–717.
32. Liu, H., Q. Cai, P. Lian, Z. Fang, S. Duan, X. Yang, X. Deng, and S. Ryu. β-tricalcium phosphate nanoparticles adhered carbon nanofibrous membrane for human osteoblasts cell culture. *Mater Lett* 64, no. 6 (2010): 725–728.
33. Bi, L., W. Cheng, H. Fan, and G. Pei. Reconstruction of goat tibial defects using an injectable tricalcium phosphate/chitosan in combination with autologous platelet-rich plasma. *Biomaterials* 31, no. 12 (2010): 3201–3211..

34. Luginbuehl, V., K. Ruffieux, C. Hess, D. Reichardt, B. Von Rechenberg, and K. Nuss. Controlled release of tetracycline from biodegradable β-tricalcium phosphate composites. *J Biomed Mater Res Part B* 92, no. 2 (2010): 341–352.
35. Dorozhkin, S. V. Biphasic, triphasic and multiphasic calcium orthophosphates. *Acta Biomater* 8, no. 3 (2012): 963–977.
36. Ellinger, R. Histological assessment of periodontal osseous defects following implantation of hydroxyapatite and biphasic calcium phosphate ceramics: A case report. *Int J Periodont Restor Dent* 6 (1986): 23–33.
37. Daculsi, G. Biphasic calcium phosphate concept applied to artificial bone, implant coating and injectable bone substitute. *Biomaterials* 19, no. 16 (1998): 1473–1478.
38. Lobo, S. E., and T. L. Arinzeh. Biphasic calcium phosphate ceramics for bone regeneration and tissue engineering applications. *Materials* 3, no. 2 (2010): 815–826.
39. Daculsi, G., S. Baroth, and R. LeGeros. 20 years of biphasic calcium phosphate bioceramics development and applications. *Adv Bioceram Porous Ceram II* (2009): 45–58.
40. Amirian, J., N. T. B. Linh, Y. K. Min, and B-T. Lee. Bone formation of a porous gelatin-pectin-biphasic calcium phosphate composite in presence of BMP-2 and VEGF. *Int J Biol Macromol* 76 (2015): 10–24.
41. He, F., W. Ren, X. Tian, W. Liu, S. Wu, and X. Chen. Comparative study on in vivo response of porous calcium carbonate composite ceramic and biphasic calcium phosphate ceramic. *Mater Sci Eng: C* 64 (2016): 117–123.
42. Ramay, R. R. Hassna, and M. Zhang. Biphasic calcium phosphate nanocomposite porous scaffolds for load-bearing bone tissue engineering. *Biomaterials* 25, no. 21 (2004): 5171–5180.
43. Thamaraiselvi, T., and S. Rajeswari. Biological evaluation of bioceramic materials-a review. *Carbon* 24, no. 31 (2004): 172.
44. Dong, J., P. Lin, N. E. Putra, N. Tümer, M. A. Leeflang, Z. Huan, L. E. Fratila-Apachitei, J. Chang, A. A. Zadpoor, and J. Zhou. Extrusion-based additive manufacturing of Mg-Zn/bioceramic composite scaffolds. *Acta Biomater* (2022).
45. Chitra, S., N. K. Mathew, S. Jayalakshmi, S. Balakumar, S. Rajeshkumar, and R. Ramya. Strategies of bioceramics, bioactive glasses in endodontics: Future perspectives of restorative dentistry. *BioMed Res Int* 2022 (2022).
46. dos Santos, G. G., L. Q. Vasconcelos, I. C. Barreto, F. B. Miguel, and R. P. C. de Araújo. Wollastonite and tricalcium phosphate composites for bone regeneration. *Res Soc Develop* 11, no. 9 (2022): e12011931662–e12011931662.
47. Cárdenas-Balaguera, C. A., and M. A. Gómez-Botero. Engineering applications of chemically-bonded phosphate ceramics. *Ingeniería e Investigación* 39, no. 3 (2019): 10–19.
48. Kamitakahara, M., C. Ohtsuki, and T. Miyazaki. Behavior of ceramic biomaterials derived from tricalcium phosphate in physiological condition. *J Biomater Appl* 23, no. 3 (2008): 197–212.
49. Wu, J., C. Ling, A. Ge, W. Jiang, S. Baghaei, and A. Kolooshani. Investigating the performance of tricalcium phosphate bioceramic reinforced with titanium nanoparticles in friction stir welding for coating of orthopedic prostheses application. *J Mater Res Technol* 20 (2022): 1685–1698.
50. Ducheyne, P., and G. W. Hastings, eds. *Metal and Ceramic Biomaterials*, vol. 2. CRC Press, Boca Raton, 1984.
51. Shackelford, J. F. Bioceramics(applications of ceramic and glass materials in medicine). In *Materials Science Forum*. Trans Tech Publications, 1999.
52. Khandan, A., M. Abdellahi, N. Ozada, and H. Ghayour. Study of the bioactivity, wettability and hardness behaviour of the bovine hydroxyapatite-diopside bionanocomposite coating. *J Taiwan Inst Chem Eng* 60 (2016): 538–546.

53. Fan, G., H. Su, J. Zhang, M. Guo, H. Yang, H. Liu, E. Wang, L. Liu, and H. Fu. Microstructure and cytotoxicity of Al_2O_3-ZrO2 eutectic bioceramics with high mechanical properties prepared by laser floating zone melting. *Ceram Int* 44, no. 15 (2018): 17978–17985.
54. Lops, D., E. Bressan, M. Chiapasco, A. Rossi, and E. Romeo. Zirconia and titanium implant abutments for single-tooth implant prostheses after 5 years of function in posterior regions. *Int J Oral Maxillofac Implants* 28, no. 1 (2013).
55. Wang, J., and R. Stevens. Surface toughening of TZP ceramics by low temperature ageing. *Ceram Int* 15, no. 1 (1989): 15–21.
56. Merolli, A., P. Tranquilli Leali, P. L. Guidi, and C. Gabbi. Comparison in in-vivo response between a bioactive glass and a non-bioactive glass. *J Mater Sci: Mater Med* 11, no. 4 (2000): 219–222.
57. Siriphannon, P., Y. Kameshima, A. Yasumori, K. Okada, and S. Hayashi. Formation of hydroxyapatite on CaSiO3 powders in simulated body fluid. *J Eur Ceram Soc* 22, no. 4 (2002): 511–520.
58. Campbell, A. A. Bioceramics for implant coatings. *Mater Today* 6, no. 11 (2003): 26–30.
59. Finšgar, M., A. P. Uzunalić, J. Stergar, L. Gradišnik, and U. Maver. Novel chitosan/diclofenac coatings on medical grade stainless steel for hip replacement applications. *Sci Rep* 6, no. 1 (2016): 1–17.
60. Li, L., W. Bai, X. Wang, C. Gu, G. Jin, and J. Tu. Mechanical properties and in vitro and in vivo biocompatibility of aC/aC: Ti nanomultilayer films on Ti6Al4V alloy as medical implants. *ACS Appl Mater Interfaces* 9, no. 19 (2017): 15933–15942.
61. Wang, L., X. Zhao, M. H. Ding, H. Zheng, H. S. Zhang, B. Zhang, X. Q. Li, and G. Y. Wu. Surface modification of biomedical AISI 316L stainless steel with zirconium carbonitride coatings. *Appl Surf Sci* 340 (2015): 113–119.
62. Fu, Q., Q. Liu, L. Li, X. Li, H. Gu, and B. Sheng. Effect of doping different Si source on Ca-P bioceramic coating fabricated by laser cladding. *J Appl Biomater Funct Mater* 18 (2020): 2280800020917322.
63. Fukada, E., and I. Yasuda. Piezoelectric effects in collagen. *Jpn J Appl Phys* 3, no. 2 (1964): 117.
64. Bandyopadhyay, A., S. Bernard, W. Xue, and S. Bose. Calcium phosphate-based resorbable ceramics: Influence of MgO, ZnO, and SiO2 dopants. *J Am Ceram Soc* 89, no. 9 (2006): 2675–2688.
65. Shen, X., Y. Zhang, Y. Gu, Y. Xu, Y. Liu, B. Li, and L. Chen. Sequential and sustained release of SDF-1 and BMP-2 from silk fibroin-nanohydroxyapatite scaffold for the enhancement of bone regeneration. *Biomaterials* 106 (2016): 205–216.
66. Rintoul, L., E. Wentrup-Byrne, S. Suzuki, and L. Grøndahl. FT-IR spectroscopy of fluoro-substituted hydroxyapatite: Strengths and limitations. *J Mater Sci: Mater Med* 18, no. 9 (2007): 1701–1709.
67. Shamoradi, F., R. Emadi, and H. Ghomi. Fabrication of monticellite-akermanite nanocomposite powder for tissue engineering applications. *J Alloys Comp* 693 (2017): 601–605.
68. Najafinezhad, A., M. Abdellahi, H. Ghayour, A. Soheily, A. Chami, and A. Khandan. A comparative study on the synthesis mechanism, bioactivity and mechanical properties of three silicate bioceramics. *Mater Sci Eng: C* 72 (2017): 259–267.
69. Sharafabadi, A. K., M. Abdellahi, A. Kazemi, A. Khandan, and N. Ozada. A novel and economical route for synthesizing akermanite ($Ca_2MgSi_2O_7$) nano-bioceramic. *Mater Sci Eng: C* 71 (2017): 1072–1078.
70. Kazemi, A., M. Abdellahi, A. Khajeh-Sharafabadi, A. Khandan, and N. Ozada. Study of in vitro bioactivity and mechanical properties of diopside nano-bioceramic synthesized by a facile method using eggshell as raw material. *Mater Sci Eng: C* 71 (2017): 604–610.

71. Karamian, E., M. Abdellahi, A. Khandan, and S. Abdellah. Introducing the fluorine doped natural hydroxyapatite-titania nanobiocomposite ceramic. *J Alloys Comp* 679 (2016): 375–383.
72. Razavi, M., M. Fathi, O. Savabi, D. Vashaee, and L. Tayebi. Improvement of biodegradability, bioactivity, mechanical integrity and cytocompatibility behavior of biodegradable Mg based orthopedic implants using nanostructured bredigite (Ca7MgSi4O16) bioceramic coated via ASD/EPD technique. *Ann Biomed Eng* 42, no. 12 (2014): 2537–2550.
73. Hench, L. L. *An Introduction to Bioceramics*, vol. 1. World scientific, 1993.
74. Yang, K., and Y. Ren. Nickel-free austenitic stainless steels for medical applications. *Sci Technol Adv Mater* (2010).
75. Choi, S. R., J. W. Kwon, K. S. Suk, H. S. Kim, S. H. Moon, S. Y. Park, and B. H. Lee. The Clinical Use of Osteobiologic and Metallic Biomaterials in Orthopedic Surgery: The Present and the Future. *Materials* 16, no. 10 (2023): 3633.
76. Jiang, J., D. Xu, T. Xi, M. Babar Shahzad, M. Saleem Khan, J. Zhao, X. Fan, C. Yang, T. Gu, and K. Yang. Effects of aging time on intergranular and pitting corrosion behavior of Cu-bearing 304L stainless steel in comparison with 304L stainless steel. *Corros Sci* 113 (2016): 46–56.
77. Thostenson, E. T., Z. Ren, and T-W. Chou. Advances in the science and technology of carbon nanotubes and their composites: A review. *Compos Sci Technol* 61, no. 13 (2001): 1899–1912.
78. Lee, C., X. Wei, J. W. Kysar, and J. Hone. Measurement of the elastic properties and intrinsic strength of monolayer graphene. *Science* 321, no. 5887 (2008): 385–388.
79. Zhao, C., W. Liu, M. Zhu, C. Wu, and Y. Zhu. Bioceramic-based scaffolds with antibacterial function for bone tissue engineering: A review. *Bioact Mater* (2022).
80. Zhi, W., X. Wang, D. Sun, T. Chen, B. Yuan, X. Li, et al. Optimal regenerative repair of large segmental bone defect in a goat model with osteoinductive calcium phosphate bioceramic implants. *Bioact Mater* 11 (2022): 240–253.
81. Zhou, L., Y. Wang, and G. Cao. Estimating the elastic properties of few-layer graphene from the free-standing indentation response. *J Phys: Condens Matter* 25, no. 47 (2013): 475301.
82. Dorozhkin, S. V. Calcium orthophosphates as bioceramics: State of the art. *J Funct Biomater* 1, no. 1 (2010): 22–107.
83. Anderson, J. M., A. Rodriguez, and D. T. Chang. Foreign body reaction to biomaterials. *Seminars Immunol* 20, no. 2: 86–100. Academic Press, 2008.
84. Roveri, N., and M. Iafisco. Evolving application of biomimetic nanostructured hydroxyapatite. *Nanotechnol Sci Appl* 3 (2010): 107.
85. Dorozhkin, S. V. Calcium orthophosphate-based bioceramics. *Materials* 6, no. 9 (2013): 3840–3942.
86. Antoniac, I. V., ed. *Handbook of Bioceramics and Biocomposites.* Springer, Berlin, Germany, 2016.
87. Eliaz, N., and N. Metoki. Calcium phosphate bioceramics: A review of their history, structure, properties, coating technologies and biomedical applications. *Materials* 10, no. 4 (2017): 334.
88. Ebrahimi, M., M. G. Botelho, and S. V. Dorozhkin. Biphasic calcium phosphates bioceramics (HA/TCP): Concept, physicochemical properties and the impact of standardization of study protocols in biomaterials research. *Mater Sci Eng: C* 71 (2017): 1293–1312.
89. Kumar, P., B. S. Dehiya, and A. Sindhu. Bioceramics for hard tissue engineering applications: A review. *Int J Appl Eng Res* 13, no. 5 (2018): 2744–2752.
90. Yu, Q., J. Chang, and C. Wu. Silicate bioceramics: From soft tissue regeneration to tumor therapy. *J Mater Chem B* 7, no. 36 (2019): 5449–5460.

91. Saba, J. N., D. A. Siddiqui, L. C. Rodriguez, S. Sridhar, and D. C. Rodrigues. Investigation of the corrosive effects of dental cements on titanium. *J Bio- Tribo-Corros* 3, no. 2 (2017): 1–7.
92. Baláž, M. Ball milling of eggshell waste as a green and sustainable approach: A review. *Adv Colloid Interf Sci* 256 (2018): 256–275.
93. Ulfyana, D., F. Anugroho, S. H. Sumarlan, and Y. Wibisono. Bioceramics synthesis of hydroxyapatite from red snapper fish scales biowaste using wet chemical precipitation route. *IOP Conf Ser: Earth Environ Sci* 131, no. 1: 012038 (2018). IOP Publishing.
94. About, I. Recent trends in tricalcium silicates for vital pulp therapy. *Curr Oral Health Rep* 5 (2018): 178–185.
95. Weisleder, R. Is vital pulp therapy fiction or a real treatment option?
96. Alegre, A. C., S. A. Verdú, J. I. Z. López, E. P. Alcina, J. R. Climent, and A. P. Sabater. Intratubular penetration capacity of HiFlow bioceramic sealer used with warm obturation techniques and single cone. *Heliyon* (2022): e10388.

15 Reuse, Reduce and Recycling Standards of Bioceramics

R. Senthilkumar
REVA University

G.V. Swarnalatha
Rayalaseema University

15.1 INTRODUCTION

The science and technology knowledge is essential for biomaterials to exhibit a good performance in a living system. Biological, clinical and material sciences' knowledge is important for the interaction between the living body and material that is implanted. True examples for multidisciplinary field are biomaterials. These may be natural or synthetic with biomedical applications (Bose and Bandyopadhyay 2013). These biomaterials are also used in drug delivery systems. These biomaterials are developed by engineers and material scientists. The expertise biologists and physicians validate the function of biomaterials in human body. Identification of specific need starts the procedure of biomaterials. Then, specific implants were created and inserted into the body of a patient. This lengthy process has to be verified at different stages. These include design, synthesis, manufacturing and biomedical evaluation. Regulatory requirements are necessary for a potential biomaterial. Physical state of these biomaterials is solid. Biomaterials are broadly of two types, that is, synthetic and biological (Ratner et al. 2004), Collagen and chitin are biological ones. Bioceramics, biometals, biopolymers and biocomposites are the four major groups of synthetic biomaterials based on their composition and nature (Figure 15.1). All four types of biomaterials are playing a crucial role in regeneration and replacement of tissues in humans.

Out of these four, bioceramics gained a lot of development as these are in connection with the living tissues. Research is focusing on generation of bioceramic materials especially toward the repair and regeneration of damaged bones during any trauma. Increase in the number of aging populations is also one of the reasons for this as they are in huge medical needs. Key role of bioceramics is in dentistry and orthopedics that is in clinical field. The use of ceramics has been evolving along with human evolution. The evolution of ceramics from storage to structural and finally to functional is observed along with human evolution. Finally, they are evolved as biomaterials toward the improvement of human life with regard to length and quality.

DOI: 10.1201/9781003258353-19

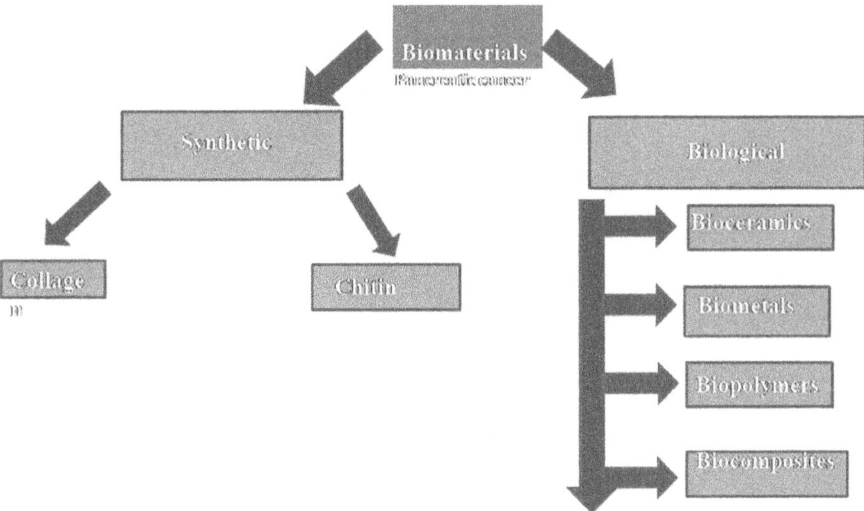

FIGURE 15.1 Types of biomaterials.

Most of the biomaterials are involved in skeletal muscle repair and also in replacements of components of cardiovascular system (Hench 2013). Ceramics are resistant to heat, nonmetallic, chemically stable and inorganic. Ceramics are used for hard tissue replacements. Ceramics are nonimmunogenic.

First-, second- and third-generation bioceramics are there. First generations are used in orthopedics, implants and in dental work as they are inert biologically or chemically. Zirconia, titanium and alumina come under first generation (Dorozhkin 2018). Bioresorbable and bioactive biomaterials are second-generation materials. A bioresorbable substance changes according to its tissue type and slowly dissolves in human body. Wollastonite ($CaSiO_3$), calcium silicate, tricalcium phosphate (TCP, $Ca_3(PO_4)_2$), polylactic–polyglycolic copolymeric acid, calcium carbonate ($CaCO_3$) and calcium oxide (CaO) are examples for second generation (Jeong et al. 2019). Third-generation biomaterials form scaffolds and porous structures with three dimensions which allow the natural processes of living cells. Mesoporous silica, mesoporous glasses, calcium phosphate (Sabudin et al. 2019) and organic/inorganic hybrids are examples for third generation (Cerqueira et al. 2021). Different types of bioceramics are represented in Figure 15.2.

Bioceramics are used in medical field and also in optical, energy and electronic fields. First bioceramic material used was porcelain for crown treatment during 18th century. Then, in 19th century, plaster of Paris was used for treating dental disorders. In 20th century, medical field-related bioceramics usage is increased rapidly with the improvement in technology (Chevalier and Gremillard 2009). Bioceramics are good biocompatible and have high mechanical strength and show low degradation. Thus, bioceramics are friendly substitutes of the body (Rieger 2001). Some of the bioceramic interactions with tissues and their responses are shown in Table 15.1 (Shekhawat et al. 2020).

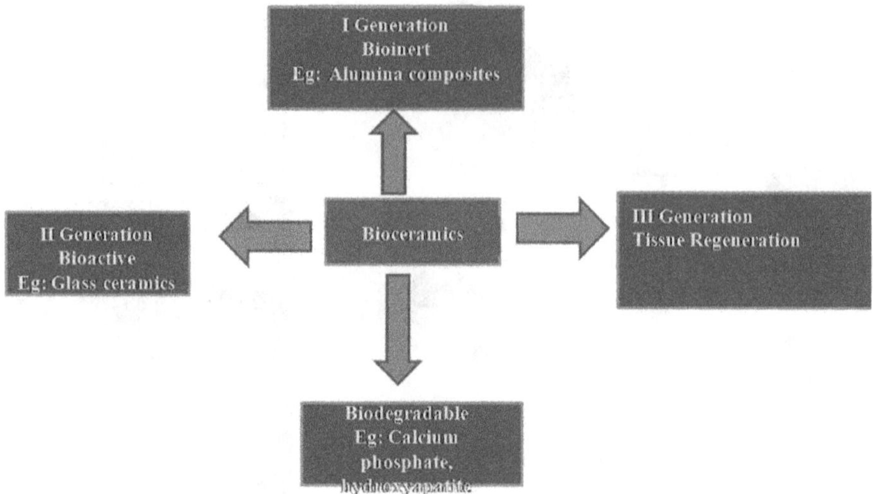

FIGURE 15.2 Types of bioceramics.

TABLE 15.1
Interaction and Response of Bio Ceramic Implants

Interaction	Response
Implant surrounding tissues are killed	Toxic
Implant dissolution and replacement by tissues	Dissolution
Implant dissociation from host	Inert
Implant and tissue interface bonding	Bioactive

Bioactive, bioinert and bioresorbable are the three subclasses of bioceramics which exhibit both advantages and disadvantages (Kalita et al. 2007). Synthetic and traditional bioceramics are used in tissue engineering by which they allow the enhancement of implants life. Research is needed to improve further toward quality and stability of implants. Mechanical properties of bioactive glass ceramics and hydroxyapatite are to be focused further. Bioglass forms apatite layer when encountered with body fluid which allows bone regeneration. Combination of zirconia and alumina enhances the biological characteristics. Dental applications are provided through hydroxyapatite (Pawan Kumar et al. 2018). Science and technology is developing to find the route of producing value-added products from waste. Waste from any sector like agriculture, industry, etc. is accumulating more and more in the environment. The effective usage of these wastes prevents the issues of environment. Abundancy of any waste definitely affects the plants, livestock and human ecosystems. The waste is varying across the countries. Environmental protection and health quality is enhanced through efficient handling of these wastes. Hence, the waste can be recycled and reused and further produce value-added products from them. Sustainable development is achieved through the proper use of waste.

Thus, ceramic research is focusing on recycling of waste, industrial by-products into useful ceramics. This chapter provides the information on different types of wastes used for ceramics including bioceramics, recycle and reuse of those ceramics in different fields.

15.2 BIOCERAMICS PROPERTIES

Bioceramics exhibit mechanical and biological properties.

a. Mechanical properties
   The different bioceramics have different properties based on that they are used in different medical applications. Anticorrosive and biological behavior is the two considerable factors for these applications. One good thing with the glass ceramics is easy bone bonding without toxicity and also, they are risk-free. The inert and bioactive properties make these composites very useful (Pawan Kumar et al. 2018). The mechanical properties of different bioceramics are given in Table 15.2.
b. Biological properties
   1. Bone similarity
      The factors that are responsible for bone similarity are porosity, composition and mechanical properties. Macroporous (100–500 µm) and microporous (1–10 µm) structures of bioceramics show different functions. Macro allows the tissue growth, and micro allows blood vessels in growth. Hydroxyapatite that is obtained by natural resources showed similar properties of bone interconnecting pore structure. Progenitors are used for creating pores in synthetic bioceramics. The crystal size and pores of some synthetic bioceramics are controlled by sintering temperature. The acidic conditions of cell also favor degradation of these bioceramics. Structural composition, surface area, crystallinity and porosity are the factors that cause degradation of either natural or synthetic bioceramics under both in vitro and in vivo conditions. New bone formation depends on inductivity and osteoconductivity of a bioceramic scaffold. These scaffolds show similar mechanical properties of bone. Particle size and composition of scaffold are important for its success (Livingstone and Daculsi 2003). The potential uses of calcium silicate-based bioceramics are shown in Figure 15.3.

TABLE 15.2
Mechanical Properties of Different Bioceramics (Pawan kumar et al. 2018)

Mechanical properties	Alumina	Zirconia	Bioglass	Hydroxyapatte	Glass Ceramic
Density (g/cm^3)	3.3–3.9	6.0	2.5	3.1	2.8
Bond Strength (GPa)	300–400	200–500	50	120	215
Compressive strength (MPa)	2000–5000	2000	600	500–1000	1080
Young Modulus (GPa)	260–410	150–200	75	73–117	118

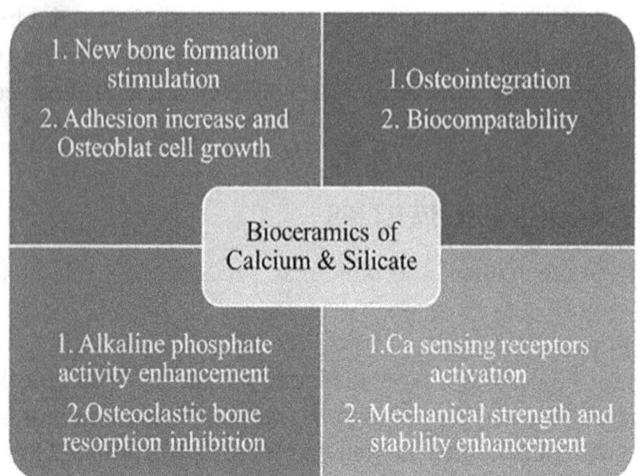

FIGURE 15.3 Potential uses of calcium silicate-based bioceramics.

2. Bioactivity
   Each bioceramic has different bonding ability with the tissues of body; in simple, they show different bioactivities. In a similar way, the immune response is also different. Morphological and physicochemical factors are important for bioactivity. The bioceramics may be synthetic or biological and are almost inorganic materials. Autografts and allografts are best substituted by calcium phosphate and bioglass bioactive. Material bioactivity is affected by composition and fabrication of bioceramics. Bone deposition enhancement is due to formation of calcium phosphate crystal by the release of calcium and phosphate from bioceramics. Four different types of tissue attachment are exhibited by bioceramics: (i) inert and dense surface, (ii) porosity, (iii) dense phase, and (iv) easy integration and brake down. All these allow easy surface attachment, chemical bonding and tissue growth. Biomaterial assimilation with biological agents enhances clinical potential by enhancing the bioactivity. The surface chemistry of ceramic materials is changed by biological agents. The bioceramic and inorganic bone part phase similarity forms a strong bond between those two interfaces (LeGeros et al. 2008).
3. Biodegradation
   In vivo degradation by osteoclasts and phagocytic cells is due to cell-mediated acidic condition. The composition, crystallinity, porosity, structure and surface area are the influencing factors for dissolution of materials. Physiological factors like temperature, pH, ion and protein concentration and buffer capacity also influence for dissolution. Stimulated body fluid provides the information of degradation of ceramics and formation of apatite. Some serum proteins resist the apatite formation. The biomaterials' surface charge chemistry is modified by dissolution and precipitation reactions; as a result, the strength,

reactivity and degradation rate of these materials are enhanced (Heymann et al. 2001).

4. Biotoxicity
   Advanced bioceramics may increase the toxicity risk in human body. Thus, biocompatibility and performance testing should be done at cellular level prior to clinical applications of bioceramics. For material implantation, a complete data of physiological, morphological, biological and mechanical tests are essential. This favors the long life of implant without toxicity. Before clinical trials, the bioceramic toxicity evaluation is important. Ethical and legal issues are also to be considered for these trials. Three different toxicity analyses, namely cyto, histo and genotoxicities, are to be done. Good laboratory practice regulations prevent the biocompatibility issues (Thrivikraman et al. 2014)

15.3 BIOCERAMICS AS NANOBIOMATERIALS

The different states of bioceramics act as nanobiomaterials like coatings, nanotubes, fiber and particle. The usage of bulk material form of bioceramics is observed in scaffolds and graft substitution of tissue engineering. The degradation ability of resorbable bioceramics is common in those applications. For adipose-derived stem cell culturing, HA bioceramics scaffolds are used. Osteogenesis and angiogenesis of mesenchymal stem cells need calcium silicate scaffolds. Calcium phosphate scaffolds are used in synthetic bone graft substitutes.

Bioceramics are also used for coating several medical devices and constructs of tissue engineering. This improves the long-term stability, osteoconductivity and biocompatibility. Artificial heart valves are coated with carbon, calcium phosphate and hydroxyapatite. Particle states of bioceramics are used in drug delivery systems. The therapeutic agents like isotopes and chemotherapeutics of cancer treatment, antigens, antibodies and enzymes are carried by insoluble and porous glass beads. Newly studied bioceramics with graphene oxide nanoparticles are also suitable for drug delivery. Risedronate is a drug for treatment of osteoporosis is a surface HA nanoparticle. Hallowed micro- and nanotubes are also one of the forms of bioceramics. These are widely used in tissue regeneration and drug and gene delivery systems due to their high surface area and thermal and electrical conductivity. Hyperthermia therapy is also proved. The analogues of carbon nanotubes are graphene oxide nanotubes and are predominantly used in biosensors. Halloysite is a novel aluminosilicate nanotube structure which is used as a nanoreactor. This is used for loading of drugs that are poorly soluble. Titanium oxide nanotubes also exhibit various potential biomedical applications. Fiber form of bioceramics emits far infrared waves. Magneto therapeutic devices are fabricated by bioceramic-woven fibers. Zirconium dioxide is hollow fiber bioceramics (Kiaie et al. 2017).

15.4 REUSE, REDUCE AND RECYCLING OF BIOCERAMICS

Skeletons of marine organisms produce various natural ceramics. These are often useless and also considered as pollutants for environment. The reason for this is due to favoring of growth of different bacteria. In undeveloped and developed societies, this kind of waste is more common through hatching farms, food processing units and egg baking units. In the field of biomedical and tissue engineering, these are used as source for calcium as calcium phosphate ceramics (Ibrahim et al. 2013, Kamalanathan et al. 2014). Some of the calcium phosphate phases used for biomedical applications are shown.

This is the turning point during European societal challenge where the waste is converted to potential material and used in patients with bone disease. The common chemical synthesis of calcium phosphates has several backdrops. Thus, substitution of chemical synthesis with natural products like bioceramics is crucial for environmental protection and recycling of waste materials. The simulating solutions form the bone-like apatite, that is, calcium phosphate bioceramics on surface of multiphasic calcium phosphate. In vivo, the same process occurs; thus, it allows direct bonding of bone with host bones. This multiphasic calcium phosphate adsorbs bioorganic molecules, cells and proteins. In bone tissue engineering, the multiphasic calcium phosphate bioceramics are used as carrier systems where they possess the properties like osteoinductive and osteopromotive that leads toward the formation of new bone. Multiphase calcium phosphate bioceramics have bioactivity control. They release orthophosphate and calcium ions that pave the path for new bone formation. Carbonate apatite is formed by multiphasic calcium phosphate. These have wide range of applications in orthopedic, dental, drug delivery and in stem cell tissue engineering (Dorozhkin 2016). Eggshell is studied to produce calcium phosphate foams with a natural bone porosity similarity. This technology is effective with respect to quality, reproducibility and cost (Sandra et al. 2017). Coal ignition produces a waste called fly ash (FH). FH contains the mineral silicoaluminum which is a potential raw material for ceramics (Fan et al. 2018). Solid waste is also generated during the process of ceramics in ceramic industry. The favorable waste management method is the recycling of this ceramic waste to same industry (Ke et al. 2016). Bayer's alumina extraction produces red mud (RM) waste. This RM is utilized in ceramic industry (Xu et al. 2019). Coffee husk ashes (CHA) act as substitutes for ceramic formulations (Acchar et al. 2013). The general waste-derived ceramics are shown in Figure 15.4, and the potential waste raw materials used for bioceramic production are shown in Figure 15.5.

Glass ceramic products are produced from solid waste, that is, FA, filter dust slag from steel plant, etc. (Holanda 2017). Wollastonite is a calcium silicate mineral with widespread properties. This plays crucial role, and around 30%–40% is used in ceramic industries around the world. Synthetic wollastonite is synthesized from the ingredients of low-cost waste (Palakurthy et al. 2019).

One more ceramic material is hydroxyapatite with extensive bioactivity, biocompatibility and osteoconductivity. It is widely used in humans toward teeth and bone repairs. It will neither show inflammation nor toxicity in the body. So, it has wide range of biomedical applications (Hendi 2017). It is not only used in biomedical but also in soil treatment plant, adsorption of heavy metals and an absorbent material in wastewater (Nzihou and Sharrock 2010). Holanda (2017) reviewed and stated that the nanostructured hydroxyapatite is produced from recycling of calcium rich wastes. Thus, waste materials with rich calcium serve as good components

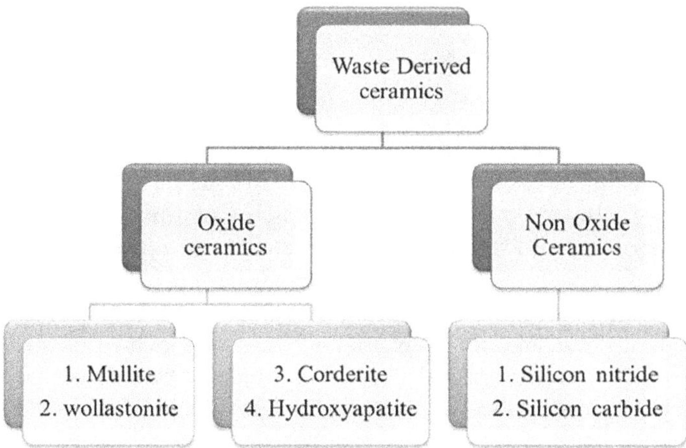

FIGURE 15.4 General waste-derived ceramics.

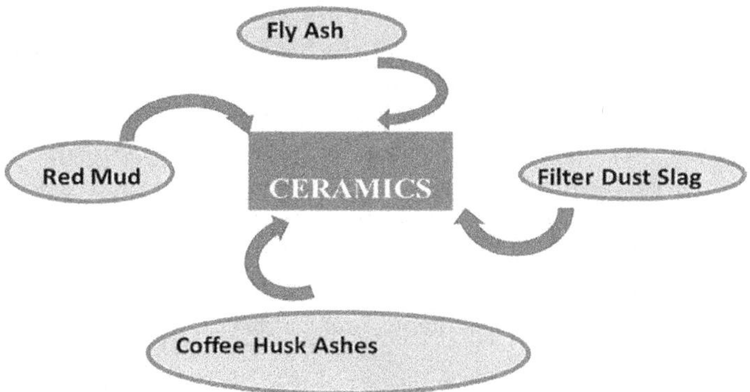

FIGURE 15.5 Potential waste raw materials for bioceramics.

for bioceramics production. In separation technology, membrane technology has a prominent role. In this ceramic, membranes are gaining more importance than other membranes. This may be due to the factors that they have good thermos mechanical stability, resistance, antifouling, etc.; one limitation is that they are expensive. Research is focusing on production of cost-effective ceramic membranes (Hossain and Roy 2020). The bones contain 70% of hydroxyapatite, and the rest is collagen and water. Removal of biological parts from bones results in only hydroxyapatite, a complete mineral structure with trace elements also. The ceramics with calcium phosphate hydroxyapatite act as a substitute for broken bone and also used in implants. Natural hydroxyapatite possesses all essential elements and has a good ease to access those raw materials. Fish bones, which are considered to be waste and useless, are rich source of calcium phosphate. Salmon fish bone wastes are used for production of natural hydroxyapatite. Thus, waste reuse of these fishbone reduces the environmental pollution with good biomedical applications. A cost-effective and environment-friendly source of calcium phosphate is through the salmon fish bones. These natural bones assure biocompatibility and osteointegration (Mahesh 2022). One of the agricultural wastes called rice straw or rice husk is providing a considerable value in bioceramics preparation. This rice straw acts as precursor for biomedical implants, various biomaterials and in surgery applications. A new dimension of using agricultural waste toward the value-added products leads to production of useful biomaterials. This may be due to the presence of active compounds in those wastes that are essential for formation of biomaterials. The low cost of availability, abundance and renewability of the agricultural wastes make them as energy potentials and efficient feedstock value-added product producers (Agwa et al. 2020). In recent years, rice husk and rice straw usage for formation of biogenic materials has gained a lot of interest (Grimm et al. 2021). Some studies reported that one of the bioceramic or bioglass safe substitutes is the agricultural waste (Palakurthy et al. 2020). They have wide range of applications from bone substitution to development, drug delivery systems and dental and medical therapies. Day-to-day bone replacement patient's number is increasing, and hence, innovative biomaterials are potent for such replacements. One important point to be noted is that all the bone implants that are used should be biocompatible. Agricultural waste products like rice straw, rice straw ash, rice husk, rice husk ash are used in production of wollastonite bioceramics and shown in Figure 15.6 (Ismail and Mohamad 2021).

Wollastonite has several biomedical applications. CERABONE is a bioactive glass ceramic implant used commercially. It is made from wollastonite apatite and shows excellent bioactive properties and hence used in human medicine toward bone replacement. Other important apatite components used are CERAVITAL, BIOVERIT, and BIOGLASS in head and throat surgery device (Höland and Beall 2012).

Calcium-containing biomaterials are synthesized from eggshells. These eggshells are rich source of calcium of around 95%, and the rest is protein with some organic

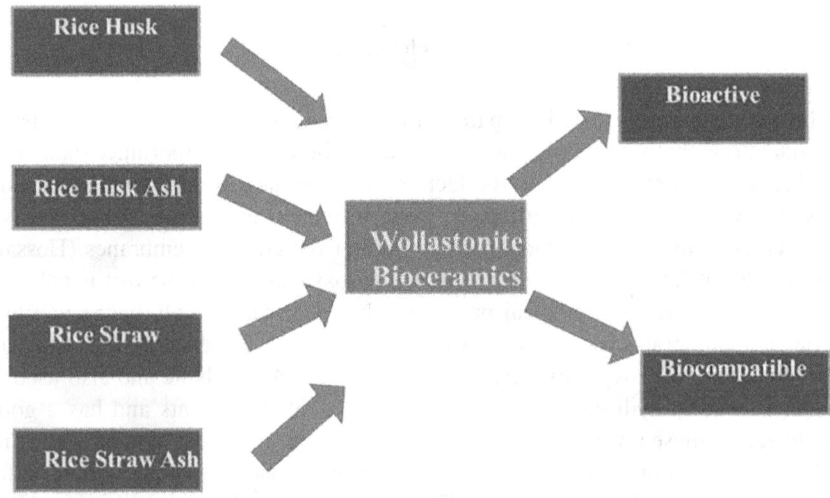

FIGURE 15.6 Agricultural wastes used for wollastonite bioceramics production.

moieties. Food processing industries are producing 2,50,000 tons of eggshells as waste by-product. These unattended eggshells were foul-smelled and forms a platform for various bacteria to grow, thus causing environmental pollution. There are other available sources of shells from snails, oyster and corbicula; even then eggshells are in high preference because of high calcium level and easy availability. The eggshells' recycling is an economic and environment-friendly strategy (Verma et al. 2012). One of the first magnesium-based bioceramic is diopside, a silicate bioceramic with bioactivity. The mechanical stability of diopside is superior to the human bone. The cell differentiation and proliferation are induced by the dissolution of silicate, Ca and Mg ions of diopside. This diopside is prepared from wastes and that too one pure form is obtained through eggshells too. Rice husk and eggshells both are used. Along with diopside, other bioceramics like wollastonite, pseudo wollastonite, cristobalite and calcite are also formed as secondary phases (Kazemi et al. 2017). Wollastonite as discussed earlier is also a calcium silicate bioceramic. Based on previous reports, it is stated that eggshells act as calcium source for wollastonite. The chicken eggshell and rice husks are used in wollastonite preparation (Anjaneyulu and Sasikumar 2014).

Forsterite is also one of the bioceramic synthesized from quartz, fly ash, silica fumes, rice husk ash and waste glass. In biomedical field, forsterite has significant importance. This is due to its mechanical strength because of magnesium oxide, stimulation of proliferation and slow degradation (Sadek et al. 2016).

15.5 CONCLUSIONS

There are many areas of biomedicine where bioceramics are crucial. They contribute to improving people's lives. There is a wide variety of bioceramics available for use in various settings. Based on the specifics of the problem, a suitable bioceramic can be selected. Therefore, the bioceramic industry is rapidly developing in order to achieve the desired properties. Mechanical and a wide range of biological properties are engineered for specific uses. Tissue regeneration and repair are made possible thanks to the scaffolding effect of bioceramics. Bioceramics' noninvasive nature makes surgical excision unnecessary. The by-products of bioceramic degradation are safe to use. The fabrication of nanoscale bioceramics is another significant use of bioceramics in the medical field. Dumping of wastes causes severe environmental damage on a global scale. There are many advantages of using trash as a resource replacement. They include saving money, preserving natural resources, and enhancing public health and safety by reducing pollution levels. The use of natural resources in ceramic production made from recycled materials is advantageous. When a relatively small amount of waste is transformed into a sizable quantity of ceramics, a correspondingly large amount of waste is absorbed. While there have been a number of studies looking into the feasibility of turning trash into ceramics, most of those studies have focused on bioceramics rather than other types of trash. In this chapter, we present the findings from the current study that pertains to the reusing, reducing and recycling of bioceramics. The transfer of technology from universities to businesses is an important step toward developing sustainable bioceramics, but this process needs more research.

REFERENCES

Acchar, W., Dultra, E.J.V., Segadães, A.M. 2013. Untreated coffee husk ashes used as flux in ceramic tiles. *Applied Clay Science*. 75–76, 141–147.

Agwa, I.S., Omar, O.M., Tayeh, B.A., Abdelsalam, B.A. 2020. Effects of using rice straw and cotton stalk ashes on the properties of lightweight self-compacting concrete. *Construction and Building Materials*. 235, 117541.

Anjaneyulu, U., Sasikumar, S. 2014. Bioactive nanocrystalline wollastonite synthesized by sol-gel combustion method by using eggshell waste as calcium source. *Bulletin of Material Science*. 37, 207–212.

Bose, S., Bandyopadhyay, A. 2013. *Introduction to Biomaterials*. Elsevier BV: Amsterdam, The Netherlands, pp. 1–9.

Cerqueira, A., Romero-Gavilán, F., García-Arnáez, I. et al. 2021. Bioactive zinc-doped sol-gel coating modulates protein adsorption patterns and in vitro cell responses. *Material Science and Engineering*. 122, 111839.

Chevalier, J., Gremillard, L. 2009. Ceramics for medical applications: A picture for the next 20 years. *Journal of European Ceramic Society*. 29, 1245–1255.

Dorozhkin, S. 2016. Multiphasic calcium orthophosphate ($CaPO_4$) bioceramics and their biomedical applications. *Ceramics International*. 42.

Dorozhkin, S.V. 2018. Current state of bioceramics. *Journal of Ceramic Science and Technology*. 9, 353–370.

Fan, F., Liu, Z., Xu, G. et al. 2018. Mechanical and thermal properties of fly ash based geopolymers. *Construction and Building Materials*. 160, 66–81.

Grimm, A.M., Dorsch, L.Y., Kloess, G.H., Enke, D., Roppertsz, A. 2021. Catalysing the combustion of rice husk and rice straw towards an energy optimized synthesis of metal modified biogenic silica. *Social Science Research Network*, 1–31.

Hench, L.L. 2013. *An Introduction to Bioceramics*. Imperial College Press: London, UK.

Hendi, A.A. 2017. Hydroxyapatite based nanocomposite ceramics. *Journal of Alloys and Compounds*. 712, 147–151

Heymann, D., Guicheux, J., Rousselle, A.V. 2001. Ultrastructural evidence in vitro of osteoclast- induced degradation of calcium phosphate ceramic by simultaneous resorption and phagocytosis mechanisms. *Histology and Histopathology*. 16, 37–44.

Höland, W., Beall, G.H. 2012. *Glass-Ceramic Technology*. Wiley: Hoboken, NJ.

Holanda, J.N.F. 2017. Nanostructured calcium phosphatebased bio-ceramics from waste materials. In: *Handbook of Ecomaterials*, L.M.T. Martínez et al. (eds.). Springer International Publishing AG.

Hossain, S.K.S., Roy, P.K. 2020. Sustainable ceramics derived from solid wastes: A review. *Journal of Asian Ceramic Societies*. 8(4), 984–1009.

Ibrahim, A., Wei, W., Zhang, D., Wang, H., Li, J. 2013. Conversion of waste eggshells to mesoporous hydroxyapatite nanoparticles with high surface area. *Materials Letters*. 110, 195–197.

Ismail, H., Mohamad, H. 2021. Bioactivity and biocompatibility properties of sustainable wollastonite bioceramics from rice husk ash/rice straw ash: A review. *Materials*. 14, 5193.

Jeong, J., Kim, J.H., Shim, J.H., Hwang, N.S., Heo, C.Y. 2019. Bioactive calcium phosphate materials and applications in bone regeneration. *Biomaterial Research*. 23, 1–11.

Kalita, S.J., Bhardwaj, A., Bhatt, H.A. 2007. Nanocrystalline calcium phosphate ceramics in biomedical engineering. *Material Science and Engineering C*. 27, 441–449.

Kamalanathan, P., Ramesh, S., Bang, L.T. et al. 2014. Synthesis and sintering of hydroxyapatite derived from eggshells as a calcium precursor. *Ceramics International*. 40, 16349–16359.

Kazemi, M., Abdellahi, A., Khajeh-Sharafabadi, A. et al. 2017. Study of in vitro bioactivity and mechanical properties of diopside nano-bioceramic synthesized by a facile method using eggshell as raw material. *Materials Science Engineering C*. 71, 604–610.

Ke, S., Wang, Y., Pan, Z. et al. 2016. Recycling of polished tile waste as a main raw material in porcelain tiles. *Journal of Cleaner Production*. 115, 238–244.

Kiaie, N., Aavani, F., Razavi, M. 2017. 2: Particles/fibers/bulk. In: *Nanobiomaterials Science, Development and Evaluation*, M. Razavi, A. Thakor (eds.). Woodhead Publishing, pp. 7–25.

LeGeros, R.Z., Daculsi, G., LeGeros, J.P. 2008. Bioactive Bioceramics. In: *Musculoskeletal Tissue Regeneration*, W.S. Pietrzak (ed.). Orthopedic Biology and Medicine, Humana Press.

Livingstone, T.L., Daculsi, G. 2003. Mesenchymal stem cells combined with biphasic calcium phosphate ceramics promote bone regeneration. *Journal of Materials Science: Materials in Medicine*. 14, 211–218.

Mahesh, G. 2022. A note on synthesis zirconia as a dental biomaterial. *Bioceramics Development and Applications*. 12, 210.

Nzihou, A., Sharrock, P. 2010. Role of phosphate in the remediation and reuse of heavy metal polluted wastes and sites. *Waste Biomass Valorization*. 1(1), 163–174.

Palakurthy, S., Reddy, K.V., Patel, S., Azeem, P.A. 2020. A cost effective SiO_2-CaO-Na_2O bio-glass derived from bio-waste resources for biomedical applications. *Progress in Biomaterials*. 9, 239–248.

Palakurthy, S., Reddy, K.V.G., Samudrala, R.K. et al. 2019. In vitro bioactivity and degradation behaviour of βwollastonite derived from natural waste. *Materials Science Engineering C*. 98, 109–117.

Pawan Kumar, B., Dehiya, S., Anil, S. 2018. Bioceramics for hard tissue engineering applications: A review. *International Journal of Applied Engineering Research*. 13(5), 2744–2752.

Ratner, B.D., Hoffman, A.S., Lemons, J.E., Schoen, F.J. 2004. *Biomaterials Science*. Elsevier: Amsterdam, The Netherlands.

Rieger, W. 2001. Ceramics in orthopedics – 30 years of evolution and experience. In: *World Tribology Forum in Arthroplasty*, C. Rieker, S. Oberholzer, U. Wyss (eds.). Hans Huber: Bern, Switzerland.

Sabudin, S., Marzuke, M.A., Hussin, Z. 2019. Effect of mechanical properties on porous calcium phosphate scaffold. *Materials Today: Proceedings*. 16, 1680–1685.

Sadek, H.E.H., Khattab, R.M., Zawrah, M.F. 2016. Preparation of porous forsterite ceramic using waste silica fumes by the starch consolidation method. *International Ceramic Review*. 65, 174–178.

Sandra, S., Mercedes V., Alfredo D. et al. 2017. Synthesis of bioceramic foams from natural products. *Journal Sol-gel Science Technology*.

Shekhawat, D., Singh, A., Banerjee, M.K., Singh, T., Patnaik, A. 2020. Bioceramic composites for orthopaedic applications: A comprehensive review of mechanical, biological, and microstructural properties. *Ceramics International*. doi:10.1016/j.

Thrivikraman, G., Madras, G., Basu, B. 2014. In vitro/in vivo assessment and mechanisms of toxicity of bioceramic materials and its wear particulates. *Royal Society of Chemistry Advances*. 4, 12763–12781.

Verma, V., Kumar, M.C., Bansal, C. 2012. Utilization of eggshell waste in cellulase production by Neurospora crassa under wheat bran-based solid fermentation. *Polish Journal of Environmental Studies*. 21, 491–497.

Xu, X., Song, J., Li, Y. et al. 2019. The microstructure and properties of ceramic tiles from solid wastes of Bayer red muds. *Construction and Building Materials*. 212, 266–274.

Section E

Challenges, Issues and Sustainability

16 Ethical Issues of Bioceramics

M. Vidya
M.S. Ramaiah College of Arts, Science and Commerce

Samer Shamshad
ICAR-National Institute of Veterinary
Epidemiology and Disease Informatics
REVA University

R. Senthilkumar
REVA University

16.1 INTRODUCTION

Bioceramics are a type of biomaterial that is used to treat, augment, repair or replace diseased or injured hard tissue in the human body. Ceramics are used in electronic, optical and energy fields in addition to medicine (Chevalier and Gremillard 2009). Since 1960, bioceramics have been used as an alternative for hard tissue engineering. A promising approach to hard tissue engineering involves merging isolated functional cells with biodegradable scaffolds made from synthetic bioceramics to restore the function of diseased or injured bone tissue. Bioceramic materials have a better chance of replacing damaged tissues and restoring function to existing tissue (Kumar et al. 2018). Currently, millions of people have implants, and the number is rapidly increasing. It is crucial to consider the ethical implications of using artificial, non-living materials to repair, restore or augment live tissues. Clinicians have recognised the necessity for joint efforts to solve humanity's biomedical problems, which has spurred not just the development of research in biomedical engineering but also the ethical concerns that have arisen as a result (Musiband and Saha 2011, Kashi and Saha 2009). Questions about clinical trials, developing technologies (such as telemedicine, stem cell research, nanobiotechnology and tissue engineering) and animal research are among the ethical concerns (Saha and Saha 1999). Many strategies and materials that were exclusively exploited or reserved for traditional engineering purposes once are increasingly implemented for biomedical applications (e.g., industrial ceramics have been modified to be used as bioceramics in dentistry and orthopaedics). When utilised for humans, these examples can raise ethical questions regarding their efficacy, safety and long-term repercussions (Bertolami 2004, Pietrzak and Vacanti 2014).

DOI: 10.1201/9781003258353-21

16.2 ETHICAL CONCERNS WITH NEW TECHNOLOGIES

Some of the most promising areas of new breakthroughs include nanotechnology in stem cell research, medicine and innovative imaging tools for better diagnosis and tissue engineering. Many conceptual scientific ideas have been proposed over the years, but given the lack of improved manufacturing, imaging tools and characterisation until recently, they could not be proven in laboratory trials (Figure 16.1). Anticipatory ethics is a new field of discussion in biomaterials research, in which researcher scientists are motivated to be able to foresee challenges related to ethics and answers to any concerns raised with technological breakthroughs (Malsch 2013). Even though this is not achievable at all times and may be difficult sometimes, the intention that people can be motivated to foresee issues posed by newer technology will be critical to understand how to address them from a societal and scientific standpoint.

In an era when technology and science are progressing at unprecedented rates, we must discover new and more efficient ways to analyse and handle these concerns. This research might point to strategies to begin an ethical examination sooner and make it a part of every researcher's basis. Addressing ethical concerns during future researcher education will not only boost their knowledge but also instil a stronger sense of accountability for how their findings may be utilised in unethical ways.

FIGURE 16.1 Ethical concerns related to biomaterials research.

16.3 BIOSAFETY AND BIOCOMPATIBILITY

Biocompatibility is defined as a biomaterial's ability to function predictably in the body without generating a persistent inflammatory response, foreign body response, undesired biological response or toxicity as a consequence of the interaction between the biomaterial and the host tissues (Freire et al. 2014). Because one or more of the components may be potentially poisonous or irritating, no single material is fully inert from a physiological standpoint. Chemical by-products formed during the material's cure or as a result of the interplay with other materials in the area may also have negative consequences (Schmalz 1997). Before using materials, all doctors must have a thorough understanding of their qualities and properties, as well as their anticipated interactions with the biological environment.

16.4 ETHICS AND STEM CELL RESEARCH

Yet another promising field of biomedical engineering and science is stem cell research, which aims to provide remedies to a variety of medical issues. Cells' origin (i.e., stem cells produced from aborted foetuses, adult stem cells from tissues of the adult, or whether these represent embryonic stem cells of 5- to 7-day-old embryos) and the viability and safety of the stem cells when utilised for human therapeutic applications are the key ethical consideration related to stem cell research. It is an interdisciplinary branch of science that will undoubtedly encounter many ethical issues in the future (Kashi and Saha 2015, Nestor et al. 2014, Stephens 2015). Collaboration between specialists such as biomedical engineers, cellular biologists, clinicians and molecular biologists will allow this scientific discipline to expand successfully. To progress with the stem cell research and develop biologically active products, such coordinated initiatives are required. As a result of these one-of-a-kind collaborations, scientists from many disciplines will be required to collaborate.

Animal experiments, however contentious, have been deemed necessary and appropriate for performing safety and effectiveness investigations (Assen et al. 2021). Education, policy and legislation all benefit from broadening the breadth of ethical thinking. The difficulty is to strike a balance between how much flexibility and education researchers should have to cope with potential ethical consequences on their own and where legislation and regulation may help (Swierstra 2015). More significantly, a better knowledge of the ethical implications of stem cell research should assist researchers and others in thinking about how to foresee, and so perhaps avoid or lessen, any future issues rather than dealing with ethical challenges as they arise (Assen et al. 2021).

In either situation (human life beginning at conception or later in development), some would argue that even though blastocysts are not yet human, they are still part of human life, and so they may believe that killing blastocysts for stem cell research is immoral. For example, U.S. President George W. Bush's Council on Bioethics declared in 2001 that "it is ethically wrong to exploit and destroy growing human life, even for good cause," referring to Kant's Categorical Imperative. Some virtue ethicists, on the other hand, believe that terminating embryo life for stem cell research is immoral. The argument they endorse is that it is our responsibility to value human life. They believe that not endangering human life is a noble human conduct based on this principle.

16.5 ETHICS AND NANO BIOLOGY

At the nanoscale, researchers have recently been able to observe, analyse and even manufacture materials (Lee and Henthorn 2012). The field of nanotechnology has the potential to solve a wide range of biomedical issues. To mention a few instances are the invention of nanoparticles that are competent enough to transport medications to specific tissues and the modification of biological substances to improve efficacy when used in vivo. In the field of nanotechnology, several ideas that were formerly considered hypothetical or theoretical have lately manifested into actual devices. This is due to advancements in imaging techniques, fabrication and material processing technologies. Although it is reasonable to believe that science is progressing at a quicker rate than its linked ethical standards, research in nanotechnology has prompted concerns about safety, particularly connected to nanomaterials when used for human uses (Florczyk and Saha 2007). The consequences of nanoparticles on the environment have not been thoroughly investigated. Academic institutions, government agencies and international organisations must consider a shortage of personnel to address ethical challenges associated with nanotechnology.

The first ethical difficulty in nanobiotechnology is connected to the medical profession (40%), followed by environmental issues (30%), health and safety issues (20%) and societal issues (10%). The majority of responders (50%–90%) feel that nanobiotechnology will offer more threats than benefits (25%) to human health. As a result, it is critical to encourage ethics researchers to submit high-quality research proposals on ethical concerns and to engage the public in discussions about the societal implications of nanotechnology. Nanomedicine has the potential to injure rather than repair the human body. Particles that cannot be seen or controlled would penetrate the body and deliver hazardous substances such as poisons. Materials utilised in nano-medical technology might be poisonous or environmentally benign. Increasing the risk of privacy violations or abuse of personal information: for example, releasing medical information (in a DNA chip) to insurance firms. Transhumanists want to alter human nature itself.

16.6 TISSUE ENGINEERING

The purpose of tissue engineering is to use laboratory-engineered tissue constructs to replace lost or diseased tissue (Mhanna and Hasan 2017). Despite the fact that tissue engineered materials have a variety of clinical purposes; their end goal is to restore lost tissue with satisfactory results. Tissue engineering, like stem cell research, is a quite interdisciplinary subject that requires consideration of increasing ethical concerns. The origin of the cells used and their safety and long-term effects in the tissue construct, expenses and intellectual property challenges are only a few of the concerns. Inadequate routes for addressing and/or resolving one or more of the previously stated problems may impede tissue engineering progress. It is unclear if the cells themselves may be patented or whether the combination of cells and artificial materials used to create tissue engineered structures. Some patients may avoid tissue products or cells derived from animals for religious reasons (Kashi and Saha 2017).

When addressing the translational progression from bench to bedside, the ethical considerations arising from tissue engineering research are similarly complex, and investigators in the field of tissue engineering act as moral agents at each step of their research along the translational pathway, from early benchwork and preclinical studies to clinical research. This article compares difficulties surrounding two translational tissue engineering technologies, the bioartificial pancreas and a tissue engineered skeletal muscle construct, to highlight the ethical implications and hurdles at each level of development. The relevant ethical issues and questions to consider at each step along the translational pathway, from the basic science bench to preclinical research to first-in-human clinical trials, should be addressed.

Maintaining data integrity, appropriate reporting and distribution of results and ensuring that experiments are intended to provide results acceptable for furthering research are all topics at the bench level. Preclinical research topics include the "modest translational distance" idea and suitable animal models. Clinical research topics encompass significant difficulties that emerge in early-stage clinical studies, such as patient-subject selection, disclosure of uncertainty and defining success. The contrast of these two methods, as well as their ethical difficulties, highlights several obstacles for translational tissue engineering research and gives recommendations for investigators working on any tissue engineering technique (Baker et al. 2015).

The best approach for toxicity evaluation is to test the substance on humans, although this is difficult due to legal and ethical concerns. The prolonged direct contact of a bioceramic implant with bodily tissue increases the danger to the point where it becomes a major concern. When compared to biologics and pharmaceuticals, the complex elements of tissue engineering and bioprinting make its therapeutic potential morally difficult (Tamminen and Vermeulen 2019). The cell source and associated processing techniques are one of the primary ethical problems in tissue engineered medical products (TEMPs) and bioprinted constructions.

To eliminate ethical concerns about cell sources, patient-specific cells might be employed as a cell source for both seeding on TEMPs and bioprinting organs. In rare situations, impregnated cells from TEMPs may migrate via the bloodstream to other regions of the host body, causing needless undesirable effects such as cancer or other unknown disorders. As per the European Commission opinion on the state of the art concerning tissue engineering microbial contamination (re-infection), inadequate sterility and toxicity due to the presence of allogenic sources, malfunction of bioactive motifs, overreactivity of growth factors and cryo-preservative toxicity are some of the typical ethical difficulties and hazards related with TEMPs. Furthermore, clinical studies using 3D bioprinted tissues necessitate a more individual-centric approach (transplantation medicine) than standard normal trials.

For example, these products must be tried in a single subject and should only be continued on further subjects once the clinical response in the first subject has been evaluated. Furthermore, 3D bioprinted tissue/organs may pose a significant danger to the individual; hence, such studies should only be conducted when there are no other options for the chosen subject (Sekar et al. 2021). Table 16.1 highlights all published, new and amended versions of the U.S., India and European Union recommendations for tissue engineered product regulation, along with a brief summary of the rules and their scope. Because the goals of these rules are to provide maximum benefits with little risk, the ethical committee, researchers and other stakeholders are constantly reliant on benefit–risk analyses (Venkatakrishnan and Ecsedy 2017). These recommendations inform researchers, industrial producers, marketing authorisation members and other stakeholders about regulatory requirements, approaches and analyses. Importantly, these guidelines must be amended, adjusted and enhanced on a regular basis if tissue engineered products are to be commercialised in the future (Table 16.1).

16.7 COST VS BENEFIT ANALYSIS

When performing research on biomedical engineering, just like any other type of study, scientists must assess if their products are cost-effective. This indicates that the research will yield answers that are both scientifically solid and cost-effective. A cost–benefit analysis shows that longer-lasting prostheses lower long-term medical healthcare expenditures and increase patients' quality of life, at the same time allowing them to be more creative and self-sufficient. As a result, the cost–benefit analysis will be favourable. On the other side, if a modern design of medical instruments lasts the same or even less time than an existing device, the newer design will have a negative cost-to-benefit ratio, which will not be a primary motivating factor to continue researching. Even though some scientists, ethicists and physicians claim that comparing health and health care relating to costs and benefits is problematic, others believe that it is necessary. To ensure that meaningful studies are carried out in the future, careful budget allocation for healthcare research is critical (Kashi and Saha 2017). Research projects aimed towards attaining better awareness of the disease and leads to new medical therapies should be motivated. Future scientists will benefit from this in making discoveries in biomedical research and, as a result, improve our country's health.

16.8 ETHICS AND AUTHORSHIP

While biomedical engineering research is becoming highly interdisciplinary, various ethical issues across engineering and medicine disciplines may arise as a consequence (Gordon and Gordon 2010). While biomedical engineering research is becoming highly interdisciplinary, various ethical issues between engineering field and medicine field may appear as a result. Concerns about article authorship and conflicts of interest when undertaking research are among them. When it comes to authorship, various criteria must be considered, including the intellectual contributions of distinct individuals. Conflicts of interest are an important part of research that might lead to bias in the reporting of results. This can sometimes involve the financing source (e.g., industry) and the final ownership of the study results. Nonfinancial and financial conflicts of interest (e.g., publishing data in advance to promote one's profession) and institutional conflicts of interest (e.g., universities holding conflicts with a faculty's activity) are some of the other types of conflicts of interest. To eliminate

TABLE 16.1
Compilation of the Published Guidelines Applicable to US, India and European Union for Tissue-Based Product Regulations

	Year	Guidance/published guidelines	Scope
U.S (FDA) (U.S. Food and Drug Administration (FDA))*	2020	Regulatory considerations for human cells, tissues, and cellular and tissue-based products: minimal manipulation and homologous use	For better understanding of terms minimal manipulation and homologous use criteria for tissue engineered product to regulate under 21 CFR Part 1271 and specifically the 21 CFR 1271.10(a)(1) criterion of minimal manipulation and the 21 CFR 1271.10(a)(2) criterion of homologous use
	2019	Enrichment strategies for clinical trials to support approval of human drugs and biological products	Exploring the strategies of patient selection to improve drug potency by decreasing the clinical group variations, improving group prognostics and prediction for the drug specific treatment
	2017	Deviation reporting for human cells, tissues, and cellular and tissue-based products regulated solely under section 361 of the public health service act and 21 CFR part 1271	With respect to the technical innovation and development of tissue-based products this guideline describes the HCT/P deviations clearly for its regulation under 361 of the public health service act and 21 CFR part 1271
	2016	Investigating and reporting adverse reactions related to human cells, tissues, and cellular and tissue-based products (HCT/Ps) regulated solely under section 361 of the public health service act and 21 CFR part 1271	This provides the supplement guidance to the "guidance for industry: current good tissue practice (CGTP) and additional requirements for manufacturers of human cells, tissues, and cellular and tissue-based products (HCT/Ps)" (December 2011) for reporting adverse reactions on the study
	2011	Current good tissue practice (CGTP) and additional requirements for manufacturers of human cells, tissues, and cellular and tissue-based products (HCT/Ps)	For manufacturer's better understanding of good clinical practice requirements (facilities, environment control, equipment and reagent source criteria, processing and process control, storage, package, shipment, labeling, donor screening testing and eligibility), registration and listing requirements for HCT/Ps regulation

(*Continued*)

TABLE 16.1 (*Continued*)
Compilation of the Published Guidelines Applicable to US, India and European Union for Tissue-Based Product Regulations

	Year	Guidance/published guidelines	Scope
	2008	Certain human cells, tissues, and cellular and tissue- based products (HCT/Ps) recovered from donors who were tested for communicable diseases using pooled specimens or diagnostic tests	Donor eligibility considerations and requirements (screening and testing) to avoid or reduce the communicable disease transmission
	2007	Regulation of human cells, tissues, and cellular and tissue-based products (HCT/Ps) – small entity compliance guide	Provide details for the HCT/Ps product to register (registration form FDA 3356) and list with center for biologics evaluation and research (CBER) with regular updates and complete guidance for its safety evaluation
	2007	Eligibility determination for donors of human cells, tissues, and cellular and tissue-based products	Represent the establishment of donor safety with free of risk factors for the better availability of tissue-based products in medical care applications
	2006	Compliance with 21 CFR Part 1271.150(c)(1)— manufacturing arrangements	Provide requirements for manufacturing if there is another establishment to undertake any manufacture steps to ensure whether good tissue practice (CGTP) requirements are satisfied
	2002	Validation of procedures for processing of human tissues intended for transplantation	In order to avoid infection of any communicable diseases through HCT/Ps it is necessary to prepare all necessary procedures written with complete validation to avoid infections to patient
	1997	Screening and Testing of Donors of Human Tissue Intended for Transplantation	Provide general requirements for screening and testing procedures for the individual involved in whole procedures to ensure complete diagnosis of heath to assure absence of any diseases
India (ICMR) (Indian Council of Medical research (ICMR))**	2019	National guidelines for gene therapy product development and clinical trials	States the development of effective and safe GTP with controlled product characterization, quality assessment, preclinical evaluation, clinical study and long-term patient analysis to ensure proper medicinal application

(*Continued*)

TABLE 16.1 (*Continued*)
Compilation of the Published Guidelines Applicable to US, India and European Union for Tissue-Based Product Regulations

	Year	Guidance/published guidelines	Scope
	2017	National ethical guidelines for biomedical research involving children	Detail description about the ethical issue need to take care for biomedical research on children since there is wide variation in metabolic activity, disease, pharmacokinetics, and adverse effect to study the effect of treatment is safer for children
	2017	National ethical guidelines for biomedical and health research involving human participants	It's a latest version of guideline with incorporation of Indian cultural, social behavioral, and natural environment areas into biomedical research field involving human studies for better medical application
	2017	National guidelines for stem cell research	Guidelines providing procedures, regulatory pathways, research works on human stem cells with proper ethical and scientific considerations
	2008	Guidelines for good clinical laboratory practices (GCLP)	Provide principles and procedures for the clinical trial laboratories to produce quality results with assured human safety
	2006	Ethical guidelines for biomedical research on human participants	Earlier versions of ethical guidance for clinical research in India
EU (EMA)***	2019	Guidelines on quality, non-clinical and clinical requirements for investigational advanced therapy medicinal products in clinical trials	Provide data requirement and other need for manufacturing, development, quality control for clinical and non-clinical studies of ATMP
	2019	Guideline on the quality, non-clinical and clinical aspects of gene therapy medicinal products	Provide specific requirements for Marketing Authorization Application (MAA) and regulation of GTPs with respect to quality, nonclinical and clinical aspects
	2019	Guideline on strategies to identify and mitigate risks for first-in-human and early clinical trials with investigational medicinal products	It's a first revision published with first-in-human (FIH) and phase clinical trial (CT) study with their EU regulations to study safety, tolerability, pharmacokinetic and pharmacodynamics of the investigational medicinal product

(*Continued*)

TABLE 16.1 (*Continued*)
Compilation of the Published Guidelines Applicable to US, India and European Union for Tissue-Based Product Regulations

Year	Guidance/published guidelines	Scope
2018	Guideline on potency testing of cell-based immunotherapy medicinal products for the treatment of cancer	It provides guidance for the specific requirement of assays and procedures for the cell-based immune therapy products for quality control and manufacture
2014	Reflection paper on clinical aspects related to tissue engineered products	Guidance for the clinical testing of tissue engineered products under regulation (EC) No 1394/2007
2013	Guideline on the risk-based approach according to annex I, part IV of directive 2001/83/EC applied to Advanced therapy medicinal products	Describes the importance and assessment of risk-based approaches for the regulation of advanced therapy medicinal products (ATMP) under directive 2001/83/EC
2011	Reflection paper on stem cell-based medicinal products	Provide requirements of stem cell-based products marketing authorization application (MAA) and it should be considered along with "guideline on human cell-based medicinal products"
2008	Guidelines on safety and efficacy follow-up and risk management of advanced therapy medicinal products	Provide characteristics of ATMPs, describes requirements and procedures for the risk-based analysis, safety and efficacy study, clinical aspects and marketing of ATMPs
2008	Guidelines on human cell-based medicinal products	Provide the guidelines for the quality control, manufacturing approaches and safe human cell-based medicinal products development

Source: *U.S. Food and Drug Administration (FDA). FDA guidance documents, https://www.fda.gov/vaccines-blood-biologics/biologics-guidances/recently-issued-guidance-documents (2021).

** Indian Council of Medical Reserach (ICMR). ICMR national ethical guidelines, https://ethics.ncdirindia.org/ICMR_Ethical_Guidelines.aspx (2018).

*** European Medicines Agency, https://www.ema.europa.eu/en/human-regulatory/research-development/scientificguidelines/multidisciplinary/multidisciplinary-cell-therapy-tissue-engineering.

such biases, clinical studies should use double-blind research as the standard (Kashi and Saha 2017).

16.9 PUBLICATION ETHICS AND MALPRACTICE STATEMENT

16.9.1 Editor's Responsibilities

This Journal is always a collaborative effort. Handling challenges of research integrity and publication ethics in journals is no exception. These concerns may potentially raise or entail legal difficulties. We recommend that journals use these principles when developing policies and processes, as well as a starting point should problems emerge.

We recommend that editors, publishers and other journal team members discuss the issues highlighted as a first step in addressing any issue. We recommend that these talks take place prior to taking any further action and that legal counsel be obtained if necessary, particularly when matters involve potential defamation, violation of contract, privacy or copyright infringement.

16.9.2 Confidentiality

The editor and any editorial staff shall not reveal any information about a submitted manuscript to anybody other than the corresponding author, reviewers, potential reviewers, other editorial advisers and, as necessary, the publisher. The ICMJE code of conduct and best practise recommendations for journal editors on publishing ethics are adopted by the journal.

16.9.3 Reviewers' Responsibilities

The peer-reviewing process helps the editor and editorial board make editorial judgements and may also help the author improve the article. Any selected referee who believes they are unqualified to examine the research provided in a paper or understands that timely review is impossible should tell the editor and resign from the review process. Manuscripts for review must be treated as confidential documents. They must not be divulged or discussed with anybody else unless specifically permitted by the editor.

16.9.4 Standards of Objectivity

Reviews should be carried out objectively. Personal assaults on the author are unacceptable. Referees should express their opinions clearly and give evidence to back them up. Reviewers should look for instances when relevant published work mentioned in the manuscript has not been cited in the reference section. They should specify if observations or arguments obtained from other publications are accompanied by a citation. Reviewers will alert the editor if there is a significant resemblance or overlap between the article under consideration and any other published paper about which they are personally aware.

16.10 REPORTING STANDARDS

Authors of original research reports should offer an accurate overview of their work as well as an objective assessment of its significance.

The underlying data should be appropriately reported in the study. A paper should provide enough detail and references to allow others to reproduce the work. Fraudulent or wilful incorrect remarks are unethical and must be avoided.

16.11 DATA ACCESS AND RETENTION

Authors may be required to provide the raw data from their study along with the manuscript for editorial review, and they should be willing to make the data publicly available if possible. In any case, authors should ensure that such data are accessible to other competent professionals for at least 10 years after publication (preferably through an institutional or subject-based data repository or other data centre), provided that participant confidentiality can be protected and legal rights to proprietary data do not preclude their release.

16.12 CONCLUSION

For all groups of patients, there is a rising demand for the various materials used to be aesthetically acceptable, functionally biocompatible and economically viable. It is consequently critical to develop improved techniques to evaluate, predict and assess material safety issues at both industrial and consumer levels. Researchers must report impacts of all novel biomaterials *in vitro* and *in vivo* and newer technologies to treat disease in an ethical manner. To ensure that any emerging items or technology undergo thorough and systematic clinical review, clinically driven research networks and practitioner organisations should be participating in ethically driven patient studies. Reliable research methodologies will ensure that all biosafety concerns are addressed, and the frequency of adverse responses is reduced. Despite the introduction of human induced pluripotent stem cells (hiPSCs), the usage of stem cells remains a contentious issue. While their creation does not need the loss of an embryo, as it does with Embryonic Stem Cells (ESCs), issues over how they should be employed remain. Relevant legislation must be created to address all of the difficulties raised in this article and to guarantee that hiPSCs are not abused or used unethically. Perhaps the recent NIH session might serve as an example for proactive policy review (NIH VideoCast – Workshop on Animals Containing Human Cells 2015; NOT-OD-15–158: NIH Research Involving Introduction of Human Pluripotent Cells into Non-human Vertebrate Animal Pre-gastrulation Embryos).

If stem cell scientists, bioethicists and policymakers can keep an open discourse regarding the present status of research, possible ethical difficulties can be addressed ahead of time. Such a strategy would allow hiPSCs for human therapy to be appropriately controlled while not impeding essential scientific advancements that would benefit everybody.

These case studies raise several important ethical questions. To begin, it is critical to seek iterative value in tissue engineering research. Because tissue engineering research methods are complicated and heterogeneous, incorporating surgery, cells and a wide range of assisting technologies in animal and human trials, innovation is nearly unavoidable. Researchers should plan for the possibility of innovation and build and explain methods that can not only accommodate but also turn innovations into study characteristics when necessary.

To achieve the correct balance of flexibility and reproducibility in tissue engineering protocols and to update procedures when necessary to capture methodological advancements, clear and close contact with research oversight organisations is therefore crucial.

Secondly, data openness is crucial, despite the fact that it is becoming increasingly difficult to accomplish in today's highly competitive financing climate. It is very important for tissue engineering research that researchers (i) register clinical studies and (ii) publish clear and comprehensive data from well-designed research at all phases, particularly data from early clinical trials, regardless of success or failure. Transparency in the study plan and trial outcomes allows scientists in earlier stages of development to identify what is and is not working with the model in use. This will drive and motivate future effort and enhance the discipline overall.

Furthermore, making the results of these trials publicly available and understandable to the general public will inform those considering trial participation and help set the stage for understanding not only the benefits and limitations of trial participation but also any future treatments and new research directions that may emerge from the research. Thirdly, much work is required to increase parsimony in animal research while identifying the best animal models and developing effective adjuncts and alternatives to animal use. As previously stated, research into body-on-a-chip and organ-on-a-chip has significant potential in this area.

Fourthly, it is becoming clear that the FDA's traditional phase designations are not necessarily compatible with innovative trial designs and the more complex factors driving subject selection. Early bone alkaline phosphatase (BAP) clinical studies, like Phase I cancer trials and most, if not all, first in human (FIH) and early-stage clinical trials of innovative biotechnologies, must involve patients as subjects rather than the "healthy volunteers" normally enrolled in Phase I pharmacological trials. These studies' first patients may be older or younger, experienced or treatment-naive. The ethical issues that occur in early clinical trials of innovative biotechnologies such as tissue engineering are more directly related to design and subject selection than to phase designation.

Thus, regardless of phase classification, adequate rationale for a clinical trial's design and aims is vital, as is full consideration of the consequences of enrolling patients as subjects, including but not limited to the permission form and process. A similar difficulty emerges when complicated tissue engineering intervention experiments cannot be randomised and blinded adequately or ethically. Patients who are potential subjects may have strong and justifiable preferences for one arm over another since their objectives and quality of life evaluations may differ in ways that influence their participation decisions. In addition to the "gold standard" randomised controlled trials, it may be good to explore whether there are new techniques to acquire relevant information in research.

In clinical research, "patient-centred outcomes" are receiving a lot of attention. Incorporating patients' viewpoints into research may be appropriate for some tissue engineering studies and may enhance enrolment. In novel trial designs, methodological issues will surely develop, including not just statistical design and data credibility but also a possible increase in the possibility of TM in investigators, the media, the public and patient-subjects. These problems may be effectively handled by providing clear and transparent reasoning for each phase of the research pathway, as well as excellent preclinical evidence, exceptional communication and the innovation those results from collaborative scholarship along the translational journey. As a result, intervention profiles that are more predictive of clinical outcomes may emerge.

Finally, not only is long-term follow-up required but researchers may also play an important role in aiding translation to clinical treatment for these complicated tissue engineering therapies. Even after successful research, clinical usage of cell-seeded scaffolds and capsules in TESM and BAP will face obstacles. One issue that is frequently disregarded is the infrastructure required for practical and effective tissue engineering therapies. Hospital workers, for example, would need to be educated in the handling cell-scaffold products, and hospitals would need to be affiliated with a good manufacturing process facility for the fabrication of the constructs. The FDA has authorised many biomaterial scaffolds for a range of purposes, including rotator cuff repair, hernia repair and heart valve patches. Familiarity with these devices may assist future adoption and usage of tissue engineering technologies, provided that researchers and sponsors plan ahead of time for efficient clinical translation (Baker et al. 2015).

The advances in 3D bioprinting have made the fabrication of individualised live tissues/organs more feasible and practical. Despite the availability of adaptable regulations, only a small number of TEMPs have been licenced. Despite frequent revisions to scientific prerequisites and preclinical standards, numerous tissue engineered products and 3D bioprinted tissues have yet to undergo clinical trials. This might be related to the fact that present legislation and regulations are better adapted for traditional drugs/therapeutics and hence fail to properly absorb the complexities inherent in the most recent medical items.

As a result, it is critical for regulatory agencies and other parties to understand the history of these sophisticated goods and adapt to the intricacies in order to expedite the approval procedures for TEMPs and 3D bioprinted tissues. Furthermore, physicians and researchers should study the scientific requirements, history of standards and essential clinical trials and GMP training to meet the demands of regulatory bodies and worldwide standards. In conclusion, given the nature of TEMPs and 3D bioprinted tissues, it is critical to ensure that the goods remain within their shelf-life at the time of clinical procedure and assessed using well-planned procedures to determine safety and long-term efficacy (Sekar et al. 2021).

16.13 FUTURE PERSPECTIVES

Researchers in the area of biomaterial research and practise, which involves the development and evaluation of new materials, must follow the highest ethical standards possible. They must adhere to the guidelines outlined in numerous engineering societies' codes of ethics, Recommendations for Clinical Research and the Declaration of Helsinki, and other comparable papers.

Before beginning the trials, the research protocols must be approved by the appropriate Institutional Animal Care and Use Committee (IACUC for animal studies) and Institutional Review Boards (IRB for human studies). For the use of investigational devices and implants, scientists and biomedical engineers must collaborate with clinicians and other appropriate healthcare professionals who possess suitable expertise in the treatment of patients. Such a study should only be carried out after a careful review of the benefits and risks to the patient and others. The findings of the study should be beneficial to the clinical trial populations. Clinical trial participants must be informed about the study's objectives and techniques. It is important to talk about any potential benefits and risks to the subjects. Any possible conflicts of interest and the sources of financing should be made public.

To forecast the future of 3D bioprinting, it is critical to first comprehend the present advances made in the effective production of 3D bioprinted tissues such as skin, cartilage and bone. This accomplishment is related to the therapeutically most significant properties achieved in printing these constructions, such as appropriate cell kinds, desirable functional aspects, biomimetic resolution and other essential accompanying cues, such as vascular network. As a result, effectively 3D bioprinted skin, cartilage and bone might serve as a yardstick for predicting the future prospects and commercialisation potential of manufactured constructions. The advancements in 3D bioprinted tissues/organs and bioprinted organ-on-chip technologies, as well as the anticipated years for regulatory approval and commercialisation. Given the medical device, ISO and other laws in place, we predict that bioprinted 3D tissue models and organ-on-chip will not face rigorous requirements and may be commercialised over the next 5–8 years. Due to the complicated structure and varied biological components of 3D bioprinted tissues and organs, economic opportunities are expected only in the future decades. Furthermore, a coordinated approach between academics, industry and hospitals is required to develop regulatory changes that might facilitate the clinical translation of bioprinted organs/tissue.

Despite the advantages of generating completely functional soft and hard 3D tissues with neovascularisation capabilities, additional hurdles include culturing the printed heterogeneous tissues/organs in a bioreactor or with any particular set up for maturation. Furthermore, the maximum printing resolution of a laser bioprinter is 20 m, whereas tiny capillaries in original tissues are 3 µm in diameter. Aside from the modest diameter, 3D bioprinting has yet to accomplish the complexity of vascular networks with a multi-level structure.

Other challenges include the need for a large number of cells for printing, extensive research on optimal bioink composition with good printing capabilities and shape fidelity and the need for sophisticated advanced complementary bioprinting technologies with guaranteed environment and sterility maintenance. Furthermore, for effective translation into clinical practise, bioprinting faces a number of challenges, including ethical concerns, safety, price and large-scale production. Marketing permission for bioprinted tissue equivalents has yet to be obtained under current/revised regulations; hence, attention must be directed toward these directions in order to use 3D bioprinting for the construction of organs for transplantation purposes.

The 4D bioprinting idea, with the added potential of printed objects to undergo full maturation in response to dynamic external stimuli, may be followed as a next-generation manufacturing technique. Several extensive studies on smart bioink compositions with excellent shape fidelity, feasibility for large-scale production, potential bioprinters and faster clinical trials for the developed bioprinted constructs will undoubtedly pave the way for clinical success and commercialisation scope when compared to conventional TEMPs in the future.

REFERENCES

Assen LS, Jongsma KR, Isasi R, Tryfonidou MA, Bredenoord AL. 2021. Recognizing the ethical implications of stem cell research: A call for broadening the scope. *Stem Cell Reports*. 16(7):1656–1661.

Baker HB, McQuilling JP, King NM. 2015. Ethical considerations in tissue engineering research: Case studies in translation. *Methods*. 99:135–144.

Bertolami CN. 2004. Why our ethics curricula don't work. *Journal of Dental Education*. 68(4):414–425.

Chevalier J, Gremillard L. 2009. Ceramics for medical applications: A picture for the next 20 years. *Journal of the European Ceramic Society*. 29:1245–1255. doi: 10.1016/j.jeurceramsoc.2008.08.025.

Florczyk SJ, Saha S. 2007. Ethical issues in nanotechnology. *Journal of Long-Term Effects of Medical Implants*. 17(3):271–280. doi: 10.1615/jlongtermeffmedimplants.v17.i3.90.

Freire W, Fook M, Barbosa E, Araújo C, Barbosa R, Pinheiro Í. 2014. Biocompatibility of dental restorative materials. *Materials Science Forum*. 805:19–25. doi: 10.4028/www.scientific.net/MSF.805.19.

Gordon SL, Gordon LE. 2010. Ethical dilemmas in medicine in the 21st century. 1(2):101–106. doi: 10.1615/EthicsBiologyEngMed.v1.i2.30.

Kashi A, Saha S. 2009. Ethics in biomaterials research. *Journal of Long-Term Effects of Medical Implants*. 19:19–30. doi: 10.1615/JLongTermEffMedImplants.v19.i1.30.

Kashi A, Saha S. 2015. Chapter 64: Ethical/legal aspects of tissue engineered products." In *Stem Cell Biology and Tissue Engineering in Dental Sciences*, edited by Ajaykumar Vishwakarma, Paul Sharpe, Songtao Shi and Murugan Ramalingam, 865–870. Boston: Academic Press.

Kashi A, Saha S. 2017. Chapter 11: Ethical issues in biomaterials research. In *Materials for Bone Disorders*, edited by Susmita Bose and Amit Bandyopadhyay, 493–503. Academic Press.

Kumar P, Dehiya BS, Sindhu A. 2018. Bioceramics for hard tissue engineering applications: A review.

Lee S, Henthorn DB. 2012. Materials in biology and medicine.

Malsch I. 2013. Nanotechnology and Human Health.

Mhanna R, Hasan A. 2017. Introduction to tissue engineering: Regenerative medicine, smart diagnostics and personalized medicine. 1–34.

Musiband M, Saha S. 2011. Ethical considerations in biomaterials research and development. 483–492.

Nestor MW, Artimovich E, Wilson RL. 2014. The ethics of gene editing technologies in human stem cells. *Ethics in Biology, Engineering and Medicine: An International Journal*. 5(4):323–338.

Pietrzak WS, Vacanti CA. 2014. *Musculoskeletal Tissue Regeneration: Biological Materials and Methods*. Humana Press.

Saha S, Saha P. 1999. Ethical issues of animal and human experimentation in the development of medical devices.

Schmalz G. 1997. Concepts in biocompatibility testing of dental restorative materials. *Clinical Oral Investigations*. 1(4):154–162. doi: 10.1007/s007840050027.

Sekar MP, Budharaju H, Zennifer A, Sethuraman S, Vermeulen N, Sundaramurthi D, Kalaskar DM. 2021. Current standards and ethical landscape of engineered tissues-3D bioprinting perspective. *Journal of Tissue Engineering*. 12:20417314211027677.

Stephens N. 2015. Ethics and emerging laws in stem cell science. 855–863.

Swierstra T. 2015. Identifying the normative challenges posed by technology's 'soft' impacts. *Etikk i praksis*. 9:5–20.

Tamminen S, Vermeulen N. 2019. Bio-objects: new conjugations of the living. *Sociologias*. 21:156–179.

Venkatakrishnan K, Ecsedy JA. 2017. Guideline on strate- gies to identify and mitigate risks for first-in-human and early clinical trials with investigational medicinal prod- ucts. *Clinical Pharmacology & Therapeutics*. 101:99–113.

17 Opportunities, Challenges and Future of Bioceramics

M. Azam Ali and M. L. Gould
University of Otago

17.1 INTRODUCTION

Ceramics are known largely as engineering materials and have been used in the electronic industry, automobile sectors, optical fields, etc. for centuries. But bioceramics also exhibit vital biomaterials in the biomedical field, which have been used for the development of resorbable implants for the repair or replacement of damaged bone tissues since the 1960s (Kumar et al. 2018). Biomaterials are non-drug-based substances appropriate for inclusion in a living system that augment or act as a substitute for body tissues or organs. Changes in the biological response and the design of these devices have been the direct consequence of the progression in material science. The most basic attempt in reparation of hard tissue using biomaterials was to provoke a biological response from the physiological surrounding microenvironment. Alternatively, bioceramics remain inactive when simply inserted as a foreign body to serve as just a mechanical load bearing device at the implantation site. These materials are classified as 'bioinert' as any material introduced into the physiological environment will normally elicit a response, although bioinert is considered the minimal level of response from the host tissue as the implant becomes encapsulated within a fine protective non-adherent fibrous layer; however, because of the lack of a toxic response, this is considered a positive result.

As of late, interest has refocussed towards the benefits of ceramics including their high compressive strength and chemical and wear resistance, particularly in the fields of dental and bone engineering to enhance individual well-being. Additive manufacturing (AM), more commonly known as three-dimensional (3D) printing, is a technique generally utilised to process bioceramics into scaffolds using a layered well-defined technique. The application of AM in bioceramics has the ability to produce commercial scaffolds specifically customised towards clinical applications, but most importantly, these scaffolds have the potential to endure hard tissue fabrication.

Bone tissue has a complex, yet elegant, construction. Bone is composed of an organic base, principally collagen, where the inorganic matrix of calcium-containing crystals are embedded. In the human body, bones play a critical role in support and locomotion but also provide protection for vital organs. But bones are liable to fracture from injury or to lose integrity as a result of degenerative disease. Bone is

DOI: 10.1201/9781003258353-22

also considered an anisotropic material due to the variety of mechanical properties that occur in numerous anatomical locations with diverse loading directions. Furthermore, the outer section of bone or cortical bone is dense, and the inner section of bone known as cancellous or trabecular bone is less dense, and this too can create further complications in design. Scaffolds produced using AM have a similar histomorphometry to bone, and thus, it is used for the repair of impaired bone tissue.

Bone implants fabricated from biomaterials can be considered as (i) first, (ii) second or (iii) third generations. First-generation implants are simply bioinert, whereas second-generation implants are biodegradable, as well as bioinert. Third-generation implants have developed even further, emphasising promoting tissue regeneration by stimulating the cells towards bone regeneration (Gautam et al. 2022). For example, first-generation ceramics were inert, targeted as a natural bone substitute; second-generation ceramics targeted mirroring biomineralisation and with the third generation. These bioceramics provide not only a functional scaffolding system but also, more importantly, facilitates bone cell osteogenesis (Figure 17.1).

Tissue bioengineering also focusses on the regeneration of dental tissue, specifically the damaged tooth, as the AM-generated scaffolds in their different forms meticulously mimic the natural components and framework of dental tissues, including enamel, dentin and periodontal tissues. In particular, biopolymer hybrid particles have gained notice as unique materials for clinical situations, such as dental regeneration, bone repair or surgery.

This chapter will discuss the opportunities and applications of bioceramic-based AM in hard tissue engineering and furthermore focus on, AM processing of common

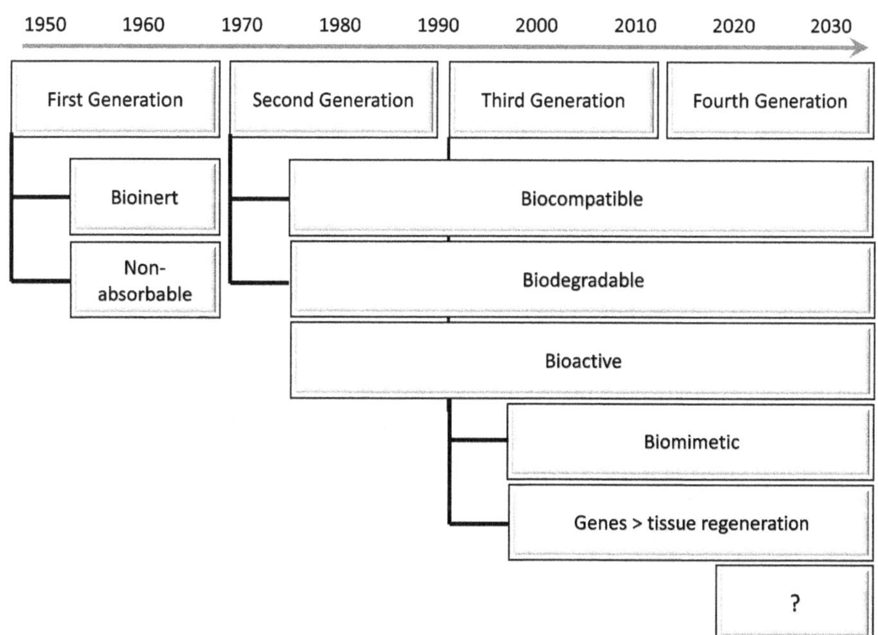

FIGURE 17.1 Evolution of biomaterials through the generations.

materials utilised for scaffolds, bone cement as dental or tooth tissue regeneration, and the complexities, challenges and future trends related to these materials have been expounded.

17.2 COMPOSITES AND BIOCOMPOSITES

What is a composite? In the most basic form, when two or more different materials are combined, the resultant material is a composite. The modern epoch of composites commenced with the development of plastics in the early 1900s, when vinyl, polystyrene and polyester were invented (Figure 17.1). These new synthetic materials outperformed the previous plant-based resins. However, plastic alone was not strong enough for some structural applications, and hence, reinforcement was required for additional strength. In 1935, glass fibres were combined with a plastic polymer to create an incredibly strong lightweight structure named fibreglass. This paved the way for the fibre-reinforced polymers industry. Carbon fibre was developed in 1958 and increasingly, over time, has replaced parts formerly made of steel. In the 1970s, the composites industry began to mature, providing more advanced plastic resins and reinforcing fibres. For example, the DuPont company produced an aromatic polyamide (aramid) fibre, commonly known as Kevlar, used in body armour because of its high tensile strength and lightweight properties. The composites industry is still evolving, constantly pushing the limits on size moving down into nanosized materials and bio-based polymers.

Hence, a biocomposite is a composite formed from a matrix, possibly in resin form with reinforcement for strength or biocompatibility, for example, in the form of natural fibres. In these times of global warming and environmental concerns, the cost and lack of recyclability of synthetic fibres have led to the basis of using natural fibre instead as the reinforcement in polymeric composites. The matrix phase protects the fibres from environmental degradation and mechanical damage, holds the fibres together and takes the mechanical loads. The fibres derived from biological origins are, for example, fibres from crop foods (flax, cellulose or hemp) and crop processing waste products. Advocates of biocomposites, such as we are, state that these materials have a place to use as scaffolds within the human body and are environmentally superior.

17.3 CLASSIFICATION OF VARIOUS BIOCERAMICS

Ceramic materials are inorganic non-metallic compounds composed of polymers of long chains of carbon atoms interconnected with other atoms such as H, O, C or N. However, ceramics are not carbon-based materials, but instead are inorganic. Most ceramic materials are oxides, including magnesia (MgO), silica (SiO_2), alumina (Al_2O_3) and zirconia (ZrO_2). They can also be carbon-based when combined with an element of either lower or comparable electronegativity using either a metal or metal oxide; these include silicates such as zirconium silicate ($ZrSiO_4$), magnesium metasilicate ($MgSiO_3$), silicon carbide (SiC) and nitrides (Si_3N_4) (Figure 17.2). Those compounds not containing carbon, classified as inorganic particles, are composites of fillers and additives combined together into a polymer matrix that may

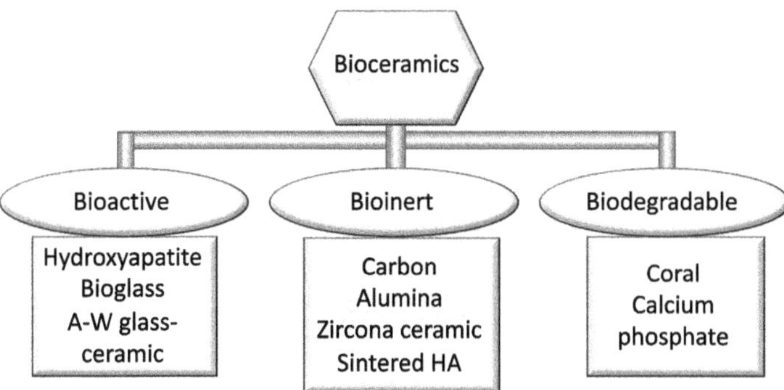

FIGURE 17.2 Classification of bioceramics according to bioinertness, biodegradability and bioactivity.

be antimicrobial or antiviral, but definitely are bioactive when balanced with strong mechanical properties.

Bioceramics, on the other hand, can be composed of hydroxyapatite, bioglass or ceramics that take the form of nanocrystals, nanoparticles, pastes, amongst others, and these have been exploited in the regeneration of both soft and hard tissue in the human body. Over the last few years, bioceramics have been used for both restoration and complete replacement, but ultimately, the aim remains the same, complete regeneration of injured tissues by providing a fundamental function in the integration, jointly, into both soft and hard tissue (Figure 17.3). The use of bioceramics such as hydroxyapatite, bioactive glasses and bioinert substances as resorbable implants demonstrated the potential for hard tissue applications in various studies. It has also been documented that bioceramic biomaterials and implants are vital to the medical needs of the ageing population that is rapidly growing throughout the world, and in particular for developing countries. However, the clinical success of the uses of bioceramics as medical implants in a clinical setting are still challenging and inconclusive. The major challenges identified relate to the failure of osteoinduction and osteoconduction of bioceramics implants, often requiring patients to endure many surgeries. Furthermore, this provides an added cost to the patients and a burden for the healthcare system (Hench and Thompson 2010).

17.4 CHARACTERISATION OF BONE BIOCOMPOSITES USING BIOCERAMICS

Biomineralisation or biomimetic mineralisation is the activity by which living organisms produce minerals, thus hardening existing tissues and leading to the construction of hierarchically structured inorganic–organic composites with the inorganic hydroxyapatite and calcium precipitated within the supportive organic matrix. In general, bioceramics for skeletal repair and reconstruction exhibit extraordinary biological and osteoinductive properties that generate increased proliferative and adhesive properties in bone tissue regeneration (Blokhuis and Arts 2011).

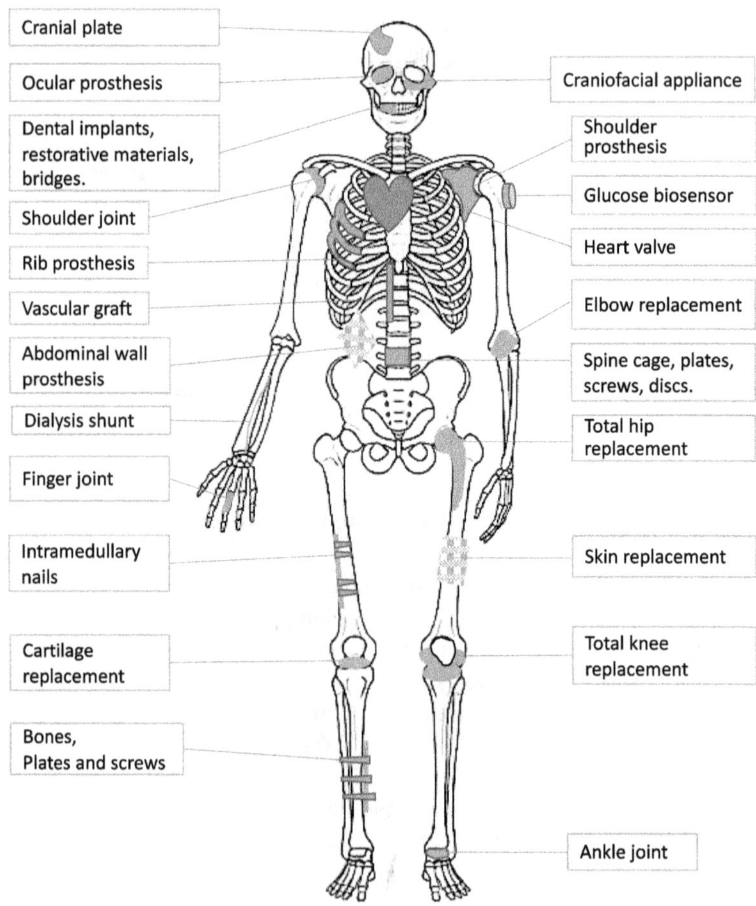

FIGURE 17.3 Application of bioceramic biomaterials as implants in the human body.

17.4.1 Hydroxyapatite

In 1926, de Jong observed a resemblance between bone X-ray diffraction patterns and the calcium phosphate mineral, named hydroxyapatite (HA) (Zhao et al. 2011). Decades later, the existence of HA and carbonate in tooth and bone was identified using infrared spectroscopy (Zhao et al. 2011). It was only later in 1965 that the use of bioactive ceramics incorporating HA was fully recognised.

Initially, the dental or bone implants composed of HA entailed comprehensive optimisation of the properties guaranteeing stability and mechanical strength. Consequently, stoichiometric, highly crystalline, high temperature sintered HA demonstrating decreased porosity and high crystallinity was produced, but the sintered HA was too brittle for load bearing. Unfortunately, the product was also non-biodegradable and non-bioactive. Daculsi et al. (1989) increased the degradation rate by combining HA with ceramic base β-TCP:biphasic CaP (BCP), making it more soluble.

HA ($Ca_{10}(PO_4)_6(OH)_2$) as the primary mineral component of natural teeth and bone in vertebrates, comprising approximately 65% of complete bone mass, is recognised as being bioactive, biocompatible, non-inflammatory, non-toxic and non-immunogenic but most importantly is osteogenic and osteoconductive (Swetha et al. 2010). HA does have increased stability and high crystallinity and therefore is challenging when it comes to *in vivo* degradation (Shuai et al. 2013). Hence, HA exhibits exceptional biocompatibility and is frequently used as a supportive substance for bone tissue engineering owing to its similarity to natural bone. However, its brittle temperament makes it problematic to fabricate into different sizes and shapes. But a combination of HA with other biodegradable biopolymers enhances the binding interactions and mechanical properties of each material as both materials benefit from each other's positive properties.

HA originates from either chemical or natural resources, such as eggshells (Derkus et al. 2016), bovine bones (Ratnayake et al. 2017) or fish scales (Wijedasa et al. 2020), and when taken from natural resources, HA is more cost-effective with higher metabolic activity and increased bioactivity compared to their synthetic equivalents (Ratnayake et al. 2017). Furthermore, when HA is synthesised, the choice of method can create tailor-made HA nanoparticles (Varadavenkatesan et al. 2021).

A combination of HA with either inorganic or organic compounds can augment bioactivity, biocompatibility and mechanical strength. For example, polyurethane membranes electrospun with HA and titanium dioxide (TiO_2) have a higher water contact angle, indicating low wetting or hydrophobicity, increased mechanical strength and stiffness and better cell adhesion, viability and proliferation. Incorporation of HA into dental tissue provides increased osteoinduction as seen by increased calcium content and alkaline phosphatase activity (Prieto et al. 2015; Watcharajittanont et al. 2020).

Laser stereolithography (SLA) has produced complex HA-containing structures. For example, Barry et al. in 2008 used helium–cadmium (HeCd) laser technology to fabricate HA-based oligocarbonate dimethacrylate (OCM-2) scaffolds. By using laser machining, any residual toxic by-products can be effectively removed via supercritical carbon dioxide ($scCO_2$), making the scaffolds biocompatible and showing cell attachment in both *in vitro* and *in vivo* studies (Barry et al. 2008). A bioink prepared by combining nanohydroxyapatite and deproteinised bovine bone with collagen created a biomimetic porous structure via 3D printing (Lopez-Heredia et al. 2017). This bioink was shown to be biocompatible and stimulate osteogenic differentiation of human bone marrow-derived mesenchyme stem cells (MSCs)(Li et al. 2019).

17.4.2 Calcium Phosphate

Calcium phosphate (Schoche et al. 2017) has long existed as a synthetic bone graft. Calcium phosphate ($Ca_3(PO_4)_2$) is a family of minerals containing calcium ions along with inorganic phosphate anions. Biomimetic CaPs produced at low temperatures mimic the composition of the bone minerals, by modulating the material composition, and then textural properties such as nanopores are formed, which balance material resorption and bone formation. The first publication on CaP stimulating bone regeneration was in 1920, when Albee described a faster healing rate when 'triple CaP' was injected into defects created surgically in canine bones (Albee 1920). The original CaP synthetic bone grafts achieved commercial production in the 1970s, and their usage increased in the 1990s because of an accumulating apprehension behind disease transmission by xeno- and allografts. This was related to the transmission of diseases like acquired immunodeficiency syndrome and bovine encephalitis through grafts (Mathe 1994, Fushimi et al. 2002, Mochizuki et al. 2003).

Calcium phosphate is the basis of the inorganic phase of bone and so logically is useful as a filler for dental polymer composites (Zhao et al. 2011) as it provides the required biocompatible properties, as well as osteoinductivity and osteoconductivity. When calcium phosphate makes contact with biological media, the mineral experiences a solution facilitated surface reaction that is partially accountable for new bone construction. Many crystalline forms of calcium phosphate exist, including HA and tricalcium phosphate (TCP) (Safronova et al. 2020) that form a diverse range of shapes, including spheres, zero-dimensional (0D) particles, 1D rods, fibres, wires and whiskers, 2D plates, disks, sheets, belts, ribbons or flakes and 3D hollow and yet porous biomimetic structures similar to biological tooth and bone. The dimensions, such as shape or size, of the calcium phosphate crystals significantly influence biological responses such as stimulating osteoblast proliferation, encouraging cellular actions, cell death and macrophage activity (Lin et al. 2014). By modifying the concentration and the precursor Ca/P ratio, the predicted shape and size of calcium phosphate particles can be modified to fit the need (Piazza et al. 2020). Gelatin/strontium-substituted CaP composites constructed using a coprecipitation method synthesised homogeneous unidirectional scaffolds with an orientated porous microtubular structure, creating a scaffold capable of guiding bone tissue regeneration (Wu et al. 2015).

Kovach et al. in 2015 manufactured a supramolecular CaP ball by dripping a sodium phosphate solution into a calcium chloride solution containing chitosan and gelatin at 90°C; this action grew a flower-like construction of thin CaP platelets. With gelatin alone, large rhombic crystal aggregates formed because at the molecular level, gelatin alone could not regulate the generation of thin platelets, whereas with chitosan alone, thin calcium phosphate platelets were generated to form equilateral cubes as chitosan directed calcium phosphate crystal growth. The combination of gelatin/chitosan manipulated the supramolecular ordering to guide calcium phosphate crystal growth, consequently forming supramolecular nanoporous structures. *In vitro*, when acidic fibroblast growth factor (aFGF) was loaded into those shelled nanocarrier scaffolds, the aFGF discharged over 4 weeks at a highly maintainable rate, which caused increased proliferation of osteoblastic precursor cells. When implanted into the rat, a significantly increased invasion of fibroblasts displaying a homogeneous population was seen, indicating a possible therapeutic 3D matrix for tissue engineering.

17.4.3 Bioactive Glass

Bioactive glass, also called bioglass, is a group of surface-reactive glass-ceramic biomaterials, in particular. Bioglass is composed of sodium oxide, silicon dioxide, calcium oxide, phosphorous silica and phosphates, all of which have the potential to promote bone regeneration. Silicates in particular are a critical component in the biocompatibility of bioglass. Bioglass implanted within a living organism develops a hydroxyapatite layer across the surface, mimicking the natural bone structure and stimulating growth factor release, which will in turn aid in osteoblast proliferation and differentiation proposed to fill large bone defects (Zhao et al. 2019). In one example, bioceramic glass of silica and titanium oxide promoted the human osteosarcoma (MG63) cell line proliferation, maintaining the pH and ion concentrations of the environment (Chen et al. 2019).

Bioglass particles are mesoporous structures that are supernanoporous with pore diameters between 2 and 50 nm and are often used as drug delivery systems for dental applications. Bioglass in the presence of boron will increase its bioactivity by providing a more porous structure, and this has been beneficial for successful antibiotic delivery (Huang et al. 2020). When nanocarriers formed from mesoporous silica nanospheres were supported within a collagen foam matrix within polylactic acid, biopolymers or polyethylene glycol shell formed utilising electrospraying, the use of a shell significantly protracted the release period of the model protein, bovine serum albumin (Kim et al. 2014).

Bioglass 45S5 or calcium sodium phosphosilicate is a bioactive glass composed of 45 wt% SiO_2, 24.5 wt% Na_2O, 24.5 wt% CaO and 6.0 wt% P_2O_5. Bioglass 45S5 was fabricated by Hench in the late 1960s (Rahaman et al. 2011) using a sol–gel method. By controlling environmental factors, such as concentration, pH and temperature, the solution forms a sol as the colloidal particles amass to form a 3D gel. The advantage of the sol–gel is the maintenance of bioactivity as the synthesis temperature is low and the homogeneity of the particle size can be regulated, obtaining high-purity samples with high uniformity.

Silicon dioxide/calcium oxide (SiO_2/CaO) binary glass scaffolds fabricated using the sol–gel process formed fibrous silica-based scaffolds based on bacterial cellulose fibres, which exhibited mesopores of 10.6 nm. Additionally, this type of bioglass was utilised in soft tissue implants because of its superior bonding capability.

Bioglass may be useful for upregulating osteogenesis; nevertheless, their utilisation in load bearing bone defects were limited because of the reduced fracture toughness, increased brittleness and low mechanical strength (Baino et al. 2016, Fernandes et al. 2017).

At the same time that Bioglass® was developed, Kokubo et al. developed a new glass-ceramic material called apatite-wollastonite (A-W) glass/ceramic to be used as a bone substitute. A-W glass-ceramic contains tiny apatite flakes reinforced by the presence of wollastonite, a calcium inosilicate mineral that contains low quantities of magnesium, iron and manganese that can substitute calcium. A-W glass-ceramic combines increased bioactivity with high mechanical properties. Fracture toughness, bending strength and Young's modulus of the A-W glass-ceramic were higher than glass ceramic or bioactive glass and hence is utilised in prostheses undergoing high compression/load bearing applications, such as the vertebrae (Kokubo et al. 1987, 1990, Kitsugi et al. 1989, Juhasz et al. 2002).

The osteoinductive properties of 3D 45S5 bioglass scaffolds with a gelatin coating, or a coating of cross-linked gelatin with poly(3-hydroxybutyrate-co-3-hydroxyvalerate) polymers were examined by seeding the scaffolds with human mesenchymal stem cells (hMSCs) that were then placed in immunodeficient mice. Histomorphometry and μ-computed topographical analysis showed that each bioglass scaffold displayed visible bone regeneration with the gelatin-coated bioglass scaffold having increased cell incorporation (Westhauser et al. 2016).

17.4.4 Tricalcium Phosphate

Beta-tricalcium phosphate (β-TCP) is used for craniofacial defects because it displays high wear resistance and increased biodegradability and forms enhanced chemical bonds with bone tissue under every tested load bearing condition. The difficulty of β-TCP comes with a sintering temperature of 1,100°C because at temperatures above this, β-TCP metamorphosises to alpha-tricalcium phosphate (α-TCP), which is chemically unstable but is soluble. Generally, the calcium (Ca)-to-phosphate (P) ratio changes the dissolution; a lower Ca-to-P ratio has high solubility, a more acidic nature and a more soluble calcium phosphate phase in comparison to calcium phosphate with an increased Ca-to-P ratio, as with HA (Koutsoukos et al. 1980).

β-TCP with the chemical formula of $Ca_3(PO_4)_2$ is a biodegradable bioceramic calcium salt of phosphoric acid. β-TCP bioceramic scaffolds have been used as bone graft replacements in orthopaedic and dental applications. A ceramic stereolithographic method produced a β-TCP/collagen scaffold by combining stereolithography and gel casting. In this case, they used histological information to design the scaffold to define the morphological characteristics of bone and cartilage. This acquired evidence was used in designing the biomimetic/biphasic scaffolds (Bian et al. 2012).

Sphingosine-1-phosphate (S1P)-coated β-TCP scaffolds were tested by using rat bone marrow stromal cells or macrophages, which showed that the scaffold stimulated cell compatibility and osteogenesis and also regulated the immune response when compared to a conventional artificially contrived scaffold (Cao et al. 2019).

3D-printed β-TCP scaffolds incorporating Fe^{3+} and Si^{4+} dopes were placed into a rat distal femur. The scaffold was fabricated from synthesized β-TCP powder using the binder jetting technique. The Fe^{3+} added to the TCP formed a scaffold, which increased early-stage bone restoration by increasing collagen type I. The TCP scaffold also exhibited neovascularisation, indicating that this ceramic powder-based scaffold has a potential for bone defect restoration (Bose et al. 2018).

17.5 THE USE OF BIOCERAMICS IN 3D BIOPRINTING TECHNOLOGY

Additive manufacturing (AM), also called 3D printing, has caught increased attention in manufacturing products aimed at tissue engineering applications, specifically hard tissue regeneration. 3D printing can be tailored by utilising computer-aided design (Mihalache et al. 2020) to produce the perfect scaffold with the required measurements, materials and porosity coupled with interconnected structures using a time-efficient and low-cost material. Although this technique has shown a momentous capacity for hard tissue engineering, some specific difficulties still exist.

Clinically, AM ceramic scaffold designs supply a unique technique for fast consistent production of a tissue replica that mimics an individual's imperfections. AM is either acellular or cellular, meaning that this will include the printing of living cells.

Binder jetting techniques are used for printing powder materials where large particle size was preferred due to providing increased flowability along with a small surface area, but less surface roughness has been reported by using a fine powder in the binder jetting AM printers (Snelling et al. 2017, Miyanaji et al. 2018).

Direct energy deposition-based AM is a process used to create a project by delivering materials and energy simultaneously. Energy can be focussed on a particular area using electricity or a laser beam to lay down, melt and solidify materials, usually ceramics, metals or a composite simultaneously. Laser-directed energy deposition uses elemental powders titanium (Ti), niobium (Chu et al. 2002) and zirconium (Zr) to produce niobium-based biomedical β-type titanium alloys (Ti-Nb) or zirconium (Ti-Zr-Nb) alloys. Laser-deposited Ti-Nb and Ti-Zr-Nb alloys do not show any defects or pores, but instead have a low elastic modulus but are significantly harder than titanium. Furthermore, corrosion resistance was exceptional because of a stable passive protective oxide film forming on the material surface. Additionally, they also possessed cytocompatibility using osteosarcoma Saos-2 cells (Arias-Gonzalez et al. 2021).

Direct energy deposition (DED), also called electron beam AM, delivers ceramic flecks to the laser focal point that then melts and solidifies in layers. DED provides a range of viscosities, increases resolution and allows control of the cell-to-cell adhesions. An example of this is the use of cobalt chromium (CoCr)-based alloys for increasing strength and wear, but concerns remain due to the high stiffness of CoCr, which results in stress shielding and bone reabsorption. 3D printed CoCr/Ti6Al4V implants that have an interconnected open porous architecture fabricated using electron beam melting were implanted into adult sheep femora. Bone formation densification was seen surrounding the implant with a slow ingrowth into the porous structure, with osteocytes directly contacting the Ti6Al4V and CoCr. However, DED too has its challenges, including a low speed, high cost and increased complexity; hence, the use of DED is restrictive as more research is required to augment productivity.

Fused deposition lays down an uninterrupted layer-by-layer ceramic-loaded paste to fabricate 3D entities. Dense ceramic particles in a thermoplastic filament are partially melted then extruded layer-by-layer creating complex 3D objects. FDM using plasticised poly (3-hydroxybutyrate)/poly(d, l-lactide) with a tricalcium phosphate bioceramic modification including the plasticiser, Syncroflex (oligomeric adipate ester). Samples with Syncroflex demonstrated non-cytotoxicity and biocompatibility including viability, proliferation and the osteogenic differentiation of hMSCs (Melcova et al. 2020).

Extrusion-based bioprinting has more capacity in terms of printing speed and consequential deposition when compared to other AM techniques, and some complexities do exist, including decreased resolution and a lower shear stress put upon cells. Methacryl-modified gelatin (GM), acetylated GM (GMA) and non-modified gelatin (G) were investigated in tailoring the bioink towards printability and vascular network construction. Human dermal microvascular endothelial cells (HDMEC) and human adipose-derived stem cells (Moreira et al. 2021) grown in co-culture were investigated for biocompatibility using both a vascular element and an osteogenic element to fabricate a bone bioink using extrusion-based bioprinting. Vascular-like structures were formed and maintained by depositing bone matrix-related protein (Leucht et al. 2020).

Powder bed fusion, also known as selective laser sintering (SLS), melts ceramic powders by using laser energy, which sinters the powder close to melting point to fabricate individual layers in a 3D design. This is the main hindrance with using SLS, the high temperatures that inhibit the inclusion of cells. SLS constructed a 3D bone scaffold embedded with microspheres of polycaprolactone (PCL) and HA. Both *in vivo* and *in vitro* investigations demonstrated that the scaffolds promoted cell proliferation and adhesion and encouraged cell differentiation. Furthermore, the scaffolds showed histocompatibility inducing vascularisation of the newly formed tissue (Du et al. 2015).

Vat polymerisation, also called SLA, is used to construct tissue scaffolds for use in the regenerative medicine field due to its high fabrication accuracy. The use of the SLA technique provides excellent advantages when it comes to constraints in producing porosity, pore size and interconnectivity. There are also disadvantages, such as difficulties in producing scaffolds at the micron level due to layer thickness (Melchels et al. 2010). SLA was used to prepare porous hydroxyapatite-based scaffolds to investigate bone regeneration in Yucatan minipig mandibles. Two different pore network architectures, either orthogonal or radial channels, were used, in the orthogonal design, HA with bone formed an interporous matrix, whilst in the radial model, bone regenerated at the centre of the scaffold, indicating the significance of pore inclusion in the scaffold designed for *in vivo* bone regeneration (Chu et al. 2002).

3D printing technologies using bioceramics are advantageous over other methods, when it comes to designing personalised bone implants designed specifically for the individual patient (Bose et al. 2013, Kumar et al. 2016). An accurate scan of the geometry of the wound, based on X-ray computed tomography (CT) would provide a precise fit between the scaffold and the anatomical defect. The development of CaP inks composed of reactive CaP such as α-TCP, that self-harden at low temperatures, forego the sintering step making them extremely convivial for incorporating biological molecules such as growth factors or drugs (Ginebra et al. 2012, Maazouz et al. 2017). The low temperatures expand the application to provide patient-specific bone scaffolds, allowing surgeons to readily obtain bone grafts within a few hours, revolutionising surgical treatment by fine-tuning the geometry and microstructure of the scaffold (Roohani-Esfahani et al. 2016).

17.5.1 Chitosan

Chitosan has become the optimal product for scaffold production because it has biodegradability and antibacterial properties; furthermore, this product is non-toxic after degradation. When chitosan is added into bioglass, the alkaline antibacterial environment of bioglass can persist, and the mechanical characteristics of hardness and elasticity are increased, so the biocomposite still maintains true mechanical strength in the humid acidic environment of the mouth. Moreover, bioglass precipitates phosphorus and calcium on the implant surface, forming a layer of hydroxy-carbonated apatite when submerged within biological fluids to enhance protein adsorption to the implant surface and incorporation with the surrounding tissue (Yu et al. 2018). The most critical trait of this material is the formation of the carbonate/apatite layer on the surface following scaffold implantation.

In one example, nanohydroxyapatite (Lopez-Heredia et al. 2017) particles incorporated into scaffolds composed of chitosan, gelatin and nanoHA (Cs/Gel/nHa) without exogenous addition growth factors promoted the accumulation of mineralised tissue within the dental pulp stem cell-seeded Cs/Gel/nHa in the guided regeneration of the barrier membrane. This product induced odontogenesis, biomineralisation and differentiation with the prospect of guiding bone regeneration, thereby promoting bone formation for periodontal tissue regeneration (Vagropoulou et al. 2021).

Chitosan can also form micro- or nanoparticles, fibres, films, sponges or gels that are used for the treatment of tooth caries and periodontitis or in endodontic therapies. Chitosan can sustain drug release at therapeutic concentrations in the extracted tooth pocket for extended periods of time. Drug-loaded chitosan nanoparticles have low toxicity and enhanced controlled release in the acidic oral environment. But chitosan is mechanically weak, and this property can be improved by composite formulations.

Calcium phosphate cement (CPC) self-hardens when implanted into living tissue and forms HA when combined with 20% chitosan; composites with increased mechanical strength and washout endurance can be fabricated with a flexible strength three-fold than that of CPC alone (Sun et al. 2007).

Biodegradable chitosan films formed via condensation utilising the amidation reaction between native chitosan and synthetic poly[(maleic anhydride)-alt-(vinyl acetate)] are aimed at the delivery of efficient concentrations of the local anaesthetic bupivacaine, which provides prolonged anaesthetic treatment and relieves pain after 15 minutes (Mihalache et al. 2020).

The advantage of hybrid hydrogels, is that each component provides its own benefits, for example, a hybrid hydrogel composed of poly(N-isopropylacrylamide) copolymer combined with graphene oxide (GO) and chitosan showed that GO increased the mechanical strength of the hydrogel whereas chitosan combined with GO increased osteogenic differentiation and mineralisation. The engineered fusion gel presented desirable potential for bone tissue engineering and also upregulated the Runt-related transcription factor 2 expression, which is involved in osteoblastic differentiation and skeletal morphogenesis and osteocalcin in the human dental pulp stem cells cultivated in both standard and osteogenic media (Amiryaghoubi et al. 2020).

17.5.2 Hyaluronic Acid

Hyaluronic acid composed of glycosaminoglycans of polysaccharide and carbohydrate is another biomolecule used in dental applications. Natural hyaluronic acid is located within the oral tissues gingiva and periodontal ligaments. The elasticity, hydrophilicity and viscosity of hyaluronic acid can be altered, thereby increasing the opposition to biodegradation. A combination of hyaluronic acid gelatin hydrogel (Nejadshafiee et al. 2019) polymers, β-TCP or biphasic CaP bioceramics fabricated using a freeze-drying method as a plug for bone grafting into an extracted tooth socket had high osteoconductivity and biocompatibility, and the plugs possessed increased haemostatic ability in the artery of the rabbit ear. The HG/TCP/BCP scaffold plug showed promising bone formation potential after implantation in a rabbit femur defect. This mix of biopolymer and bioceramic provided bone deposition through the osteocytes, without any detrimental effects on the tissue (Kang et al. 2020).

17.6 BIOCERAMIC DENTAL CEMENTS

Although considerable publications are available on the first-generation resin-based composites for dentistry, the literature available for bioceramic-based restorative materials is sadly lacking.

Since the first dental adhesive bonded resin directly onto the tooth enamel without dentin adhesion, the use of bioceramics in the dental field has made a remarkable evolution. Consequently, dental adhesives bond directly to dentin to seal dentin margins whilst still improving adhesion to the enamel. The more recent intensified demand to discontinue amalgam- and mercury-free dentistry has accelerated progressions in composite technology to now provide a dental resin that is more similar to a natural tooth and is composed of a healthier mix of plastics and glass that is aesthetically pleasing and will endure (Figure 17.4). But the drawbacks of resin composites are major: they absorb water and substances from the surrounding environment but possibly more importantly release substances out into the surroundings, which may be detrimental for the patient. For example, bioglass may promote tissue regeneration but essentially delivers alkaline substances into the environment, which slows bacterial proliferation. However, the composite of poly(lactide-co-glycolide) (PLGA), which is a stable, biodegradable, biocompatible and mechanically strong material with bioglass, sounds like an ideal combination; however, when PLGA degrades, the by-product is acidic, which neutralises the alkaline by-product from the bioglass, changing the pH of the environment to become more neutral that was unable to reduce the bacterial colony forming units (Hild et al. 2013).

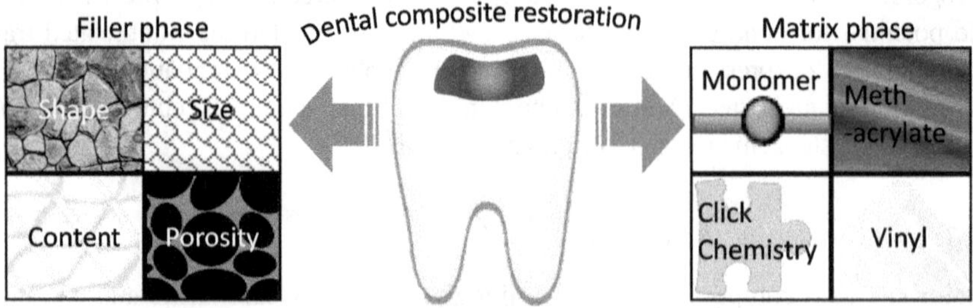

FIGURE 17.4 Filler and matrix resin characteristics in dental applications.

17.7 THE FUTURE CHALLENGES AND OPPORTUNITIES FOR BIOCERAMICS

The last few decades have seen a positive evolution in both the availability of the number of ceramics utilised in clinical applications and also in the superiority of bone repair offered; of course, there is still a huge scope for progressive advances to be made. The mechanical performance and stability could be improved in the existing bioactive ceramics. Of necessity, enhanced bioactivity, in terms of the ability of biomedical coatings to deliver biological agents, are paramount in developing smart materials that sense bioactivity in the development of improved biomimetic composites.

17.7.1 New Opportunites Do Exist in the Field of Bioceramics

As previously stated, ceramics are naturally brittle, displaying poor mechanical properties thus making them unsuitable for loadbearing applications. The combination of ceramics with polymers would improve the lack of mechanical properties, but to date, no biodegradable ceramic composites have been produced (Johnson and Herschler 2011). Many clinical treatments could indeed be improved, and for this reason, this area of research has attracted extensive attention from researchers, and thus, this the field is expected to remain active snd productive.

17.7.2 There Will Always Be Challenges

Biomimetic bioceramics could be utilised for drug delivery because they are formed at low temperatures, which could provide a platform for encapsulating drugs whilst preserving their activity for controlling drug release kinetics by adjusting the porosity of the scaffold. Drugs that could be integrated include antiinflammatory antibiotics or anticancer drugs. Indeed, a suture developed by our group incorporated diclofenac (Deng et al. 2021). However, control over release kinetics requires monitoring, initially *in vitro* and then *in vivo* over an appropriate time period in different locations throughout the body to ensure a reproducible release of a low drug concentration over a long period of time.

Similarly, delivery of active metal ions including cobalt, copper, zinc, strontium, silicon or silver at the injury site could stimulate a specific biological response. Some ions stimulate osteogenesis, whilst others are antibiotics. The purpose of a synthetic bone graft with both antimicrobial and osteogenic characteristics would provide a critical innovation for the clinical field. Indeed, in our hands, silver nanoparticles (AgNPs) as antimicrobial agents were added into a chitosan/keratin polymer base and electrospun into a 3D nanofibrous scaffold (3DENS). Approximately half of the total AgNPs were released during the first 6 hours after immersion before reaching an equilibrium. Therefore, this study could be useful in understanding the mechanism of release from nanofibrous polymeric products in biomedical application (Islam et al. 2019).

For cell-based therapies, hybrid composites providing a combination of porous ceramic scaffolds loaded with molecular signals and with either bone marrow cells or MSCs could be utilised in injuries that lack sufficient osteogenic potential such as diseased or damaged bone tissue as the scaffolds deliver the precise niche that cells prefer to attach and proliferate. An injectable porous biomaterial that hardens *in situ* within the human body at 37°C would benefit clinical practice. Injectable bioceramics would allow the development of more minimally invasive surgical techniques providing a product that geometrically fits the defect. The presence of nanopores would provide not only superior drug delivery capability but also the possibility of being colonised by the tissues. Our group has developed an injectable thermoplastic chitosan-based gel that solidifies at 37°C in the brain for stroke.

Bioceramics could be used as a growth delivery system. Growth factors that enhance the bone graft natural osteogenic potential could be also be incorporated. Bioceramic beta-tricalcium phosphate (β-TCP) could logically be utilised as a bone graft substance and drug carrier for treating bone deficiences in the oral and maxillofacial regions because of its biocompatibility and osteoconductivity. But β-TCP has a low mechanical strength and low osteoinductivity, restricting bone regeneration. Therefore, a biomimetic incorporation of bone morphogenetic protein-2 (BMP-2) into the bioceramic was proposed. Where the biosilicification and collagen layer on the β-TCP surface allowed incorporation of BMP-2 without bioactivity loss. Continuous release of BMP-2 from the composite stimulated osteogenic behaviour of preosteoblasts and furthermore increased the bone regeneration within the rodent calvarial bone defect (Vagropoulou et al. 2021).

Owing to their uniquely beneficial properties, bioceramic materials might just become the new generation of materials for medical applications. Although various types of biomaterials have been extensively studied for biomedical purposes, many problems still exist. In this respect, the new generation of bioceramics and their potential applications in tissue engineering should be studied in respect to the biological system and its interactions.

Modern techniques face certain limitations and to achieve a long-lasting implant with acceptable bioactivity, there is still a vital need to improve the range of bioceramics and devise new manufacturing approaches.

17.7.3 What Does the Future Hold for Bioceramics?

Even though several clinical studies are currently being undertaken because this therapy is comparatively new, at this time, there is insufficient reported data available to allow for the comparison and correlation of results (Cho et al. 2022, Hayashi et al. 2022, Zhou et al. 2022).

A further obstruction to moving forward is the lack of understanding of the biological system and its interaction with bioceramics. More information on the bond formation between bone minerals and collagen is desperately required, particularly if the repair mechanism instigated by bioceramics is a consequence of the increased mineralisation or is there an additional multifaceted signalling activity involved.

One future line of research could be to design a scaffold that is ceramic base using a biomimetic methodology. The fabrication of a scaffold to replicate the hierarchical structures of bone from the molecular to the macroscopic scale, from the femur with a hard cortical bone with compact structures and low porosity to the trabecular bone with an interconnected network, down to the osteoblasts, down to the molecular structure of HA just might mimic each level, consequently forming tunable combinations of strength, toughness, density and porosity augmenting the mechanical properties.

Understanding the fundamentals of bone in response to the presence of bioceramics and the specific signals that are activated would provide a prime position to fabricate optimal bioactive ceramics for our future.

17.8 CONCLUSION

This chapter largely gathers the information available at present regarding attributes of bioactive ceramic products critical for hard tissue including dental tooth regeneration. Furthermore, information regarding bioceramics in AM is discussed with comprehensive instances in the use of inorganic particles, including hydroxyapatite, calcium phosphate and bioactive glass and also the biopolymers of chitosan, gelatin and collagen. By writing this chapter, we hope to inspire the invention of novel biopolymer/bioceramic hybrid particles considering their biocompatibility and bioactivity for use in hard tissue regeneration.

Bioceramics are uniquely placed as biomedical materials. Their development and revolutionary manufacturing techniques have expanded their diversity within the human body. Bioceramics have a positive future in the development of novel formulations with innovative methods of manufacture. From a clinical point of view, bioceramics containing hydroxyapatite is osteoconductive, non-toxic and bioactive and, for some time, was the primary component in bone implants due to the bioactivity and physicochemical properties. Bioactive glass is also currently used in clinical practice due to its highly biocompatible properties. Additionally, chitosan could be utilised as an antibacterial product or a carrier system that is biodegradable and therefore destined for drug delivery systems. Native hyaluronic acid present in oral tissues is also discussed. The use of all these products, therefore, represent new directions for the modelling of biopolymer bioactive ceramics to promote tissue regeneration in both bone and teeth.

Bone is capable of self-regeneration, but in some clinical situations, due to injury or degenerative disease, the self-regeneration mechanism fails. In order to stimulate bone formation above the capability of the host tissue, materials with a biomimetic composition and structure close to the native bone state represents a hopeful approach to expand the balance between bone creation and scaffold reabsorption. Currently, the development of new techniques such as AM manufacturing provides increased optimal manipulation in the scaffold design, able to be customed for each situation, and for each patient. For the future, focus needs to shift to increasing our comprehension of the biological interaction between the cells/tissue and the scaffold as this information is critical to further the development of successful implantation at the bony site (Table 17.1).

TABLE 17.1
Commercially Available Implants

	Brand Name	Company	Composition	Application	Reference
Collagen	Bio-Gide-membrane	Geistlich, Wolhusen, Switzerland	A compact collagenous layer covered by a very dense film	Designed to prevent invasion of soft tissue in a membrane-protected bone defect	Schlegel et al. (1997)
	BioMend-membrane	Zimmer, Warsaw, IN, USA	Freeze-dried fascia lata and cross-linked bovine type I collagen	Guided bone regeneration	McGinnis et al. (1998)
	BCP 4Bone	Biomatlante, Nantes, France	Biphasic calcium phosphate	Bioceramic graft material	Ezirganli et al. (2013)
	Vitala	Osteogenics	Collagen-based membrane	Guided bone regeneration	Coelho et al. (2012)
	Collprotect-membrane	Botiss biomaterials GmbH	Porcine dermis-derived collagen membranes	Transmembranous vascularisation	Barbeck et al. (2015)
Bioceramics	Maxresorb	Botiss biomaterials GmbH	Calcium phosphate-base	Bone substitute	Konermann et al. (2014)
	Bio-Oss	Geistlich, Wolhusen, Switzerland	Hydroxylapatite ceramic	Bone implant	Mandelkow et al. (1990)
	ChronOS	Synthes, PA, USA	99% β-TCP, traces of α-TCP and HA	Bone substitute	Gorla et al. (2015)
	OsteoSet	Wright Medical Technology	Calcium sulfate pellets	Bone graft	Wilkins et al. (1999)
	Pro-Osteon	Biomet, Warsaw, IN, USA	Coral-derived	Bone substitute	Jensen et al. (1996)
	MasterGraft	Medtronic, Minneapolis, MN, USA	Collagen-ceramic sponge	Bone graft substitute	Kraiwattanapong et al. (2005)

(Continued)

TABLE 17.1 (*Continued*)
Commercially Available Implants

	Brand Name	Company	Composition	Application	Reference
Composite bioceramics/collagen	Collagraft	Zimmer, Warsaw, IN, USA	Porous beads of 60% hydroxyapatite and 40% tricalcium phosphate ceramic and fibrillar collagen	Bone graft substitute	Cornell et al. (1991)
	Collapat/Collapat II	Symatese, Chaponost, France	Collagen scaffolds containing hydroxyapatite cross-linked with glutaraldehyde or EDC/NHS	Osteochondral defects	Chajra et al. (2008)
	Healos	DePuy, Raynham, MA, USA	Collagen-Hydroxyapatite Matrix	Injectable bone graft that mimics natural bone	Kitchel (2006)
	MinerOss X Collagen	Biohorizon Birmingham, AL, USA	Freeze-dried bone allograft	Guided tissue regeneration	Sato et al. (2012)
	OsteoGen	Impladent, NY, USA	Resorbable hydroxylapatite graft	Repair of a large mandibular lesion	Wagner (1989)
Marine	Biocoral®	Biocoral Inc, Saint-Gonnery, France	Natural calcium carbonate from coral	Bone graft	Doherty et al. (1994)
	Frios®Algipore®	Dentsply Sirona, NC, USA	Algae-derived hydroxyapatite	Bone augmentation following tooth extraction	Schopper et al. (2003)
	Pro-Osteon®200R Pro-Osteon®500R	Zimmer, Warsaw, IN, USA	Coral calcium carbonate matrix covered by calcium phosphate	Bone defects	Thalgott et al. (1999), Jensen et al. (1996)
	Interpore	Interpore, CA, USA	hydroxyapatite by purification of natural coral - corraline hydroxyapatite	Skeletal stabilisation	Ebraheim et al. (1997)

(Continued)

TABLE 17.1 (*Continued*)
Commercially Available Implants

	Brand Name	Company	Composition	Application	Reference
β-TCP	Bioresorb®	Sybron Implant Solutions, Germany	99% β-TCP, traces of α-TCP and calcium pyrophosphate	Bioactive, bioresorbable bone replacement material,	Wiedmann-Al-Ahmad et al. (2007)
	Vitoss®	Orthovita, PA, USA	98.8% β-TCP, traces of calcium pyrophosphate, and 1.2% organic bone matrix	Bone viod filler	Resnick (2002)
	Calciresorb C35	Ceraver, France	Macroporous biphasic calcium phosphate ceramic granules	Bone reconstruction	Frasca et al. (2017)
	JAX	Smith and Nephew Orthopaedics USA	Beta tricalcium phosphate granules with carboxymethylcellulose (CMC) gel	Bone healing	Clarke et al. (2007)
	Osferion	Olympus Terumo Biomaterials, Japan	Highly purified β-tricalcium phosphates	Filling substitute for defects following curettage of small bone tumours	Seto et al. (2013)
	PepGen® P15	CeraMed Dental, Lakewood, CO, USA	100% HA coated with a synthetic peptide (P-15) that mimics the cell-binding domain of Type I collagen	Bone replacement graft	Krauser et al. (2000)
Hydroxyapatite	Endobon® Cerabones®	Zimvie Dental, Canada	100% HA, traces of calcium oxide	Bovine bone	Jensen et al. (1996)
	Algipore®	Dentsply Sirona, North Carolina, USA	95% HA, 2.4% organic matrix, 2.3% $CaCO_3$	Bone graft	Bieniek (1990)
	Cementek	Teknimed, France	Calcium phosphate cement	Thoracolumbar spine support	Vanderschot et al. (2002)
	Actifuse	ApaTech, UK	Silicate-substituted calcium phosphate	Bone void filler	Hardenbrook and Lombardo (2006)
	Apaceram	Pentax, Japan	Synthetic hydroxyapatite	Implanted ceramic prostheses	Fukuta et al. (1992)
	ApaPore	ApaTech, UK	Hydroxyapatite bone graft substitute	Bone graft substitute	Coathup et al. (2008)
	Bonefil	Pentax, Japan	70% calcium hydroxyapatite	Bone fill	Rao et al. (2010)

Source: Taken from Pubmed and Web of Science.

REFERENCES

Albee, F. H. (1920). "Studies in bone growth – Triple calcium phosphate as a stimulus osteogenesis." *Ann Surg* **71**: 32–39.

Amiryaghoubi, N., N. N. Pesyan, M. Fathi and Y. Omidi (2020). "Injectable thermosensitive hybrid hydrogel containing graphene oxide and chitosan as dental pulp stem cells scaffold for bone tissue engineering." *Int J Biol Macromol* **162**: 1338–1357.

Arias-Gonzalez, F., A. Rodriguez-Contreras, M. Punset, J. M. Manero, O. Barro, M. Fernandez-Arias, F. Lusquinos, F. J. Gil and J. Pou (2021). "In-situ laser directed energy deposition of biomedical Ti-Nb and Ti-Zr-Nb alloys from elemental powders." *Metals* **11**(8).

Baino, F., S. Fiorilli and C. Vitale-Brovarone (2016). "Bioactive glass-based materials with hierarchical porosity for medical applications: Review of recent advances." *Acta Biomaterialia* **42**: 18–32.

Barbeck, M., J. Lorenz, A. Kubesch, N. Bohm, P. Booms, J. Choukroun, R. Sader, C. J. Kirkpatrick and S. Ghanaati (2015). "Porcine dermis-derived collagen membranes induce implantation bed vascularization via multinucleated giant cells: A physiological reaction?" *J Oral Implantol* **41**(6): e238–251.

Barry, J. J. A., A. V. Evseev, M. A. Markov, C. E. Upton, C. A. Scotchford, V. K. Popov and S. M. Howdle (2008). "In vitro study of hydroxyapatite-based photocurable polymer composites prepared by laser stereolithography and supercritical fluid extraction." *Acta Biomaterialia* **4**(6): 1603–1610.

Bian, W. G., D. C. Li, Q. Lian, X. Li, W. J. Zhang, K. Z. Wang and Z. M. Jin (2012). "Fabrication of a bio-inspired beta-Tricalcium phosphate/collagen scaffold based on ceramic stereolithography and gel casting for osteochondral tissue engineering." *Rapid Prototyp J* **18**(1): 68–80.

Bieniek, K. W. (1990). "[Algipore–new material for repair of bony periodontal pockets]." *ZWR* **99**(1): 24–25.

Blokhuis, T. J. and J. J. C. Arts (2011). "Bioactive and osteoinductive bone graft substitutes: Definitions, facts and myths." *Injury-Int J Care Inj* **42**: S26–S29.

Bose, S., D. Banerjee, S. Robertson and S. Vahabzadeh (2018). "Enhanced in vivo bone and blood vessel formation by iron oxide and silica doped 3D printed tricalcium phosphate scaffolds." *Ann Biomed Eng* **46**(9): 1241–1253.

Bose, S., S. Vahabzadeh and A. Bandyopadhyay (2013). "Bone tissue engineering using 3D printing." *Mater Today* **16**(12): 496–504.

Cao, Y. X., L. Xiao, Y. F. Cao, A. Nanda, C. Xu and Q. S. Ye (2019). "3D printed beta-TCP scaffold with sphingosine 1-phosphate coating promotes osteogenesis and inhibits inflammation." *Biochem Biophys Res Commun* **512**(4): 889–895.

Chajra, H., C. F. Rousseau, D. Cortial, M. C. Ronziere, D. Herbage, F. Mallein-Gerin and A. M. Freyria (2008). "Collagen-based biomaterials and cartilage engineering. Application to osteochondral defects." *Biomed Mater Eng* **18**(1 Suppl): S33–45.

Chen, I. H., M. J. Lian, W. Fang, B. R. Huang, T. H. Liu, J. A. Chen, C. L. Huang and T. M. Lee (2019). "In vitro properties for bioceramics composed of silica and titanium oxide composites." Applied Sciences-Basel **9**(1).

Cho, S. H., K. K. Shin, S. Y. Kim, M. Y. Cho, D. B. Oh and Y. T. Lim (2022). "In situ-forming collagen/poly-gamma-glutamic acid hydrogel system with mesenchymal stem cells and bone morphogenetic protein-2 for bone tissue regeneration in a mouse calvarial bone defect model." *Tissue Eng Regener Med.*

Chu, T. M. G., D. G. Orton, S. J. Hollister, S. E. Feinberg and J. W. Halloran (2002). "Mechanical and in vivo performance of hydroxyapatite implants with controlled architectures." *Biomaterials* **23**(5): 1283–1293.

Clarke, S. A., N. L. Hoskins, G. R. Jordan, S. A. Henderson and D. R. Marsh (2007). "In vitro testing of advanced JAX bone void filler system: Species differences in the response of bone marrow stromal cells to beta tri-calcium phosphate and carboxymethylcellulose gel." *J Mater Sci Mater Med* **18**(12): 2283–2290.

Coathup, M., N. Smith, C. Kingsley, T. Buckland, R. Dattani, G. P. Ascroft and G. Blunn (2008). "Impaction grafting with a bone-graft substitute in a sheep model of revision hip replacement." *J Bone Joint Surg Br* **90**(2): 246–253.

Coelho, P. G., G. Giro, W. Kim, R. Granato, C. Marin, E. A. Bonfante, S. Bonfante, T. Lilin and M. Suzuki (2012). "Evaluation of collagen-based membranes for guided bone regeneration, by three-dimensional computerized microtomography." *Oral Surg Oral Med Oral Pathol Oral Radiol* **114**(4): 437–443.

Cornell, C. N., J. M. Lane, M. Chapman, R. Merkow, D. Seligson, S. Henry, R. Gustilo and K. Vincent (1991). "Multicenter trial of collagraft as bone graft substitute." *J Orthop Trauma* **5**(1): 1–8.

Daculsi, G., R. Z. Legeros, E. Nery, K. Lynch and B. Kerebel (1989). "Transformation of biphasic calcium-phosphate ceramics invivo – Ultrastructural and physicochemical characterization." *J Biomed Mater Res* **23**(8): 883–894.

Deng, X. X., M. Gould and M. A. Ali (2021). "Fabrication and characterisation of melt-extruded chitosan/keratin/PCL/PEG drug-eluting sutures designed for wound healing." *Mater Sci Eng C – Mater Biol Appl* **120**.

Derkus, B., Y. E. Arslan, K. C. Emregul and E. Emregul (2016). "Enhancement of aptamer immobilization using egg shell-derived nano-sized spherical hydroxyapatite for thrombin detection in neuroclinic." *Talanta* **158**: 100–109.

Doherty, M. J., G. Schlag, N. Schwarz, R. A. Mollan, P. C. Nolan and D. J. Wilson (1994). "Biocompatibility of xenogeneic bone, commercially available coral, a bioceramic and tissue sealant for human osteoblasts." *Biomaterials* **15**(8): 601–608.

Du, Y. Y., H. M. Liu, J. Q. Shuang, J. L. Wang, J. Ma and S. M. Zhang (2015). "Microsphere-based selective laser sintering for building macroporous bone scaffolds with controlled microstructure and excellent biocompatibility." *Colloids Surf B – Biointerfaces* **135**: 81–89.

Ebraheim, N. A., A. O. Mekhail and M. Darwich (1997). "Open reduction and internal fixation with bone grafting of clavicular nonunion." *J Trauma* **42**(4): 701–704.

Ezirganli, S., S. Polat, E. Baris, I. Tatar and H. H. Celik (2013). "Comparative investigation of the effects of different materials used with a titanium barrier on new bone formation." *Clin Oral Implants Res* **24**(3): 312–319.

Fernandes, J. S., P. Gentile, R. A. Pires, R. L. Reis and P. V. Hatton (2017). "Multifunctional bioactive glass and glass-ceramic biomaterials with antibacterial properties for repair and regeneration of bone tissue." *Acta Biomaterialia* **59**: 2–11.

Frasca, S., F. Norol, C. Le Visage, J. M. Collombet, D. Letourneur, X. Holy and E. Sari Ali (2017). "Calcium-phosphate ceramics and polysaccharide-based hydrogel scaffolds combined with mesenchymal stem cell differently support bone repair in rats." *J Mater Sci Mater Med* **28**(2): 35.

Fukuta, S., H. Niwa and N. Yanagita (1992). "[Magnified tomography for identification of ceramic prostheses]." *Nihon Jibiinkoka Gakkai Kaiho* **95**(4): 535–540.

Fushimi, M., K. Sato, T. Shimizu and H. Hadeishi (2002). "PLEDs in Creutzfeldt-Jakob disease following a cadaveric dural graft." *Clin Neurophysiol* **113**(7): 1030–1035.

Gautam, G., S. Kumar and K. Kumar (2022). "Processing of biomaterials for bone tissue engineering: State of the art." *Mater Today-Proc* **50**: 2206–2217.

Ginebra, M. P., C. Canal, M. Espanol, D. Pastorino and E. B. Montufar (2012). "Calcium phosphate cements as drug delivery materials." *Adv Drug Deliv Rev* **64**(12): 1090–1110.

Gorla, L. F., R. Spin-Neto, F. B. Boos, S. Pereira Rdos, I. R. Garcia-Junior and E. Hochuli-Vieira (2015). "Use of autogenous bone and beta-tricalcium phosphate in maxillary sinus lifting: A prospective, randomized, volumetric computed tomography study." *Int J Oral Maxillofac Surg* **44**(12): 1486–1491.

Hardenbrook, M. A. and S. R. Lombardo (2006). "Silicate-substituted calcium phosphate as a bone void filler after kyphoplasty in a young patient with multiple compression fractures due to osteogenesis imperfecta variant: Case report." *Neurosurg Focus* **21**(6): E9.

Hayashi, K., X. Fang, H. Ueda, A. Miwa, T. Naka and H. Tsuchiya (2022). "Bone regeneration using autologous adipose-derived stem cell spheroid complex." *J Biomater Tissue Eng* **12**(6): 1216–1223.

Hench, L. L. and I. Thompson (2010). "Twenty-first century challenges for biomaterials." *J R Soc Interface* **7 Suppl 4**: S379–391.

Hild, N., P. N. Tawakoli, J. G. Halter, B. Sauer, W. Buchalla, W. J. Stark and D. Mohn (2013). "pH-dependent antibacterial effects on oral microorganisms through pure PLGA implants and composites with nanosized bioactive glass." *Acta Biomater* **9**(11): 9118–9125.

Huang, C. L., W. Fang, B. R. Huang, Y. H. Wang, G. C. Dong and T. M. Lee (2020). "Bioactive glass as a nanoporous drug delivery system for teicoplanin." *Appl Sci-Basel* **10**(7).

Islam, M., A. Ali, M. McConnell, R. Laing and C. Wilson (2019). "Mechanisms and kinetics of silver nanoparticle release from polyvinyl alcohol/keratin/chitosan electrospun nanofibrous scaffold." Proceedings of the 19th World Textile Conference, Autex(2B4_0290): 1–5.

Jensen, S. S., M. Aaboe, E. M. Pinholt, E. Hjorting-Hansen, F. Melsen and I. E. Ruyter (1996). "Tissue reaction and material characteristics of four bone substitutes." *Int J Oral Maxillofac Implants* **11**(1): 55–66.

Johnson, A. J. W. and B. A. Herschler (2011). "A review of the mechanical behavior of CaP and CaP/polymer composites for applications in bone replacement and repair." *Acta Biomaterialia* **7**(1): 16–30.

Juhasz, J. A., M. Kawashita, N. Miyata, T. Kokubo, T. Nakamura, S. M. Best and W. Bonfield (2002). "Apatite-forming ability and mechanical properties of glass-ceramic A-W-polyethylene composites." *Bioceramics* **218–220**: 437–440.

Kang, H. J., S. S. Park, T. Saleh, K. M. Ahn and B. T. Lee (2020). "In vitro and in vivo evaluation of Ca/P-hyaluronic acid/gelatin based novel dental plugs for one-step socket preservation." *Mater Design* **194**.

Kim, T. H., M. Eltohamy, M. Kim, R. A. Perez, J. H. Kim, Y. R. Yun, J. H. Jang, E. J. Lee, J. C. Knowles and H. W. Kim (2014). "Therapeutic foam scaffolds incorporating biopolymer-shelled mesoporous nanospheres with growth factors." *Acta Biomaterialia* **10**(6): 2612–2621.

Kitchel, S. H. (2006). "A preliminary comparative study of radiographic results using mineralized collagen and bone marrow aspirate versus autologous bone in the same patients undergoing posterior lumbar interbody fusion with instrumented posterolateral lumbar fusion." *Spine J* **6**(4): 405–411; discussion 411–402.

Kitsugi, T., T. Yamamuro, T. Nakamura and T. Kokubo (1989). "Bone bonding behavior of Mgo-Cao-Sio2-P2o5-Caf2 glass (mother glass of A.W-glass-ceramics)." *J Biomed Mater Res* **23**(6): 631–648.

Kokubo, T., S. Ito, Z. T. Huang, T. Hayashi, S. Sakka, T. Kitsugi and T. Yamamuro (1990). "Ca, P-rich layer formed on high-strength bioactive glass-ceramic a-W." *J Biomed Mater Res* **24**(3): 331–343.

Kokubo, T., S. Ito, M. Shigematsu, S. Sakka and T. Yamamuro (1987). "Fatigue and lifetime of bioactive glass ceramic a-W containing apatite and wollastonite." *J Mater Sci* **22**(11): 4067–4070.

Konermann, A., M. Staubwasser, C. Dirk, L. Keilig, C. Bourauel, W. Gotz, A. Jager and C. Reichert (2014). "Bone substitute material composition and morphology differentially modulate calcium and phosphate release through osteoclast-like cells." *Int J Oral Maxillofac Surg* **43**(4): 514–521.

Koutsoukos, P., Z. Amjad, M. B. Tomson and G. H. Nancollas (1980). "Crystallization of calcium phosphates. A constant composition study." *J Am Chem Soc* **102**(5): 1553–1557.

Kovach, I., S. Kosmella, C. Prietzel, C. Bagdahn and J. Koetz (2015). "Nano-porous calcium phosphate balls." *Colloids Surf B* **132**: 246–252.

Kraiwattanapong, C., S. D. Boden, J. Louis-Ugbo, E. Attallah, B. Barnes and W. C. Hutton (2005). "Comparison of Healos/bone marrow to INFUSE(rhBMP-2/ACS) with a collagen-ceramic sponge bulking agent as graft substitutes for lumbar spine fusion." *Spine (Phila Pa 1976)* **30**(9): 1001–1007; discussion 1007.

Krauser, J. T., M. D. Rohrer and S. S. Wallace (2000). "Human histologic and histomorphometric analysis comparing OsteoGraf/N with PepGen P-15 in the maxillary sinus elevation procedure: A case report." *Implant Dent* **9**(4): 298–302.

Kumar, A., S. Mandal, S. Barui, R. Vasireddi, U. Gbureck, M. Gelinsky and B. Basu (2016). "Low temperature additive manufacturing of three dimensional scaffolds for bone-tissue engineering applications: Processing related challenges and property assessment." *Mater Sci Eng R-Rep* **103**: Iii–+.

Kumar, P., B. S. Dehiya and A. Sindhu (2018). "Bioceramics for hard tissue engineering applications: A review." *Int J Appl Eng Res* **13**(5): 2744–2752.

Leucht, A., A. C. Volz, J. Rogal, K. Borchers and P. J. Kluger (2020). "Advanced gelatin-based vascularization bioinks for extrusion-based bioprinting of vascularized bone equivalents." *Sci Rep* **10**(1).

Li, Q., X. X. Lei, X. F. Wang, Z. G. Cai, P. J. Lyu and G. F. Zhang (2019). "Hydroxyapatite/collagen three-dimensional printed scaffolds and their osteogenic effects on human bone marrow-derived mesenchymal stem cells." *Tissue Eng Part A* **25**(17–18): 1261–1271.

Lin, K. L., C. T. Wu and J. Chang (2014). "Advances in synthesis of calcium phosphate crystals with controlled size and shape." *Acta Biomaterialia* **10**(10): 4071–4102.

Lopez-Heredia, M. A., A. Lapa, A. C. Mendes, L. Balcaen, S. K. Samal, F. Chai, P. Van der Voort, C. V. Stevens, B. V. Parakhonskiy, I. S. Chronakis, F. Vanhaecke, N. Blanchemain, E. Pamula, A. G. Skirtach and T. E. L. Douglas (2017). "Bioinspired, biomimetic, double-enzymatic mineralization of hydrogels for bone regeneration with calcium carbonate." *Mater Lett* **190**: 13–16.

Maazouz, Y., E. B. Montufar, J. Malbert, M. Espanol and M. P. Ginebra (2017). "Self-hardening and thermoresponsive alpha tricalcium phosphate/pluronic pastes." *Acta Biomaterialia* **49**: 563–574.

Mandelkow, H. K., K. K. Hallfeldt, S. B. Kessler, M. Gayk, M. Siebeck and L. Schweiberer (1990). "[New bone formation following implantation of various hydroxyapatite ceramics. Animal experiment with bore hole models of the sheep tibia]." *Unfallchirurg* **93**(8): 376–379.

Mathe, G. (1994). "Graft versus host and auto-immune reactions may explain the discontinuity and severity of HIV-1-AIDS disease." *Biomed Pharmacother* **48**(1): 1–2.

McGinnis, M., P. Larsen, M. Miloro and F. M. Beck (1998). "Comparison of resorbable and nonresorbable guided bone regeneration materials: A preliminary study." *Int J Oral Maxillofac Implants* **13**(1): 30–35.

Melchels, F. P. W., J. Feijen and D. W. Grijpma (2010). "A review on stereolithography and its applications in biomedical engineering." *Biomaterials* **31**(24): 6121–6130.

Melcova, V., K. Svoradova, P. Mencik, S. Kontarova, M. Rampichova, V. Hedvicakova, V. Sovkova, R. Prikryl and L. Vojtova (2020). "FDM 3D printed composites for bone tissue engineering based on plasticized poly(3-hydroxybutyrate)/poly(d,l-lactide) blends." *Polymers* **12**(12).

Mihalache, C., D. M. Rata, A. N. Cadinoiu, X. Patras, E. V. Sindilar, S. E. Bacaita, M. Popa, L. I. Atanase and O. M. Daraba (2020). "Bupivacaine-loaded chitosan hydrogels for topical anesthesia in dentistry." *Polym Int* **69**(11): 1152–1160.

Miyanaji, H., M. Orth, J. M. Akbar and L. Yang (2018). "Process development for green part printing using binder jetting additive manufacturing." *Front Mech Eng* **13**(4): 504–512.

Mochizuki, Y., T. Mizutani, N. Tajiri, T. Oinuma, N. Nemoto, S. Kakimi and T. Kitamoto (2003). "Creutzfeldt-Jakob disease with florid plaques after cadaveric dura mater graft." *Neuropathology* **23**(2): 136–140.

Moreira, M. S., G. Sarra, G. L. Carvalho, F. Goncalves, H. V. Caballero-Flores, A. C. F. Pedroni, C. A. Lascala, L. H. Catalani and M. M. Marques (2021). "Physical and biological properties of a chitosan hydrogel scaffold associated to photobiomodulation therapy for dental pulp regeneration: An in vitro and in vivo study." *Biomed Res Int* **2021**: 6684667.

Nejadshafiee, V., H. Naeimi, B. Goliaei, B. Bigdeli, A. Sadighi, S. Dehghani, A. Lotfabadi, M. Hosseini, M. S. Nezamtaheri, M. Amanlou, M. Sharifzadeh and M. Khoobi (2019). "Magnetic bio-metal-organic framework nanocomposites decorated with folic acid conjugated chitosan as a promising biocompatible targeted theranostic system for cancer treatment." *Mater Sci Eng C – Mater Biol Appl* **99**: 805–815.

Piazza, R. D., T. A. G. Pelizaro, J. E. Rodriguez-Chanfrau, A. A. La Serna, Y. Veranes-Pantoja and A. C. Guastaldi (2020). "Calcium phosphates nanoparticles: The effect of freeze-drying on particle size reduction." *Mater Chem Phys* **239**.

Prieto, E. M., A. D. Talley, N. R. Gould, K. J. Zienkiewicz, S. J. Drapeau, K. N. Kalpakci and S. A. Guelcher (2015). "Effects of particle size and porosity on in vivo remodeling of settable allograft bone/polymer composites." *J Biomed Mater Res Part B – Appl Biomater* **103**(8): 1641–1651.

Rahaman, M. N., D. E. Day, B. S. Bal, Q. Fu, S. B. Jung, L. F. Bonewald and A. P. Tomsia (2011). "Bioactive glass in tissue engineering." *Acta Biomater* **7**(6): 2355–2373.

Rao, D. V., M. Swapna, R. Cesareo, A. Brunetti, T. Akatsuka, T. Yuasa, T. Takeda and G. E. Gigante (2010). "Synchrotron-based scattered radiation from phantom materials used in X-ray CT." *J Xray Sci Technol* **18**(3): 327–337.

Ratnayake, J. T. B., M. L. Gould, A. Shavandi, M. Mucalo and G. J. Dias (2017). "Development and characterization of a xenograft material from New Zealand sourced bovine cancellous bone." *J Biomed Mater Res B Appl Biomater* **105**(5): 1054–1062.

Resnick, D. K. (2002). "Vitoss bone substitute." *Neurosurgery* **50**(5): 1162–1164.

Roohani-Esfahani, S. I., P. Newman and H. Zreiqat (2016). "Design and fabrication of 3D printed scaffolds with a mechanical strength comparable to cortical bone to repair large bone defects." *Sci Rep* **6**.

Safronova, T. V., Selezneva, II, S. A. Tikhonova, A. S. Kiselev, G. A. Davydova, T. B. Shatalova, D. S. Larionov and J. V. Rau (2020). "Biocompatibility of biphasic alpha,beta-tricalcium phosphate ceramics in vitro." *Bioact Mater* **5**(2): 423–427.

Sato, E. Y., T. Svec, B. Whitten and C. M. Sedgley (2012). "Effects of bone graft materials on the microhardness of mineral trioxide aggregate." *J Endod* **38**(5): 700–703.

Schlegel, A. K., H. Mohler, F. Busch and A. Mehl (1997). "Preclinical and clinical studies of a collagen membrane (Bio-Gide)." *Biomaterials* **18**(7): 535–538.

Schoche, S., N. Hong, M. Khorasaninejad, A. Ambrosio, E. Orabona, P. Maddalena and F. Capasso (2017). "Optical properties of graphene oxide and reduced graphene oxide determined by spectroscopic ellipsometry." *Appl Surf Sci* **421**: 778–782.

Schopper, C., D. Moser, A. Sabbas, G. Lagogiannis, E. Spassova, F. Konig, K. Donath and R. Ewers (2003). "The fluorohydroxyapatite (FHA) FRIOS Algipore is a suitable biomaterial for the reconstruction of severely atrophic human maxillae." *Clin Oral Implants Res* **14**(6): 743–749.

Seto, S., K. Muramatsu, T. Hashimoto, Y. Tominaga and T. Taguchi (2013). "A new beta-tricalcium phosphate with uniform triple superporous structure as a filling material after curettage of bone tumor." *Anticancer Res* **33**(11): 5075–5081.

Shuai, C. J., P. J. Li, J. L. Liu and S. P. Peng (2013). "Optimization of TCP/HAP ratio for better properties of calcium phosphate scaffold via selective laser sintering." *Mater Charact* **77**: 23–31.

Snelling, D. A., C. B. Williams, C. T. A. Suchicital and A. P. Druschitz (2017). "Binder jetting advanced ceramics for metal-ceramic composite structures." *Int J Adv Manuf Technol* **92**(1–4): 531–545.

Sun, L., H. H. Xu, S. Takagi and L. C. Chow (2007). "Fast setting calcium phosphate cement-chitosan composite: Mechanical properties and dissolution rates." *J Biomater Appl* **21**(3): 299–315.

Swetha, M., K. Sahithi, A. Moorthi, N. Srinivasan, K. Ramasamy and N. Selvamurugan (2010). "Biocomposites containing natural polymers and hydroxyapatite for bone tissue engineering." *Int J Biol Macromol* **47**(1): 1–4.

Thalgott, J. S., K. Fritts, J. M. Giuffre and M. Timlin (1999). "Anterior interbody fusion of the cervical spine with coralline hydroxyapatite." *Spine (Phila Pa 1976)* **24**(13): 1295–1299.

Vagropoulou, G., M. Trentsiou, A. Georgopoulou, E. Papachristou, O. Prymak, A. Kritis, M. Epple, M. Chatzinikolaidou, A. Bakopoulou and P. Koidis (2021). "Hybrid chitosan/gelatin/nanohydroxyapatite scaffolds promote odontogenic differentiation of dental pulp stem cells and in vitro biomineralization." *Dental Mater* **37**(1): E23–E36.

Vanderschot, P., E. Schepers, A. Vanschoonwinkel and P. Broos (2002). "A newly designed vertebral replacement implant to reconstruct the thoracolumbar spine anteriorly." *Acta Chir Belg* **102**(1): 37–45.

Varadavenkatesan, T., R. Vinayagam, S. Pai, B. Kathirvel, A. Pugazhendhi and R. Selvaraj (2021). "Synthesis, biological and environmental applications of hydroxyapatite and its composites with organic and inorganic coatings." *Prog Organ Coat* **151**.

Wagner, J. R. (1989). "Clinical and histological case study using resorbable hydroxylapatite for the repair of osseous defects prior to endosseous implant surgery." *J Oral Implantol* **15**(3): 186–192.

Watcharajittanont, N., M. Tabrizian, C. Putson, P. Pripatnanont and J. Meesane (2020). "Osseointegrated membranes based on electro-spun TiO2/hydroxyapatite/polyurethane for oral maxillofacial surgery." *Mater Sci Eng C Mater Biol Appl* **108**: 110479.

Westhauser, F., C. Weis, M. Prokscha, L. A. Bittrich, W. Li, K. Xiao, U. Kneser, H. U. Kauczor, G. Schmidmaier, A. R. Boccaccini and A. Moghaddam (2016). "Three-dimensional polymer coated 45S5-type bioactive glass scaffolds seeded with human mesenchymal stem cells show bone formation in vivo." *J Mater Sci – Mater Med* **27**(7).

Wiedmann-Al-Ahmad, M., R. Gutwald, N. C. Gellrich, U. Hubner and R. Schmelzeisen (2007). "Growth of human osteoblast-like cells on beta-tricalciumphosphate (TCP) membranes with different structures." *J Mater Sci Mater Med* **18**(4): 551–563.

Wijedasa, N. P., S. M. Broas, R. E. Daso and I. A. Banerjee (2020). "Varying fish scale derived hydroxyapatite bound hybrid peptide nanofiber scaffolds for potential applications in periodontal tissue regeneration." *Mater Sci Eng C Mater Biol Appl* **109**: 110540.

Wilkins, R. M., C. M. Kelly and D. E. Giusti (1999). "Bioassayed demineralized bone matrix and calcium sulfate: Use in bone-grafting procedures." *Ann Chir Gynaecol* **88**(3): 180–185.

Wu, Y. C., W. Y. Lin, C. Y. Yang and T. M. Lee (2015). "Fabrication of gelatin-strontium substituted calcium phosphate scaffolds with unidirectional pores for bone tissue engineering." *J Mater Sci Mater Med* **26**(3): 152.

Yu, Y., Z. Bacsik and M. Eden (2018). "Contrasting in vitro apatite growth from bioactive glass surfaces with that of spontaneous precipitation." *Materials (Basel)* **11**(9).

Zhao, B. S., H. P. Xu, Y. Gao, J. Z. Xu, H. M. Yin, L. Xu, Z. M. Li and X. R. Song (2019). "Promoting osteoblast proliferation on polymer bone substitutes with bone-like structure by combining hydroxyapatite and bioactive glass." *Mater Sci Eng C – Mater Biol Appl* **96**: 1–9.

Zhao, J., Y. Liu, W. B. Sun and H. Zhang (2011). "Amorphous calcium phosphate and its application in dentistry." *Chem Cent J* **5**: 40.

Zhou, M. Y., M. Guo, X. C. Shi, J. Ma, S. T. Wang, S. Wu, W. Q. Yan, F. Wu and P. B. Zhang (2022). "Synergistically promoting bone regeneration by Icariin-incorporated porous microcarriers and decellularized extracellular matrix derived from bone marrow mesenchymal stem cells." *Front Bioeng Biotechnol* **10**.

Index

Note: **Bold** page numbers refer to tables and *italic* page numbers refer to figures.

For Product Safety Concerns and Information please contact our EU representative GPSR@taylorandfrancis.com
Taylor & Francis Verlag GmbH, Kaufingerstraße 24, 80331 München, Germany

www.ingramcontent.com/pod-product-compliance
Lightning Source LLC
Chambersburg PA
CBHW060746220326
41598CB00022B/2343

* 9 7 8 1 0 3 2 1 9 2 5 2 9 *